Petra Perner (Ed.)

Case-Based Reasoning on Images and Signals

Studies in Computational Intelligence, Volume 73

Editor-in-chief
Prof. Janusz Kacprzyk
Systems Research Institute
Polish Academy of Sciences
ul. Newelska 6
01-447 Warsaw
Poland
E-mail: kacprzyk@ibspan.waw.pl

Further volumes of this series can be found on our homepage: springer.com

Vol. 54. Fernando G. Lobo, Cláudio F. Lima and Zbigniew Michalewicz (Eds.)
Parameter Setting in Evolutionary Algorithms, 2007
ISBN 978-3-540-69431-1

Vol. 55. Xianyi Zeng, Yi Li, Da Ruan and Ludovic Koehl (Eds.)
Computational Textile, 2007
ISBN 978-3-540-70656-4

Vol. 56. Akira Namatame, Satoshi Kurihara and Hideyuki Nakashima (Eds.)
Emergent Intelligence of Networked Agents, 2007
ISBN 978-3-540-71073-8

Vol. 57. Nadia Nedjah, Ajith Abraham and Luiza de Macedo Mourella (Eds.)
Computational Intelligence in Information Assurance and Security, 2007
ISBN 978-3-540-71077-6

Vol. 58. Jeng-Shyang Pan, Hsiang-Cheh Huang, Lakhmi C. Jain and Wai-Chi Fang (Eds.)
Intelligent Multimedia Data Hiding, 2007
ISBN 978-3-540-71168-1

Vol. 59. Andrzej P. Wierzbicki and Yoshiteru Nakamori (Eds.)
Creative Environments, 2007
ISBN 978-3-540-71466-8

Vol. 60. Vladimir G. Ivancevic and Tijana T. Ivacevic
Computational Mind: A Complex Dynamics Perspective, 2007
ISBN 978-3-540-71465-1

Vol. 61. Jacques Teller, John R. Lee and Catherine Roussey (Eds.)
Ontologies for Urban Development, 2007
ISBN 978-3-540-71975-5

Vol. 62. Lakhmi C. Jain, Raymond A. Tedman and Debra K. Tedman (Eds.)
Evolution of Teaching and Learning Paradigms in Intelligent Environment, 2007
ISBN 978-3-540-71973-1

Vol. 63. Wlodzislaw Duch and Jacek Mańdziuk (Eds.)
Challenges for Computational Intelligence, 2007
ISBN 978-3-540-71983-0

Vol. 64. Lorenzo Magnani and Ping Li (Eds.)
Model-Based Reasoning in Science, Technology, and Medicine, 2007
ISBN 978-3-540-71985-4

Vol. 65. S. Vaidya, L.C. Jain and H. Yoshida (Eds.)
Advanced Computational Intelligence Paradigms in Healthcare-2, 2007
ISBN 978-3-540-72374-5

Vol. 66. Lakhmi C. Jain, Vasile Palade and Dipti Srinivasan (Eds.)
Advances in Evolutionary Computing for System Design, 2007
ISBN 978-3-540-72376-9

Vol. 67. Vassilis G. Kaburlasos and Gerhard X. Ritter (Eds.)
Computational Intelligence Based on Lattice Theory, 2007
ISBN 978-3-540-72686-9

Vol. 68. Cipriano Galindo, Juan-Antonio Fernández-Madrigal and Javier Gonzalez
A Multi-Hierarchical Symbolic Model of the Environment for Improving Mobile Robot Operation, 2007
ISBN 978-3-540-72688-3

Vol. 69. Falko Dressler and Iacopo Carreras (Eds.)
Advances in Biologically Inspired Information Systems: Models, Methods, and Tools, 2007
ISBN 978-3-540-72692-0

Vol. 70. Javaan Singh Chahl, Lakhmi C. Jain, Akiko Mizutani and Mika Sato-Ilic (Eds.)
Innovations in Intelligent Machines-1, 2007
ISBN 978-3-540-72695-1

Vol. 71. Norio Baba, Lakhmi C. Jain and Hisashi Handa (Eds.)
Advanced Intelligent Paradigms in Computer Games, 2007
ISBN 978-3-540-72704-0

Vol. 72. Raymond S.T. Lee and Vincenzo Loia (Eds.)
Computational Intelligence for Agent-based Systems, 2007
ISBN 978-3-540-73175-7

Vol. 73. Petra Perner (Ed.)
Case-Based Reasoning on Images and Signals, 2008
ISBN 978-3-540-73178-8

Petra Perner
(Ed.)

Case-Based Reasoning on Images and Signals

With 132 Figures and 30 Tables

Springer

Dr. Petra Perner
Institute of Computer Vision
and Applied Computer Sciences, IBaI
Arno-Nitzsche-Str.43
04277 Leipzig
Germany
E-mail:- pperner@ibai-institut.de

ISSN print edition: 1860-949X
ISSN electronic edition: 1860-9503
ISBN 978-3-642-09221-3 e-ISBN 978-3-540-73180-1

Springer is a part of Springer Science+Business Media
springer.com

Cover design: deblik, Berlin

Printed on acid-free paper

Preface

This book is the first edited book that deals with the special topic of signals and images within case-based reasoning (CBR).

Signal-interpreting systems are becoming increasingly popular in medical, industrial, ecological, biotechnological and many other applications. Existing statistical and knowledge-based techniques lack robustness, accuracy, and flexibility. New strategies are needed that can adapt to changing environmental conditions, signal variation, user needs and process requirements. Introducing CBR strategies into signal-interpreting systems can satisfy these requirements. CBR can be used to control the signal-processing process in all phases of a signal-interpreting system to derive information of the highest possible quality. Beyond this CBR offers different learning capabilities, for all phases of a signal-interpreting system, that satisfy different needs during the development process of a signal-interpreting system.

In the outline of this book we summarize under the term "signal" signals of 1-dimensional, 2-dimensional or 3-dimensional nature.

The unique data and the necessary computation techniques require extraordinary case representations, similarity measures and CBR strategies to be utilised.

Signal interpretation (1D, 2D, or 3D signal interpretation) is the process of mapping the numerical representation of a signal into logical representations suitable for signal descriptions. A signal-interpreting system must be able to extract symbolic features from the raw data e.g., the image (e.g., irregular structure inside the nodule, area of calcification, and sharp margin). This is a complex process; the signal passes through several general processing steps before the final symbolic description is obtained.

The structure of the book is divided into a theoretical part and into an application-oriented part.

The first chapter gives an introduction to case-based reasoning and describes the special problems associated with signal-interpreting systems and why CBR is especially appropriate to solve the well-known bottleneck when developing signal-interpreting systems. At the end of this chapter an outlook to new developments is given.

The chapter is followed by a chapter on similarity. This chapter gives a fresh view to similarity and describes how important the concept of similarity is for many application including information retrieval, pattern recognition, computer vision and technical diagnosis. It reviews similarity under the aspects of using similarity for reasoning and explains the advantages of the different properties of the different similarity measures. While doing that it also takes into account the different possible data representations and the requirements following from that to the similarity measures.

Although similarity is a widely used concept in human reasoning it is often not clear how to assess the similarity. Therefore learning of similarity is important aspect to achieve the expected performance of a system. Chapters 3 and 4 describe two different methods for learning the similarity. While Chap. 3 describes a new approach for learning the weight of the attributes in Chap. 4 induction of similarity is proposed.

Structural similarity is an important concept in computer vision and design. A lot of effort has been put into the development of different structural similarity measures and into the development of computational efficient algorithm. Chapter 5 reviews structural similarity measures and gives a classification of the different measures.

Memory organization plays an important role in CBR. Different memory structure have been developed that help to organize a large case base so that a case can be efficiently retrieved from the case base. Chapter 6 gives an excellent overview and classification about what has been achieved so far.

Performance evaluation also under sparse data set is another important aspect. A method for performance evaluation that bridges between CBR and statistics is given in Chap. 7

In the second part of the book special signal-related applications are described and how subtasks of a CBR system such as retrieval, image segmentation or memory organization are solved for this kind of applications.

Chapter 8 describes an agent-based approach for environmental monitoring. The interpretation of 1-dimensional medical signals is described in Chap. 9. The application of prototypical cases for medical diagnosis is described in Chap. 10. The use of the case-based reasoning process for building the model of a watershed-based image segmentation system is described in Chap. 11. The last three chapters deal with the different aspects of image retrieval that is one subtask of CBR. Methods for image retrieval of biomedical images are described in Chap. 12. Chapter 12.8 faces on the special problems when dealing with different multimedia sources and Chap. 14 uses a dissimilarity-based representation for the retrieval.

The book is made for scientist and computer science experts from industry, medicine, and biotechnology who like to work on the special topics of CBR for signals and images. Although CBR is often not a standard lecture at universities we hope we can inspire PhD students to deal with this topic.

The hope is that in the next edition of that book we will see more inspiring work dealing with the challenging topic of CBR.

The main conferences on CBR are the International Conference on Case-Based Reasoning, ICCBR and the European Conference on Case-Based Reasoning, ECCBR. Special emphasis on CBR for signals and images is given at the International Conference on Machine Learning and Data Mining in Pattern Recognition, MLDM and the Industrial Conference on Data Mining ICDM-Leipzig. We are curious to see what new ideas you can present for CBR on signals and image. Looking forward to welcome you at one of our next event.

Leipzig, April 2007 Petra Perner

Contents

1

Introduction to Case-Based Reasoning for Signals and Images

P. Perner

Institute of Computer Vision and Applied Computer Sciences Leipzig, Arno-Nitzsche-Str. 43, 04277 Leipzig, Germany

Summary. Case-based reasoning (CBR) is used when generalized knowledge is lacking. The method works on a set of cases formerly processed and stored in the case base. A new case is interpreted based on its similarity to cases in the case base. The closest case with its associated result is selected and presented as output of the system. Signal-interpreting systems for 1-d, 2-d, or 3-dimensional signals are becoming increasingly popular in medical and industrial applications. New strategies are necessary that can adapt to changing environmental conditions, user needs, and process requirements. Introducing CBR strategies into signal-interpreting systems can satisfy these requirements. We describe in this chapter the basics of CBR and review what has been done so far in the field of signal-interpreting systems.

1.1 Introduction

Signal-interpreting systems for 1-d, 2-d, or 3-dimensional signals are becoming increasingly popular in medical and industrial applications. The existing statistically and knowledge-based techniques lack robustness, accuracy, and flexibility. New strategies are necessary that can adapt to changing environmental conditions, user needs, and process requirements. Introducing case-based reasoning (CBR) [1–3] strategies into signal-interpreting systems can satisfy these requirements. CBR provides a flexible and powerful method for controlling the signal-processing process in all phases of a signal-interpreting system to derive information of the highest possible quality. Beyond this CBR offers different learning capabilities, for all phases of a signal-interpreting system, that satisfy different needs during the development process of a signal-interpreting system. Therefore, they are especially appropriate for signal interpretation.

Although all this has been demonstrated in various applications [4–10, 47, 52–54, 60, 68], case-based signal-interpreting systems are still not well established in the computer-vision and pattern-recognition community. One reason might be that CBR is not very well known within this community. Also, some relevant activities have been shied away from developing large complex systems in favor of developing special algorithms for well-constrained

P. Perner: *Introduction to Case-Based Reasoning for Signals and Images*, Studies in Computational Intelligence (SCI) **73**, 1–24 (2008)
www.springerlink.com

tasks (e.g., texture, motion, or shape recognition). A recent overview about case-based signal-interpreting system can be found in [68].

In this chapter, we will show that a CBR framework can be used to overcome the modeling burden usually associated with the development of image-interpretation systems.

We seek to increase attention for this area and the special needs that signal processing tasks require.

In Sect. 1.2 we describe the CBR. Section 1.3 explains the developmental burden relating to signal-interpreting systems. The process model of CBR is explained in Sect. 1.4. In Sect. 1.5 we explain the knowledge containers of a CBR system. Design considerations of a CBR system are given in Sect. 1.6. The formal case description is explained in Sect. 1.7. The cognitive and the mathematical aspect of similarity are described in Sect. 1.8. The overview of methods for organization of a case base is given in Sect. 1.9. The different learning methods in a CBR system are described in Sect. 1.10. We give an outlook on further developments of CBR based signal-interpreting systems in Sect. 1.11 and, finally, we summarize the chapter in Sect. 1.12.

1.2 Case-Based Reasoning

Rule-based systems or decision trees are difficult to utilize in domains where generalized knowledge is lacking. However, often there is a need for a prediction system, even though there is not enough generalized knowledge. Such a system should (a) solve problems, using the already stored knowledge and (b) capture new knowledge, making it immediately available for solving the next problem. To accomplish these tasks, CBR is useful. CBR explicitly uses past cases from the domain expert's successful or failing experience.

Therefore, CBR can be seen as a method for problem solution, and also as a method to capture new experience. It can be seen as a learning and knowledge-discovery approach, since it can capture from new experience some general knowledge, such as case classes, prototypes, and some higher-level concepts.

To point out the differences between a CBR learning system and a symbolic learning system, which represents a learned concept explicitly, e.g., by formulas, rules, or decision trees, we follow the notion of Wess et al., [11]: "A CBR system describes a concept C implicitly by a pair (CB, sim). The relationship between the case base CB and the measure sim used for classification may be characterized by the equation:

$$\boxed{\text{Concept} = \text{Case Base} + \text{Measure of Similarity}}$$

This equation indicates in analogy to arithmetic that it is possible to represent a given concept C in multiple ways, i.e., there exist many pairs $C = (CB_1, sim_1), (CB_2, sim_2), \ldots, (CB_i, sim_i)$ for the same concept C. Furthermore, the equation gives a hint how a case-based learner can improve its classification ability. There are three possibilities to improve a case-based system. The system can

1. Store new cases in the case base *CB*
2. Change the measure of similarity *sim*
3. Change *CB* and *sim*

During the learning phase a case-based system gets a sequence of cases $X_1, X_2, \ldots, X_i$ with $X_i = (x_i, class(x_i))$ and builds a sequence of pairs $(CB_1, sim_1), (CB_2, sim_2), \ldots, (CB_i, sim_i)$ with $CB_i \subseteq \{X_1, X_2, \ldots, X_i\}$. The aim is to get in the limit a pair $(\mathrm{CB_n}, \mathrm{sim_n})$ that needs no further change, i.e., $\exists \mathrm{n}\ \forall \mathrm{m} \geq \mathrm{n}\ (\mathrm{CB_n}, \mathrm{sim_n}) = (\mathrm{CB_m}, \mathrm{sim_m})$, because it is a correct classifier for the target concept C".

1.3 Development Concerns for Signal-Interpreting Systems

Several factors influence the quality of the final result of a signal-interpreting system, including environmental conditions, the selected signal acquisition device, noise, number of observations from the task domain, and the chosen part of the task domain. Often these cannot all be accounted for during system development, and many of them will only be discovered during system execution. Furthermore, the task domain cannot even be guaranteed to be limited. For example, in defect classification for industrial tasks, new defects may occur because the manufacturing tool that had been used for a long period suddenly causes scratches on the surface of the manufactured part. In optical character recognition, imaging defects (e.g., heavy print, light print, or stray marks) can occur and influence the recognition results. Rice et al. [12] attempted to systematically overview the factors that influence the result of an optical character-recognition system, and how different systems respond to them. However, it is not yet possible to observe all real-world influences, or provide a sufficiently large sample set for system development and testing.

A robust signal-interpreting system must be able to deal with such influences. It must have intelligent strategies on all levels of a signal-interpreting system that can adapt the processing components to these new requirements. A strategy that seems to satisfy these requirements could be CBR. CBR does not rely on a well-formulated domain theory, which is, as we have seen, often difficult to acquire.

This suggests that we must consider different aspects during system development that are frequently studied CBR issues. Because we expect that users will discover new aspects of the environment and the objects during system usage, an automatic signal-interpreting system should be able to incrementally update the system's model, as illustrated in Fig. 1.1. This requires knowledge maintenance and learning. The designated lifetime of a case [66] also plays an important role. Other aspects are concerned with system competence. The

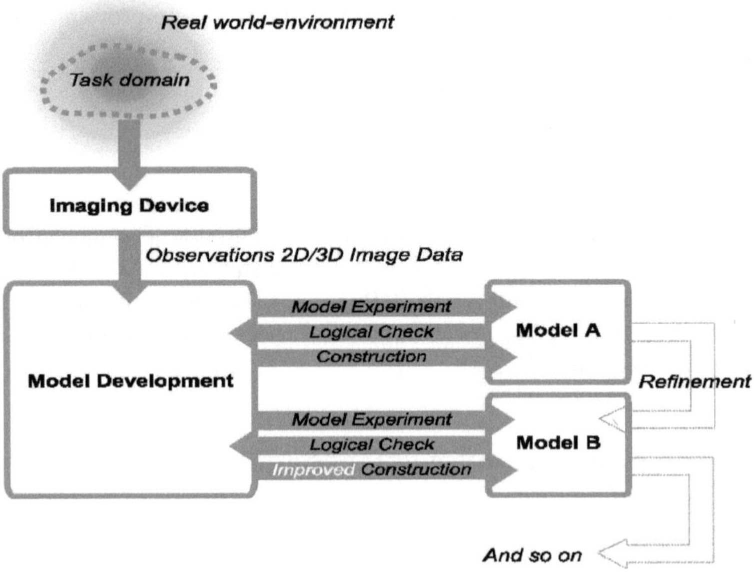

Fig. 1.1. Model development process

range of target problems, which a given system or algorithm can solve, is often not quite clear to the developer of the signal-interpreting system. Often researchers present to the community a new algorithm that can, for example, recognize the shape of an object in a particular signal and then claim that they have developed a model. Unfortunately, all too often another researcher inputs a different signal to the same algorithm and finds that it fails. Did the first researcher develop a model or did he/she instead develop a function? Testing and evaluation of algorithms and systems is an important problem in computer vision [13], as is designing the algorithm's control structure, so that it fits best to the current problem. CBR strategies can help to solve this problem in signal-interpreting systems.

1.4 The Process Model of CBR

1.4.1 The CBR Reasoning Process

At the highest level of generality, a general CBR cycle may be described by four tasks [1]: Retrieve the most similar case or cases, reuse the information and knowledge in that case to solve the problem, revise the proposed solution, and retain the parts of this experience likely to be useful for future problem solving (Fig. 1.2).

The four tasks each involve a number of more specific subtasks.

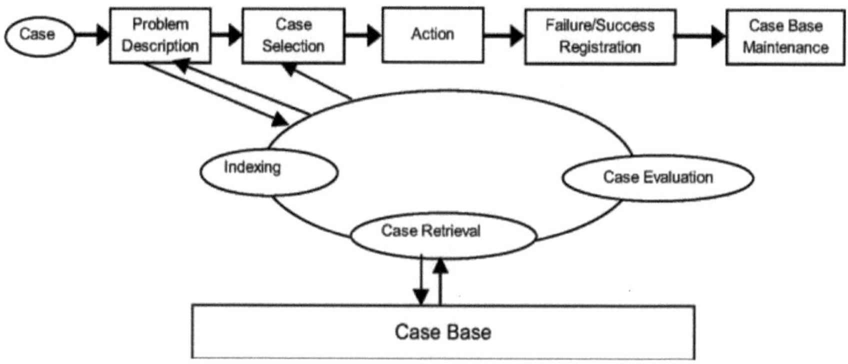

Fig. 1.2. Case-based reasoning process

The current problem is described by some keywords, attributes or any abstraction that allow describing the basic properties of a case. In pattern recognition and signal interpretation this can be attributes describing the basic properties of an object/signal or an attributed graph describing a scene. Based on this case description a set of close cases is retrieved from the case base (retrieve task).

The closest case is evaluated and the solutions associated to the closest case are given as output to the user.

The problem solution associated to the current case is applied to the current problem and the result of the test is observed and evaluated by a teacher (revise task). When the user is not satisfied with the result, or no similar case could be found in the case base, then case-base maintenance (retain tasks) starts. The revise and retain task can be viewed as a supervised learning task, which is necessary for improving the performance of the system.

Retrieval can be based on an index. The index can be a structure, such as for example a classifier, or any hierarchical organization of the case base. This index is built and updated during the retain task.

1.4.2 CBR Maintenance

CBR maintenance (see Fig. 1.3) will operate on new cases as well as on cases already stored in the case base.

If a new case has to be stored into the case base, it means that there is no similar case in the case base. The system has recognized a gap in the case base. A new case has to be incorporated into the case base, in order to close this gap. From the new case a predetermined case description has to be extracted, which should be formatted into the predefined case format. Afterwards, the case can be stored into the case base.

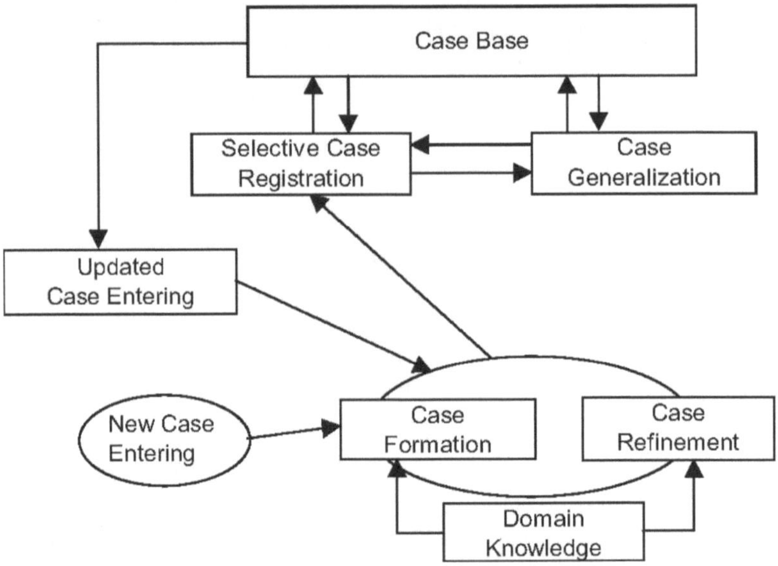

Fig. 1.3. CBR maintenance

Selective case registration means that no redundant cases will be stored into the case base and that the case will be stored at the right place, depending on the chosen organization of the case base. Similar cases will be grouped together or generalized by a case that applies to a wider range of problems. Generalization and selective case registration ensure that the case base will not grow too large and that the system can find similar cases fast.

It might also happen that too many cases not applicable to the current problem would be retrieved from the case base. Then it might be wise to rethink the case description, or to adapt the similarity measure. For the case description, more distinguishing attributes should be found that allow to filter out cases that do not apply to the current problem. The weights in the similarity measure might be updated, in order to retrieve only a small set of similar cases.

CBR maintenance is a complex process and works for all knowledge containers (vocabulary, similarity, retrieval, case base) [14] of a CBR system. Consequently, architectures and systems have been developed that support this process [15–17].

1.5 Knowledge Containers in a CBR System

The notion of knowledge containers has been introduced by Richter [14]. It gives a helpful explanation model or view on CBR systems. A CBR system has four knowledge containers which are:

1. The underlying vocabulary (or features)
2. The similarity measure
3. The solution transformation (adaptation knowledge)
4. The cases

The first three represent compiled knowledge, since this knowledge is more stable. A raw description of a situation has to be transformed into the case description, based on the selected vocabulary and the underlying domain knowledge (see Fig. 1.3).

The cases are interpreted knowledge. As a consequence, newly added cases can be used directly. This enables a CBR system to deal with dynamic knowledge.

In addition, knowledge can be shifted from one container to another. For instance, in the beginning a simple vocabulary, a rough similarity measure, and no knowledge on solution transformations are used [2]. However, after a large number of cases have been collected, over time, the vocabulary can be refined and the similarity measure defined in higher accordance with the underlying domain. In addition, it may be possible to reduce the number of cases, because the improved knowledge within the other containers now enables the CBR system to better differentiate between the available cases.

1.6 Design Consideration

The main problems concerned with the development of a CBR system are:

1. What is the right case description?
2. What is an appropriate similarity measure for the problem?
3. How to organize a large number of cases for efficient retrieval?
4. How to acquire and to refine a new case for entry in the case base?
5. How to generalize specific cases to a case that is applicable to a wide range of situations?

1.7 Case Description

There are different opinions about the formal description of a case. Each system utilizes a different representation of a case. Formally, we like to understand the following definition for a case:

Definition 1. *A case C is a triple $(P; E; L)$ with a problem description P; an explanation of the solution E and a problem solution L.*

For signal-related tasks we have two main different types of information that make up a case consisting of signal-related information and a nonsignal-related information. Signal-related information could be the 1D, 2D, or 3D signals of the desired application. Nonsignal-related information could be information about the signal acquisition, such as the type and parameters of the sensor, and information about the objects or the illumination of the scene. Which type of information should be taken into consideration for the interpretation of the signal depends on the type of application. In the case of the medical CT image segmentation described in [6], we used patient-specific parameters, such as age and sex, slice thickness and number of slices. Jarmulak [4] took into consideration the type of sensor for the railway-inspection application. Based on this information, the system controls the type of case base that the system is using during reasoning.

How the 2D or 3D image matrix is represented depends on the purpose, and rarely on the developer's point of view. In principle it is possible to represent an image by one of various abstraction levels. An image may be described by the pixel matrix itself or by parts of this matrix (pixel-based representation). It may be described by the objects contained in the image and their features (feature-based representation). Furthermore, it may be described by a more complex model of the image scene, composed of objects and their features, as well as of the spatial relation between the objects (attributed graph representation or semantic networks).

Jarmulak [4] has solved this problem by a four-level hierarchy for a case and different case bases for different sensor types. At the lowest level of the hierarchy are stored the objects described by features, such as their location, orientation, and type (line, parabola, or noise) parameters. The next level consists of objects of the same channel within the same subcluster. In the following level the subcluster is stored and at the highest level the whole image scene is stored. This representation allows him to match the cases on different granularity levels. Since the whole scene may have distortions caused by noise and imprecise measurements, he can reduce the influence of noise by retrieving cases on these different levels.

Grimnes and Aamodt [5] developed a model-based image-interpreting system for the interpretation of abdominal CT images. The image content is represented by a semantic network, where concepts can be general, special cases, or heuristic rules. Not well-understood parts of the model are expressed by cases and can be revised during the usage of the system by the learning component. The combination of the partially well-understood model with cases helps them to overcome the usual burden of modeling. The learning component is based on failure-driven learning and case integration. Nonimage information is also stored such as sex, age, earlier diagnosis, social condition, etc.

Micarelli et al. [7] have also calculated image properties from their images and stored them into the case base. They use the Wavelet transform, since it is scale-independent.

In all this work, CBR is only used for the high-level unit. We have studied different approaches for the different processing stages of an image interpretation system. For the image-segmentation unit [6] we studied two approaches:

(1) a pixel-based approach and (2) a feature-based approach that described the statistical properties of an image. Our results show that the pixel-based approach can give better results for the purpose of image segmentation. For the high-level approach of an ultra-sonic image interpretation system, we used a graph-based representation [18].

However, if we do not store the image matrix itself as a case, but store the representation of a higher-level abstraction instead, we will lose some information. An abstraction means that we have to make a decision between necessary and unnecessary details of an image. It might happen that, not having seen all the objects at the same time, we might think that one detail is of no interest, since our decision is only based on a limited number of objects. This can cause problems later on. Therefore, keeping the images themselves is always preferable, but needs a lot of storage capacity. The different possible types of representation require different types of similarity measures.

1.8 Similarity Relation

An important point in CBR is the determination of similarity between case A and case B. We need an evaluation function that gives us a measure for similarity between two cases. This evaluation function reduces each case from its case description to a numerical similarity measure *sim*. These similarity measures show the relation to other cases in the case base.

The concept of similarity is important for many tasks; therefore, we will briefly review the cognitive aspect of similarity, the basic properties of similarity, and the design criteria for similarity.

1.8.1 Formalization of Similarity

The problem with similarity is that it has no meaning, unless one specifies the kind of similarity. Smith [19] distinguishes five different kinds of similarity:

1. Overall similarity
2. Similarity
3. Identity
4. Partial similarity
5. Partial identity

Overall similarity is a global relation that includes all other similarity relations. All colloquial similarity statements are subsumed here.

Similarity and identity are relations that consider all the properties of the objects at once and where no single part is left unconsidered. A red ball and a red ball are similar, a red ball and a red car are dissimilar. The holistic relation's similarity and identity are different in the degree of the similarity. Identity describes objects that are not significantly different. All red balls are identical and, therefore, they are also similar. Similarity contains identity and is a more general concept.

Partial similarity and partial identity compare the significant parts of objects. One aspect or attribute can be marked. Partial similarity and partial identity are different with respect to the degree of similarity. A red ball and a pink cube are partially similar, but a red ball and a red cube are partially identical.

The described similarity relations are in connection in many respects. Identity and similarity are unspecified relations between objects. Partial identity and similarity are relations between single properties of objects. Identity and similarity are equivalence relations, this means they are reflexive, symmetrical, and transitive. For partial identity and similarity, these relations do not hold. From identity follows similarity and partial identity. From that follows partial similarity and general similarity.

It seems advisable to require from a similarity measure the reflexivity, which means an object is similar to itself. Symmetry should be another property of similarity. However, Bayer et al. show that these properties are not bound to belong to similarity in colloquial use. Let us consider the statements "A is similar to B" or "A is the same as B." We notice that these statements are directed and that the roles of A and B cannot be exchanged. People say: "A circle is like an ellipse." but not "An ellipse is like a circle." or "The son looks like the father." but not "The father looks like to the son." Therefore, symmetry is not necessarily a basic property of similarity. However, in the above examples it can be useful to define the similarity relation to be symmetrical. The transitivity relation also must not necessarily hold. Let us consider the block world: a red ball and a red cube might be similar; a red cube and a blue square are similar; but a red ball and a blue square are dissimilar. However, a concrete similarity relation might be transitive.

Similarity and identity are two concepts that strongly depend on the context.

The context defines the essential attributes of the objects that are taken into consideration when similarity is determined. An object "red ball" may be similar to an object "red chair" because of the color "red". However the objects "ball" and "chair" are dissimilar. These attributes may be relevant, depending on whether they are given priority or saliency in the considered problem.

1.8.2 General Remarks on Similarity

This little example shows that the calculation of similarity between the attributes must be meaningful. It makes no sense to compare two attributes that do not make a contribution to the considered similarity. That brings us to the vocabulary or feature-elicitation problem.

The homogeneity of the data matrix must be assumed. Attributes should have equal scale level. Metrical attributes should have a similar variance. Attribute weighting is only possible if the class structure will not be blurred.

Observations containing mixed data types, such as numerical and categorical attributes, require special distance measure.

1.8.3 Properties of a Distance Measure

A distance $d(x, y)$ between two vectors x and y is a function for which the following condition can be required:

$$\begin{aligned}
&d(x,y) \geq 0 \text{ (Nonnegativity)} \\
&d(x,x) = 0 \text{ (Identity, Reflexivity)} \\
&d(x,y) = 0 \text{ if and only if } x = y \text{ (Identity of Indiscernibles,} \\
&\qquad\qquad \text{Strong Reflexivity)} \\
&d(x,y) = d(y,x) \text{ (Symmetry)}
\end{aligned}$$

We call the distance $d(x, y)$ a metric if the triangle inequality holds:

$$d(x,y) \leq d(x,z) + d(z,y) \text{ (subadditivity / triangle inequality).}$$

If we require the following condition, then we call $d(x, y)$ an ultra metric:

$$d(x,y) \leq \max\{d(x,z), d(z,y)\}.$$

1.8.4 Distance Measures for Metrical Data

A well-known distance measure is the Minkowski metric:

$$d_{ii}^{(p)} = \left[\sum_{j=1}^{J} |x_{ij} - x_{i'j}|^p\right]^{1/p}$$

The choice of the parameter p depends on the importance we give to the differences in the summation.

If we choose $p = 2$, then we give special emphasis to big differences in the observations. The measure is invariant to translations and orthogonal linear transformations (rotation and reflection). It is called Euclidean distance:

$$d_{ii} = \sqrt{\sum_{j=1}^{J} |x_{ij} - x_{i'j}|^2}$$

If we choose $p = 1$, the measure gets insensitive to outliers, since big and small differences are equally treated. This measure is also called the City-Block_Metric:

$$d_{ii} = \sum_{j=1}^{J} |x_{ij} - x_{i'j}|$$

If we choose $p = \infty$, we obtain the so called Max Norm:

$$d_{ii} = \max_{j} |x_{ij} - x_{i'j}|$$

This measure is useful if only the maximal distance between two variables among a set of variables is of importance, whereas the other distances do not contribute to the overall similarity.

The disadvantange of the measures described above is that these measures require the stastical independence of the attributes. Each attribute is considered to be independent and isolated. A high correlation between attributes can be considered as a multiple measurement of an attribute. This means that the measures described above give this feature more weight than an uncorrelated attribute. The Mahalanobis distance:

$$d_{i'}^2 = (\overline{x_i} - \overline{x_{i'}})S^{-1}(\overline{x_i} - \overline{x_{i'}})$$

takes into account the covariance matrix of the attributes.

1.8.5 Using Numerical Distance Measures for Categorical Data

Categorical attributes require special similarity measures. Otherwise the symbolical representation has to mapped to a numerical scale. Let us suppose that we have an attribute color with green, red, and blue as attribute values. Let us suppose further that we have an observation 1 with red color and an observation 2 also with red color, then the two observations are identical. Therefore, the distance gets zero. Suppose now, we have an observation 3 with green color and we want to compare it to the observation 2 with red color. The two attribute values are different; therefore, the distance gets the value one. If we want to express the degree of dissimilarity, we have to assign levels of dissimilarity to all the different combinations between the attribute values.

If $\mathrm{a} \in \mathrm{A}$ is an attribute and $\mathrm{W_a} \subseteq \mathrm{W}$ is the set of all attribute values, which can be assigned to a, then we can determine for each attribute a a mapping:

$$\mathrm{distance_a} : \mathrm{W_a} \rightarrow [0, 1].$$

The normalization to a real interval is not absolutely necessary, but advantageous for the comparison of attribute assignments.

For example, let a be an attribute a=spatial_relationship and

$$W_a = \{\text{behind\_right, behind\_left, in front\_right}, \dots\}.$$

Then we could define:

$$\text{distance}_a(\text{behind\_right, behind\_right}) = 0$$
$$\text{distance}_a(\text{behind\_right, in front\_right}) = 0.25$$
$$\text{distance}_a(\text{behind\_right, behind\_left}) = 0.75.$$

Based on such distance measure for attributes, we can define different variants of distance measure as mapping:

$$\text{distance} : B^2 \rightarrow R^+$$

($R^+ \dots$ set of positive real numbers) in the following way:

$$\text{distance(x,y)} = 1/D \sum_{a \in D} \text{distance}_a(x(a),\ y(a))$$

with $D = \text{domain}\ (x) \cap \text{domain}(y)$.

1.8.6 Distance Measure for Nominal Data

For nominal attributes special distance coefficients have been designed. The basis for the calculation of these distance coefficients is a contingency table, see Table 1.1. The value N_{00} in this table is the frequency of observation pairs (i, j) that do not share the property, neither in the one observation nor in the other. The value N_{01} is the frequency of observation pairs, where one observation has the property and the other does not have this property.

Given that, we can define different distance coefficients for nominal data:

$$d_{ii} = 1 - (N_{11} + N_{00})/(N_{11} + N_{00} + 2(N_{10} + N_{01})$$

Whereas the similarity is increased by the value of N_{00} and the value of N_{11}, the noncorrespondence N_{10} and N_{01} gets double weight.

$$d_{ii} = 1 - N_{11}/(N_{11} + N_{10} + N_{01})$$

The nonexistence of a property N_{00} is not considered in the Jaccard coefficient.

Table 1.1. Contigency table

Status of the observation i	**Status of the observation j′**	
	0	1
0	N_{00}	N_{01}
1	N_{10}	N_{11}

1.8.7 Contrast Rule

This measure has been developed by Tversky [20]. It describes the similarity between a prototype A and a new example B *as*:

$$S(A,B) = \frac{|D_i|}{\alpha\,|D_i| + \beta\,|E_i| + \chi\,|F_i|} \qquad \alpha = 1, \beta, \chi = 0.5$$

with D_i being the features that are common to both A and B; E_i the features that belong to A but not to B; and F_i the features that belong to A, but not to B.

1.8.8 Similarity Measures for Images

Images can be rotated, translated, different in scale, or may be different in contrast and energy, but they might be considered as similar. In contrast to this two images may be dissimilar, since the object in one image is rotated by 180°. The concept of invariance in image interpretation is closely related to that of similarity. A good similarity measure should take this into consideration.

The classical similarity measures do not allow this. Usually the images or the features have to be preprocessed in order to be adapted to the scale, orientation, or shift. This process is a further processing step that is expensive and needs some a priori information, which is not always given. Filters, such as matched filters, linear filters, Fourier or Wavelet filters, are especially useful for invariance under translation and rotation, which has also been shown by Micarelli et al. [7]. There has been a lot of work done to develop such filters for image interpretation in the past. The best way to achieve scale invariance from an image is by means of invariant moments, which can also be invariant under rotation and other distortions. Some additional invariance can be obtained by normalization (reduces the influence of energy).

Depending on the image representation (see Fig. 1.4), we can divide similarity measures into:

1. Pixel (iconic)-matrix based similarity measures
2. Feature-based similarity measures (numerical or symbolical or mixed type)
3. Structural similarity measures [21–27, 57, 67].

Since a CBR image interpretation system has also to take into account nonimage information, such as information about the environment or the objects, etc., we need similarity measures that can combine nonimage and image information. We have shown a first approach to this in [6].

To better understand the concept of similarity, systematic studies on the different kinds of image similarity have to be done. Zamperoni and Starovoitov [21] studied how pixel-matrix based similarity measures behave under different real world influences, such as translation, noise (spikes, salt,

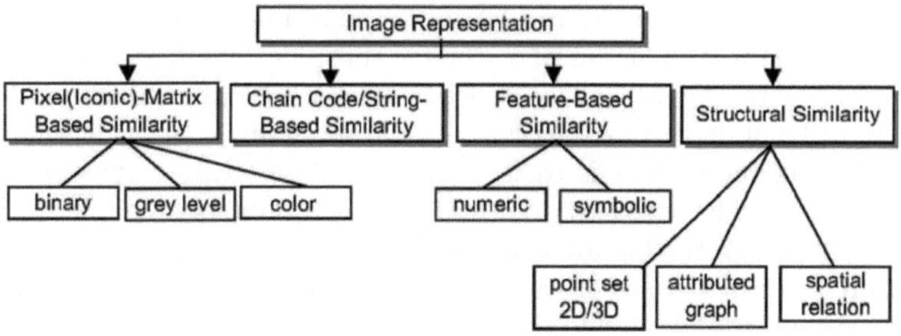

Fig. 1.4. Image representations and similarity measure

and pepper noise), different contrast, and so on. Image feature-based similarity measures have been studied from a broader perspective by Santini and Jain [22]. Those are the only important works that we are aware of. Otherwise, at every new conference on pattern recognition [28–33], new similarity measures are proposed for specific purposes and the different kinds of image representation, but a more methodological work is missing.

1.9 Organization of a Case Base

Cases can be organized into a flat case base or in a hierarchical fashion. In a flat organization we have to calculate the similarity between the problem case and each case in memory. It is clear that this will take time, especially if the case base is very large. Systems with a flat case-base organization usually run on a parallel machine, to perform retrieval in a reasonable time, and do not allow the case base to grow over a predefined limit. Maintenance is done by partitioning the case base into case clusters and by controlling the number and size of these clusters [34].

To speed up the retrieval process, a more sophisticated organization of the case base is necessary. This organization should allow separating the set of similar cases from those cases that are not similar to the actual problem at the earliest stage of the retrieval process. Therefore, we need to find a relation p that allows us to order our case base:

Definition. A binary relation p on a set CB is called a partial order on CB, if it is reflexive, antisymmetric, and transitive. In this case the pair $\langle CB, p\rangle$ is called a partially ordered set or poset.

The relation can be chosen depending on the application. One common approach is to order the case-base based on the similarity value. The set of cases can be reduced by the similarity measure to a set of similarity values. The relation p over these similarity values gives us a partial order over these cases. The derived hierarchy consists of nodes and edges. Each node in this hierarchy contains a set of cases that do not exceed a specified similarity value. The edges show the similarity relation between the nodes. The relation between two successor nodes can be expressed as follows: Let z be a node and x and y two successor nodes of z; then x subsumes z and y subsumes z. By tracing down the hierarchy, the space gets smaller and smaller, until finally a node will not have any successor. This node will contain a set of close cases. Among these cases is to be found the closest case to the query case. Although we still have to carry out matching, the number of matches will have decreased through the hierarchical ordering. The nodes can be represented by the prototypes of the set of cases assigned to the node. When classifying a query through the hierarchy, the query is only matched with the prototype. Depending on the outcome of the matching process, the query branches right or left of the node.

Such a kind of hierarchy can be created by hierarchical or conceptual clustering [35, 49], k–d trees [36] or a decision tree [37]. There are also set-membership-based organizations known, such as semantic nets [5], relational structure [63], and object-oriented representations [38].

1.10 Learning in a CBR System

CBR management is closely related to learning [62]. It aims to improve the performance of the system.

Let X be a set of cases collected in a case base CB. The relation between each case in the case base can be expressed by the similarity value *sim*. The case base can be partitioned into n case classes $C : CB = \bigcup_{i=1}^{n} C_i$ such that the intracase class similarity is high and the intercase class similarity is low. The set of cases in each class C_i can be represented by a representative who generally describes the cluster i. This representative can be the prototype, the mediod, or an a-priori selected case, whereas the prototype implies that the representative is the mean of the cluster that can easily be calculated from numerical data. The mediod is the case whose sum of all distances to all other cases in a cluster is minimal. The relation between the different case classes C_n can be expressed by higher-order constructs expressed e.g., as super-classes that gives us a hierarchical structure over the case base.

There are different learning strategies that can take place in a CBR system:

1. Learning takes place when a new case x has to be stored into the case base such that: $CB_{n+1} = CB_n \cup \{x\}$. That means that the case base is incrementally updated according to the new case.
2. It may incrementally learn the case classes and/or the prototypes representing the class [35, 39–41].
3. The relationship between the different cases or case classes may be updated according the new case classes [35].
4. The system may learn the similarity measure [42–44].

1.10.1 Learning New Cases and Forgetting Old Cases

Learning new cases means just adding cases into the case base upon some notification. Closely related to case adding is case deletion or forgetting cases [45], which have shown low utility. This should control the size of the case base. There are approaches that keep the size of the case base constant and delete cases that have not shown good utility within a fixed time window [46]. The failure rate is used as utility criterion. Given a period of observation of N cases, if the CBR component exhibits M failures in such a period, we define the failure rate f as $f = M/N$. Other approaches try to estimate the "coverage" of each case in memory and by using this estimate to guide the case memory revision process [47].

The adaptability to the dynamics of the changing environment that requires storing new cases in spite of the case base limit is addressed in [34]. Based on intraclass similarity it is decided whether a case is to be removed from a cluster stored into it.

1.10.2 Learning of Prototypes

Learning of prototypes has been described in [39] for a flat organization of a case base and for a hierarchical representation of a case base in [35]. The prototype or the representative of a case class is the most general representation of a case class. A class of cases is a set of cases sharing similar properties. The set of cases does not exceed a boundary for the intraclass dissimilarity. Cases that are on the boundary of this hyperball have a maximal dissimilarity value. A prototype can be selected a priori by the domain user. This approach is preferable when the domain expert knows for sure the properties of the prototype. The prototype can be calculated by averaging over all cases in a case class or the median of the cases is chosen. If only a few cases are available in a class and subsequently new cases are stored in the class, then it is preferable to incrementally update the prototype according to the new cases.

Learning the concept and representatives by fuzzy clustering and relevance feedback has been described in [58]. Learning prototypes for graph-based representation has been described in [65].

1.10.3 Learning of Higher-Order Constructs

The ordering of the different case classes gives an understanding of how these case classes are related to each other. For two case classes, which are connected by an edge, similarity relation holds. Case classes that are located at a higher position in the hierarchy, apply to a wider range of problems than those located near the leaves of the hierarchy. By learning how these case classes are related to each other, higher-order constructs are learnt [35].

1.10.4 Learning of Similarity

By introducing feature weights, we can put special emphasis on some features for the similarity calculation. It is possible to introduce local and global feature weights. A feature weight for a specific attribute is called local feature weight. A feature weight that averages over all local feature weights for a case is called global feature weight. This can improve the accuracy of the CBR system. By updating these feature weights, we can learn similarity [43, 44]. An approach using clustering to supervise learning the distance function has been described in [55]. Improving the similarity by linear feature transformation is described in [56].

1.11 Outlook

New work is being done into the direction of case mining. The scope of case mining is to search a large case base for relevant data and summarize these data into a more abstract representation, such as prototypes or concepts. An approach for literature mining has been proposed in [49]. Mining case classes and learning prototypes for 2D shapes has been proposed based on conceptual clustering in [64].

The feature selection problem for CBR systems has been studied in [69].

Signal-interpreting systems require special graphical user support to manage the case description and case base. The work of O'Sullivan et al. [58] has dealt with this problem.

Other problems go along with the evaluation and the lifecycle [66] of a case-based signal-interpreting system for which special architectures are required.

The main problem with image analysis by itself is that a person is often not able to sufficiently describe the image content. Unfortunately, for most image retrieval systems it is hard to narrow down the subset of retrieved images to the most relevant ones, unless the image is represented in a content-rich, high-level representation, which is only possible in a domain-specific environment. Associated free text documents with images would enable the construction of an ontology, which could be used to describe high-level features. In general, a system that can retrieve documents based on textual features, as well as on features obtained from the image content, could significantly improve its performance and user-friendliness. To develop retrieval methods based on conversational CBR (CCBR) for images seems to be a challenging topic to make an image-retrieval system more applicable. The CCBR should guide the user through visual content, based on a flexible interactive conversational strategy [72]. Based on the feedback from the user, the flexible dialogue strategy can narrow down the retrieval path to the desired group of documents, based on the image content.

In CCBR, a query describing a target problem is incrementally, and often incompletely, elicited in an interactive dialogue with the user [70, 71].

Alternatively, these dialogues are often initiated when the user provides a free-text description of the problem (or query), which may be incrementally extended. On each cycle of the problem-solving process, the user is shown the cases that are most similar to the current query and invited to select from a list of questions, ranked in order of expected usefulness for solving the problem, or presented a number of possible solutions, upon which to make recommendations. Unless terminated by the user, the dialogue continues until some predefined termination criteria are satisfied. Throughout this process the system highlights any cases whose descriptions, up to this point, have an above-threshold similarity to the query, and the user can select any of the displayed case solutions. Upon termination, some CCBR systems present the solution of the currently most similar case as the proposed solution to the target query. Much of the research has focused on CCBR as an approach to interactive problem solving in domains, such as fault diagnosis, help-desk support, and product recommendation. The usage of CCBR for multimodal retrieval has not been intensively studied. The only contribution to this area is that of case knowledge acquisition and performance improvement, solely in the area of CCBR for image retrieval [73].

1.12 Conclusion

We surveyed special topics associated with a case-based signal-interpreting system. From our point of view, case-based signal interpretation differs in many aspects from other CBR applications that require further investigation. First, more systematic work on special similarity measures is needed that investigates the measures under different influences that may occur. Next, case representations are required for all the different abstraction levels of an image. Finally, the maintenance and learning strategies must be defined, so that they can help to improve the system performance and discover the range of target problems that the system can solve.

In this chapter are described special topics for similarity assessment, memory organization, similarity learning, and applications of CBR.

References

1. Aamodt, A., Plaza, E., 1995. Case-based reasoning: foundational issues, methodological variations and system approaches. AI Communication 7(1), 39–59.
2. Althoff, K.-D., 2001. Case-based reasoning. In: Chang, S.K. (Ed.), Handbook of Software Engineering and Knowledge Engineering, Vol. I. World Scientific, Singapore.
3. Briscoe, G., Caelli, T., 1996. A Compendium of Machine learning: Symbolic Machine Learning, Vol. 1. Ablex Publishing Corporation, Norwood, NJ.

4. Jarmulak, J. (1998). Case-based classification of ultrasonic B-Scans: Case-base organisation and case retrieval. In B. Smyth & P. Cunningham (Eds.) *Advances in Case-Based Reasoning* (pp. 100–111). Berlin: Springer Verlag.
5. Grimnes, M. & Aamodt, A.(1996). A two layer case-based reasoning architecture for medical image understanding, In I. Smith & B. Faltings (Eds.) *Advances in Case-Based Reasoning* (pp. 164–178). Berlin: Springer Verlag.
6. Perner, P. (1999). An architecture for a CBR image segmentation system. *Journal of Engineering Application in Artificial Intelligence, 12*(6), 749–759.
7. Micarelli, A. Neri, A., & Sansonetti, G. (2000). A case-based approach to image recognition, In E. Blanzieri & L. Portinale (Eds.) *Advances in Case-Based Reasoning* (pp. 443–454). Berlin: Springer Verlag.
8. Venkataraman, S., Krishnan, R., & Rao, K.K. (1993). A rule-rule-case based system for image analysis. In M.M. Richter, S. Wess, K.D. Althoff, & F. Maurer (Eds.) *First European Workshop on Case-Based Reasoning* (Technical Report SFB 314) (pp. 410–415). Kaiserslautern, Germany: University of Kaiserslautern.
9. Ficet-Cauchard, V., Porquet, C., & Revenu, M. (1999). CBR for the reuse of image processing knowledge: A recursive retrieval/adaption strategy. In K.-D. Althoff, R. Bergmann, & L.K. Branting (Eds.) *Case-Based Reasoning Research and Development* (pp. 438–453). Berlin: Springer.
10. Cheetham, W. & Graf, J. (1997), Case-Based Reasoning in Color Matching, In: Leake, D.B. & Plaza, E. (Eds.) *Case-Based Reasoning Research and Development,* (pp. 1–12). Berlin, Springer Verlag
11. Wess, St., Globig, Chr., 1994. Case-based and symbolic classification. In: Wess, St., Althoff, K.-D., Richter, M.M. (Eds.), Topics in Case-Based Reasoning. Springer, Berlin, pp. 77–91.
12. Rice, S.V., Nagy, G., & Nartker, T.H. (1999). *Optical character recognition: An illustrated guide to the frontier.* London: Kluwer.
13. Klette, R., Stiehl, H.S., Viergever, M.A., & Vincken, K.L. (2000). *Performance characterization in computer vision.* London: Kluwer
14. Richter MM (1998) Introduction to Case-Based Reasoning. In: M. Lenz, B. Bartsch-Spörl, H.-D. Burkhardt, S. Wess (Eds.), Case-based Reasoning Technology: from Foundations to Applications, Springer Verlag 1998, lnai1400, p. 1–16
15. Perner, P. (1998). Different learning strategies in a case-based reasoning system for image interpretation. In B.Smyth & P. Cunningham (Eds.) *Advances in Case-Based Reasoning* (pp. 251–261). Berlin: Springer Verlag.
16. Heister F, Wilke W (1998) An Architecture for Maintaining Case-Based Reasoning Systems, In: B. Smyth and P. Cunningham (Eds.), Advances in Case-Based Reasoning, lnai 1488, Springer Verlag, p. 221–232
17. Lluis Arcos, J., Plaza, E., 1993. A reflective architecture for integrated memory-based learning and reasoning. In: Wess, St., Althoff, K.-D., Richter, M.M. (Eds.), Topics in Case-Based Reasoning. Springer, Berlin, pp. 289–300.
18. Perner, P. (1998). Using CBR learning for the low-level and high-level unit of a image interpretation system. In S. Singh (Ed.) *Advances in Pattern Recognition* (pp. 45–54). Berlin: Springer Verlag.
19. Smith LB (1989) From global similarities to kinds of similarities: the construction of dimensions in development. In: St. Vosniadou and A. Ortony (Eds.), Similarity and Analogical Reasoning, Cambridge University Press, 1989
20. Tversky, A. (1977). Feature of Similarity. Psychological Review 84 (4), 327–350.

21. Zamperoni, P. & Starovoitov, V. (1995). How dissimilar are two gray-scale images? In *Proceedings of the Seventeenth DAGM Symposium* (pp.448–455). Berlin: Springer Verlag.
22. Santini, S. & Jain, R. (1999). Similarity measures *IEEE Transactions on Pattern Analysis and Machine Intelligence,. 21*(9), 871–883.
23. Horikowa, Y. (1996). Pattern recognition with invariance to similarity transformations based on third-order correlations. In *Proceedings of Internatinal Confernce on Pattern Recognition'96*, IEEE Computer Society Press (pp 200–204)
24. Leitao, F. (1999). A study of string dissimilarity measures in structural clustering. In: S. Singh (Ed.) *Advances in Pattern Recognition* (pp. 385–394). Berlin: Springer Verlag.
25. Mehrotra, G. (1993). Similar shape retrieval using a structural feature index. Information Systems, 18 (5), 525–537.
26. P. Perner and W. Paetzold. An Incremental Learning System for Image Interpretion, In: Shape, Structure and Pattern Recognition, D. Dori and A. Bruckstein (Eds.), World Scientific Publishing Co. 1994, pp. 311–323
27. Messmer, B., & Bunke, H. (2000). Efficient subgraph isomorphism detection: A d composition approach. *IEEE Trans. on Knowledge and Data Engineering, 12*(2), 307–323.
28. Horikowa, Y. (1996). Pattern recognition with invariance to similarity transformations based on third-order correlations. In *Proceedings of Internatinal Confernce on Pattern Recognition'96*, IEEE Computer Society Press (pp 200–204).
29. Crouzil, A., Massipo-Pail, L., & Castan, S. (1996). A new correlation criterion based on gradient fields similarity. In *Proceedings of Internatinal Confernce on Pattern Recognition'96*, IEEE Computer Society Press, (pp. 632–636).
30. Moghadda, Nastar, & Pentland (1996). A Bayesian similarity measure for direct image matching. In *Proceedings of Internatinal Confernce on Pattern Recognition'96*, IEEE Computer Society Press, (pp. 350–358).
31. Moghadda, Jebra, & Pentland (1998). Efficient MAP/ML similarity matching for visual recognition, In *Proceedings of Internatinal Confernce on Pattern Recognition'98*, IEEE Computer Society Press, (pp. 876–881).
32. Wilson, D.L., Baddely, A.J., & Owens R.A. (1997). A new metric for gray-scale image comparison, *International Journal of Computer Vision, 24*(1), 5–19.
33. van der Heiden, A., & Vossepoel, A. A landmark-based approach of shape dissimilarity. In *Proceedings of the International Conference on Pattern Recognition'99*, IEEE Computer Society Press, (pp. 120–124).
34. Surma, Tyburcy J (1998) A Study on Competence-Preserving Case Replacing Strategies in Case-Based Reasoning, In: B. Smyth and P. Cunningham (Eds.), Advances in Case-Based Reasoning, lnai 1488, Springer Verlag 1998, p. 233–238
35. P. Perner, Case-base maintenance by conceptual clustering of graphs, Engineering Applications of Artificial Intelligence, vol. 19, No. 4, 2006, pp. 381–295.
36. Wess, St., Althoff, K.-D., Derwand, G., 1993. Using k-d trees to improve the retrieval step in case-based reasoning. In: Wess, St., Althoff, K.-D., Richter, M.M. (Eds.), Topics in Case-based Reasoning. Springer, Berlin, pp. 167–182.
37. McSherry, D., 2001. Precision and recall in interactive case-based reasoning. In: Aha, D.W., Watson, I. (Eds.), Case-Based Reasoning Research and Development, LNAI 2080. Springer, Berlin, pp.392–406.
38. Bergmann, R., Stahl, A., 1998. Similarity measures for object-oriented case representations. In: Smith, B., Cunningham, P. (Eds.), Proceedings: Advances in Case-Based Reasoning, LNAI 1488. Springer, Berlin, pp. 25–36.

39. Uehara, K., et al., 1993. PBL: prototype-based learning algorithms, topics in case-based reasoning, workshop EWCBR-93. Springer, Berlin, pp. 261–273.
40. Peresdes, R., Vidal, E. Weighting prototypes a newediting approach. Proceedings of the ICPR200, Vol. 2. IEEE Press, London, pp. 25–28.
41. Huang, Y.S., et al. 2000. Construction optimized prototypes for nearest neighbor classifier. Proceedings of the ICPR200, Vol. 2. IEEE Press, London, pp. 17–20.
42. Jarmulak, J., Craw, S., Rowe, R., 2000. Genetic algorithm to optimise CBR retrieval. In: Blanzieri, E., Portinale, L. (Eds.), EWCBR 2000, LNAI 1898. Springer, Berlin, pp. 136–147.
43. Aha, D., 1998. Feature weighting for lazy learning algorithms. In: Lui, H., Motoda, H. (Eds.), Feature Extraction, Construction anSelection: A Data Mining Perspective. Kluwer, Norwell, MA.
44. Bonzano, A., Cunningham, P., 1998. Learning feature weights for CBR global versus local.
45. Leake, D.B., & Wilson, D.C. (2000). Remembering why to remember: performance-guided case-base maintenance, In E. Blanzieri & L. Portinale (Eds.) *Advances in Case-Based Reasoning* (pp. 161–172). Berlin: Springer Verlag.
46. Portinale, L., Torasso, P., & Tavano, P. (1999). Speed-up, quality, and competence in multi-modal reasoning In K.-D. Althoff, R. Bergmann, & L.K. Branting (Eds.) *Case-Based Reasoning Research and Development* (pp. 303–317). Berlin: Springer.
47. Smyth, B., & McKenna, E. (1998). Modeling the competence of case-bases. In B. Smyth & P. Cunningham (Eds.) *Advances in Case-Based Reasoning* (pp. 208–220). Berlin: Springer Verlag.
48. St. Montani, L. Portinale, R. Bellazzi, G. Leonardi: RHENE: A Case Retrieval System for Hemodialysis Cases with Dynamically Monitored Parameters. In: P. Funk, P. A. González-Calero (Eds.): Advances in Case-Based Reasoning, lncs 3155, Springer Verlag 2004, p. 659–672
49. I. Bichindaritz, E. Kansu, K.M. Sullivan: Case-Based Reasoning in CARE PARTNER: Gathering Evidence for Evidence-Based Medical Practice. In: B. Smyth and P. Cunningham (Eds.): Advances in Case-Based Reasoning, lncs 1488, Springer 1998, p. 334–345
50. Bichindaritz, S. Akkineni: Concept Mining for Indexing Medical Literature. In: P. Perner and A. Imiya (Eds.): Machine Learning and Data Mining in Pattern Recognition, lncs 3587, Springer 2005, p. 682–691
51. R. Schmidt, O. Vorobieva: Adaptation and Medical Case-Based Reasoning Focusing on Endocrine Therapy Suppor, In: S. Miksch, J. Hunter, E.T. Keravnou (Eds.): Artificial Intelligence in Medicine, lncs 3581 Springer 2005, p. 300–309.
52. R. Schmidt, L. Gierl, Temporal Abstractions and Case-Based Reasoning for Medical Course Data: Two Prognostic Applications. In: Petra Perner (Ed.): Machine Learning and Data Mining in Pattern Recognition, lncs 2123 Springer 2001, p. 23–34
53. E. Olsson, P. Funk, M. Bengtsson: Fault Diagnosis of Industrial Robots Using Acoustic Signals and Case-Based Reasoning. In: P. Funk, P.A. González-Calero (Eds.): Advances in Case-Based Reasoning, lncs 3155, Springer 2004, p. 686–701
54. M. Nilsson, P. Funk: A Case-Based Classification of Respiratory Sinus Arrhythmia. In: P. Funk, P.A. González-Calero (Eds.): Advances in Case-Based Reasoning, lncs 3155, Springer 2004, p. 673–685

55. Ch. F. Eick, A. Rouhana, A. Bagherjeiran, R. Vilalta: Using Clustering to Learn Distance Functions for Supervised Similarity Assessment. In: P. Perner, A. Imiya (Eds.): Machine Learning and Data Mining in Pattern Recognition, lncs 3587, Springer 2005, p. 120–131
56. L. Bobrowski, M. Topczewska: Improving the K-NN Classification with the Euclidean Distance Through Linear Data Transformations. In: Petra Perner (Ed.): Advances in Data Mining, Applications in Image Mining, Medicine and Biotechnology, Management and Environmental Control, and Telecommunications, lncs 3275 Springer 2004, p. 23–32
57. J.D. Carswell, D.C. Wilson, M. Bertolotto, Digital Image Similarity for Geo-spatial Knowledge Management. In: S. Craw, A.D. Preece (Eds.): Advances in Case-Based Reasoning, lncs 2416, Springer 2002, p. 58–72
58. D. O'Sullivan, E. McLoughlin, M. Bertolotto, D.C. Wilson: A Case-Based Approach to Managing Geo-spatial Imagery Tasks, In: P. Funk, P.A. González-Calero (Eds.): Advances in Case-Based Reasoning, lncs 3155, Springer 2004, p. 702–712
59. B. Bhanu, A. Dong: Concepts Learning with Fuzzy Clustering and Relevance Feedback. Petra Perner (Ed.): Machine Learning and Data Mining in Pattern Recognition, lncs 2123, Springer 2001, p. 102–116
60. J.M. Corchado, E. Corchado, J. Aiken: An IBR System to Quantify the Ocean's Carbon Dioxide Budget. In: P. Perner (Ed.): Advances in Data Mining, Applications in Image Mining, Medicine and Biotechnology, Management and Environmental Control, and Telecommunications, lncs 3275, Springer 2004, p. 33–41
61. G. Giacinto, F. Roli: Dissimilarity Representation of Images for Relevance Feedback in Content-Based Image Retrieval. In: P. Perner, A. Rosenfeld (Eds.): Machine Learning and Data Mining in Pattern Recognition, lncs 2734, Springer 2003, p. 202–214
62. S. Craw, Introspective Learning to Build Case-Based Reasoning (CBR) Knowledge Containers. In: P. Perner and A. Rosenfeld (Eds.): Machine Learning and Data Mining in Pattern Recognition, lncs 2734, Springer 2003, p. 1–6.
63. M.S. Costa, L.G. Shapiro: Relational Indexing. In: P. Perner, P. Shen-Pei Wang, A. Rosenfeld (Eds.): Advances in Structural and Syntactical Pattern Recognition, lncs 1121 Springer 1996, p. 130–139
64. Jänichen, S. & Perner, P. (2006), Conceptual Clustering and Case Generalization of 2-dimensional Forms, *Computational Intelligence*, Volume 22, Number 3/4, 2006, p. 177–193.
65. Jiang X, Muenger A, Bunke H (2001) On median graphs: properties, algorithms, and applications. IEEE Trans. on Pattern Analysis and Machine Intelligence, 23:1144–1151
66. M. Minor, and A. Hanft. The Life Cycle of Test Cases in a CBR System. In E. Blanzieri, and L. Portinale, (Eds.), Advances in Case-Based Reasoning, lnai 1898, Springer 2000, p. 455–466.
67. E. Blanzieri, F. Ricci: Probability Based Metrics for Nearest Neighbor Classification and Case-Based Reasoning. In: K.-D. Althoff, R. Bergmann, K. Branting (Eds.): Case-Based Reasoning and Development, Third International Conference, lncs 1650 Springer 1999, p. 14–28
68. P. Perner, A. Holt, M. Richter, Image Processing in Case-Based Reasoning, The Knowledge Engineering Review, Vol. 20:3, 311–314
69. N. Arshadi, I. Jurisica, Feature Selection for Improving Case-Based Classifiers on High-Dimensional Data Sets. FLAIRS Conference 2005: 99–104.

70. Aha, D.W. Breslow, L & Muñoz-Avila, H. Conversational Case-Based Reasoning. Applied Intelligence 14(1): 9–32.
71. Shimazu, H. ExpertClerk: A Conversational Case-Based Reasoning Tool for Developing Salesclerk Agents in E-Commerce Webshops. Artif. Intell. Rev. 18(3–4): 223–244 (2002)
72. 21. Schmitt, S., Dopichaj, P., Domínguez-Marín, P. 2002. Entropy-based vs. Similarity-influenced: Attribute Selection Methods for Dialogs Tested on Different Electronic Commerce Domains. SLNAI 2416, Springer Verlag.
73. Perner, P., Perner H., and B. Müller. Similarity Guided Learning of the Case Description and Improvement of the System Performance in an Image Classification System. In: S. Craw and A.Preece (Eds.), Advances in Case-Based Reasoning, ECCBR2002, Springer Verlag, LNAI 2416, pp. 604–612, 2002.

2

Similarity

M.M. Richter

Department of Computer Science, University of Calgary, 2500 University Dr., Calgary, AB, T2N 1N4, Canada mrichter@cpsc.ucalgary.ca

Summary. In this chapter, we analyze and discuss the concept of similarity. Similarity plays an important role in many computer applications. These are often tasks that have either no precise input description or where the solution can only be approximated. We consider two main methodologies, case-based reasoning (CBR) and pattern recognition (PR). The specific tasks we deal with are mainly classification, diagnosis, image understanding, and information retrieval. The way similarity enters such scenarios is quite diverse and therefore we will present a relatively broad investigation of similarity relations and measures. When solving a task, similarity is not the only concept that is used; therefore, the overall process and its methods have to be considered. That means, besides the syntax of similarity concepts we discuss also the meaning and the semantics. For this, the semantics is ultimately reduced to utility.

In order to solve problems some knowledge is necessary. This knowledge has different sources and can be used in different ways. Therefore, we put some emphasis on the question what kind of knowledge is needed, what is contained in a measure and how is it entered into a system. This is quite different in CBR and PR. Here we make use of the concept of knowledge containers and of the local–global principle.

For illustration, we also present many examples of problems, measures, and application types. Because of the diverging terminology we are also heading toward a unified view on the subject.

2.1 Introduction and History

Similarity research, often in the form of its dual distances, is an established area in mathematics, for example, in topology and approximation theory. This is somewhat different in computer science and in computer applications. In computer science, approximation is often used in a nonsystematic and ad-hoc manner. In this context, the essential concept of similarity occurs in many forms, interpretations, and many applications.

The concept of similarity has many owners and appears in many shapes. Despite the differences, they all have in common, similarity is used for comparing two (or more) objects, two situations, two problems, etc. for various

M.M. Richter: *Similarity*, Studies in Computational Intelligence (SCI) **73**, 25–90 (2008)
www.springerlink.com

reasons. There is always some purpose for such a comparison, because a subsequent action is taken and ultimately a certain problem has to be solved. For that reason, the two objects to be compared play different roles. The first object is under actual consideration and is called the *problem* object. The second object is already known and stored; often it is called a *prototype* or a *case*.

Similarity is used indirectly in the problem solving process. Prominent indirect uses are analogy-based methods, case-based reasoning (CBR), and pattern recognition (PR). These are related to each other and there is no clear boundary between analogy and the other methods. Here we adopt the view that analogy relates objects across domains, while CBR and PR use similarity within some domain.

A basic difference to analogy is that in CBR often (but not always) objects described in the same description language and in the same terminology are considered, while in analogy completely different theories can be considered. These theories are compared in the structural mapping theory (SMT), see [1]. Historically, analogy was already used by Leibniz [2]. In CBR, early work adopted this view and was presented in [3] and [4].

Similarity is always used for describing something like "closely related." The dual notion is *dissimilar* or distant for describing "far away." These are not crisp notions; they have a number of possible degrees. In addition, these concepts do not have an universal definition like equality but rather very different interpretations. In our examples we will frequently switch between the concepts of similarity and distance.

CBR is a very general way for solving problems by making use of previous experiences. These experiences are recorded in a database called case base. The underlying idea of employing similarity for reusing previous experiences was quite naïvely formulated in the

CBR – Paradigm: If two problems are similar then they have similar solutions.

CBR has as a basic assumption that experiences are available. Under this condition, CBR can be applied to almost all types of applications. Often one has very many experiences stored and an essential aspect is to find useful experiences rapidly (the retrieval problem). Later on we will describe some extensions of CBR that make use of the similarity techniques and do not need experiences.

In **Databases**, similarity is related to search, too, and there is some relation to CBR. Mostly, in databases exact match is required. Similarity measures play a role in some special databases like spatial or geo-databases.

Pattern Recognition is also a very general problem and methodology area that studies the operation and design of systems that recognize patterns in data. Because such patterns are not always identical, the concept of similarity often plays a crucial role.

Both, CBR and PR are methodologies that have developed powerful techniques. In the remaining applications we consider some problem types in these areas. They often use of these methodologies in a different way. Many variations and new methods have been developed.

In **Classification and Cluster Analysis**, similarity is used for classifying objects: Similar objects belong to the same class or cluster and dissimilar objects belong to different ones.

Diagnosis is an extension of classification: One does not only want to know to which class an object belongs, but one also wants to design a subsequent therapy or a repair.

In **Image Interpretation** images are interpreted with respect to their meaning and they are compared. For example, an actual medical image and a certain nonpathological one are compared; the similarity between these images is used to tell whether the actual image contains a pathology or not. Image identification falls into this area too.

There are other application areas in which similarity plays a role but we will not investigate them in detail, see Sect. 2.11.

In **Cognitive and Social Psychology**, similarity played a role in connection with memory since about 1920 when the schema theory came up. It was used, for example, to describe recognizing earlier structures in the memory, see [5]. Important is that similarity is something subjective. It refers to how closely attitudes, values, interests, and personality match between people. There are different kinds of psychological models of similarity and four very prominent ones are geometric, featural, alignment-based, and transformational.

In **Engineering Disciplines**, in particular in mechanical engineering, similarity reasoning is considered as a methodology, see [6]. It is mainly used in connection with analogy. There are two main considerations:

1. The first one considers physical processes with respect to similarity under the assumption that the participating magnitudes are the same.
2. The second one considers physical processes under the assumption that the describing differential equations are the same or closely related, while the participating magnitudes can be different.

The latter two applications are very much of historical interest because they show that similarity was already a research topic many decades ago. However, we will not investigate them here furthermore.

The aspects have been refined in many ways and some will be discussed below. We do not intend to introduce the reader into the different application and mathematical areas; in fact we assume some familiarities with them. We only will point out what role is played by similarity.

We regard similarity as a concept in some more general problem solving process and, therefore, we are interested in which way similarity makes use of problem solving knowledge.

One of our purposes is to relate the different appearances and uses of similarity to each other. Therefore, we identify several underlying general principles and show their different instantiations. For this we use two abstract principles: The local–global principle and the knowledge containers.

The most important aspect is related to knowledge. Without knowledge one cannot solve problems. The question is how and where knowledge is represented and used. It can be represented in the similarity concept or measure, but it can also be represented at other places. The representation of knowledge will distinguish, for instance, CBR and Pattern Recognition.

2.2 Mathematical and Computational Models

Besides the aspects from psychology, design and art, computer science is interested to make use of similarity in computational models. The goal is aimed at a computer-supported problem solving environment. Here we encounter a confrontation between expressive power and efficiency. The more knowledge a similarity concept can carry, the less efficient its computational processing usually will be.

In this section, we first present principal aspects, and in the next sections examples will be shown.

2.2.1 Relations vs. Functions

In order to understand the scope of the computational models it is helpful to think of two origins of the similarity concept (later on we show other influences), as shown in Fig. 2.1

From both ancestors the similarity concept has inherited not only properties of intuitive character but, in addition, certain precise notions, axioms, and techniques for applications. In the past, both have rarely been discussed together. For investigating similarity, this view is appropriate, however. From fuzzy sets similarity inherits its vague character and from metrics the aspects of being close together. A short summary of axiomatic properties gives Table 2.1.

In the transitivity axiom, min can be replaced by any other t-norm. Often transitivity is seen as a counterpart of the triangle inequality. But these concepts are not logically equivalent. In fact, first it has to be made clear what equivalence should mean precisely in an abstract way.

These axioms have been questioned and will be discussed below.

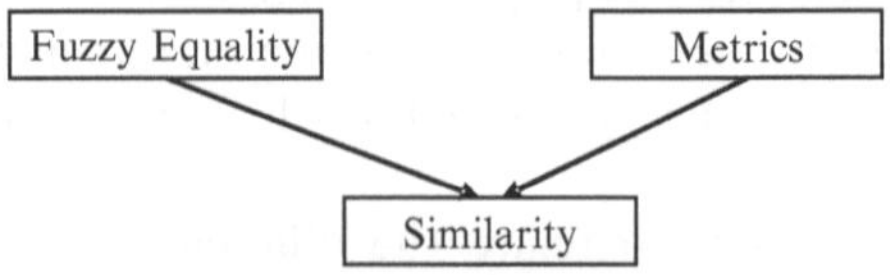

Fig. 2.1. Ancestors of similarity

Table 2.1. Possible axioms

	Fuzzy equality	Metric
Reflexivity	E(x, x) = 1	d(x, x) = 0
Symmetry	E(x, y) = E(y, x)	d(x, y) = d(y, x)
Transitivity	$E(x, y) \geq \sup \{(\inf (E(x, z),\ E(z, y))\|\}$	?
Triangle inequality	?	$d(x, y) \leq d(x.z) + d(z, y)$
Substitution axiom	$(s(a) = t(a) \wedge a = b) \rightarrow (s(a) = t(b))$	?

Basically, there are two ways to represent similarity in computational models:

(a) As a relation
(b) As a function

(a) Relational models:
There are three kinds of relational models:
(i) A binary similarity predicate:
SIM(x,y) ⇔ "x and y are similar"
(ii) A binary dissimilarity predicate:
DISSIM(x,y) ⇔ "x and y are dissimilar"
(iii) Similarity as a partial order relation:
R(x,y,z) ⇔ "x is at least as similar to y as x to z"
For simplicity, we assume that all elements are taken from the same underlying set. Obviously, the similarity and the dissimilarity predicate are definable from each other. To view similarity as a binary predicate looks, however, too simple in order to grasp the intuition. On the other hand, if similarity is the basis of a binary decision, this view is essential. The step to obtain a binary predicate from a more complex as introduced below can be regarded as a defuzzification step.
The predicate in (iii) is more powerful in expressiveness, because different degrees of similarity can be distinguished. A generalization of (iii) is
(iv) S(x,y,u,v) ⇔ "x is at least as similar to y as u is to v."

This allows to define R as R(x,y,z):↔ S(x,y,x,z).

Most important, with relation R one can define the nearest neighbor concept:

For some fixed x each y that satisfies R(x,y,z) for all z is called a *nearest neighbor* of x.

Notation: NN(x,y). This means NN is a relation and the nearest neighbor is not necessarily uniquely defined; there may be several such (equally similar) elements.

A slight extension is to look at the first k nearest neighbors what is denoted as the k-NN methods.

In applications, the nearest-neighbor relation often is the most important one. The reason is that *closer* has in general the meaning of *better*; a nearest neighbor is the *best*. We will discuss this in the section on semantics. The computation of the nearest neighbor is known as the retrieval problem and it is a big problem area in its own. We will not discuss this in detail, however, and remark only that similarity computation should be efficient.

There are several possible properties ("axioms") that the relation S can enjoy:

For all x, y, z, u, v, s, t holds:

(1) Reflexivity:

$S(x, x, u, v)$: Every object is at least as similar to itself as two other arbitrary objects could be to each other.

(2) Symmetry:

$$S(x, y, u, v) \leftrightarrow S(y, x, u, v) \leftrightarrow S(x, y, v, u)$$

(3) Transitivity:

$$S(x, y, u, v) \wedge S(u, v, s, t) \rightarrow S(x, y, s, t)$$

These axioms are intuitively plausible but we will question them later on.

(b) Functional Models

Similarity functions ("measures") are used to refine the relational approaches by assigning a degree of similarity to two objects. They are defined on ordered pairs:

$$\mathrm{sim}: \ \mathrm{U} \times \mathrm{U} \rightarrow [0, 1].$$

This notion can be extended to relate elements of different sets to each other:

$$\mathrm{sim}: \ \mathrm{U} \times \mathrm{V} \rightarrow [0, 1].$$

Hence, similarity functions can be considered as fuzzy sets of ordered pairs. From fuzzy equality similarity functions have inherited four possible axioms:

(1) $\mathrm{sim(x,\ x)} = 1$ (reflexivity)
(2) $\mathrm{sim(x,\ x)} = \mathrm{sim(y,\ y)}$ (constancy of self-similarity)
(3) $\mathrm{sim(x,\ y)} = \mathrm{sim(y,\ x)}$ (symmetry)
(4) $\mathrm{sim(x,y)} \geq \sup \{(\mathrm{inf(sim(x,z),\ sim(z,x))}\}$

The requirement that similarity functions are bounded to [0,1] is some kind of restriction. It may lead to difficulties and may be given up.

Similarity measures have besides the ordinal information also a quantitative statement about the degree of similarity, as illustrated in Fig. 2.2.

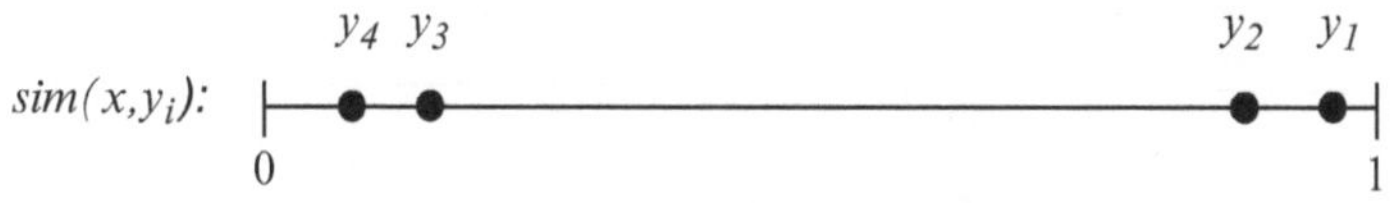

Fig. 2.2. Degrees of dissimilarity

The dual concepts for similarity measures are distance functions that are supposed to represent dissimilarity.

$$d : U \times V \rightarrow [0, 1].$$

Often the range of distance functions is not bounded.

For pseudo-metrics besides reflexivity and symmetry, an additional axiom is common:

(5) $d(x, z) \leq d(x, y) + d(y, z)$ (triangle inequality).
For metrics the reflexivity is sharpened to
(6) $d(x, y) = 0 \leftrightarrow x = y$.

One often encounters the problem that there are very many nearest neighbors (or almost nearest neighbors) to some object that are not close to each other. The selection principle is aimed at determining a maximal set of objects that are dissimilar to each other with a minimum degree of dissimilarity (diversity problem), see [7]. This problem is quite involved.

From a principal point of view, similarity measures and distance functions are equivalent: "very close" is the negation of "far away." There are, however, two kinds of difficulties:

1. What is easily definable for similarity functions may be difficult to express in terms of distances and vice versa. For instance, the triangle axiom cannot be easily expressed in terms of similarity measures.
2. Different kinds of techniques have been developed for similarities and distances. For instance for similarity measures, techniques for finding nearest neighbors were cultivated, while for distances one was more interested in finding objects that were far away from each other.

Both, similarity measures sim and distance functions d induce similarity relations S_{sim} and S_d in an obvious way. There is a natural relation between them:

A similarity measure *sim* and a distance measure *d* are called compatible if and only if

- For all x, y, u, v $\in$ M : $S_d(x, y, u, v) \leftrightarrow S_{sim}(x, y, u, v)$ (the same similarity relation is induced)
  Transformation of measures: If a *bijective, order inverting* mapping f: $[0, 1] \rightarrow [0, 1]$ exists with
- $f(0) = 1$
  $f(d(x, y)) = sim(x, y)$
  then *sim* and *d* are *compatible.*

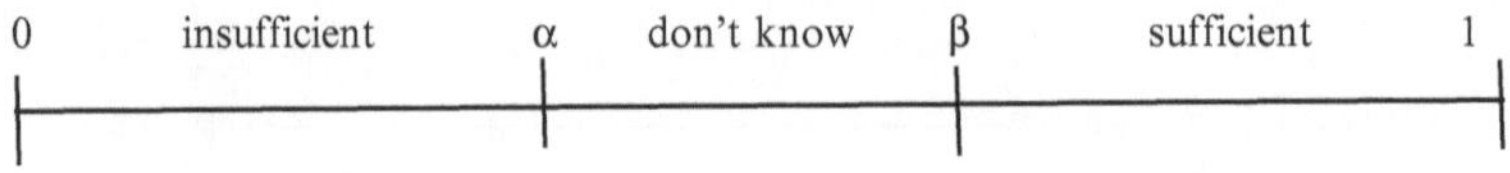

Fig. 2.3. Uncertainty area

If one puts the focus on a fixed object x (usually the current problem) we have additional useful relations induced $\geq_x$, $>_x$, $\sim_x$ defined by:

1. $y \geq_x z \Leftrightarrow R(x, y, z)$ "y is at least as similar to x as to z"
2. $y >_x z \Leftrightarrow (y >_x z) \wedge \neg(z \geq_x y)$ "y is more similar to x than to z" (strict part of $\geq_x$)
3. $y \sim_x z \Leftrightarrow (y \geq_x z) \wedge (z \geq_x y)$ "y and z are indistinguishable"

If only a few objects are given, it is easier for a human to provide a relational similarity than a quantitative one. For a large number of objects, a similarity measure is more useful. However, one has to be careful: If one looks at the induced relation of a measure, then it may very well happen that it disagrees with the relational intuition which one had in mind, and the measure has to be improved.

By its very nature, similarity contains an essential element of uncertainty. On the other hand, if similarity is involved in some process, for example, for reaching a decision, one often encounters a "yes or no" problem. That means, some defuzzification has to take place. This can be done by introducing thresholds and proceed in the spirit of Rough Set Theory.

Example (see Fig. 2.3): Decisions based on sufficient similarity: We introduce two thresholds α and β that define three areas:

If the similarity is in the "insufficient" area the solution is not accepted, in the "sufficient" area it is accepted. The "don't know" area is the *uncertainty area* and one needs additional criteria for a decision.

2.2.2 An Abstract View: The Local–Global Principle

In the sequel, we will discuss many examples for similarity relations and measures. In order to relate different representations and functions to each other we need some structural principle that is shared by them and allows discovering common aspects as well as differences.

In order to formulate a systematic and general approach, we introduce a general structural representation principle (see [8]). It is based on the view that (complex) objects to be compared are built up in a systematic way; the structure of the similarity measure is "parallel" to the structure of the objects. The objects can be thought as very general ones like machines, the human body, images, etc.

This structure is called the *local–global principle* for complex object descriptions; it says:

(1) There are *local* (atomic) description elements; for simplicity, we assume that these are attributes
(2) Each object or concept A is *(globally)* described by some construction operator C from the local elements:

$$A = C(A_i | i \in I).$$

Here I is some index set for the atomic elements (i.e., the attributes).

The local–global principle is also formulated for similarities and is stated as:

There are *local measures* sim_i on the attributes A_i and there is some amalgamation function F such that for a, b $\in$ U, U being the universe under consideration, $a = (a_i | i \in I),\ b = (b_i | i \in I)$

$$sim(a,\ b) = F(sim_i(a_i, b_i) | i \in I).$$

The local measures sim_i compare values of individual attributes and sim is the global measure. The amalgamation function F is the construction operator and it is important that the two construction operators for objects and measures are closely related. In fact, the definition of the measure should follow the structure of the objects. We will discuss this principle below for various examples.

The principle gives rise to two tasks:

(a) The decomposition task: Break the object or concept down into atomic parts. This task often occurs because the object may be presented globally and the parts are initially unknown.
(b) The synthesis task: Compose an object or concept from simpler parts.

Both tasks play a role for relating the concepts we are interested in. For the objects under investigation, the local–global principle has very different realizations. We will discuss it for representations of symbolic character as well as for approximation-oriented representations.

The claim is not that the principle itself is very innovative, in fact, it is quite standard. The point is the unified use of the principle in order to allow a systematic treatment of the different techniques. For this purpose we will briefly introduce these techniques and discuss them from this point of view.

There is also the term *structural similarity*; it is used in two ways. On the one hand, it means to consider structural aspects of the problems or cases in the sense of the local–global principle but it can also mean to consider the similarity of a whole set of cases [9].

2.2.3 Semantics

It is important to observe that the axioms do not carry any specific knowledge about the domain or the problem. They reflect properties, which hold in all applications of the concepts.

In the subsequent sections, we will discuss how these axioms have partially to be given up due to the different interpretations of the similarity concept.

The axioms also do not in any way determine the specific choice of the measure; they only restrict the choice. It is also not clear when such axioms are acceptable.

This depends on the intended meaning of the measure, i.e., on the semantics.

Our starting point is that determining the semantics of a similarity between two objects needs to consider the overall process, as for example, classification, diagnosis, or planning. The similarity is supposed to be useful for the process and therefore the similarity measure must "know" something about this process. This knowledge should be reflected in the semantics of the measure.

The local–global principle splits the semantics of a measure into two parts:

- The local measures reflect the intentions of the specific attributes of the domain and therefore local domain knowledge.
- The global measure puts the local measures into a global relation and reflects usually, for example, the importance of the attributes for the task under investigation.

The views on similarity have undergone several changes in the last 20 years, in particular in CBR. Basically, one can observe three phases that are not clearly separated. For a detailed description of the phases, see [10]. In the first phase, a very naïve and intuitive view was dominating that was slowly changed to more abstract views. This was leading to a much broader scope of applications, but has, on the other hand, preserved the computational properties developed initially.

The Naïve Period

In this period, two objects were considered as *similar* if they *looked similar*. For classification, the interpretation was: Objects that look similar are in the same class. In diagnosis, it meant that similar looking observations lead to the same diagnosis and similar looking images describe the same situation. Of course, looking similar is something subjective and this was discussed in cognitive psychology.

The CBR-paradigm then reads as

Two problems that look similar have similar solutions.

The property of "*looking similar*" was taken naïvely and not questioned very much. The practical interpretation of the CBR paradigm then was: Problems, which look similar, can be attacked with the same or almost the same solution method.

The applications at that time did not need much knowledge. They were in principle quite simple but nevertheless useful because they saved humans much problem solving time.

The crucial topics and techniques of CBR have been formulated in the naïve period. In a nutshell, they are similarity-based matching and the nearest neighbor search. The use of the similarity measure in using the experience base is to select the most similar experience, i.e., the nearest neighbor to the actual problem. Because the actual problem and the selected problem will still differ, an adaptation of the solution may become necessary. But even then there is no guarantee for getting a correct solution: CBR is an approximation technique. This differs essentially from the database view.

The Sophisticated Period: Utility Orientation

Over time applications became more involved and the naïve view turned out to be more and more insufficient. In the beginning of the nineties, demands from applications asked to pay more attention to the specific task and to use more involved knowledge, in particular, when defining measures. It was no longer sufficient or necessary that objects *looked similar* but to find out what is really important for certain purposes. One had to investigate:

- What influences a classification, a diagnosis, or a more general problem solution?
- Are all *important* description elements listed? What means important?
- Which description elements are important for what?
- Which relations and dependencies of interest exist between them?

The term *important* can be interpreted in different ways. The uses of this term in CBR can essentially be interpreted in the sense of usefulness, i.e., in the sense of utility theory. This means the degree of similarity is the degree of utility of the solution for a previous problem for the actual problem (see [11]). This aspect was generalized later on, see below. In this view, similarity may be far away from the intuitive notion of the naïve period and one may ask whether the term "similar" is still justified. This is in fact questionable but the term was kept, because all computational properties and all tools needed for CBR were still applicable.

A crucial theoretical point is that the reduction to utility is a way to provide a rigorous foundation for the semantics of the concept of similarity. The reason is that, at least in principle, utility can be evaluated by an outside and independent experiment. It also explains the origins of the many variations of similarity. Truth is an absolute concept (at least in logic) while utility has to be defined in any single situation. This is related to the fact that there is only one equality notion, but there are infinitely many inequalities. In addition, the theoretical concepts and results of utility theory can be used. A practical aspect is that the similarity assessment now has a guideline.

In this sense, similarity is an a-priori approximation of utility and a goal of similarity modeling is to provide a good approximation of the intended utility. On the relational level, similarity then compares to preference relations: More similar cases are preferred.

From utility theory, we need the following notions and concepts:

Utility Functions and Preference Relations

In order to describe the utility orientation of the similarity concept we discuss utility itself briefly. The relational version is the preference relation expressing what one finds more useful. Utility functions are the functional versions of preference relations; they assign real numbers as values to the elements of the domain:

$$u : U \rightarrow \mathrm{IR}.$$

If $u(a) > 0$, we call it the benefit, otherwise the cost of a. Mostly u is considered as bounded; in this case, we can assume without loss of generality that the values of u are in the interval $[-1, 1]$.

The weaker relational formulation for utility functions uses preference relations. Both, utility functions and preference relations usually are complex. The local–global principle demands that they are defined on objects $A = C(A_i | i \in I)$.

The *local–global principle for utility functions* u (and analogously for preference relations) says that u can also be represented as

$$u(a) = G(u_i(a_i) | i \in I)$$

where the u_i are called the local utilities. It is desirable to model the situation in such a way that the global utility is a linear combination of local utilities:

$$u(a) = u_w(a_1, ..., a_n) = \Sigma(w_i \cdot u(a_i), 1 \leq i \leq n).$$

The vector $w = (w_1, \ldots, w_n)$ of real-valued coefficients is called the weight vector of the representation. Generally, the global utility is in the center of interest to the user and is given in the first place. In order to represent it formally, one has to find the influence factors, i.e., to perform a sensitivity analysis that finally breaks the utility function down to local components; this is the decomposition task. This task is often quite difficult. The point is that it has to be solved before the similarity can be defined.

For defining the similarity measure, one often has to go in an opposite direction starting with the elements that can be influenced directly, i.e., one has to perform a synthesis task. That means, one has to define the local similarities first and then to define the construction operator. The local similarity measures are defined in such a way that they are in correspondence to the local utilities. The construction of the global similarity should reflect among others importance that is present in the utility function. In this construction,

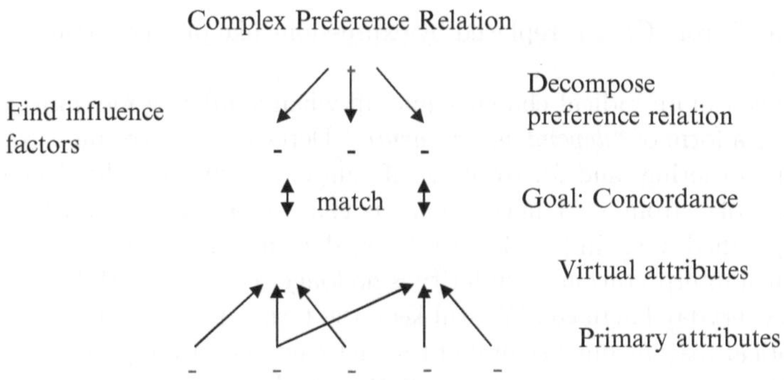

Fig. 2.4. Preference and similarity

Table 2.2. General partners

Observations	Class
Symptoms	Therapy
Knowledge needed	Documents
Questions	Answers
Functionalities	Machines
Workers	Coworkers
Images	Meaning
Desired products	Available products

new attributes called *virtual attributes* (as opposed to the original primary ones, see Sect. 2.4.3), are introduced. These are attributes that are of major importance and can be defined in terms of the given primary attributes.

An exact match between similarity and utility cannot be expected here. A weaker and more practical demand is that utility and similarity are *concordant*; this means, one function increases or decreases if and only if the other one does, see Sect. 2.3. Figure 2.4 shows how the decomposition of preferences and the synthesis of attributes are related.

The Generalized Period

The view of the sophisticated period was generalized extensively and the most recent extensions (starting at the end of the nineties) could be named as "*partnership measures.*" This leads to connecting similarity and utility even stronger. Because the possible partners may come from different sets U and V, the domain of sim has to be generalized to U × V. The intention of partnership is that both objects cooperate more or less well as partners. In the extended view, the measure compares objects like the ones seen in Table 2.2.

This view is today dominating in information retrieval (IR) and many e-commerce applications. An immediate observation is that, as a consequence, the axioms of reflexivity and symmetry have to be given up. Also, the term "case base" from CBR is replaced by expressions like product base, document base, etc.

An even more radical change came up when similarity measures were regarded as a form of "*dependency measure*." Dependency introduces some kind of partial ordering, and for example, if sim(x, y) expresses the degree of dependency of y from x. A major point is that symmetry for sim is again no longer justified, i.e., sim(x, y) = sim(y, x) does not hold any more.

In such interpretations, similarity is no longer coherent with the use of this term in everyday language. We still keep the term similarity not only because of historical reasons but because of the fact that the techniques developed for similarity reasoning apply here as well. Such techniques cover mainly retrieval algorithms, the structure of the database, the assessment of measures, and maintenance operations.

The leading and unchanged motivation in this (ongoing) period was still the view that similarity approximates or imitates utility; more or less all other aspects were on stake. In particular, and most importantly, CBR was not anymore restricted to the use of previous experiences. Important aspects subject to generalization were:

1. Scope of applications (the driving force)
2. Objects to be compared
3. Knowledge contained and the process of filling in knowledge
4. Properties and axioms

2.3 Basic Examples

In mathematics, metrics are discussed in detail. There is little domain knowledge involved in the mathematical functions; for applications, one has to find out when functions are useful and with which other knowledge can be combined. For our purposes, the role of knowledge is crucial and we will have a close eye on it. We will start with some elementary examples.

2.3.1 Global Measures With Numerical Arguments

A basic similarity measure is the *Hamming measure*, coming from coding theory. It applies to attribute-value representations with Boolean-valued attributes:

$$\mathrm{H}((\mathrm{a}_1,\ldots,\mathrm{a}_\mathrm{n}),(\mathrm{b}_1,\ldots,\mathrm{b}_\mathrm{n})) = \Sigma(\mathrm{i}|\mathrm{a}_\mathrm{i} = \mathrm{b}_\mathrm{i}, 1 \leq \mathrm{i} \leq \mathrm{n}).$$

The *scalar product* is a variation of the Hamming measure. It is defined as

$$S(x,y) = \sum_{i=1}^{n} x_i \cdot y_i$$

The scalar product is often used in PR.

For binary attributes we distinguish two cases:

(a) The values are 0 and 1: Then only values 1 contribute to the similarity.
(b) The values are −1 and +1: Then, in addition, nonagreeing values give a penalty to the measure.

Both, the Hamming measure and scalar product are of very simple character: The only knowledge they contain is the number of coinciding coordinates.

In applications, the disadvantage is that all attributes in the measure are of equal importance. To overcome this difficulty the weighted Hamming measure was introduced:

$$\mathrm{H_w}((\mathrm{a}_1, \ldots, \mathrm{a}_n), (\mathrm{b}_1, \ldots, \mathrm{b}_n)) = \Sigma(\mathrm{w_i} | \mathrm{a_i} = \mathrm{b_i}, 1 \leq \mathrm{i} \leq \mathrm{n})$$

where $\mathrm{w} = (\mathrm{w}_1, \ldots, \mathrm{w}_n)$ with $\sum \mathrm{w_i} = 1$ is a weight vector of nonnegative coefficients. The local measures are still simple (only test of equality) while the global measure reflects importance of attributes.

A further generalization is obtained by allowing arbitrary ranges of the attributes and arbitrary local similarity measures. If $\mathrm{sim} = (\mathrm{sim}_1, \ldots, \mathrm{sim}_n)$ is such a measure vector, the *generalized Hamming measure* is defined as:

$$\mathrm{H_{w,sim}}((\mathrm{a}_1, \ldots, \mathrm{a}_n), (\mathrm{b}_1, \ldots, \mathrm{b}_n)) = \Sigma(\mathrm{w_i} \cdot \mathrm{sim_i}(\mathrm{a_i}, \mathrm{b_i}) | 1 \leq \mathrm{i} \leq \mathrm{n}).$$

In the same way, weighted Euclidean measures can be defined that read familiar in the form of distances:

$$d(x, y) = \sqrt{\sum_{i=1}^{n} w_i \cdot (x_i - y_i)^2}$$

For the real case, another type of axiom is often used that runs under *invariance properties.* A typical example is *scale invariance*: If T is a linear transformation then d(x,y) = d(Tx, Ty); in a similar way rotation invariance is defined that is important for image understanding.

A measure frequently used in cognitive psychology runs under the name *Tversky feature model or contrast model* (see [12]).

A feature is a Boolean-valued attribute. The basic idea is that the similarity increases if a feature is common to both objects and decreases if it is observed only in one object. In this context, the different types of objects (problem objects p and case objects (prototypes) c) play a role. The features are split into three subsets (depending on p and c):

- Com: The set of features observed for p and c; a = |Com|.
- P-C: The set of features observed for p but not for c (redundant features); b = |P-C|.
- C-P: The set of objects observed for c but not for p (missing features); c = |C-P|.
- NO: Features not observed for p and for c; d = |NO|.

The contrast model assigns a numerical value sim(p,c) for the similarity on the basis of these observations. It is defined by a real-valued function h on sets of features and three positive real numbers α, β, and γ:

$$\mathrm{sim}_{\mathrm{contr}}(\mathrm{p},\mathrm{c}) = \gamma \mathrm{h}(\mathrm{Com}(\mathrm{p},\mathrm{c})) - \alpha \mathrm{h}(\mathrm{P\text{-}C}(\mathrm{p},\mathrm{c})) - \beta \mathrm{h}(\mathrm{C\text{-}P})(\mathrm{p},\mathrm{c}))$$

If the function h is a weighted sum like in the weighted Hamming measure we get

$$\begin{aligned}\mathrm{sim}_{\mathrm{contr}}(\mathrm{p},\mathrm{c}) = \; & \gamma\Sigma(\mathrm{w_i}|\mathrm{i} \in \mathrm{Com}(\mathrm{p},\mathrm{c})) - \alpha\Sigma(\mathrm{w_i}|\mathrm{i} \in (\mathrm{P} - \mathrm{C}(\mathrm{p},\mathrm{c})) \\ & -\beta\Sigma(\mathrm{w_i}|\mathrm{i} \in (\mathrm{C} - \mathrm{P}(\mathrm{p},\mathrm{c}))\end{aligned}$$

In experiments with humans [13], it is reported that humans prefer $\alpha > \beta$, i.e., the similarity increases with the number of features in the actual problem that are also in the prototype while features observed in the problem object only play a lesser role. In [13] it is also reported that for images diverging features are higher rated while in verbal descriptions of the images features in Com are higher rated.

Another slight computational variation of the Tversky measure is the *Jaccard coefficient*. In [14] an extension of the Tversky approach is described. There the local measures on the features are no longer binary (indicating presence or absence of features) but real values; these measures are called Fuzzy Features Contrast Models in [14].

There are many other variations of such measures. They have many applications in databases, PR, image interpretation, and applications in biology. There is much experience when to use which measure but there is no general theory. Next we give a short overview that contains some additional measures besides the ones introduced above.

An early and very comprehensive reference for simple measures is in [15].

There is a huge number of simple similarity measures. Often, they differ only the choice of certain coefficients. This choice is motivated by the needs of some application. Mostly, the background is a deep insight into the application. However, the choice is mostly ad-hoc and there is no principle behind it.

A summary (without details) of some simple measures is shown in Table 2.3.

2.3.2 Simple Measures and Databases

Despite the simplicity of the Hamming measure or the Euclidean distance they have many applications, as in [13] experiments with humans it is reported that humans prefer $\alpha > \beta$ i.e., the similarity increases with the number of features in the actual problem that are also in the prototype while features observed in the problem object only play a lesser role. In [13] it is also reported that for images diverging features are higher rated while in verbal descriptions of the images features in Com are higher rated.

A summary (without details) of some simple measures is shown in Table 2.3.

Table 2.3. Examples of simple measures

	Formula	Comments
Euclidean distance	SQRT(b+c)	The square root of the sum of discordant objects, minimum value is 0, and it has no upper limit
Squared Euclidean distance	b+c	The sum of discordant cases, minimum value is 0, and it has no upper limit
Size difference		An index of asymmetry. It ranges from 0 to 1.
Pattern difference	bc/(n**2)	Dissimilarity measure for binary data that ranges from 0 to 1
Variance	(b+c)/4n	Ranges from 0 to 1
Simple matching	(a+d)/n	This is the ratio of matches to the total number of values. Equal weight is given to matches and nonmatches
Dice	2a/(2a + b + c)	This is an index in which joint absences are excluded from consideration, and matches are weighted double. Also known as the Czekanowski or Sorensen measure
Lance and Williams	(b+c)/(2a+b+c)	Range of 0 to 1. (Also known as the Bray-Curtis nonmetric coefficient)
Nei & Lei's genetic distance	2a/[(a+b)+(a+c)]	
Yule coefficient	(ad−bc)/(ad+bc)	A function of the cross-ratio; has range [−1, 1]. Also known as the coefficient of colligation

Another slight computational variation of the Tversky measure is the *Jaccard coefficient.* In [14] an extension of the Tversky approach is described. There the local measures on the features are no longer binary (indicating presence or absence of features) but real values; these measures are called Fuzzy Features Contrast Models in [14].

There are many other variations of such measures. They have many applications in databases, PR, image interpretation, and applications in biology. There is much experience when to use which measure but there is no general theory. Next we give a short overview that contains some additional measures besides the ones introduced above.

An early and very comprehensive reference for simple measures is in [15].

2.3.3 Simple Measures and Data Bases

Despite the simplicity of the Hamming measure or the Euclidean distance they have many applications, for example, in database search and PR. One reason is that they are simple to compute. This plays a central role in high dimensions. The knowledge in the measures is again of simple character. If the vectors are very long and all arguments are of little importance, then these measures are basically of the same expressive power as the measures equipped with weights.

However, it was discovered that these measures often did not lead to the desired results because the intended semantics was not covered. This was due to the fact that the roles of the local measures as well as of the local–global principle were not sufficiently well understood. Instead, (sometimes very sophisticated) weightings and query expansions (see Sect. 2.9) have been introduced.

Another method to overcome such difficulties is to introduce *elliptic queries.* Traditionally, the nearest neighbors are computed by using circles around the point of interest (the query). If the circles are no longer adequate, the approach uses hyper-elliptic surfaces in order to represent relevant coordinates better. However, this is nothing than introducing weights; it is in one-to-one correspondence to choose the axes of the ellipses.

The drawback of this method as well as taking weights is that the method is trapped into a local optimum, especially because each vector is wrapped with many irrelevant coordinates.

2.3.4 Local Measures With Numerical Arguments

Next we come to measures that can contain more knowledge. Local measures contain knowledge coming from the distribution of the values of the attributes. That means, local measures can contain specific information about the attributes.

Example

Water temperature is an attribute wt with the domain, for example, dom(wt) = [−100, +200]. In numerical domains, we can separate regions in which the similarity undergoes little changes by using landmarks that separate the real axis into intervals. For water temperature we identify three regions: [−100, 0], [0, 100], and [100, 200]. These landmarks reflect facts from physics and not personal relevancies. The local similarity should not be uniform and reflect these facts too: numbers inside of a region should be quite similar but numbers in different regions should have a low degree of similarity. Hence, the measure should perform some kind of qualitative reasoning.

The relevance of temperature for some problem is reflected by a weight.

For attributes A with integer or real values similarity often based on the difference of attribute values and not on the attribute values themselves, for example:

- linearly scaled value ranges: $sim_{\mathrm{A(x,y)}} = f(\mathrm{x} - \mathrm{y})$
- exponentially scaled value ranges: $sim_{\mathrm{A(x,y)}} = f(\log(\mathrm{x}) - \log(\mathrm{y}))$

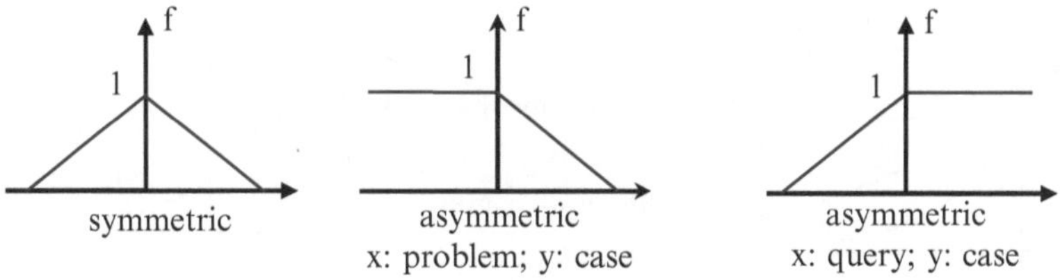

Fig. 2.5. Symmetry and asymmetry

where $f:$ IR $\rightarrow [0..1]$ or Z $\rightarrow [0..1]$. ($f(0) = 1$ indicates reflexivity)

Often one assumes that f(x) is monotonously decreasing for x > 0 or monotonously increasing for x < 0. Symmetric and asymmetric measures are shown in Fig. 2.5.

Discussion

There is a lot of knowledge contained in the measures. The first asymmetric measure says "smaller is better (y $<$ x)" while the second measure says "greater is better (x $<$ y)." As an example, we consider the attribute price (for an object, a therapy, etc.). In the problem x, the desired price is presented and it is perfectly matched in the case for y = x. If in the first example the price to pay and in the second example the price obtained is meant then there are two utilities involved. In the first situation, the utility is still optimal if one has to pay less and in the second example it is still optimal if one gets more. This justifies the shape of the curves. However, there is a subtle point because the utility function may not be bounded. That means, to obtain even higher prices will still increase the utility and this is not reflected by the measure. From the viewpoint of fuzzy sets, these measures would be regarded as fuzzy preferences.

Instead of linear functions, more involved ones for the decrease can occur that contain more knowledge about the problem. Convexity is shown in Fig. 2.6.

- Downward convexity: A small difference between two values causes a big decrease of similarity
- Upward convexity: A small difference between two values causes only a small decrease of similarity
- Combined

Such knowledge is often decisive for technical machines as well as for economic decisions.

If we tolerate differences between the two values even to a defined distance x_1 we can use the following similarity function as shown in Fig. 2.7 (S-Function) what gives rise to an even more sophisticated local measure:

$$Sim = \exp\left((-ax)^3\right) \text{ with } a \in IR$$

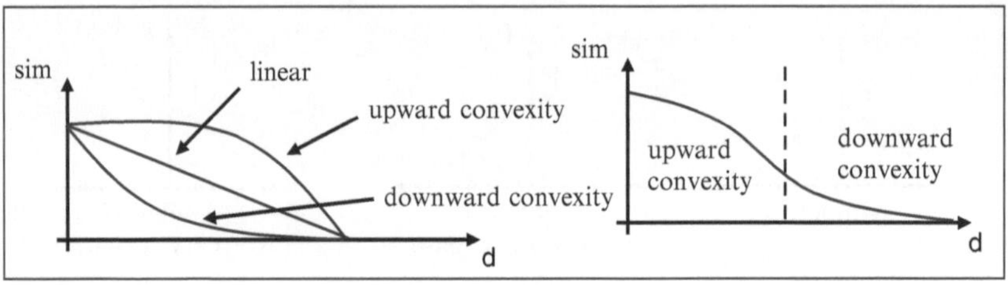

Fig. 2.6. Convexity

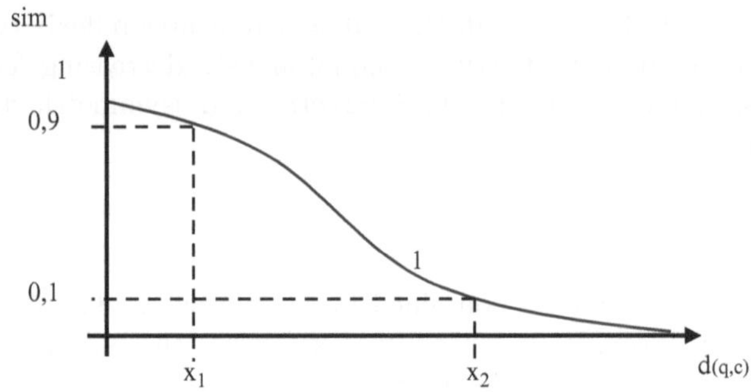

Fig. 2.7. S-Function

As a result, one can say that a local attribute can contain important knowledge for the problem solution. Because the knowledge can be of arbitrary nature there is no general method to represent it in the measure. This depends on the familiarity with the domain and with the representation methods.

These similarity measures reflect very much the utility orientation of the sophisticated and the generalized period. In particular, similarity reflects preference. This shows again that the generalization is useful because all the algorithms and tools can still be used; it is one of the reasons for commercial success of CBR.

In specific situations, one has, of course, to define precisely the intended semantics.

2.3.5 Symbolic Arguments

Symbolic arguments often are coded by numerical values. Symbolic values can simply be sets or they can carry a structure. This can be, for example, an ordering or a taxonomy. These structures are reflected by measures. Other possibilities are interval-based values. Symbolic arguments are often coded by numerical values. Such a coding is sometimes justified but not always. If one encodes symbolic values into numerical values this increases efficiency at run time. On the other hand, some transparency is lost because one looses the connection to the original sources of the values. In particular,

Table 2.4. Similarity table

s[x,y]	v_1	v_2	...	v_k
v_1	s[1,1]	s[1,2]		s[1,k]
v_2	s[2,1]	s[2,2]		s[2,k]
...				
v_k	s[k,1]	s[k,2]		s[k,k]

this creates difficulties in similarity assessment and maintenance, because the original motivations are no longer visible.

For attributes A with unstructured symbolic values $\{v_1, v_2, \ldots, v_k\}$, there is no other way for defining measures than using tables; they are also called similarity matrices.

A similarity table $sim_{A(x,y)} = s[x, y]$ is shown in Table 2.4.

The more structure is defined on the symbolic values the more systematic the similarity measures can be defined.

For reflexive similarity measures diagonal entries are 1 and for symmetric similarity measures we have: upper triangle matrix = lower triangle matrix.

For ordered symbols, an ordering is on the value range. Example: Qualitative values:

$$\{\text{small, medium, large}\}$$

where small < medium < large.

We want to use local similarity measures like for numeric types. Therefore, we define an integer value for each symbol so that the ordering is preserved. Example:

- Small $\rightarrow$ 1
- Medium $\rightarrow$ 2
- Large $\rightarrow$ 3

On the ordinal scale itself there are no differences defined. The mapping to numbers, however, allows to model differences between the values, for example, if the gap between small and medium is smaller than the one between medium and large; it is a simple way to represent this knowledge in the measure. So, for example,

- Small $\rightarrow$ 1
- Medium $\rightarrow$ 5
- Large $\rightarrow$ 6

would indicate such different gaps, namely that large is closer to medium that medium to small.

In case of attributes like *name* where the values are strings, the Levenshtein distance [16] often used. It counts the number of changes needed to change one string into another one. The possible change actions are insertion, deletion, and modification. If these steps have different costs then more similar means cheaper. This measure is of a very simple character but it is sometimes useful for correcting misprints. Often it is extended by making use of the fact that certain orderings of letters do not or rarely occur in a language; this is of course language dependent.

This principle of using modifications can also be used in global measures. For example, if one compares technical devices a measure is given by the number (or cost) of changes that transform one device into the other one.

In symbolic representations the local–global principle is quite standard. Examples are component-oriented descriptions of complex objects like machines, processes, or image descriptions. They very often employ a part-of (or part-whole) hierarchy. The underlying language elements can be of different nature. Quite common are description logics; here we will restrict ourselves to attribute-value representations, see [17].

This gives rise to a taxonomic structures (trees or graphs) that can be used to define similarities for representing knowledge about the taxonomies.

The simplest measure would count the number of nodes to the deepest common predecessor what defines a graph distance.

A more refined method is seen in the next example.

The principle definition of the similarity for leaf nodes is:

- Assignment of a similarity value to each inner node
- Similarity values for successor nodes become larger
- Similarity between two leaf nodes is computed by the similarity value at the deepest common predecessor

Example:

We consider a small example of a hierarchy of computer tools (Fig. 2.8).

Taxonomies are also widely used in order to investigate the similarity of concepts. This lead to the investigation of the similarity of objects in an ontology, see [18]. There the approach of just counting the number of steps to the deepest common ancestor was refined. The difficulty here is that not all links may have the same weights and counting the links will not suffice. Among the

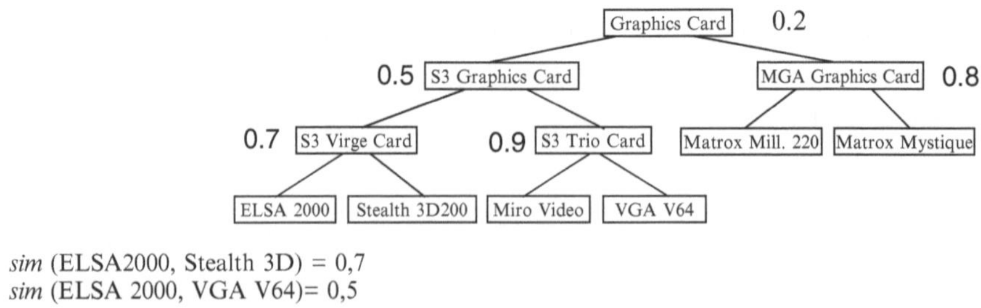

Fig. 2.8. Taxonomy

generalizations one finds information content measures or additional information about the part-of or the is-a relation, see Sect. 2.3.5. An interesting point is that one can use context information too. Concepts may occur in several ontologies, e.g., in a chemical as well as in a biological ontology. The context is the specific ontology that can lead to different distances, see Sect. 2.3.5. One can regard this as a utility orientation because an ontology is built for a specific purpose.

These ideas are extended to nonhierarchical relations. A running system is the lexical database WordNet, see [19].

2.3.6 More General Arguments

Besides attributes there may be many other kinds of expressions that can serve as arguments for a similarity measure. We will discuss four of them:

1. Real-functions
2. Fuzzy functions
3. Random variables
4. Combinations of probabilities and taxonomies

We give some examples.

1. Real valued functions or sets in IR^n: This will be discussed in Sect. 2.8.2 (Hausdorff and Frechet measures)
2. Fuzzy memberships functions as arguments (see [20]; for a general discussion of similarity and fuzzy sets see [21])

In order to compare fuzzy membership functions, two methods became popular:

(a) Crisp Method:
Select some a_i for which μ_i (a_i) maximal, $i = 1, 2$; put $d(\mu_1, \mu_2) := |a1 - a2|$
(b) Integral Method:
F_i =Area between μ_i and x - axis
$d\ (\mu_1,\ \mu_2) = \text{Size}\ (F_1 \Delta F_2)$ (Symmetric difference, Fig. 2.9)

Distances now compare fuzzy membership functions!

Problems:

(a) Two fuzzy functions with disjoint areas have always the same distance.
(b) The shape of the curves in the crisp method does not play a role.
A combined method is as follows:

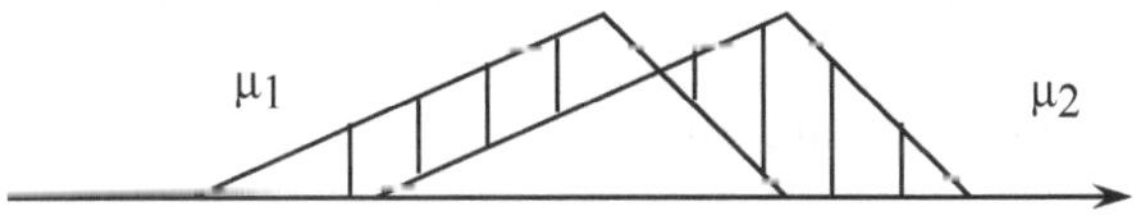

Fig. 2.9. Symmetric difference

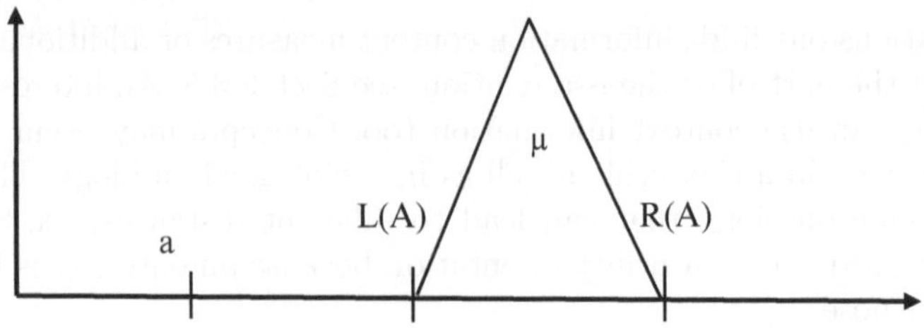

Fig. 2.10. Points and fuzzy functions

- If the areas are not disjoint, apply the integral method
- If the areas are disjoint, add to the distance obtained by the integral method the distance between the two points where both curves reach zero.

It is also of interest to compare numbers and fuzzy functions, for example, "How similar is 43 years of age to young?"; see [20].

Suppose $y \in Y \subseteq IR$ and there is a fuzzy predicate A with membership function μ that has its maximum at m(A), see Fig. 2.10.

Requirements:

If $a < b < m(A)$ then $sim(b, \mu) \leq sim(a, \mu)$;
If $m(A) < a < b$ then $sim(a, \mu) \leq sim(b, \mu)$;
For $a < L(A)$ or $R(A) < a$ $sim(a, \mu)$ decreases monotonically in $|a-L(A)|$ or $|R(A)-a|$

3. Random variables as arguments

For random variables, there exist different methods to define similarity measures.

Let $(X, Y)^T$ be a vector of random variables. The linear correlation coefficient for $(X, Y)^T$ is

$$\rho(X, Y) = \frac{Cov(X, Y)}{\sqrt{Var(X)} * \sqrt{Var(Y)}}$$

$\rho(X, Y)$ is a similarity measure in the sense that it measures linear dependencies and it is quite natural for elliptical distributions. For other types of distributions, the situation is, however, different.

An extension provides a stochastic version of the concept "having the same monotonicity behavior":

Let (x^T, y^T) and $(\underline{x}^T, \underline{y}^T)$ be two observations of the continuous random variables (X^T, Y^T).

(x^T, y^T) and $(\underline{x}^T, \underline{y}^T)$ are called

- *concordant* if $(x - \underline{x})(y - \underline{y}) > 0$
- *discordant* if $(x - \underline{x})(y - \underline{y}) < 0$

This has a probabilistic version (see [22]) where one considers independent vectors (X^T, Y^T) and ($\underline{X}^T$, $\underline{Y}^T$) of continuous random variables. Then between these two vectors

- the probability of concordance is $\mathrm{Prob}((X-\underline{X})(Y-\underline{Y})>0)$
- the probability of discordance is $\mathrm{Prob}((X-\underline{X})(Y-\underline{Y})<0)$
- the order difference is $Q=\mathrm{Prob}((X-\underline{X})(Y-\underline{Y})>0)-\mathrm{Prob}((X-\underline{X})(Y-\underline{Y})<0)$

This give rise to define a similarity measure called *measure of concordance* between random variables X and Y.

A simple example is the measure called *Kendall's tau* (the order difference from above). Suppose ($\underline{X}$, $\underline{Y}$) is an independent copy of (X, Y). The we define

$$\tau(X,Y):=Q=\mathrm{Prob}((X-\underline{X})(Y-\underline{Y})>0)-\mathrm{Prob}((X-\underline{X})(Y-\underline{Y})<0)$$

Hence τ measures simply the difference of the probabilities for concordance and disconcordance, which is intuitively clear for risk analysis if one assumes that X and Y represent costs.

The measure can be computed according to the following formula:

$$\tau(X,Y)=Q(C,\underline{C})=4\times\iint\underline{C}(u,v)dC(u,v)-1$$

where the integral is taken over $[0,\ 1]^2$. The factor 4 is due to the fact that the range of the measure is $[-1,1]$ instead of [0,1].

There is a reason why such measures are of interest in our context. What one wants is to avoid unwanted financial consequences. For example, one does not want to become bankrupt because all shares in a portfolio go down simultaneously. One knowledge source is implicit and hidden in the numerical data. The concordance measure takes care of this. In risk analysis, the statistical methods based on copulas and measures of concordance play an essential role. On the other hand, there are several symbolic data of importance where no statistical data may exist. It was emphasized several times that qualitative, i.e., symbolic attributes are also important for describing risks. Such data may be the type of the company, political stability, or type of the products sold.

How can these two data types be integrated? The problem is the lack of integration with numerical attributes. We suggest here that the similarity-based approach provides a possibility for such an integration method. The idea is to have a combined vector of attributes, one part contains random variables and the other one symbolic attributes.

In the context of risk analysis for investment it has to be avoided or minimized that several assets of a portfolio go down drastically at the same time, i.e., that they are not concordant with respect to going down. We assume here that no distribution functions or statistical evidences for the relevancies of the values of the symbolic attributes are available. Example attributes are:

- Company type; range = {steel, military, energy, tourism}. For the local similarity sim_{CT} with respect to concordance it is reasonable to assume sim_{CT} (steel, tourism) < sim_{CT} (steel, military).
- Area; range = {EU, USA, Middle East, South America, ...}. For the local similarity sim_A with respect to concordance it is reasonable to assume sim_A (EU, South America} < sim_A (EU, USA).

These attributes have to be associated with weights. The local measures and the weights reflect domain knowledge in the same way as distribution functions (if available) do.

4. Combining Probabilities and Taxonomies

This plays a role when sequences (like DNA) are stored with a large amount of "annotation." While the text is accessible by computer applications, it is not easy to interpret the text computationally. This is one of the reasons for the growing interest in ontologies within bioinformatics.

In ontologies, terms can have multiple parents as well as multiple children along the "is-a" relationships. One way is to base measurements on the information content of each term. This is defined as the number of times each term, or any child term, occurs in the corpus. This can be expressed as a probability in such a way that the information content of each node increases monotonically toward the root node, which will have an information content of 1. Given these probabilities, there are several possible measures of similarity.

For two objects o_1 and o_2, $P(o_1\ o_2)$ is the set of parental concepts shared by both o_1 and o_2. We take the minimum probability p(o), denoted by p, if there is more than one shared parent. The minimum of these probabilities is $p_m(o_1,\ o_2) := \min(p(o)|o \in P((o_1\ o_2))$.

Possible similarity measures are:

1. $sim1(o_1,\ o_2) = -\ln(p_m(o_1,\ o_2))$ (the *Resnik measure*)
2. $sim2(p_m(o_1,\ o_2) = 2\ \ln(p_m(o_1,\ o_2))/(\ln(p_m(o_1) + \ln(p_m(o_1))$. This is known as the *Lin measure*. It uses both the information content of the shared parents and that of the query terms

For more applications, see Wordnet [19].

2.4 Case-Based Reasoning

CBR is a general problem solving method that can be applied to all problem types discussed here. The basic idea is to make use of previous experiences when solving problems. CBR has been developed in a systematic way over the periods naïve, sophisticated, and generalized, discussed in Sect. 2.2.3. The developments were always driven by applications. We will first discuss the general principles of CBR. As a practical introduction into CBR, we recommend the tool CBR Works [23].

2.4.1 Introduction into CBR

A case is an ordered pair (p,s) where p is a problem and s is its solution (that is assumed to be satisfactory). The solution may contain additional information, e.g., an explanation or an advice how to use the solution. A case base CB is a finite set of cases.

Formally, a case-based system is a quadruple (CB, L, sim, Ad) where L is the representation language for the cases from the case base CB, sim a similarity measure between problems, and Ad an adaptation operator that transforms solutions into solutions.

The basic way a case-based system works is described in Fig. 2.11.

A new problem P is presented and a case is selected from the case base where its problem part is a nearest neighbor to P. The solution part of the case is either taken over directly or improved by the adaptation operator. Taking the nearest neighbor comes from the assumption that this gives the best solution; it is analogous to the maximum likelihood principle in probability. This causes a demand for the similarity measure: The selected solution should in fact be the most useful one from the available ones. In the naïve period this was not investigated very much. In general, the process of defining such a measure is often quite involved and called similarity assessment (see [24]).

The retrieval of the nearest neighbor is a problem in itself. It is more complex than in databases and hence the similarity should be computed fast. We will not discuss the retrieval problem here and we remark only that this is one of the reasons why a measure with a weighted sum is desirable.

The utility view is shown in Fig. 2.12. As a consequence of this view the basic CBR paradigm is now seen as a demand on the similarity measure: If a new problem p is considered as similar to an old problem p_i then a useful solution for p_i should also be useful for p.

In other words, the similarity between problems should approximate the utility of the solutions.

We have now two views. The classical CBR approach (Fig. 2.11):

The reuse step either takes over the solution from the case or it modifies it. This is called a solution transformation or adaptation. It can be useful, for

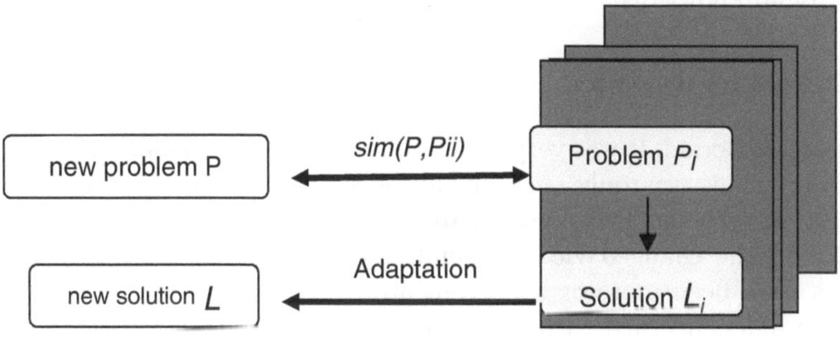

Fig. 2.11. CBR principle

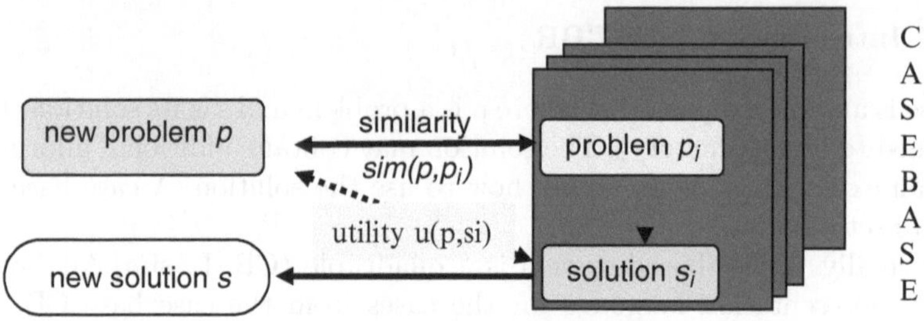

Fig. 2.12. Utility in CBR

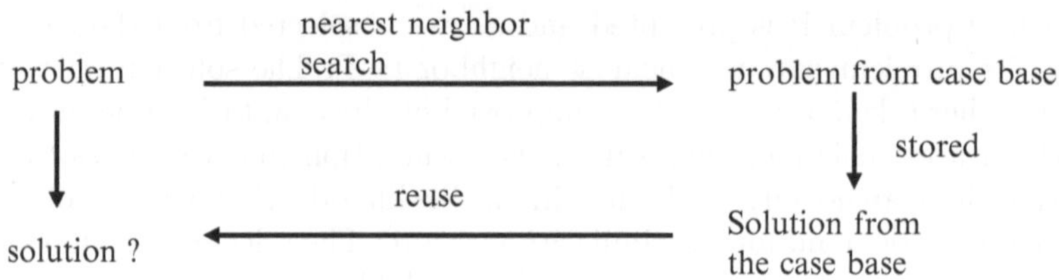

Fig. 2.13. Classical CBR

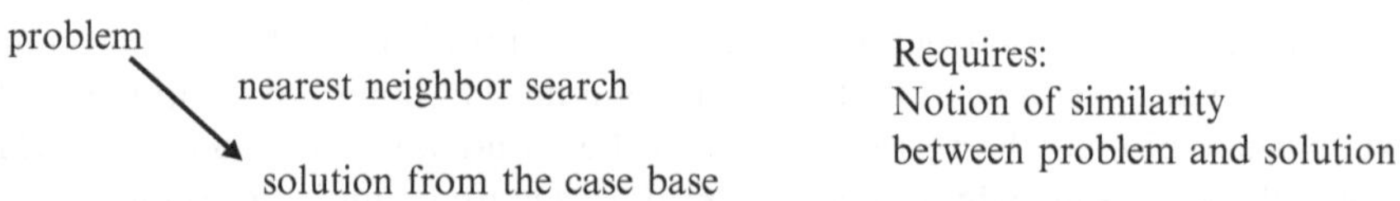

Fig. 2.14. Direct solution search

example, if the object descriptions contain much symmetry: One stores only one case from a set of symmetric ones and reintroduces the symmetry by an adaptation.

The generalized approach establishes directly a relation between the problems and the possible solutions, as stated in Fig. 2.14.

This has the advantage that no stored cases are needed anymore, only solutions. This is of particular interest if one has no experiences but much background knowledge about the relation between problems and solutions.

2.4.2 Axioms Revisited

In the naïve period, it was very intuitive to assume axioms like reflexivity and symmetry. This was only occasionally questioned as in cognitive science in connection with the Tversky measure.

Due to the changed view on similarity measures in the last years, such axioms have been more or less given up, at least in general. Originally the measure was concerned with the similarity of problem situations in order to apply previous experiences in some kind of analogical reasoning. The idea for similarity between objects was that *they looked similar*. In this view similarity was thought as a form of fuzzy equality. Most axioms discussed above survived for specific applications, only but not in general:

- From this point of view, the transitivity axiom of equality was abandoned because small errors do add up.
- The symmetry axiom fails often when similarity measures are intended to reflect utility.
  For example, the degree of usefulness of A for B is often not the same as the usefulness of B for A.
- If different sets U and V constitute the domain of the measure the axiom of reflexivity $\text{sim}(x, x) = 1$ had to be given up.

So, are we running out of axioms and a similarity will be nothing then an arbitrary real-valued binary function? There is another axiom that is not inherited from fuzzy sets and metrics. It is connected with the local–global principle:

The Monotonicity Axiom:

If $\text{sim}(a, b) > \text{sim}(a, c)$, then there is at least one $i \in I$ such that $\text{sim}_i(a_i, b_i) > \text{sim}_i(a_i, c_i)$.

This axiom can be regarded as a partial order form of the substitution axiom (substituting equals by equals) for equalities. The importance of the axiom is twofold, it allows in general more efficient computations and it simplifies the assessment of utility functions and similarity measures. The point is that local improvements of measures also lead to a global improvement. One can make use of this in learning similarity measures.

Many measures introduced above satisfy this axiom. The question arises whether the axiom can always be satisfied by a suitable choice of the measure. An example where it fails is the well-known XOR classification problem that is described in Fig. 2.15.

Suppose the case base contains already three elements that are correctly classified, then there is no weighted Hamming measure for classifying the fourth element correctly using the nearest neighbor method:

If the monotonicity axiom fails, the question is to replace the measure by one where it is valid and that is compatible with the given one (see Sect. 2.4.1)

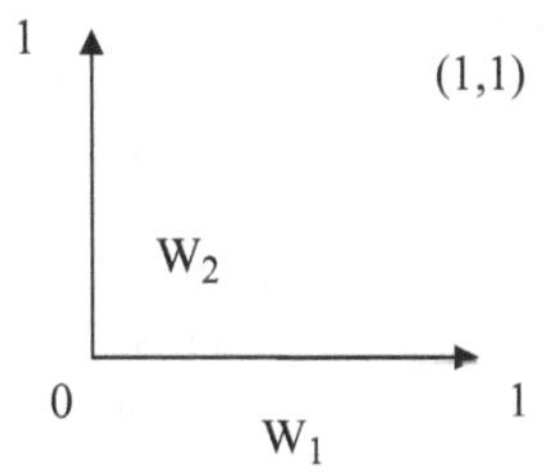

Two classes: {(0.0), (1,1)}, {(0,1), (1,0)} are given (representing the Boolean function "exclusive or")

Fig. 2.15. Violation of monotonicity

or that has at least the same nearest neighbor relation. The XOR example shows that this is not always possible unless the vocabulary is extended. The reason is that there are some dependency relations between local attribute values.

A way out is to introduce additional *virtual* (definable) attributes in addition to the original primary ones. In case of the XOR-problem, the attribute XOR (x,y) will suffice. Such additional attributes are called *virtual attributes.* The purpose of introducing virtual attributes is to shift nonlinear dependencies between attributes from the measure into the definition of virtual attributes. There is a connection with neural nets where the XOR-example shows that with a single neuron only one cannot compute all Boolean functions. The introduction of virtual attributes corresponds to the insertion of new nodes in the neural net. When virtual attributes are introduced and the measure is extended correspondingly one has to verify that all the computations for the values of the new attributes have also been performed by the old measure. However, there is still a foundational problem: Can one always achieve a measure that is monotonic?

From a practical point of view, the virtual attributes have another advantage: They can more directly reflect the utilities and therefore it is easier to determine the weights.

In Sect. 2.5, PR methods will be considered. In PR, one is less interested in the introduction of virtual attributes and takes care of dependencies in a different way.

2.4.3 Knowledge Containers in CBR

CBR is more than just making use of similarity measures. It allows symbolic reasoning as well and can be improved by learning procedures. For describing this we introduce the concept of *knowledge containers.*

The XOR example in the last section already shows that there is a relation between the similarity measures and the representation language. That means, the measure is only one part of the problem solving capacity of a CBR system. We call such parts of the problem solving process *knowledge containers* as described in Fig. 2.16, (see [25], [26]). They are not submodules of a system because they do not solve any subproblem. They are rather description elements that can be filled with knowledge elements. They also play a role in many other problem solving methodologies.

In CBR we identify four major knowledge containers, Fig. 2.16.

There is an interaction between the containers:

A simple observation is that any of the four containers can in principle contain essentially all relevant solution knowledge. In practice, however, this is usually impossible and in addition such containers would contain little efficiency knowledge:

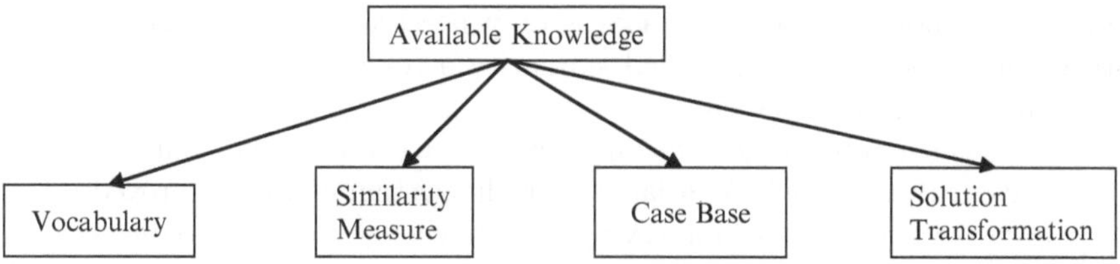

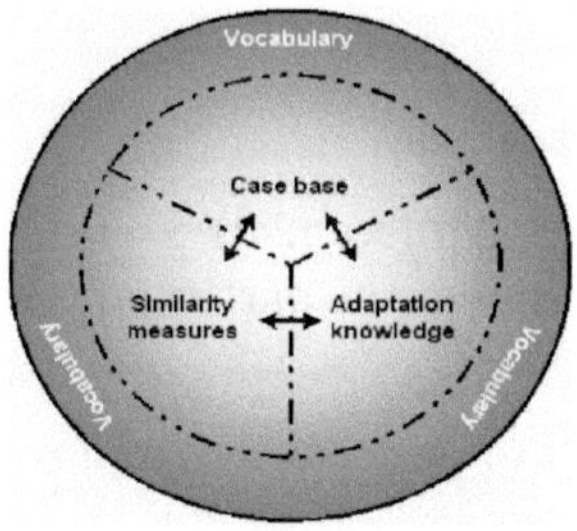

Fig. 2.16. Knowledge containers

- The vocabulary: We need for each problem description p simply an additional (ideal) virtual attribute *sol* which has as domain all solutions and where s = sol(p) gives the correct solution.
- The similarity measure: For an ideal measure sim_{ideal}, we could have sim_{ideal} (q, p) = 1 if the case (p,s) provides some best solution s and sim_{ideal} (q, p) = 0 otherwise.
- The case base: An ideal case base in this sense would simply contain all possible cases.
- The solution transformation: One can simply ignore the cases and construct the solution from scratch using the adaptation rules.

In practice of the CBR applications, this container is of little importance, it is only used in order to reduce the number of stored cases because the remaining ones can be obtained by straightforward adaptations. The situation is, as we will see, somewhat different in IR.

This shows that the similarity measure is strongly related to the other containers. The containers play a role when a CBR system is constructed; then the containers have to be filled. We distinguish it into two phases:

(a) The compilation phase: That contains everything that happens before an actual problem is presented
(b) The run time phase

For ordinary programs and knowledge-based systems, one has to understand all the knowledge that enters the system. In a CBR system the cases need not to be understood at compile time, they are just given to the case base container. Understanding the cases is only necessary at run time. The construction of the measure is called similarity assessment; the measure has to be understood at compile time.

This has the advantage that one can start with a system that works, but not necessarily very well. At a later time the system can be improved.

There are two forms of improving the knowledge in the containers:

- Improving the knowledge for the individual containers separately
- Shifting knowledge between containers (e.g., by deleting cases and improving the measure)

These two aspects are important for:

- Development of a CBR system (see [24])
- Maintenance of a CBR system, in particular as a reaction to changing contexts (see e.g., [27])

There also basically two ways to perform the improvements:

- Improvements by humans
- Improvements by machine learning techniques

A common way to improve the system is to add new (virtual) attributes: This allows defining simpler similarity measures, as was clear from the XOR problem. In fact, virtual attributes are quite common in practice and they are by no means restricted to CBR; we will present some examples.

Medicine

Suppose we have the two primary attributes height and weight of a person. In order to judge the physical condition, it is useful to introduce the body-mass index that allows comparing the physical condition of two persons. A comparison on the basis of primary attributes only would force the similarity measure to perform the computation of the body-mass index during the similarity computation. This has to be done at run time and is in addition much less transparent.

Mechanical Engineering

In making use of analogies virtual attributes play a major role. They run in this field under the name "similarity criteria." They cover wide ranges as in dynamics, thermodynamics, phase transitions, and processes in chemistry.

Robotics

Often the primary data, i.e., the digital signals coming from sensors, are not very informative, for example, for determining the location of a robot because the values of some virtual attributes need to be computed.

Economics

The financial sections of newspapers are full of economical indices that are relevant for economical decisions. Here not only few primary data are participating but in fact there are very many. The virtual data are often computed by statistical methods and in this context they run under the name metadata.

It is a frequent observation that the relation between the vocabulary container and the similarity container is often overlooked and neglected. Finding virtual attributes means to identify influence factors and belongs therefore to the area of sensitivity analysis. That should be done before or at least parallel to similarity assessment.

An extension of introducing virtual attributes is defining a whole new language. This is necessary if the two objects to be compared are of very different character and if the possible objects are not enumerated but represent all instances of certain descriptions in a language.

These may be such objects as

- Functionalities and products: Electronic products that can be composed respecting some constrains and possible actions that one can be performed with them
- Observations, therapies, etc.

This occurs if one wants to classify products with respect to a certain functionality. The description uses not the same vocabulary for products and functionalities and the techniques for building measures used so far cannot be applied.

There is a gap between objects described in different languages that has to be bridged in order to define a similarity measure.

An example from electronic commerce will illustrate this. Suppose we want to deal with electronic switches and suppose that we have a description of the. Intended functionality; then we have a description of the technical details of the products, as in Fig. 2.17.

The bridge attributes are those that occur on both sides. If all attributes are bridge attributes, one could perform the comparison directly. In a situation as here (only one bridge attribute), one has to introduce a whole new bridge level, i.e., a language to which both sides can be mapped. In this case, this is the high level machine instruction to which both, functions and products, can be mapped.

2.4.4 Learning Similarity Measures

The basic idea in CBR is to represent all knowledge and everything that can be taken into account explicitly. However, there is knowledge that is not easily available because it is hidden in the cases. In order to improve similarity measure, this knowledge has to be made explicit what is done by learning procedures. Learning can be done with or without feedback. For pure

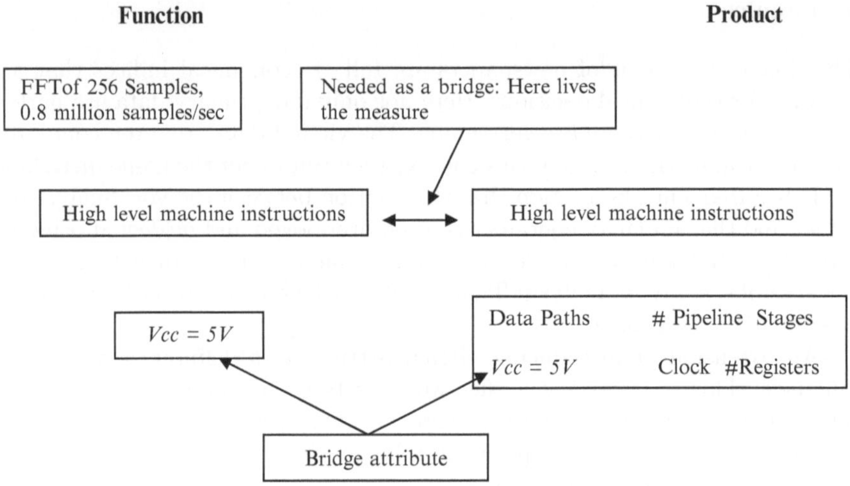

Fig. 2.17. Bridge language

attribute-value representations ideally weighted hamming measures can be sufficient. In this case, all potential nonlinearities are hidden in the (possibly virtual) attributes. As pointed out above, this goal cannot be reached by just defining suitable measures; it also requires the introduction of virtual attributes. To find virtual attributes is a topic in symbolic learning and seems presently to be out of scope for machine learning techniques for complexity reasons. Hence, we keep the vocabulary fixed and restrict ourselves to learning measures.

The measures are to a large degree determined by the domain under consideration because they should reflect utility. One can learn two things:

- Local measures
- The constructor function of the global measure

Learning of local measures has been paid relatively little attention. An approach using genetic algorithms can be found in [28].

Learning global aspects has found considerable interest for weighted Hamming measures. For this reason, we concentrate on learning procedures for the global aspects, i.e., on the weight functions.

We again consider attribute-value representations. However, we have to distinguish the different ways in which weights can be represented. In the simplest case, weights are numbers.

The weights are global in the sense that they do not depend on the arguments. In a more refined version, the weights depend on the specific arguments:

$$sim(p, s) = \sum_{i=1}^{n} w_{p,s} \cdot sim_i(p_i, s_i)$$

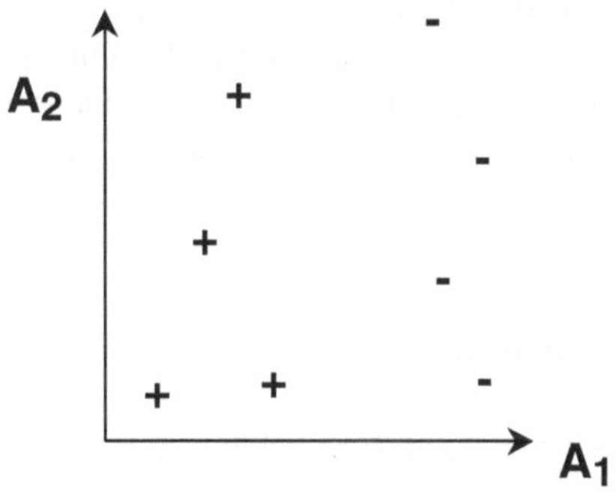

Fig. 2.18. Learning without feedback

Weight learning can take place with or without feedback and for both types a number of variants is known, see [29], [30].

Learning weights for the Tversky feature representation took place in the PATDEX system (see [31]) where a supervised version of the competitive learning from neural nets was used (for classification purposes). The basic procedure was:

- If class was correct, then increase weight for features with value 1 and decrease weights for features with value 0.
- If class was false, then do the opposite.
- Normalize the weights finally

Learning without feedback studies the distribution of the objects in order to determine the relevancies (Fig. 2.18). An example with two attributes:

Learning with feedback corrects errors in nearest neighbor search or more general in the qualitative partial ordering $\geq_x$.

The learning procedure employs a correction of the form: $w_{ik} := w_{ik} + \Delta w_{ik}$

Such a learning rule also be applied if the measure behaves correctly. One simple approach for binary attributes is described as follows (see [32]):

1. Feedback = positive (for correct behavior):
   - Weight of attributes with the same values for p and s is increased
   - Weight of attributes with different values for p and s is decreased
2. Feedback = negative (for incorrect behavior):
   - Weight of attributes with the same values for p and s is decreased
   - Weight of attributes with different values for p and s is increased

In all cases, Δw_{ik} remains constant.

More advanced methods use techniques from neural nets for weight learning as described in [33], [30], and [34]. Many theoretical results for learnability can be found in [35]; there it is investigated what can theoretically be learned and what cannot be learned.

For learning virtual attributes, the first idea is to use concept learning. Unfortunately, this is out of scope for complexity reasons. In PR, this problem is circumvented by learning numerical decision surfaces, see Sects. 2.5 and 2.6.

Learning methods have also been applied to improve other knowledge containers like the case base.

2.5 Pattern Recognition

Pattern recognition is a very general problem area that studies the operation and design of systems that recognize patterns in data. We do not intend to describe PR in general, we are only interested in aspects that are connected with similarity.

The data orientation is important for PR. These data can be, however, of a very general nature, for example numeric or symbolic as in CBR. Because such patterns are not always identical, the concept of similarity often plays a crucial role and we encounter similarity measures frequently. They are, however, accompanied by many other techniques, which we will not discuss in detail. Some will be mentioned later. A major difference to CBR is that symbolic reasoning methods do not play such a role. This restricts the scope of possible applications. In fact, the main applications are in classification and we will discuss some of this in Sect. 2.6.

The area encloses subdisciplines like discriminant analysis, feature extraction, error estimation, cluster analysis (together sometimes called statistical PR), grammatical inference, and parsing (sometimes called syntactical pattern recognition). Important application areas are image analysis, character recognition, speech analysis, man and machine diagnostics, person identification, and industrial inspection. However, in these disciplines special methods have been developed that do not apply in other disciplines. On the other hand, all these approaches are in some sense different from the CBR approach.

A typical PR system makes its "decisions" by simply looking at one or more feature vectors *fed* as input. Pattern recognition aims to classify data (patterns) based on either a priori knowledge or on statistical information extracted from the patterns. The patterns to be classified usually groups of measurements or observations, defining points in an appropriate multidimensional space. The strength of this approach is that it can leverage a wide range of mathematical tools ranging from statistics, to geometry, to optimization techniques. There are two tasks:

- The representation task: How to represent the objects in a suitable way?
- The recognition task: How to classify the represented objects?

The first step is a preprocessing step that represents the patterns to be classified. The representation defines certain attributes that are important for the task. This is sometimes quite simple, but can also be very involved, depending on the task. Sometimes, this step requires much knowledge about the objects and the task. This is particularly difficult for dealing with images. Mostly, the values of the attributes (the patterns) are represented as real-valued vectors; these vectors code features or attributes of interest. This step requires much knowledge about the objects and the task and it can essentially be viewed as the definition of virtual attributes. This is particularly difficult for dealing with images and often requires deep domain knowledge.

An important point is that the representation phase is a closed phase; after it is finished it cannot be taken up again. This phase takes not place at run time and therefore efficiency issues are of minor importance.

Examples are discussed in the sections of classification and image understanding.

After this step is performed, there is a crucial assumption:

- All remaining knowledge is contained in the data of the examples, i.e., no background knowledge can be used afterwards.
  The knowledge in the data is concerned with:
  a) The dependencies between the attributes: Here often Bayesian networks are employed.
  b) The determination of the decision surfaces of the classes.

For the recognition task, statistical techniques have been widely used. Statistical classifiers include linear discriminant function (LDF), quadratic discriminant function (QDF), Parzen window classifier, nearest-neighbor (1-NN), and k-NN rules, etc. Under the assumption of multivariate Gaussian density for each class, the quadratic discriminant function is obtained by using Bayes theory. The modified QDF (MQDF) proposed by Kimura et al. [37] aims to improve the computation efficiency and classification performance of QDF via eigenvalue smoothing, which has been used successfully in the handwriting recognition. The difference from the QDF is that the eigenvalues of minor axes are set to a constant. The motivation behind this is to smooth the parameters for the compensation of the estimation error on finite sample size.

Because of this, in general very simple similarity measures are used; an overview was given in Table 2.3 in Sect. 2.3.1. For specific applications, the measures are often enriched by special tricks, motivated by special situations. There is, however, no general domain-independent principle behind them. Improving measures by learning plays only a little role.

Special attention was paid to high dimensions of attributes vectors. Some kind of opposite to defining useful features is removing irrelevant data, i.e., the reduction to lower dimensions. From the many methods, we mention an information-oriented one. Suppose the patterns are given as functions $S \rightarrow U$.

Any classifier (see below) $C\colon U^S \rightarrow \{1, \dots, n\}$ and any subset $X \subseteq S$ define an equivalence relation $=_{C,X}$ over U^S by putting for $\alpha_X,\ \beta_X \in U^X$:

$$\alpha_X =_{C,X} \quad \beta_X \Leftrightarrow \text{for all } \gamma_Y \in U^Y : C(\alpha_X \cup \gamma_Y) = C(\beta_X \cup \gamma_Y), \text{where } Y = S/X$$

The influence potential is then defined by

$$\nu_C(X) := |U^S / =_{C,X}|$$

The purpose of the influence potential is to see how far a subset of values of the input pattern determines the class of the pattern. Subsets with low influence potential can be omitted, see [36]. This reduction simplifies the computation of the similarity measure. In CBR, it simplifies also the similarity assessment.

Because each attribute of high-dimensional data records only contains a very small amount of information, the Euclidean distance of two high-dimensional records may not always correctly reflect their real similarity. So, a multidimensional query may have a k-nearest-neighbor set, which contains only few relevant records. To address this issue, one can use an adaptive pattern discovery method to search high-dimensional data spaces both effectively and efficiently, see [38]. The user is allowed to participate in the database search by labeling the returned records as relevant or irrelevant. By using user-labeled data records as training samples, the method employs an adaptive pattern discovery technique to learn the distribution patterns of relevant records in the data space. From the reduced data the approach returns the top-k nearest neighbors of the query to the user – this interaction between the user and the DBMS can be repeated multiple times.

As a consequence of using the simple similarity measures, the decision surfaces are quite difficult; they are in particular far from being linear. If the surfaces are under control then there is no need to invent virtual attributes or sophisticated similarity measures. Everything is shifted to identify the decision surface.

The determination of the decision surface is regarded as a learning task. This learning is mostly unsupervised. The view is that the classes are clusters that have to be identified. This is considered in the next section.

The success depends very much on the complexity of the objects and the amount and quality of available data. If there are too many aspects in the local attributes and too many dependencies one would need too many data. Even if there is background knowledge about dependencies between the available data, this knowledge could not be represented easily and could therefore not be used.

A problem is that in many real-world applications the objects are not naturally representable in terms of a vector of features. For example, graph representations lack a canonical order or correspondence between nodes. Furthermore, even if a vector mapping can be established, the vectors will be of variable length, hence would not belong to a single vector space. On the other hand, it is quite often possible to obtain a measure of the similarity/dissimilarity of the objects to be classified. It is therefore tempting to design a pattern recognizer which, unlike traditional systems, accepts as input a matrix containing the similarities between objects and produces class labels as output.

2.6 Similarity, Classification, and Clustering

2.6.1 Classification

In classification one deals with a universe U and subsets K_i, $i \in I$, of U called classes. A classifier is a function

$$f : U \rightarrow I$$

where $f(x) = i$ implies $x \in K_i$. In a *cost sensitive classification*, a certain cost is associated with each false classification and hence not all errors are alike. This is of particular importance if the classification has an insecure base (as often in CBR).

A simple way for a utility function is Prob(x is classified correctly for $x \in U$); for cost sensitive classification, it would be the expected cost.

Another classification task is what we call *dynamic classification*. Here there is not a fixed number of classes but only a language to describe classes and the particular classes can be dynamically generated. An example was given above when products and functionalities have been compared.

The classification contains two tasks as mentioned above:

- The representation task: How to represent the objects in a suitable way?
- The recognition task: How to classify the represented objects?

The representation task is in particular difficult for images where one has to extract relevant features from the image. For fingerprints, this uses e.g., arch, tended arch, left loop, right loop, and whorl. For face recognition, very different features are used, see Sect. 2.8. If the relevant features are known in advance and can be completely listed, PR techniques dominate here. See also Sect. 2.8.4.

In a situation where such a listing is not achievable as often in medicine where no listing of the pathologies but only a description language for the object exists one is more tempted to use techniques from CBR.

In all methodologies, one has the choice between two strategies for performing the recognition task:

- Similarity orientation
- Distance (dissimilarity) orientation

The first strategy favors to find out whether two objects are in the same class while the second is interested in approving that two objects are in different classes. A discussion of this question is given in [38].

In CBR, one takes a symbolic representation with attributes reflecting directly the relevant aspects. PR, all features are represented numerically in a preprocessing step. This uses essentially the same knowledge as in the symbolic representation but it is hidden from the user.

Classifiers also be represented in different ways. A case *based classifier* is essentially of the form (CB, sim) where CB is a case base and sim is a similarity measure on $U \times CB$ (to be more precise on $U \times$ (Problem parts(CB)). Therefore the class of the elements in CB is known; the class of $x \in U$ is now computed using sim, for example, by

$$NN(x) \in K_i \rightarrow x \in K_i$$

where NN(x) is the (or some) nearest neighbor of x in CB. Of course, the nearest neighbor notion may be refined by considering the first k-nearest neighbors (k-NN).

The classifier may, in addition to just providing the class of an element, give additional pieces of information. If the objects of the universe are represented as attribute value vectors, these could contain e.g., the attributes that are responsible for vector being close to its nearest neighbor.

In a more general form, a case-based classifier is of the form

$$(CB, sim, T)$$

where T is a solution transformation. A typical situation where this occurs is when symmetries are present. Suppose e.g., we have left-right symmetries in a medical situation where left means "left ear" and right means "right ear", then we can make use of this in the following way:

$$\text{if } x^{left} \text{ is in } K^{left} \text{then } x^{right} \text{ is in } K^{right}.$$

(x is an observation and K is a diagnosis). If now $NN(z) = x^{right}$, x^{left} is not in CB but closer to z than x_i^{right} (because z is an object "of left type" then the solution transformation has to map it: $T(K_i^{right}) = K_i^{left}$.

This approach is sometimes criticized because it can be computationally expensive to compute the distance to all given examples while omitting examples may loose accuracy.

In PR, there is much attention paid on the representation task. After that learning procedures take place that extract classification knowledge from the data and hence comparatively simple measures can be used, as the Euclidean distance, the Hamming distance, or the mean-square error. However, the classes are learnt in a complex process that needs many examples and statistical and learning methods.

There are quite a number of improvements and we will mention one as an example (see [40]). If one represents independent objects, the distribution of summation-based distances is normally distributed in the limit, according to the central limit theorem. Therefore, Bayesian classifiers assuming normal distributions can be used. The approach in [40] allows that the NN-search operates directly on the distances and not anymore on the objects themselves.

2.6.2 Clustering

Clustering is in some sense the opposite of classification: Given is a similarity measure and a set of objects, how to define classes? Examples (data files)

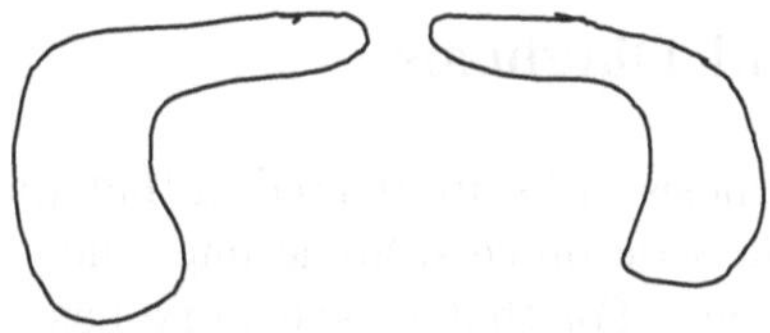

Fig. 2.19. Surfaces

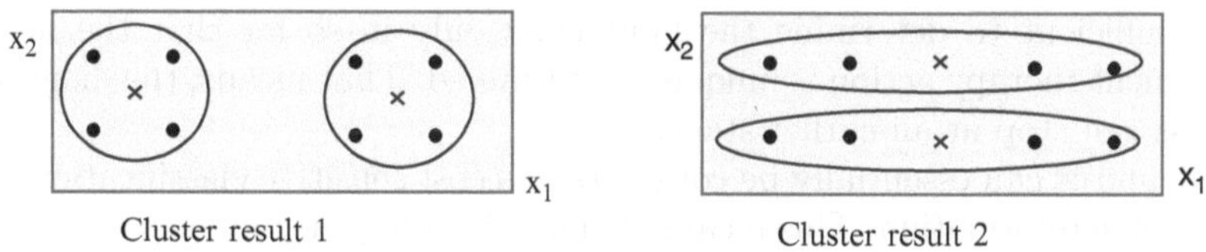

Fig. 2.20. Different cluster results

are presented without relation to a class. The task of clustering is to discover classes and definitions of classes. The examples are clustered on the basis of similarity/distance.

This process is not uniquely determined and the notion of "a correct class" does not apply because we encounter here unsupervised learning. However, the results can be tested and on the result of these tests the learning can be improved.

Density clustering is especially useful for finding clusters that are far away from circles. As an example we may get e.g., the surfaces in Fig. 2.19, each representing a cluster.

The obtained clusters depend heavily on the similarity measures as can be seen in Fig. 2.20.

For the Euclidean distance:

- Cluster result 1 is better

For the weighted Euclidean distance with weights $w_1 = 0$ and $w_2 = 1$:

Cluster result 2 is better.

Here we see that the weights reflect importances. We also see that elliptic queries are the same as introducing weights.

With respect to learning, kernel-based learning machines, as support vector machines, kernel principal component analysis, and kernel Fisher discriminant analysis have got much interest in the field of PR too.

The basic idea of kernel methods is finding a mapping ϕ such that, in new space, problem solving is easier (e.g., linear). But the mapping is left implicit. The kernel represents the similarity between two objects defined as the dot-product in this new vector space. Thus, the kernel methods can be easily generalized to a lot of dot-product (or distance)-based PR algorithms. Some methods can also be seen as dot-product methods by eigen-decomposition of the covariance matrix.

2.7 Similarity and Diagnosis

Classification and diagnosis tasks are related in that way that classification is the first step in a diagnostic process. Sometimes these phases can clearly be separated, but not always. For that reasons, one has to distinguish between the diagnostic process and the resulting diagnosis; the diagnostic process is an interactive process. In medical contexts or in the context of machine faults, often sufficient to determine the fault class only in so far that the needed subsequent therapy action is uniquely determined. That means, the diagnostic process can stop at an earlier stage.

Diagnosis can essentially be considered as cost sensitive classification with incomplete information. This means that establishing the diagnosis is only the final step of a diagnostic process. It does not mean that this is the ultimate goal, because it has to be followed by (or is even interleafed with) a subsequent therapy or repair operation.

The data on which the diagnosis is based are values of certain attributes that are acquired by observations, questions, or general tests. Such observations may not directly lead to attribute values (like temperature); the attribute values often have to be exhibited from more complex observations. Typical examples are images what is discussed in Sect. 2.8.

Hence, the diagnostic process splits into test selection and diagnosis (in the proper sense). Test selection is a problem in itself and can again be guided by cases (such cases are often called *strategic cases*). The goal is to ask as few (or cheap) queries as possible; or more precise to ask questions with a minimal amount of cost (where cost can e.g., be *pain*).

In addition, both the diagnostic as well as the strategic cases can provide additional information as e.g., explanations.

In diagnosis, background knowledge plays an essential role. This is concerned with aspects like fault probabilities, dependencies between observations, temporal behavior, or the structure of the investigated object (for example, a human or a machine). Because the diagnostic process is complex, experiences are useful. The computer support of the diagnostic process, in particular when radiological images are involved was traditionally quite limited. In CBR applications, it was, however, extensively in help-desk support.

In case of machine faults, one takes advantage of particular background knowledge, the taxonomic structure of the machine. The taxonomy refers to the part-of hierarchy and the different types of the parts. As an example we consider PC Trouble-Shooting in Fig. 2.21.

Diagnosis in medicine follows partially the same principle as machine diagnosis. There are, however, two kinds of essential differences (besides ethical aspects):

1. In medicine, images often play a role, e.g., in radiological clinics
2. Often large data structures are involved (for example, concerning bacterial analysis, etc.)

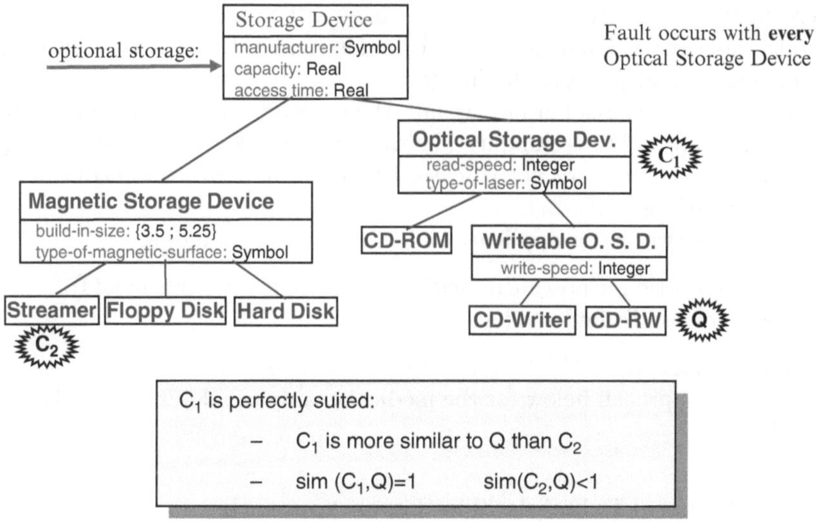

Fig. 2.21. Diagnosis with taxonomies

The first aspect falls into the category of image analysis and the second one is an aspect of PR. This will be discussed in the next sections.

The cases are previous diagnostic situations (fault descriptions) and the problem descriptions are observed symptoms for the current problem.

2.8 Similarity and Images

One distinguishes between image processing and image understanding. Image processing algorithms manipulate images without being concerned with the meaning of the images. Often this is sufficient to grasp the essentials of an image. Many sophisticated tools have been developed in medicine for this purpose. These tools are concerned with very special situations in order to highlight specific aspects. The purpose of image understanding is different. Image interpretation uses such algorithms but has a much broader scope. It has not only to deal with the images as data structures but also have to take into account the domain of objects occurring on the images. Therefore, image interpretation makes a reference to the outside world. This world is represented as a context.

The meaning of the term *understanding* has a wide range of interpretations. In recognition tasks like identification of fingerprints or faces, it simply means to perform this identification in a correct way. Here, the problem types are on the borderline of image processing and image understanding. This also shows that the borderline between image understanding and PR is not sharp.

A different problem occurs in a scenario with different kinds of objects that may occur or may be missing. In addition, for the occurring objects there may be no fixed location in the image. The only thing we have is a certain description language and the understanding task is to transfer the image into a description of the image.

Similarity measures fall mostly into the category of image understanding. There are many other techniques like the ones mentioned here (for example, regarding color) that are not discussed here because we concentrate on the role of similarities.

There are two major applications for image understanding:

1. For diagnostic purposes (in medicine or technical environments)
2. For archives: To find a relevant image

For an overview over CBR, image understanding, and processing see [41]. The specific approach below for the medical example is described in [42], [43], and [44].

2.8.1 A General Principle: The Level Approach

Because similarity measures operate on representations we have to first discuss the possible representations of images and how they are organized according to the local–global principle.

Images are in the first place represented by pixels. There are different kinds of pixels:

1. Values are only 0 or 1
2. Values are real numbers from [0, 1] (grey values)
3. Color values

For simplicity, we restrict on binary values. The increasing expressive power, in particular, when using color values is, however, of great practical importance.

For describing the meaning of images, humans use natural language or a high-level formal language that is easy to understand by the user. Such a formal system could use e.g., logic-oriented languages like predicate logic or attribute value representations with a well-defined semantics for formulating a formal version of the meaning. A major problem is now how such a high-level language can be reached from the (low) level of pixels.

For this purpose, we will introduce ordered language levels as an instance of the local–global principle as seen in Fig. 2.22.

At the highest level, the meaning of the image is formulated. The expressions of each level are defined in terms of expressions of lower levels. To obtain the expressions of the higher level from a lower level, certain constructor functions have to be defined that respect the constraints of the knowledge base. As an example we use images from the medical domain.

Similarity measures can be and have to be defined at each of these levels; they are combined according to the local–global principle.

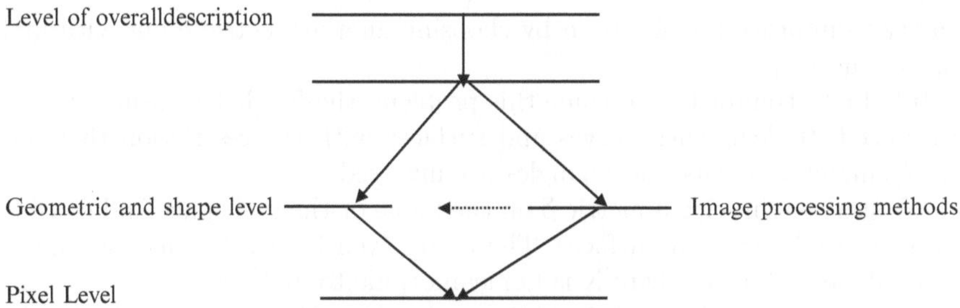

Fig. 2.22. The level approach

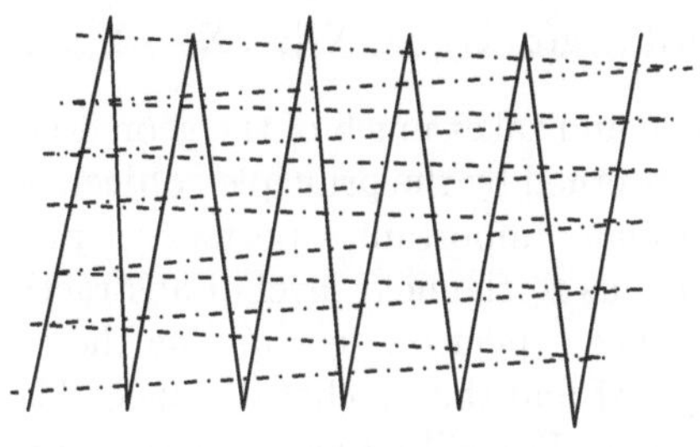

Fig. 2.23. Where Hausdorff metric fails

2.8.2 The Lowest Level

This is the (given) pixel level. Because we restrict ourselves to binary values, the pixels can be regarded as points in the real plane. There are many pixel-based measures and we restrict ourselves to some that are of principle mathematical interest.

Suppose a metric δ on the plane is given. In order to compare sets A and B of points, a common metric is the Hausdorff metric δ_H [45]:

(i) $\delta_H(x, A) := \inf(\delta(x.a)|a \in A)$
(ii) $(\delta_{asym}(A, B) := \sup(\delta_H(a, B)|a \in A)$
(iii) $(\delta_H(A, B) := \max((\delta_{asym}(A, B), (\delta_{asym}(B, A))$

The Hausdorff metric is mostly intuitive as well as useful for convex sets. Otherwise, several very strange phenomena occur as we can see from the example of the range of to curves (dotted and not dotted) where δ is the Euclidean distance, Fig. 2.23:

Intuitively, the two lines are completely different and they will play a different role in applications. On the other hand, the Hausdorff distance is relatively small. What is the reason? It disregards the shapes. The point is that not only the individual distances between points play a role but also some higher-order features; these are completely neglected here.

For this reason, the Hausdorff measure in particular and the pixel level in general is insufficient for grasping the meaning of an image. Of course, one can try to improve the situation by choosing another metric δ but with little hope of success.

In a first attempt to overcome this problem, the Frechet measure [46] was introduced. It deals with curves and surfaces with the restriction that only finitely many segments and triangles are involved.

We start again with a metric δ on the plane or the real space and consider parameterized curves or surfaces. These are given by continuous mappings f: $X \to IR^n$, $n = 2$ or 3, where X is homeomorphic to $[0, 1]^k$.

If now two such objects f: $X \to IR^n$ and g: $Y \to IR^n$ are given, the Frechet distance is defined by:

$$\delta_F(f, g) = \inf(\sup(\delta(f(x), g(\sigma(x))|x \in X)|\sigma : X \to Y \text{ is a homeomorphism})$$

This is a pseudo-metric and takes much of the geometry into account. On the other hand, it is still governed by the principle "objects are similar if they look similar." That means many important aspects, e.g., in medical diagnosis, are not grasped. This leads us to the next level of abstraction in Sect. 2.8.3.

Much research has been spent on determining the complexity of the computation of the Hausdorff and the Frechet distance. The measures are sometimes difficult to compute. We will introduce a way of computing similarities between points q and sets S by transforming the task into an optimization problem, see [47]. The method has commercial application in e-commerce (where the point denotes a query and the set is a set of products). Suppose a similarity measure sim between points is given. We define:

$$\text{sim}^*(q,\ S) = \max(\text{sim}(x, y)|y \in S).$$

Suppose the set S is defined by linear or quadratic constraints. We are interested in the optimization problem:

$$\max(\text{sim}(q, x)|h(x) \leq D)$$

where h is a set of m linear or quadratic functions and D is a vector of IR^m defining the constraints, see Fig. 2.24.

The attempt to use an arbitrary optimization method for this purpose has the following limitations describing S.

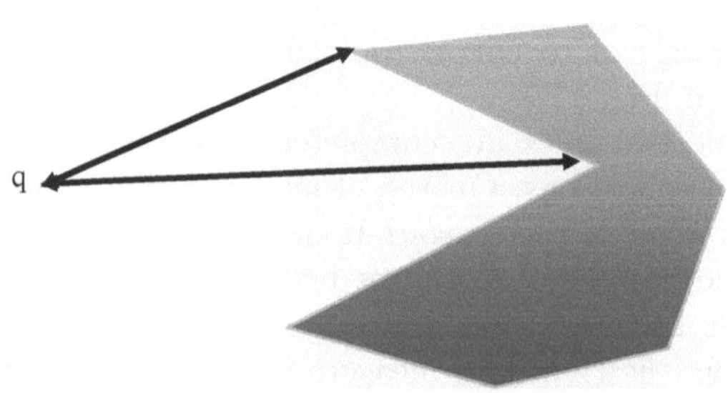

Fig. 2.24. Similarity and optimization

1. The constraints defining S are in general not convex
2. The similarity measure is in general not differentiable

For solving the optimization problem, several techniques have been presented in [47].

One method is to use Minkowski functionals (or gauges). For a compact set B, a gauge with respect to B is defined as

$$\gamma_B(x) = \inf(\lambda \geq 0 : x \in \lambda B).$$

For computing similarities, the use of gauges is to compute a distance dist(x, y) from a point x to some y by putting x in the center of B and to enlarge or shrink B until it touches y; then one sets dist(x, y) $= \gamma_B(x - y)$. This technique can be used for nonconvex sets in order to obtain upper and lower bounds for the similarity measures.

2.8.3 The Geometric and Shape Level

The elementary geometric objects are introduced as lines, curves, and areas with their boundaries and brightness, segments, etc. In addition, this level will also deal with shades, texture information, and related properties. Such geometric objects also called features and the pattern-recognition-oriented approach deals with the situation where there is a fixed set of features that have to be extracted. This is e.g., the case for fingerprint or face identification, see 8.4. A discussion of feature-based similarity measures is given in [40].

If the objects shown in the image do not have a fixed position, a demand on the measure is that it is transformation invariant for translations and rotations, see Sect. 2.3.1. In medicine, the situation is often more difficult. It has to be taken into account that such objects do not occur as mathematical objects but as the real-world objects that do not meet exactly the conditions of the mathematical definition. This leads to a wider class of problems that occur in radiological clinics e.g., for knee examination. Here a more refined approach is needed compared to the techniques of PR. Next we will present such an approach.

A first observation is that some geometrical objects are vaguely defined, e.g., "ovals." These are informal versions of ellipses. In addition, there can be new geometric objects that have not been defined before. A goal is to describe them as variations or deformation of stored objects.

A first and standard step toward a computational investigation is an approximation by polygons. Polygonal approximations are widely investigated in mathematics and image processing.

Such an approximation can be more or less precise, as seen in Fig. 2.25.

From a structural point of view, the second polygon is sufficient. In the next example (Fig. 2.26) we encounter some difficulty because of deformations of standard objects.

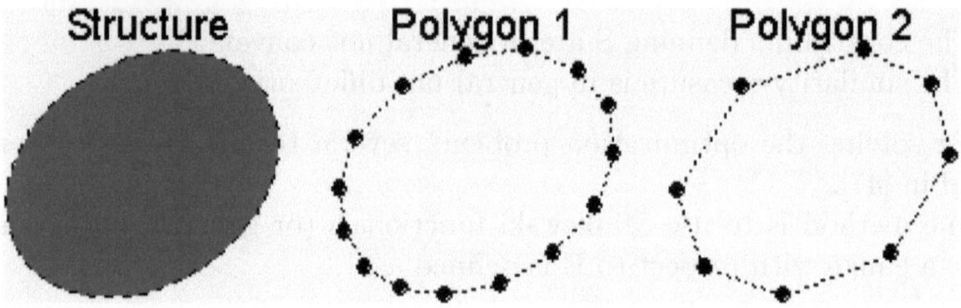

Fig. 2.25. Polygons of a oval 2D image

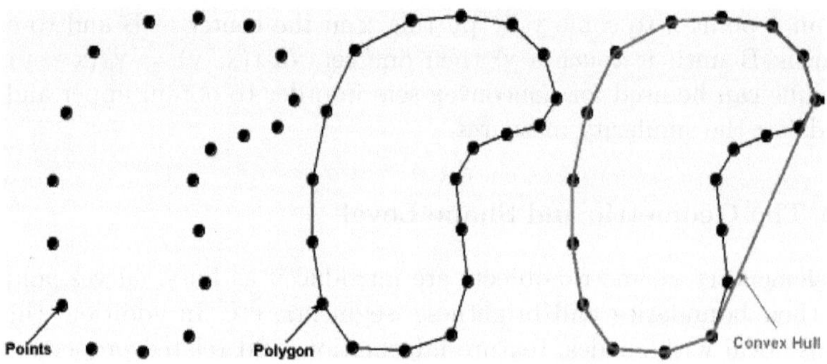

Fig. 2.26. Deformation of an oval

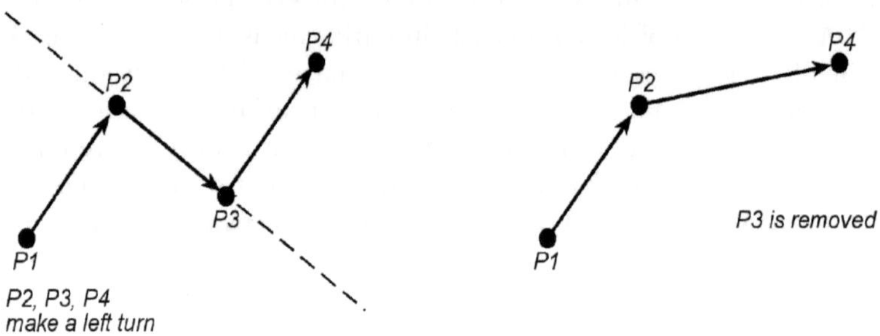

Fig. 2.27. Turns

The convex hull of the polygon does not reflect adequately the deformation of the oval. If the form of the deviation plays a role, the semantics will be missed.

In order to describe shapes we need to extract attributes from the image that are characteristic for the objects contained in the image. An example is the deviation from an oval, described by *turns*, see Fig. 2.27.

The turns can be computed and indicate the deviations from a convex oval. If the polygonal approximation is not precise enough, the phenomenon cannot be described.

Other attributes and predicates describe e.g.,

1. Convexity, concavity, direction of the bounding box, relation length/width, qualitative size, etc.
2. Relative positions are given using expressions like:
   - Right, left, up, down
   - Right of, left of, higher, lower, symmetric to
   - Bright, dark, darker then

These attributes and predicates allow describing the image in a component-oriented way according to the local–global principle. As a consequence, similarity measures can be defined for comparing them. The local similarities carry much knowledge, e.g., which deformations can be tolerated and which cannot. The weights reflect the importance of an attribute (e.g., size) for the intended purpose. This heavily uses knowledge of the intended domain as in medicine: "Left oval and right oval are in normal situations symmetric to some axis but not necessarily if pathologies are present."

We distinguish two kinds of attributes:

(i) Local image attributes: They are concerned with components, i.e., local geometric parts of the image as special form of a boundary, length of a radius, etc.
(ii) Global image attributes: They are concerned with larger parts of the image, for example, density function, center of mass, etc.

Examples:

Shape

dom(shape) = {oval, irregular, butterfly}
dom(butterfly) = {leftWing, rightWing}

Density

dom(density) = {hypodensity, middleHypodensity, middleHyperdensity, hyperdensity}

Relative density

dom(relativeDensity) = {hypodensity, hyper density}

Texture

dom(texture) = {homogeneous, heterogeneous}

We will not go into further details here.

The purpose of image processing methods in this context is to extract knowledge that is used in the higher levels. The algorithms have access to the pixel level and to the geometric level. For this a special knowledge base for image processing methods is needed. It contains a descriptions of the functionality of the algorithms and advices how to use them. It also has access to the base of the available algorithms.

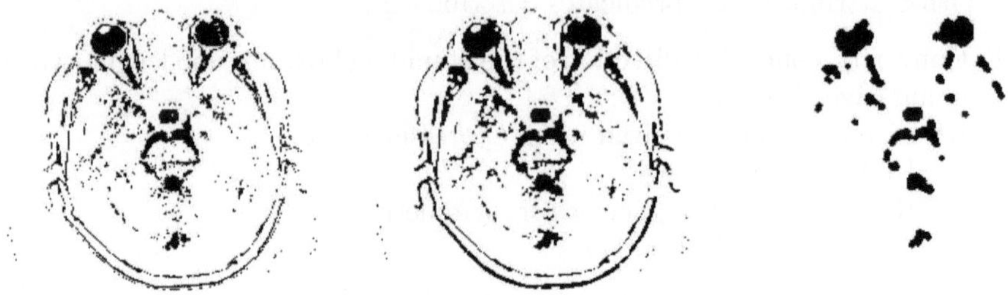

Fig. 2.28. Three processing steps

For example, in Fig. 2.28, there are three processing steps that simplify the information in the image and allow the application of geometry-describing methods. These steps are a threshold, an opening and a closing operation. The first operation omits all pixels with grey value that is not between two thresholds (980–1 003 HU). The second operation thickens the results through a morphological closing operator and the third operation eliminates small objects and noise through a morphological opening operator.

A crucial image processing method is segmentation. Perner [48] described image segmentation for CT images from the brain using CBR. As her research progressed, she used more complex case representations, reasoning and learning strategies, and data mining techniques for PR. Perner [49] proposed a system that used CBR to optimize image segmentation at the low-level stage according to changing image acquisition conditions and image quality.

2.8.4 The Domain Specific and the Overall Level

These are the levels that have a semantics in the sense of the (for example, medical) domain. At these levels, certain figures, areas, or forms are introduced together with their possible properties representing concepts of medical interest. These are concepts that medical experts use for describing a diagnosis or a fact they consider as important. Of course, the expressions of this level cannot be given for all aspects of medicine but have to be restricted to a certain area. The requirements for the vocabulary at the levels are:

(i) Each expression has to be defined in terms of expressions of the geometric level.
(ii) With this vocabulary, all properties of interest should be definable. In particular, it must be possible to distinguish pathological objects from nonpathological ones.

An example of a short informal description of the structure "skull" is:

Skull = (Shape=oval, GreatConcav=1.4, Tolerance=12%;

Density=hyperdensity, GrayLevelMean=253, Tolerance=10%;

RelativeDensity=hyperdensity, Structure=Cerebrum, GrayLevelMean=109;

Texture=Homogeneous, StandardDeviation=6.1, Tolerance=10%;

Position=Center, Centroid=260-276, Tolerance=05%;

RelativePosition=Around; Structure=Cerebrum, Centroid=259-277).

It has to be observed that these descriptions differ from those which one finds in medical textbooks. The illustrations there refer directly to the visual impression on the human. Here we have in addition to take into account that one and the same object has different representations depending on the media used (X-ray, NMR, etc.).

The Level of the Overall Description

At overall level, the meaning of the complete image is formulated in such away that it is suitable for a diagnosis. Using the object description of the domain-specific level, the position relationships of the complete description should be achieved. This may not always be possible, for example, because not all pathological variations may be known, or because they cannot be distinguished from image processing errors. In this case, such regions have no direct meaning but can be indicated to the human user as *regions of interest.* The expert will have a closer look at those regions.

At all these levels, there is a minimal set of expressions in terms that allow to define all. These expressions are the primary predicates, attributes, etc. Although the primary attributes in principal suffice, we introduced *virtual expressions* (predicates or attributes) for describing relevant properties. The specific choice of the virtual vocabulary does not so much depend on the domain but rather on the pragmatics, i.e., on the special task to perform. The definition of virtual expressions often contains important knowledge (in the vocabulary container).

Each expression at a higher level is interpreted in terms of expressions of a lower level. For the elementary geometric attributes this is different, their values are computed using image processing methods (e.g., checking convexity).

Human Interpretation

If humans interpret images, there is a degree of subjectivity involved that depends on the individual or on a group of individuals. For that reason, one distinguishes (see [14]) between a measured (or *perceived*) similarity *sim* and *judged* similarity $\underline{sim}$ that is defined as

$$\underline{sim}(\mathrm{x}, \mathrm{y}) = \mathrm{g}(\mathrm{sim}(\mathrm{x}, \mathrm{y}))$$

where g is a monotonically nondecreasing function that represents the subjective element. One should notice that only the judged similarity is accessible to experimentation. Examples for the function g are given in Sect. 2.3.3. Judged similarities play a role when images are interpreted in different hospitals or regions.

Learning

Learning methods have also been used to improve image understanding methods. As an example we mention a semisupervised fuzzy clustering method as reported in [50]. The method presented there can effectively learn class distributions from retrieval experience. The application to image retrieval was reported to be successful.

Application: Medical diagnosis

In order to use images for a medical diagnosis, we need a database of prototypes. These are images regarded by an expert as ideal instances of some concept which can either be an instance of a healthy object or of a specific pathology.

For each prototype p of some concept C, we define the fuzzy degree of membership of some object u as

$$\mu_{C,p}(u) = \mathrm{sim}(u, p) \text{ and } \mu_C(u) = \max(\mu_{C,p(u)}, p \text{ a prototype for C}).$$

If there are several concepts, then the one with the highest degree of membership is chosen. In this way one can formulate

- Some X is a Y with degree x
- Some X has the property P with degree x

If some real actions (e.g., therapeutically ones) are based on the degree of membership, the numerical values of the degree will play a role. For obtaining a binary decision the membership has to be accepted or rejected as described in Sect. 2.2.1. To achieve such a decision, we proceed in the spirit of rough set theory (see Sect. 2.2.1). We introduce two thresholds α and β with $\alpha < \beta$ which allows introducing proceed as follows:

$$\textit{membership accepted} \text{ if } \beta \leq \mu_C(u) \text{ and } \textit{not accepted} \text{ if } \mu_C(u) < \alpha, \textit{unknown} \text{ if } \alpha < \mu_C(u) < \beta.$$

The choice of the thresholds is triggered by experience and the observation of occurring risks.

In an interactive system, objects in the uncertainty area can be shown to the expert who makes the final decision.

The case base can contain nonpathological images only. Then the system is able to tell the medical expert that the situation is definitely not without pathological elements and the expert can react accordingly. Grimnes and Aamodt [51] presented a system that integrates CBR into a task-oriented model-based system for interpreting abdominal CT images. The case base therefore contains some specific pathological images. A case-based reasoner working on a segment case base contains the individual image segments. The system is based on a propose-critique-modify learning cycle.

In medical application, the judged similarity will lead to different interpretation of images depending on the specific medical expert. It reflects the utility of the diagnosis that may differ from hospital to hospital.

Application: Search in Archives

Archives store information about many patients and contain therefore general information useful for the design of therapies, etc. For many purposes it is sufficient to augment the images by a fixed textual part that can be used for searching. Because of constantly changing problems this would require a continuous updating what in general is not suitable. In addition, one cannot foresee for which purpose an image may be of interest. Therefore, it is necessary to investigate the images themselves and to interpret them during the search. In radiological clinics, much effort and time of humans is spent for this purpose.

The principle approach for using similarities is the same as for diagnosis. The main difference is in the risk of errors. The diagnosis is much more safety–critical, mistakes can be fatal. Data mining has a more statistical character and individual errors can be tolerated.

Identification of Images

In image identification, the understanding of an image is rather limited. The purpose is to select from a database of images the one that is most close to a given image. When only images of a fixed structure (i.e., a fixed number of objects on fixed positions) are considered, the situation is somewhat simpler. One does not need a complicated language; it suffices to identify some objects and evaluate certain parameters. An example is face recognition, which is shown in a simplified way in Fig. 2.29.

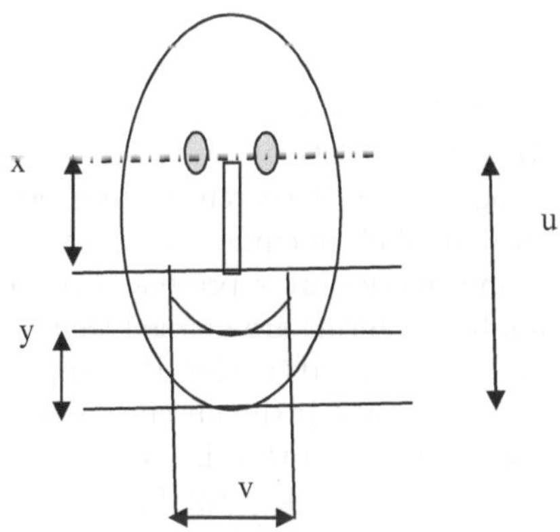

Fig. 2.29. Face

For professional face identification, these parameters are more refined. For face recognition, the components to be identified are quite obvious and intuitive. This is not always the case; it is more involved for fingerprint identification. For fingerprints, one uses arch, tended arch, left loop, right loop, and whorl for getting parameters.

In a CBR approach, the local measures contain much knowledge that can be obtained from learning processes. In the PR approach, a simple measure like the Euclidean distance may suffice; the knowledge is implicit in statistical data and used by such methods.

2.9 Information Retrieval

The task of IR has changed over the years. Originally, IR was understood as the search for a particular document in a structured set of documents. This document could be a text or an image. These types of documents require different retrieval methods. The image retrieval is closely connected to image understanding and we will here restrict ourselves to text retrieval.

Today the solution of problems is central and the retrieved documents are considered as a useful step during the problem solution. That means, IR is part of the process and knowledge management.

There may be several documents of interest (or even none) and their relevant aspects may become known only during the retrieval process itself. Typical examples are searches in the Internet and in libraries. Similarity describes the usefulness of a document within the overall process. That means, similarity does not deal with experiences in the first periods of CBR; rather belongs to the generalized period where different objects are compared. That means, we have a query and search for a useful document. In order to bridge the gap between the query and the document, both are described in the same vocabulary, namely in some terms of interest. Which terms are chosen is based on some knowledge about the domain; later on it can be extended.

A very popular model is the *vector space model* that has been widely used in the traditional IR field. This model is very suitable for comparison with the CBR approach. We will concentrate on this model, thus neglecting, for example, probabilistic models. Most search engines use similarity measures based on this model to rank Web documents. The model creates a space in which both documents and queries are represented by vectors.

As a first step, a number of terms are defined that are extracted from the documents. This can be e.g., all words except the stop words (i.e., those that do not discriminate any document). If the number of selected terms is m then an m-dimensional vector is generated that has for each term some associated weight. The weight is supposed to reflect the importance of the term for the document. This importance is mostly based on some counting, i.e., it measures a frequency.

A document vector is of the form

$$d = (w_{1j}, w_{2j}, \ldots, w_{mj})$$

and a query has the format

$$q = (w_{1i}, w_{2i}, \ldots, w_{mi})$$

There are many different ways to choose the weights. A standard method in the vector space model (VSM) associated the weights with the terms on the basis of the following two numbers:

- Term frequency, f_{ij}, the number of occurrence of term y_j in document x_i
- Inverse document frequency, $g_j = \log(N/d_j)$, where N is the total number of documents in the collection and d_j is the number of documents containing term y_i

Then, as a standard for the similarity measure the *cosine* of a query q and a document d with the *scalar product* (q,d) often used:

$$sim_{vs}(q, d) = \frac{q \cdot d}{||q|| \cdot ||d||}$$

The similarity $sim_{vs}(q, x_i)$ is supposed to reflect the relevance of the document for the query.

The documents weights w_{ij} and the query weights v_j are

$$W_{kj} = f_{kj} \cdot \log(N/d_j)$$

and

$$W_{ki} = \log(N/d_j) \text{ if } y_j \text{ is a term in q and 0 otherwise.}$$

There are two main criteria of success: Recall and precision.

- *Recall* is the ratio of the number of relevant records retrieved to the total number of relevant records in the database. It usually expressed as a percentage.
- *Precision* is the ratio of the number of relevant records retrieved to the total number of irrelevant and relevant records retrieved. It usually expressed as a percentage.

These criteria can be criticized because they do not take into account that the recall and precision may be restricted to the documents and queries the user is interested in.

The knowledge used is the weights of the terms. These weights are either of statistical nature or they are provided by the user in the hope that interesting documents are retrieved (what is rarely done). The retrieval result is, however, not necessarily convincing. Therefore, a process starts a process for improving the results. That means, the query is reformulated into an optimal query that (hopefully) finds all the relevant documents. This can be regarded as analogous to the solution transformation in CBR. A basic difference is that not the solution is transformed (one cannot rewrite a document) but rather the query. Often it is quite complex.

There are several ways to do it and a popular one is *query expansion* that has many variations. It is the process of supplementing the original query with additional terms. The idea is to add such terms to the query that are similar to the given ones. This means that the newly introduced terms are close to the old ones with respect to some similarity measure. Automatic query expansion uses a frequency measure on pairs of words; i.e., counting how often the words occur together in texts.

This can be based on user feedback, inspection of past optimal queries, or others; we will not go into further details here. In addition, the term weights can be improved, for example, by using statistical information. More refined ways like employing a thesaurus is rarely used in IR.

It should be, however, remarked that an important motivation for using ontologies and similarity measures based on them is to overcome such problems.

After the query vectors have been changed, the retrieval process starts again with the same similarity measure. This is not exactly true if one regards the weights as a part of the similarity measure (what is done in CBR); in fact, only the form of the computation remains invariant.

The knowledge of the method is mainly included in

- The terms describing the documents and queries
- The weights, these are parts of the similarity measure
- Possible statistics about former queries
- A user feedback

The vector space approach looks similar to the attribute-value representation in CBR. But there are some differences:

- In CBR, we have attributes that have a domain from which the values are taken.
- In IR, the vector coordinates are just numbers; there are no variables that could be instantiated and each coordinate is labeled by a fixed term.
- In CBR, the weights are parts of the measure; in IR, they are parts of the object representation.

It is, however, not difficult to unify the approaches. For this purpose, we introduce for each term an attribute Frequency_t(.) and take the weights as the weights of a weighted Euclidean measure. This measure is equivalent to the measure in the vector space model.

An even more sophisticated approach is to use phrases instead of words. This means, one has to define local similarity measures on phrases. First steps are described in [52].

2.10 Speech Recognition

Speech recognition is an important task in PR. Spoken language is a sequence of signals that is taken up by sensors and represented by numerical data. There are two major tasks:

1. Speaker recognition; it is the process of automatically recognizing who is speaking
2. Recognition of the words and the meaning of the spoken text, i.e., language understanding

For both tasks, the representation phase is very much involved and uses advanced methods, e.g. for filtering and denoising what we will not discuss here. The ultimate representation often uses Fourier coefficients or, more advanced, wavelet packages.

For both tasks, a similarity match has to be performed. The case base consists of

1. Recorded voice sequences of a number of speakers for speaker recognition
2. Prepared sound or language models, i.e., a base of utterances where the meaning is known

There are various ways how the case base can be achieved. One can take, for example, physiological and behavioral characteristics of the speaker, or emphasizing special vocabularies. We will not discuss this here but will remark that most of the knowledge for solving the tasks is contained in the representation phase and the structure of the case base.

The similarity measures involved can contain more or less knowledge. The simplest but widely used measure is the Euclidean distance. For speaker recognition, statistical methods are added as e.g., a normalization method for distance values using a likelihood ratio. The likelihood ratio is defined as the ratio of two conditional probabilities of the observed measurements of the utterance. The first probability is the likelihood of the given acoustic data for the claimed identity of the speaker. The second one is the given likelihood that the speaker is an imposter. The likelihood ratio normalization approximates optimal scoring in the Bayes sense.

For language understanding, most emphasis was put on efficiency. To promote higher speeds, the models must be more compact with faster similarity calculation and best word-match search processing. To make the sound models more compact, tree-type structures have been used in order to speed up the conventional similarity calculation. The method employs acoustic look-ahead functions that raise the speed of the word sequence search engine (see [53]).

The emphasis on statistics and efficiency favors certain goals but neglect others. An example would be a cost sensitive approach to language understanding. Here other, utility-oriented similarity measures seem to be needed that take care of the costs of misunderstanding an utterance.

2.11 Other Applications of Similarity in CBR

Because in CBR general knowledge about a task can be represented in a natural way, it is not limited to classification and image understanding. In fact, there is no general restriction about the applicability of the methodology. We will shortly present some examples in order to indicate how similarity can be used in other contexts. The principle richness of such applications is due to the detailed analysis of the similarity concept. It is not false to say that this has lead to a whole new area, similarity-based reasoning.

Configuration and Design. Configuration is generally understood as the construction of an artifact from a given set of components such that certain conditions are satisfied. Hence, it can be considered as a constraint satisfaction problem. Design introduces some degree of creativity because some components or even structural elements of the artifacts may not be presented a priori. Depending on this degree, one distinguishes routine design, innovative design, and creative design. In this area, the use of experience is of particular importance.

The solution to a configuration or design problem taken from a case base can almost never be used unmodified for an actual problem but has in general to be adapted. Another way to paraphrase this is that a previous solution is reused.

Planning. The term planning covers a great variety of tasks. We will restrict ourselves to *action planning.* The problem is to find a sequence of actions transforming a given initial situation into a desired goal situation. The size of the sequence is not restricted but the set of available actions are fixed. In *partially ordered* planning, the ordering of the actions is not total but only partial. Action planning in AI traditionally assumes complete knowledge what is, however, quite unrealistic for many applications.

For CBR again, the reuse aspect is dominant because the retrieved plans have to be adapted.

Decision Support. The role of *decision support* is to help the decision maker rather then to replace him. In many situations, there is no precisely stated problem but rather a complex problem structure. The output of a support system is usually not a solution but rather an advice or an useful piece of information. An example is given by considering help desks. Experience combined with the possibility of inexact matches makes CBR an important technology in this field.

Prediction. Prediction tasks occur in very different situations and require a great variety of methods. We mention just the method of extrapolating real-valued functions. Similarity enters the scenario when one makes use of recorded cases. As an example, we mention effort prediction of software development tasks as described in [54]. The method is applicable to predict effort related to any object at the requirement, feature, or project levels.

E-Commerce. The applications in e-commerce mostly refer to the generalized period of CBR. The customers express some demands and these are put directly into relation with the products. In e-commerce an uncertain and vague classification takes place. There are customer classes that are roughly described and the classification task is to assign to products that class of customers who prefer the product.

Knowledge Management and Complex Systems. Complex systems occurring in knowledge management often cannot directly be represented in feature vectors. On the one hand, this is too complex, on the other hand the objects are given in e.g., a textual form and there is no knowledge about the features that could be used for the representation. In [55] it is described that nevertheless a similarity measure can be defined that leads to successful applications as in software engineering or biology.

2.12 Comparing CBR, PR, and IR

All three areas, CBR, PR, and IR are methodologies that have developed techniques for solving problems. The scope of problems tackled is largest for CBR, see Sect. 2.11. For this reason, we will compare the methodologies for classification tasks only. IR is discussed less because it mainly uses techniques from PR and CBR. All methods deal with certain data that are not restricted in general. Because of the existence of reasoning methods CBR can deal, however, with more complex data.

In order to perform classification, knowledge is needed. There are in principle two knowledge types:

- Explicit knowledge about the problem domain
- Implicit knowledge coming from the available data

Explicit knowledge is represented in such a way that it can be directly understood and applied, mostly in the form of rules. Implicit knowledge is hidden in the data and need to be made explicit in order to be applied. In knowledge-based systems, the term "knowledge intensive systems" often occurs. This is not quite adequate because it almost always refers to explicit knowledge only.

This knowledge enters the methods of CBR and PR in different ways. The objects to be classified can originally contain data of three types:

- Numerical values of some attributes
- Symbolic values of some attributes
- Other features like geometric objects

These data come from reality and have to be represented formally. This representation step usually straightforward for numerical data.

Basically, the two views of CBR and PR can be summarized as:

- CBR view: What counts is the background knowledge because this gives the understanding of what is going on.
- The PR view: The truth is in the facts and the facts are represented in the data.

Therefore, in CBR basically one tries to represent the background knowledge while in PR the data knowledge has to be coded numerically and has later on to be made explicit by additional processes.

For more complex data like images the representation is difficult. CBR demands a symbolic representation while PR wants a numerical representation. The problem is, however, which aspects of the object (for example, the image) have to be taken into account. This requires background knowledge. For instances the features extracted for finger print detection are quite different from the features extracted for face recognition.

In CBR, the result becomes visible in the form of an explicit symbolic representation. In PR, the result is implicitly hidden in a preprocessing step that is mostly performed by humans.

After the representation is done, the two methods proceed in different ways.

1. In CBR, the knowledge is distributed over the knowledge containers. In order to define a similarity measure according to the local–global principle, one has to
   - Define virtual attributes that are independent from each other and represent influence factors
   - Define local measures
   - Define the global measure, i.e., weights

   After this is done, the classification can be obtained using the nearest principle. The classes are separated by a decision surface that is linear. For these procedures actually no data are necessary, they are only needed in order to test and improve the CBR system. The definition of the virtual attributes and the measure requires, however, a large amount of knowledge. The application can be done interactively by considering attributes step by step and the result has some explanatory character. In particular, the results are more intuitive and important aspects are more directly visible.

   On the other hand, in the case base container there is a lot of implicit knowledge. But presently there are little attempts to make use of it.
2. In PR, the representation task is done by humans, often with little computer support and without applying learning methods. From that point on, all knowledge that can be used is in the data and is presented directly to the system. The examples are n-dimensional vectors and the similarity often measures the Euclidean distance or some variation of it. The price to pay is that the decision surfaces are highly nonlinear and not easily understood by humans. What PR tries to do is to discover these surfaces. For this purpose, machine learning methods are employed, for instance clustering methods.

   To learn such surfaces, no additional background knowledge but many data are needed. An improvement is obtained by including some Bayesian learning as an intermediate step; we will not discuss this here. The

methods are based on solid mathematical and statistical methods that allow proofs for their validity. On the other hand, some tacit knowledge is hidden in the assumptions and in the preprocessing.
Even in situations where explicit knowledge is available, PR makes little use of it. As a consequence, the results, despite their accuracy, are often not very intuitive. A major reason for this is that the measures are of a simple nature comparing only simple data structures.
3. In IR the measure contains knowledge in the form of the weights and this knowledge is updated in the retrieval phase. That means in principle, that the measure is also updated but only in an ad-hoc manner. In the query expansion, much knowledge is incorporated but more in the sense of statistical than domain background knowledge.

Hence, we can summarize the different use and distribution of knowledge for CBR, PR, and IR in Table 2.5.

The black box character of the PR systems makes them easy to use for nonexperts. On the other hand, no human interaction is possible. This restricts the application to socio-technical processes where human and machine agents interact. Dynamic classification cannot be performed and one has problems with diagnostic processes. That does not exclude, however, that PR can play a useful role in diagnostics.

If one compares the approaches, the conclusion cannot be that one is superior to the other one. In fact, both have strengths and weaknesses.

CBR needs much knowledge and if this is not available then PR has an advantage. PR needs many data and if they are not available then CBR has an advantage. PR is mathematically solid (but often with some unquestioned assumptions) while CBR often of an ad-hoc character. PR operates as a black-box system while CBR is interactively usable and represents the knowledge explicitly. IR also uses mainly data but could be improved by knowledge.

Sometimes PR is criticized because of the simple character of its similarity functions. But this is compensated by the fact that the relevant knowledge is hidden in the data and extracted from them.

Both, CBR and PR, contain still "elements of art" and the scientific level needed for an engineering discipline is only partially reached.

Table 2.5. Comparison

Methodology	*Background knowledge*	*Knowledge in the data*
Case-based reasoning	In all knowledge containers	For improving measures and other containers
Pattern recognition	For the representation task	In the recognition phase
Information retrieval	In (the frequency) of terms	In the weights and the query expansion

When one is asked what to use, CBR or PR, there is no universal answer; this depends on the problem type. Unfortunately, a user gets little help for answering such a question. What can definitely be improved is the relation between the two approaches in order to benefit from the alternative. An important issue is getting more experience in the way in which the approaches can be integrated. This would improve the quality of socio-technical processes. An essential part of this is to get more insight into the question where it is better to store knowledge, in the similarity measure or at other places.

2.13 Summary

It is generally agreed that similarity and similarity-based reasoning is important (if not dominating) in many problems occurring in computer science. The reason is that one is not only concerned in problems of "yes or no" answers, but in approximations of the kind of "better or worse." In fact, such problems often are dominating, although the Boolean-valued problem types will still play an important role.

In mathematics, we observe an analogical development over the centuries: We have a systematic theory of approximation theory as well as of numerical mathematics and statistics.

In applications, similarity appears in many different ways and shapes. That does not mean that the underlying principles are totally different, in fact there are several unifying principles with, however, many diverging instances.

The approach of similarity reasoning is the attempt to make these principles clear and therefore to simplify its applications and to make the scope of applications broader, clearer, and safer.

In order to obtain a unifying view, we used two principles:

- The local–global principle that allows to establish a relation between the similarity measure and the objects to be compared.
- The knowledge container concept that allows to describe the distribution of knowledge.

We consider similarity reasoning as a special part of an overall problem solving process. To solve problems, knowledge is needed. This knowledge has to be incorporated into the parts that participate in the problem solving process. For this reason, we introduced the concept of knowledge containers.

We considered two main methodologies that use similarities, CBR and PR. It was of particular interest to observe how knowledge is used. For instance, for any task like for classification one needs some knowledge. Anybody can solve a problem only if such knowledge is used.

Essentially in this context, we have two knowledge types, explicit (background) knowledge and implicit knowledge hidden in the data. The question is how and to which extent a method makes use of the knowledge. Implicit knowledge is not directly usable and has to be made explicit and stored somewhere in the system. For this, learning and data mining techniques are used. We pointed out in which way the methodologies make use of the knowledge and which knowledge is contained in the similarity measures. In principle, both methodologies are knowledge intensive but they store the knowledge in different containers.

This is a rich area of research where many results have been achieved.

But we have also made clear that there are many isolated results and the possible improvements of synergy are still in its infancies. Two main examples are:

- If background knowledge is available, why not using it in PR systematically?
- Each case base as a whole contains a lot of implicit knowledge, why not extracting it and filling it into similarity measure?

Compared with equality the literature on similarity is relatively small, both from the philosophical as from the computational side. On the other hand, the similarity concept is much richer: There is only one equality but there are many similarities. One must admit, however, that many aspects of similarity are discussed in various other disciplines e.g., partial order or probability. Nevertheless, a unified theory of similarity would be welcome.

Acknowledgments

Many of the contributions go back to joint work with K.D. Althoff, Ralph Bergmann, Stefan Wess, and Aldo v.Wangenheim. With K.D. Althoff, Ralph Bergmann, and Stefan Wess, we developed many basic principles of CBR and similarity-based reasoning. Also, many details in the text and the diagrams come from joint lectures with Ralph Bergmann. In the cooperation with Aldo v. Wangenheim most of the techniques mentioned for image interpretation, in particular in medicine have been developed. For the present text the author is indebted to helpful discussions with Ansgar Bernardi, Theo Härder, Eyke Hüllermeyer, Wolfgang Lenski, Petra Perner, Armin Stahl, and Sandra Zilles.

References

1. Gentner, D.: Structure Mapping: A Theoretical Framework for Analogy. Cognitive Science 7 (1983); pp. 155–170
2. Leibniz, G.W.: Fragmente zur Logik. F. Schmidt Verlag 1960.
3. Kolodner, J.L. 1993. "Case Based Reasoning". Morgan Kaufmann 1993.
4. Schank, R.C. 1982. "Dynamic Memory", Cambridge University Press 1982
5. Head: Studies in Neurology. Oxford, 1920.
6. Moog, W.: Ähnlichkeits- und Analogielehre. VDI Verlag 1985.

7. McSherry, D.: Diversity-Conscious Retrieval. In: Craw, S., Preece, A. (eds.): Advances in Case-Based Reasoning LNAI 2416, pp. 219–233, Springer Verlag 2002.
8. Richter, M.M.: Logic and Approximation in Knowledge Based Systems. In: Logic versus Approximation, SLNCS 3075, ed. W. Lenski, 2004, p. 184–204.
9. Börner, K.: Konzeptbildende Analogie: Integration von Conceptual Clustering und analogem Schließen zur effizienten Unterstützung von Entwurfsaufgaben. Dissertation Kaiserslautern 1997, DISKI Verlag 177.
10. Richter, M.M.: Fallbasiertes Schließen: Vergangenheit, Gegenwart, Zukunft. In: Informatik Spektrum, Band 26 (3) (2003), S. 180–190.
11. Bergmann, R., Richter, M.M., Schmitt, S., Stahl, A., Vollrath, I.: Utility-oriented matching: A new research direction for Case-Based Reasoning. In: Professionelles Wissensmanagement: Erfahrungen und Visionen. Shaker, 2001.
12. Tversky, A.: Features of Similarity. Psychological Review 84 (1977), pp. 327–352.
13. Tversky, A., Gati, I.: Studies of Similarity. In: E. Rosch, B.B. Lloyd (Eds.), Cognition and Categorization, Lawrence Erlbaum 1978, pp. 79–98.
14. Santini, S., Jain, R.: Similarity Measures. IEEE Transaction on Pattern Analysis and Machine Intelligence 21, pp. 871–883, 1999.
15. Bock, H.H.: Automatische Klassifikation. Vandhoeck + Ruprecht 1973.
16. Levenshtein, V.I.: Binary codes capable of correction deletions, insertions and reversals. Soviet Physics Doklady 10, pp. 707–710, 1966.
17. Bergmann, R.: On the use of taxonomies for representing case features and local similarity measures. In Gierl & Lenz (Eds.) 6th German Workshop on CBR. 1998.
18. Resnik, P.: Semantic Similarity in a Taxonomy: An Information Based Measure and its Application to Problems if Ambiguity in Natural Language. Journal of Artificial Intelligence Research 11, pp. 95–130, 1999.
19. WordNet::Similarity: http://search.cpan.org/dist/WordNet-Similarity.
20. Dubitzky, W., Schuster, A., Hughes, J.G., Bell, D.A., Adamson, K.: How similar is VERY YOUNG to 43 Years of Age? Proc. IJCAI 1997, pp. 226–231.
21. Burkhard, K-D., Richter, M.M.: On the Notion of Similarity in Case Based Reasoning and Fuzzy Theory. In: Soft Computing in Case Based Reasoning (ed. Sankar K. Pal et al), Springer Verlag 2000, p. 29–46
22. Embrecht, P., Lindskog, F., McNeill, A. (01). Modeling Dependence with Copulas and Applications to Risk Management. Zurich 2001.
23. CBR-Works 4. empolis.com 2003.
24. Bergmann, R., Breen, S., Göker, M., Manago M., Wess S.: Developing Case-based Reasoning Applications: The INRECA-Methodology. Springer SNLAI 1612 (1999).
25. Richter, M.M.: Introduction to: Case-Based Reasoning Technology, ed. M. Lenz, B. Bartsch-Spörl, H.-D. Burkhard, S. Wess, LNAI 1400, 1998, S. 1–16.
26. Richter, M.M.: Knowledge Containers. To appear.
27. Roth-Berghofer, Th.: Knowledge Maintenance of Case-Based Reasoning Systems: The SIAM Methodology. Dissertation Kaiserslautern 2002. DISKI Verlag 262.
28. Stahl, A., Gabel, T.: Using Evolution Programs to Learn Local Similarity Measures. In: Proc. ICCR 03, Springer Verlag 2003.
29. Stahl, A.: Learning feature weights from case-order feedback. Proc. Of the 4th International Conference on Case-Based Reasoning, Springer 2001.

30. Wettschereck, D., Aha, D.W.: Weighting features. In: Proceedings of the 1st International Conference on Case-Based Reasoning (ICCBR 95). Springer 1995.
31. Wess, S.: Fallbasiertes Problemlösen in wissensbasierten Systemen zur Entscheidungsunterstützung und Diagnostik. Dissertation Kaiserslautern 1995. DISKI Verlag 126.
32. Bonzano, A., Cunningham, P., Smyth, B.: Using Introspective Learning to Improve Retrieval in CBR: A Case Study in Air Traffic Control. In: Proceedings of the 2nd International Conference on Case-Based Reasoning 1997 (ICCBR 97). Springer Verlag 1997.
33. Munoz-Avila, H., Huellen, J. (1996): Feature weighting by explaining case-based reasoning planning episodes. In: Proceedings of Third European Workshop on Case-Based Reasoning (EWCBR-96). Springer Verlag 1996
34. Zhang, Z., Yang, Q.: Dynamic Refinement of Feature Weights Using Quantitative Introspective Learning. In: Proceedings of the Sixteenth International Joint Conference on Artificial Intelligence (IJCAI 99), 1999.
35. Globig, C., Jantke, K, Lange, S. Sakakibara, Y.,: Krechel,D. On case-based learnability of languages. New Generation Computing 15(1), 1997, p. 57–63.
36. Li, X.: Potential Analysis for Massively Parallel Computing and its Application to Neural Networks. Dissertation Kaiserslautern 1993.
37. Kimura, F., Takashina, K., Tsuruoka, S. and Miyake, Y.: Modified quadratic discriminant functions and the application to Chinese character recognition. IEEE Transactions on Pattern Analysis and Machine Intelligence, vol 9, pp. 149–153, 1987.
38. Wu, Y., Zhang, A.: Adaptively Discovering Meaningful Patterns in High-dimensional Nearest Neighbor Search. In: IEEE International Conference on Computer Vision and Pattern Recognition (CVPR03), Baltimore 2003.
39. Perner, P.: Are case-based reasoning and dissimilarity-based classification two sides of the same coin? Journal Engineering Applications of Artificial Intelligence, 15/3. 2002. pp. 205–216.
40. Pekalska, E. Duin, R.P.W.: Automatic pattern recognition by similarity representations. Electronics Letters, Vol.37, pp. 159–160. 2001.
41. Perner, P., Holt, A., Richter, M.M.: Image processing in case-based reasoning.
42. Communello, E.: CMIIS – The Cyclops Medical Image Interpretation System. Dissertation Kaiserslautern 2004 (Dissertation.de).
43. Abdala, D., Richter, M.M., dos Santos, Th., von Wangenheim, A., Wille, P.R.: CycML – A Language to Describe Radiological Images. Proc. 16th IEEE Symposium on Computer-Based Medical Systems (CBMS 2003). IEEE Computer Society, p. 145–149, New York.
44. Krechel, D., Richter, M.M., v.Wangenheim, A.:Image Analysis and Image Interpretation with Applications to Medical Domain Problems. In preparation.
45. Hausdorff, F.: Grundzüge der Mengenlehre. Berlin, 1914. Reprint: Chelsea Publishing Company 1978.
46. Frechet, M.R.: Sur quelques points du calcul fonctionnel. Dissertation Paris 1906.
47. Mougouic, B., Richter, M.M.: Generalized Cases, Similarity and Optimization. In: Mechanizing Mathematical Reasoning, (ed. D. Hutter, W. Stephan), LNAI 2605, pp. 564–574. 2005.
48. P. Perner: An Architecture for a CBR Image Segmentation System. Engineering Applications of Artificial Intelligence Vol. 12 (6), 1999, pp. 749–759.

49. Perner, P.: Why Case-Based Reasoning Attractive is for Image Interpretation. In: D. Aha, I. Watson (Eds.): Case-Base Reasoning Research and Development, Springer LNAI 2080, 2001, pp. 27–44.
50. Bir Bhanu, Anlei Dong: Concepts Learning with Fuzzy Clustering and Relevance Feedback. 102–116. In: Proc. Machine Learning and Data Mining in Pattern Recognition, MLDM 2001 Leipzig, Germany (Ed..Petra Perner). Lecture Notes in Computer Science 2123 Springer 2001.
51. Grimnes, M., Aamodt, A.: A two layer case-based reasoning architecture for medical image understanding, In I. Smith & B. Faltings (Eds.) Advances in Case-Based Reasoning. Berlin: Springer Verlag, pp 164–178. 1996.
52. Lenski, W., Wette-Roch, E.: Structured Phrases for Indexing and Retrieval of Research Topics. In: R. Decker, W. Gaul (eds.): Classification and Information Processing at the Turn of the Millenium. Springer-Verlag Heidelberg, pp. 479–487, 2000.
53. Japan Corporate News Network, http://www.japancorp.net
54. Li, J., Ruhe, G., Al-Emran, A., Richter, M.M.: A Flexible Method for Software Effort Estimation by Analogy. Journal of Empirical Software Engineering. DOI 10.1007/s10664-006-7552-4. 2006.
55. Weber, R., Proctor, J. M., Waldstein, I., Kriete, A.: CBR for Modeling Complex Systems. In: H. Munoz, F. Ricci (Eds.), Case-Based Research and Development, LNAI 3620, Springer Verlag, pp. 625–639, 2005.

3

Distance Function Learning for Supervised Similarity Assessment

A. Bagherjeiran and C.F. Eick

Department of Computer Science
University of Houston
Houston, TX 77204-3010, USA
{abagherj,ceick}@cs.uh.edu

Summary. Assessing the similarity between cases is a prerequisite for many case-based reasoning tasks. This chapter centers on distance function learning for supervised similarity assessment. First a framework for supervised similarity assessment is introduced. Second, three supervised distance function learning approaches from the areas of pattern classification, supervised clustering, and information retrieval are discussed, and their results for two supervised learning tasks will be explained and visualized. In each of these different areas, we show how the method can be applied to areas of case-based reasoning. Finally, a detailed literature survey will be given.

3.1 Introduction

Case-based reasoning depends heavily on assessing the similarity between cases. Similarity assessment is the task of determining which cases are similar to each other and which are dissimilar. In general, defining distance functions is a difficult and tedious task.

There are several uses of distance functions in case-based reasoning all of which can benefit from distance function learning. In case-base classification, cases belonging to the same class should have a lower distance than cases in different classes. A good distance function is one that leads to a highly accurate classifier. In case base maintenance, the case base is reorganized into clusters so that search time is reduced by consulting the cluster's representative first. Cases far from the representative of the cluster can be removed to improve search time. A good distance function is one that leads to tight and cohesive clusters, effectively organizing the case base. In case-based retrieval, several cases relevant to a user's query are returned and users send feedback to improve retrieval. A good distance function is one that retrieves a variety of relevant cases for a user's query. Each of these applications places different requirements on the distance function, which makes manually tuning the distance function difficult.

A. Bagherjeiran and C.F. Eick: *Distance Function Learning for Supervised Similarity Assessment*, Studies in Computational Intelligence (SCI) **73**, 91–126 (2008)
www.springerlink.com

Due to the fact that distance functions are difficult to design for different applications, it is worthwhile to look for approaches that learn distance functions from the case base. In particular this chapter will discuss methods in the context of classification, where the goal of supervised similarity assessment is to obtain distance functions that separate cases belonging to different classes well. Consequently, the scope of the methods discussed in this chapter is limited to case bases that have class labels. Class labels either represent natural classes or capture information about the relevancy of a particular case for a particular use. Different methods are needed for unsupervised learning tasks, such as clustering. The class labels associated with different cases can be viewed as feedback that, as we will see, is instrumental for tailoring distance functions for a particular classification task.

Figure 3.1 illustrates what distance function learning for supervised similarity assessment is trying to accomplish; it depicts the distances of 13 cases, five of which belong to a class that is identified by a square and eight belong to a different class that is identified by a circle. When using the initial distance function d_{init} we do not observe much clustering with respect to the two classes. Starting from this distance function, we would like to obtain a better distance function d_{good} so that the cases belonging to the same class are clustered together. In Fig. 3.1 we can identify three clusters with respect to d_{good}, two containing only circles and one containing only squares. Why is it beneficial to find such a distance function d_{good}? Most importantly, using the learned distance function in conjunction with a k-nearest neighbor classifier [10] allows us to obtain a classifier with high predictive accuracy. For example, if we use a three-nearest neighbor classifier with d_{good} it will have 100% accuracy with respect to leave-one-out cross-validation, whereas several cases are misclassified if d_{init} is used. The second advantage is that looking

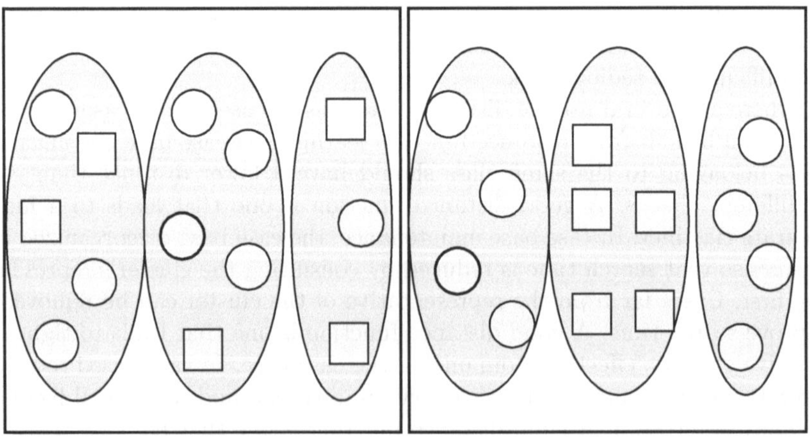

a) d_{init} b) d_{good}

Fig. 3.1. Objective of supervised distance function learning at d_{good} itself will tell us which features are important for the particular classification problem.

It is also important to understand that there are no universal distance functions for a given case base. The usefulness of a distance function is determined in the context of the task it is used for. Consequently, a distance function that does a good job in separating cancer patients from healthy patients is unlikely to be useful in separating diabetes patients from healthy patients.

This chapter surveys how distance function learning is performed in three separate areas of case-based reasoning: classification, maintenance, and retrieval. Section 3.2 reviews basic properties of distance functions and introduces a supervised distance function learning framework. Section 3.3 discusses the Relief algorithm used in case-based classification [30]. Section 3.4 discusses the inside–outside weight adjustment algorithm used as a supervised method for case-base maintenance [17]. Section 3.5 discusses a relevance feedback algorithm used in case-based retrieval [37]. In Sect. 3.6 we show how, on two example case bases, to visualize a distance function before and after learning. Section 3.7 surveys recent work in distance function learning. Section 3.8 gives a brief summary.

3.2 Distance Function Learning Model

A case x consists of a vector of features and a class label (target value). We refer to x as one case in a space of possible cases X, as $x \in X$ and $class(x)$ as the class to which x belongs.

3.2.1 Distance Functions

In general, the case need not be a vector. We define a distance function over a set of cases $x \in X$. A distance function $d : X \times X \to \mathbb{R}$ should satisfy the following two conditions for $x, y, z \in X$:

1. If $d(x, y) = 0$, then x and y are the same or as similar as possible.
2. If x is closer to y than it is to z, then $d(x, y) < d(x, z)$.

A few more restrictions are required for practical distance functions.

Definition 1. *A function $d : X \times X \to \mathbb{R}$ is a* **metric** *on the space X if for all $x, y, z \in X$ the following conditions hold:*

$$0 \le d(x, y) < \infty$$

$$d(x, y) = d(y, x) \tag{3.1}$$

$$d(x, y) = 0 \iff x = y \tag{3.2}$$

$$d(x, z) \le d(x, y) + d(x, z) \tag{3.3}$$

The pair (X, d) *is called a* **metric space.**

Condition (3.2) implies that two cases with distance 0 are equivalent. Duplicate cases occur in practice and, either because of rounding or preprocessing, they may have a distance of 0 even if the cases are not the same. A simple application of the Condition (3.3) is the saying that the shortest distance between two points is a straight line. This is because the Euclidean distance is a metric in $\mathbb{R}^2$.

Norms

A metric depends only on the definition of the distance function and not on the composition of the set. In practice, however, the set X is the vector space $\mathbb{R}^n$ where each $x \in X$ is a vector of length n:

$$x = (x_1, \ldots, x_n)$$

where $x_j \in \mathbb{R}$ for all $1 \leq j \leq n$. Each element, x_j, in the vector is referred to as a feature, and the vector x is called the feature vector. Given this vector definition of a case, we say that a distance function is imposed by a norm defined over the vector space.

Definition 2. *A function* $\|\cdot\| : \mathbb{R}^n \to \mathbb{R}_0^+$ *is a* **norm** *if for all* $x, y \in \mathbb{R}^n$ *and* $\alpha \in \mathbb{R}$*, the following conditions hold:*

$$\begin{aligned} \|x + y\| &\leq \|x\| + \|y\| \\ \|\alpha x\| &= |\alpha| \, \|x\| \\ \|x\| &> 0 \text{ for } x \neq 0 \end{aligned}$$

The first condition is known as the *triangle inequality.* Given a norm $\|\cdot\|$ over $\mathbb{R}^n$, the following distance function is a metric:

$$d(x, y) = \|x - y\|$$

where $x, y \in \mathbb{R}^n$.

We define a family of norms and distance functions.

Definition 3. *A function* $\|\cdot\|_p : \mathbb{R}^n \to \mathbb{R}_0^+$ *is the* L_p **norm** *if for all* $p \in \mathbb{R}$, $p \geq 1$*, and* $x \in \mathbb{R}^n$

$$\|x\|_p = \left(\sum_{j=1}^{n} |x_j|^p \right)^{1/p}$$

For each L_p norm, there is a corresponding L_p distance function metric.

Definition 4. *The function* $d : \mathbb{R}^n \times \mathbb{R}^n \to \mathbb{R}^+$ *is the* L_p **distance metric** *if for all* $p \in \mathbb{R}$, $p \geq 1$, *and* $x, y \in \mathbb{R}^n$

$$d(x,y) = \|x - y\|_p$$

where $\|\cdot\|_p$ *is the* L_p *norm.*

This family of distance functions is also known as the Minkowski distance. There are three common p values, which correspond to the following three distance functions:

L_1 Manhattan or city-block distance:

$$d(x,y) = \sum_{j=1}^{n} |x_j - y_j|$$

L_2 Euclidean distance:

$$d(x,y) = \sqrt{(x_1 - y_1)^2 + \cdots + (x_n - y_n)^2}$$

L_∞ Max distance:

$$d(x,y) = \max_{j=1}^{n} |x_j - y_j|$$

In the remainder of this chapter we consider only the L_1 distance because the notation is simplest and it is faster to compute than L_2.

3.2.2 Parameterized Distance Functions

In distance function learning, the objective is to find parameter values for a parameterized distance function. The most common approach to obtain a parameterized distance function is feature weighting. Feature weights are the output of the distance function learning algorithms discussed here. Weighted distance functions are a subclass of the more general linear distance functions.

Feature weights, one for each attribute, emphasize one subset of features over the others. For the L_1 distance, the weighted version is

$$d_w(x,y) = \sum_{j=1}^{n} w_j |x_j - y_j| \tag{3.4}$$

where $x, y, w \in \mathbb{R}^n$. We place a few restrictions on the vector w_j for $1 \leq j \leq n$:

$$w = (w_1, \ldots, w_n)$$
$$w_j \geq 0$$
$$\sum_{j=1}^{n} w_j = 1$$

These restrictions simplify the programming. In addition, positive weights maintain the semantics of a distance function since negative weights and distances have no clear meaning.

Linear distance functions generalize the L_2 distance by projecting the original distance vector onto a linear basis A:

$$d_A(x, y) = (x - y)^T A (x - y)$$

where x, y are column vectors and A is an $n \times n$ projection matrix. Although most definitions of linear distance function take the square root of $d_A(x, y)$, we usually ignore the root because it is expensive to compute and does not change the relative distances. When A is positive semi-definite, the distance $d_A(x, y)$ is a distance metric. When A is the identity matrix ($I_{n\times n}$), the distance reduces to the squared L_2 distance as shown:

$$\begin{aligned} d_I(x, y) &= (x - y)^T I_{n\times n} (x - y) \\ &= (x - y)^T (x - y) \\ &= \sum_{j=1}^{n} (x_j - y_j)^2 \\ &= \|x - y\|^2 \end{aligned}$$

For small- to medium-sized case bases, it is often more efficient to compute and store the distances between all pairs of cases. This forms a distance matrix for the case base X and distance function d.

Definition 5. *A matrix $D \in \mathbb{R}^{m\times m}$ is a* **distance matrix** *if and only if:*

$$D_{i,j} = d(x_i, x_j)$$

for $1 \leq i, j \leq m$, where d is a distance function, $x_i, x_j \in X$, and m is the number of cases in X.

For distance function learning, we often decompose the distance matrix into the set of matrices D_j for $1 \leq j \leq n$, storing the distance with respect to the jth feature. A weighted distance matrix for the L_1 distance can be recreated as $D = \sum_{j=1}^{n} w_j D_j$ for feature weights w. For a metric distance function, D is a symmetric matrix with zeros along the diagonal.

Examples

Example 1. A distance function without weights is assumed to have weights, but they are equal and nonzero:

$$w_j = \frac{1}{n}$$

for $1 \leq j \leq n$.

Example 2. As a linear distance function, a weighted distance function consists of a diagonal projection matrix $W = \text{diag}(w)$:

$$\begin{aligned} d_W(x, y) &= (x - y)^T W (x - y) \\ &= \sum_{j=1}^{n} w_j (x - y)^2 \end{aligned} \tag{3.5}$$

where w is a weight vector as we have previously described. Although more complex, we can obtain a linear form for the L_1 distance:

$$\begin{aligned} d_W(x, y) &= \sqrt{|x - y|}^T W \sqrt{|x - y|} \\ &= \sum_{j=1}^{n} w_j |x_j - y_j| \end{aligned} \tag{3.6}$$

where $\sqrt{\cdot}$ denotes the component-wise root. In practice, it is usually more efficient to use the form in (3.4) or (3.5).

Example 3. A common linear distance function is the Mahalanobis distance.

$$d_\Sigma(x, y) = (x - y)^T \Sigma^{-1} (x - y)$$

where Σ is the $n \times n$ covariance matrix. This is often referred to the Mahalanobis distance between x and y. It assumes that x and y are distributed according to the multivariate normal distribution $N(\mu, \Sigma)$ with mean μ. Since both vectors are in the same distribution, the effect of the mean cancels are as shown:

$$\begin{aligned} d_\Sigma(x, y) &= ((x - \mu) - (y - \mu))^T \Sigma^{-1} ((x - \mu) - (y - \mu)) \\ &= ((x - y - \mu + \mu))^T \Sigma^{-1} ((x - y - \mu + \mu)) \\ &= (x - y)^T \Sigma^{-1} (x - y) \end{aligned}$$

3.3 Case-Based Classification

In case-based classification, each case has a class label. The objective of a classification algorithm is to determine the class for an incoming case. In classification, an ideal distance function should make cases close to other cases in the same class, while making them far from cases belonging to a different class. Most distance function learning algorithms for classification find parameters for a parameterized distance function and then apply an existing distance-based classifier.

A classifier that uses distance functions is the k-nearest neighbor (k-NN) algorithm [10]. In pattern classification, we consider a case base $Y \subseteq X$ on which learning is performed. The subset Y is further divided into a training set $Tr \subset Y$ and a testing set $Te \subset Y$ such that $Tr \cup Te = Y$ and $Tr \cap Te = \emptyset$. Given a distance function d and a training set of cases Tr, the 1-NN algorithm assigns the class label to a new case in the testing set $y \in Te$ as follows based on its nearest neighbor in the training set:

$$\begin{aligned} class(y) &= class(x^\star) \\ x^\star &= \arg\min_{x \in Tr} d_w(x, y) \end{aligned}$$

The case $x^\star \in Tr$ is called the nearest neighbor to y with respect to the distance function d. Despite its simplicity, the k-NN algorithm is usually accurate and fast for medium-sized training sets.

3.3.1 Feature Evaluation

Two feature evaluation methods are common in the literature. Feature selection algorithms finds a subset of features that more compactly represent the case base. Feature evaluation algorithms assigns a score to each feature that indicates the degree to which it is useful for classification. The higher the score, the more useful a feature is. Thus, feature selection algorithms are a specialization of feature evaluation methods, in which the score is either 0 or 1. In high-dimensional case bases, such as images or signals, feature evaluation methods are used to remove features that have a low score. This sometimes improves classification accuracy and reduces computation time.

We use feature evaluation algorithms to compute weights in a distance function. We make the weights proportional to the score. Let $s = (s_1, \ldots, s_n)$ be the scores for each of the n features. We compute the weight vector w as

$$w_i = \frac{s_j}{\sum_{j=1}^{n} s_j}$$

where $s_j \geq 0$ for $1 \leq j \leq n$.

3.3.2 Relief-F

The Relief algorithm was originally designed as a feature evaluation algorithm, but it is commonly used to compute weights for distance functions [28]. The weight of a feature corresponds to how well it separates cases from different classes while not separating cases from the same class. Weights are updated for each case in the case base.

Algorithm 1 Relief-F's weight-update where $class(X)$ is the set of all classes in the space X and $p(C)$ is the proportion of cases in the training set that belong to class C. This update yields a new weight vector, and is repeated until the weights converge.

Input: Training set Tr , weight vector w, random sample size m.
Output: Updated weight vector w'

1. Select m cases $x \in T$ from Tr randomly with uniform probability.
2. $w' = w$
3. For $1 \leq i \leq m$
   a) For $C \in class(X)$
      M_i^C = Closest case to x_i in class C using old weights w
   b) $H_i = M_i^{class(x_i)}$
   c) For $1 \leq j \leq n$
      Update weight for feature j

$$w_j' = w_j' - \frac{1}{m}(x_{i,j} - H_{i,j}) + \frac{1}{m} \sum_{C \neq class(x_i)} p(C)(x_{i,j} - M_{i,j}^C) \quad (3.7)$$

Algorithm 1 shows the steps in a single iteration of the Relief-F algorithm, which is an extension of the original Relief algorithm designed to handle multiple classes [30]. From the training set $Tr \subset Y$, a random sample ($T \subseteq Tr$) of size m is selected. The weight of each feature, w_j', is updated for each case x_i in T. The update considers the nearest neighbor of x_i from each class using the last weight vector, w. The neighbor in the same class as x_i, H_i, is called the hit case. Conversely, the nearest neighbor in each class $C \neq class(x_i)$, M_i^C, is called a miss case. The update decreases the feature weight to bring x_i closer to the feature value of the hit and increases the weight to move x_i away from the feature value of the misses. This weight update procedure is repeated for the new weight vector w' until the weights converge.

Relief does not have an explicit objective function for the weights. Weights are updated based on their prediction error, which makes this an iterative hill-climbing approach. By updating weights for individual cases, Relief utilizes local information. Many other feature evaluation methods consider only global information such as the information gain (entropy with respect to class) and the χ^2 statistic (variance in the feature values) [16]. Local information is particularly useful in distance function learning.

Example 4. Consider a set Tr of four cases in two classes completely separable along the first dimension, which we will call the x-axis. For convenience, the distance matrix is decomposed:

$$D_w = w_1 D_1 + w_2 D_2$$

where D_i is the distance matrix with respect to feature i and w is the weight vector. Let D_1 and D_2 be the feature-specific distance matrices for the four cases:

$$D_1 = \begin{bmatrix} 0&0&1&1\\ 0&0&1&1\\ 1&1&0&0\\ 1&1&0&0 \end{bmatrix} \quad D_2 = \begin{bmatrix} 0&1&0&1\\ 1&0&1&0\\ 0&1&0&1\\ 1&0&1&0 \end{bmatrix}$$

Note that in D_1, the diagonal consists of 2×2 blocks of zeros and ones. This means that along the x-axis, the distance to cases in the same class is zero and the distance to cases in the other class is 1. Along the x-axis, the cases are perfectly separated. Along y-axis, they are not separated well. In this case, the ideal weight vector is $(1, 0)$, which would make the final distance matrix $D_w = D_{(1,0)} = D_1$.

We perform a single weight update as shown in Algorithm 1, starting with weights $(0.5, 0.5)$. The incrementally updated weight vector for each case is shown. We use the training set $T = Tr$ updating for all cases. The indexes of the nearest hit and miss cases are $H = (2, 1, 4, 3)$ and $M = (3, 4, 1, 2)$, where H_i denotes the hit for the ith case. The weight vector changes after considering each case as follows:

$$\begin{aligned} \text{Case } 1 &: 0.714\ 0.286 \\ \text{Case } 2 &: 0.959\ 0.041 \\ \text{Case } 3 &: 0.999\ 0.001 \\ \text{Case } 4 &: 0.999\ 0.001 \end{aligned}$$

The weight obtained after Case 4 is the new weight vector. Performing a second iteration should result in the same weight vector; thus, the weights have converged. Note that for the last two updates, the computed weights were not in $[0, 1]$. After each update, we normalize the weights to be in $[0, 1]$ and sum to 1. The learned weights converge to the weights we expected.

3.4 Case Base Maintenance

In case-based reasoning, the case base often contains irrelevant or near-duplicate cases, which degrades performance. Case base maintenance methods organize the case base by removing such cases. This is typically accomplished using clustering, where the case base is partitioned into a set of clusters. Next, a prototype of each cluster is selected and becomes the representative of the cluster. Cases that have high distance from the prototype can be removed or at least assigned a low weight. Distance function learning is often used for

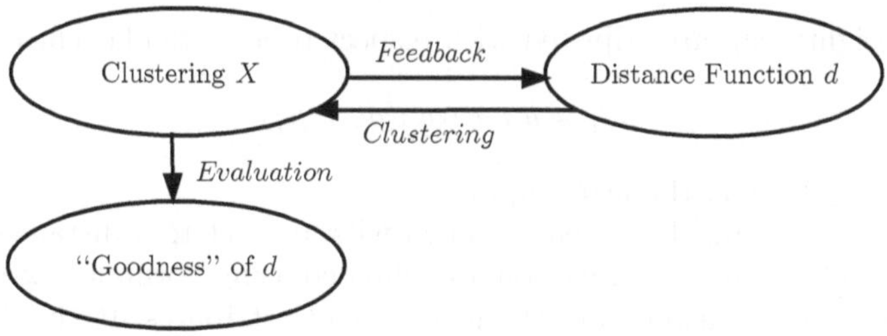

Fig. 3.2. Coevolving clusters and distance functions

case base maintenance. This creates a customized case base for the distance function.

If the case base has class labels, we seek a distance function such that the resulting clusters are more pure. Purity is defined as the average fraction of cases in a cluster that belong to a majority class. When the purity approaches 1.0, all cases in all clusters belong to a single class. The algorithm we discuss in this section, inside–outside weight updating, relies on two ideas, as depicted in Fig. 3.2. First, it uses clustering to evaluate the weighted distance function shown in (3.5). Second, it uses the local information from clusters to update the weights. We refer to this as a *coevolving* approach because the clustering and the distance function are learned together.

After the cases have been clustered, the purity of the obtained clusters is computed and is used to assess the quality of the current distance function. Any clustering algorithm can be used for this purpose; however, supervised clustering algorithms [18] that maximize cluster purity while keeping the number of clusters low are particularly useful for supervised distance function evaluation. More details on how clustering is exactly used for distance function evaluation are given in [17].

Inside–outside weight updating uses the average distance between the majority class members [1] of a cluster and the average distance between all members belonging to a cluster for the purpose of weight adjustment. More formally, let

$d_j(x, y)$ be the distance between x and y with respect to the jth attribute.

w_j be the current weight of the jth attribute.

σ_j be the average normalized distances for the cases that belong to the cluster with respect to d_j.

μ_j be the average normalized distances for the cases of the cluster that belong to the majority class with respect to d_j.

[1] If there is more than one most frequent class for a cluster, one of those classes is randomly selected to be "the" majority class of the cluster.

The weights, w_j, are adjusted with respect to a particular cluster:

$$w'_j = w_j + \alpha w_j(\sigma_j - \mu_j) \tag{3.8}$$

with $0 < \alpha \leq 1$ being the learning rate.

After a clustering[2] has been obtained with respect to a distance function the weights of the distance function are adjusted using Formula (3.8) iterating over the obtained clusters and the given set of attributes. It should also be noted that no weight adjustment is performed for clusters that are pure or for clusters that only contain a single case.

The weight adjustment formula (3.8) gives each cluster the same degree of importance when modifying the weights. Suppose we had two clusters, one with 10 majority cases and 5 minority cases and the other with 20 majority and 10 minority cases, with both clusters having identical average distances and average majority class distances with respect to a feature, the weight would have identical increases (decreases) for the two clusters. This somehow violates common sense; more efforts should be allocated to remove 10 minority cases from a cluster of size 30 than removing 5 members of a cluster that only contains 15 cases. Therefore, we add a factor λ to the weight adjustment heuristic that makes weight adjustment somewhat proportional to the number of minority cases in a cluster. Our weight adjustment formula therefore becomes

$$w'_j = w_j + \alpha \lambda w_j(\sigma_j - \mu_j) \tag{3.9}$$

with λ being defined as the number of minority cases in the cluster over the average number of minority cases for all clusters. Because the approach tends to move majority cases to the center of a cluster and nonmajority cases away from the center, it is called inside–outside weight updating.

This update rule is similar to that of the Relief algorithm as discussed in Sect. 3.3. Like this method, Relief updates the weight to bring same-class cases closer together and different-class cases further apart. Unlike this method, Relief uses simple nearest-neighbor queries instead of whole-cluster information. It does not take advantage of information from both the class labels or the result of the clustering. For example, (3.7) puts more emphasis on the cases in the same class compared to different classes. In the context of clusters, this expression is similar to the μ term from (3.9).

Example 5. Assume we have a cluster that contains six cases numbered 1 through 6 with cases 1, 2, 3 belonging to the majority class. Furthermore, we assume there are three attributes with three associated weights (w_1, w_2, w_3) which are assumed to be equal initially ($w_1 = w_2 = w_3 = \frac{1}{3}$), and the decomposed distance matrices D_1, D_2, and D_3 with respect to the three attributes are given below:

[2] Clusters are assumed to be disjoint.

$$D_1 = \begin{bmatrix} 0 & 1 & 2 & 3 & 4 & 3 \\ & 0 & 1 & 2 & 3 & 1 \\ & & 0 & 1 & 2 & 2 \\ & & & 0 & 1 & 3 \\ & & & & 0 & 1 \\ & & & & & 0 \end{bmatrix} \quad D_2 = \begin{bmatrix} 0 & 1 & 1 & 1 & 5 & 1 \\ & 0 & 1 & 1 & 5 & 1 \\ & & 0 & 1 & 5 & 5 \\ & & & 0 & 5 & 1 \\ & & & & 0 & 5 \\ & & & & & 0 \end{bmatrix} \quad D_3 = \begin{bmatrix} 0 & 3 & 3 & 2 & 2 & 3 \\ & 0 & 3 & 2 & 2 & 3 \\ & & 0 & 2 & 2 & 2 \\ & & & 0 & 1 & 3 \\ & & & & 0 & 1 \\ & & & & & 0 \end{bmatrix} \quad D = \begin{bmatrix} 0 & 1.67 & 2 & 2 & 3.67 & 2.33 \\ & 0 & 1.67 & 1.67 & 3.33 & 1.67 \\ & & 0 & 1.33 & 3 & 3 \\ & & & 0 & 2.33 & 2.33 \\ & & & & 0 & 2.33 \\ & & & & & 0 \end{bmatrix}$$

The case distance matrix D is next computed using

$$D = w_1 D_1 + w_2 D_2 + w_3 D_3$$

First, the average cluster and average inter-majority case distances for each attribute have to be computed; we obtain $\sigma_1 = 2$, $\mu_1 = 1.7$; $\sigma_2 = 2.6$, $\mu_2 = 1$; $\sigma_3 = 2.3$, $\mu_3 = 2.5$. The average distance and the average majority cases distance within the cluster with respect to d are $\sigma = 2.29$, $\mu = 1.78$. Assuming $\alpha = 0.2$, we obtain the new weights:

$$\begin{aligned} w' &= \left(\frac{1}{3}, \frac{1}{3}, \frac{1}{3}\right) * (1.06, 1.32, 0.953) \\ &= (0.353, 0.440, 0.318) \end{aligned}$$

where $*$ denotes the element-wise multiplication. After the weights have been adjusted for the cluster, the following new case distance matrix D is obtained:

$$D = \begin{bmatrix} 0 & 1.57 & 1.89 & 1.92 & 3.82 & 2.21 \\ & 0 & 1.57 & 1.60 & 3.50 & 1.57 \\ & & 0 & 1.29 & 3.19 & 3.19 \\ & & & 0 & 2.58 & 2.21 \\ & & & & 0 & 2.58 \\ & & & & & 0 \end{bmatrix}$$

The average inter-case distances have changed to $\sigma = 2.31$, $\mu = 1.64$. As we can see, the cases belonging to the majority class have moved closer to each other (the average majority class case distance dropped by 0.14 from 1.78), whereas the average distances of all cases belonging to the cluster increased very slightly, which implies that the distances involving majority cases (involving cases 1, 2, and 3 in this case) have decreased, as intended.

3.5 Case-Based Retrieval

In case-based retrieval, cases are retrieved and ranked according to their distance to a query. Unlike classification and maintenance methods, the cases are assumed not to have a class label. Offline training of a distance function is therefore not possible. A user's query, at runtime, partitions the set of cases into two classes: relevant and nonrelevant. Distance function learning uses this feedback to improve the recall and precision of subsequent retrievals. Since a new distance function is learned for each query, the learning process must be very fast, even if not very accurate.

Learning a distance function in response to a user's feedback is known as relevance feedback. One of the earliest relevance feedback algorithms was designed for text retrieval [37]. As in other fields, a case t is represented as a vector of word features:

$$\mathbf{t} = (t_1, \ldots, t_n)$$
$$t_j \geq 0$$

where each t_j is the value for the jth word from a set of all n words that occur in the case base.

Although cases are represented by feature vectors much like the other fields we have discussed, the values for the features have a special meaning. This permits several assumptions in these case bases that are otherwise not justified for the other domains we have considered. The most common form is the term-frequency inverse case frequency (TFIDF) where each term has the following form:

$$t_i = \begin{cases} \mathrm{FREQ}_i \left(1 - \frac{\log_2 \mathrm{DOCFREQ}_j}{\log_2 m}\right) & \mathrm{word}_j \in t \\ 0 & \mathrm{word}_j \notin t \end{cases}$$

where $\mathrm{DOCFREQ}_j$ is the number of cases that contain the jth word, FREQ_j is the number of times the jth word occurs in the case t, and m is the number of cases in the case base [39].

3.5.1 Distance Function Learning

The most common distance function learning algorithm for case-based retrieval using relevance feedback relies on some assumptions about relevant cases and the words they contain. An analysis of the distribution of words in text cases has shown the following [39]:

1. Relevant cases depend on a small set of relevant words (assumed to be in the query)
2. Relevant words are rare across all cases (basis for the TFIDF feature representation)
3. Relevant words are consistent across relevant cases

The most frequent words in a case base tend to be irrelevant to any particular topic, although most of these are removed during preprocessing. Within the relatively small set of relevant cases, the cases are relevant because they contain these relevant words. As a result, the relevant words are expected to occur consistently and more frequently in the set of relevant cases. Note that the query does not always contain all relevant words, as the words may have synonyms. In this case, the query may miss cases that use different words to describe the same topic.

Since we assume that relevant cases are characterized by the consistent presence of relevant words, distance function learning methods weight words by their consistency within the set of relevant cases. A common measure for inconsistency is the standard deviation of the word frequencies in the relevant cases. The weights are inversely proportional to the standard deviation:

$$w_i = \begin{cases} \frac{1}{\sigma_{i,r}} & \sigma_{i,r} \neq 0 \\ w_0 & \sigma_{i,r} = 0 \end{cases}$$

where $\sigma_{i,r}$ is the standard deviation of word i with respect to the relevant cases r and w_0 is a fixed weight when the standard deviation is close to 0. For words that consistently appear (and do not appear) in the relevant cases, the standard deviation will be low and their weight will be high. For words that do not appear consistently, the standard deviation will be high and their weight will be low.

Despite the apparent simplicity of the weight update strategy, it has been shown to be effective for text retrieval [8,37,39]. The main reason for its good performance is that it has access to the relevant cases, rather than having to predict class labels. In pattern classification and clustering, the relevance of a new case (class label) is not known at runtime, the classification process is not interactive. Information retrieval is an interactive process in which the task is slightly different than classification. Instead of asking "What is the best class of this case?" the question is "Given the relevant cases, what other cases are similar to these?" By providing cases of relevant cases interactively, the user provides significantly more information to the retriever than is available to the classifier. Since distance functions are individualized and different from each other, each distance function determines what is relevant for a particular user but not what is relevant to all users.

Evaluating the performance of these algorithms is difficult in practice. The retrieval process is evaluated in terms of precision (percent of the returned cases that were relevant) and recall (percent of relevant cases that were returned). Ideally, both precision and recall should be high. For relevance feedback algorithms, users are typically not available during experiments. A common practice is to simulate a user's feedback by assigning a class label to each case. Cases in the same class as the query are then considered relevant.

Example 6. Suppose we have a corpus (case base) of five cases with the following most frequently occurring words:[3]

[3] This example is intended to demonstrate the algorithm and not any political opinion.

1. George near a bush
2. George Herbert Walker Bush
3. George Clinton
4. The President
5. President Bush

Each case is the set of most frequently occurring words in a text document. Following the standard text preprocessing procedure, single letter words and articles ("the," "a") are removed and capitalization is ignored. The vector representation of the case base is

$$t = \begin{bmatrix} \text{bush} & \text{clinton} & \text{george} & \text{herbert} & \text{near} & \text{president} & \text{walker} \\ 0.5 & 0 & 0.8 & 0 & 0.2 & 0 & 0 \\ 0.6 & 0 & 0.6 & 0.6 & 0 & 0 & 1 \\ 0 & 0.7 & 0.6 & 0 & 0 & 0 & 0 \\ 0.1 & 0 & 0.01 & 0 & 0 & 0.8 & 0 \\ 0.6 & 0 & 0.01 & 0 & 0 & 0.9 & 0 \end{bmatrix}$$

The query of "President George W. Bush" results in the vector q:

$$q = (1, 0, 1, 0, 0, 1, 0)$$

With equal weights, cases 5, 1, 2, 4, 3 are returned in that order. The user now selects cases 4 and 5 as relevant. The normalized weight vector is then w':

$$w' = (0.17, 0.0, 0.0, 0.0, 0.0, 0.83, 0.0)$$

This assigns the highest weights to "President" and "Bush." The weight for "Bush" is less than that for "President" because "Bush" appears inconsistently across the relevant cases, in case 5 but not 4. The word "President," however, appears in both cases. The same query with the learned weights returns the cases 5, 4, 2, 1, 3 in that order. After relevance feedback, the top two cases were those marked as relevant.

3.6 Visualizing Proximity of Cases

Because distance functions are embedded in different algorithms (classification, clustering, and retrieval), it can be difficult to evaluate the performance of a distance function learning algorithm. Both to provide an intuitive understanding of distance functions and as a qualitative evaluation, we discuss several visualization methods for distance functions. They show how learning a distance function changes the spatial relationship among the cases, and try to visualize how good distance functions can be distinguished from bad ones.

3.6.1 Voronoi Diagrams

A Voronoi diagram partitions cases in a case base into mutually exclusive regions called Voronoi cells (See [33] for more details). Each cell c contains exactly one case x from the case base $Y \subseteq X$. The cell is defined such that the case x is the nearest case in the case base to all cases in the cell. Cases along the border are equally distance to the cases in the cells.

The weights change the shape of the cells in the Voronoi diagram. Figure 3.3 shows two Voronoi diagrams of a case base with two classes. On the left the cells appear to be wider but compressed vertically. This is because our original case base is skewed along one dimension. We change the weights to place more "emphasis" on the x-axis such that its weight is larger than that of the y-axis. As a result, the cells on the right diagram are stretched along the vertical dimension and compressed along the horizontal axis. As the weight for x increases towards 1, the cells become vertical line segments.

3.6.2 Multidimensional Scaling

In the 2D case, we saw that a slight change of the weights significantly changed the neighborhood of the cases. In the Voronoi diagram, this is easily seen as changing the shape of the Voronoi cells. In higher dimensions, however, computing the Voronoi diagram is computationally expensive.

Multidimensional scaling (MDS) (See [6] for more details) finds a p-dimension representation for n-dimension cases, where $1 \leq p < n$. For visualization, we assume that $p = 2$. A key benefit of MDS for visualization is that it preserves the distance between cases such that

$$d(\phi(x), \phi(y)) \approx d(x, y)$$

where x, y are cases in the original case base and $\phi : \mathbb{R}^n \rightarrow \mathbb{R}^p$ projects the input cases to the lower dimension.

Most MDS algorithm operate only on the distance matrix of the cases. In classical multidimensional scaling, the projection is a decomposition of the distance matrix. Since the purpose of distance function learning is to change the distance function and thus the distance matrix, cases change their spatial relationship as the weights change. For many MDS algorithms, the change in position is so drastic that it is hard to compare the two figures before and after changing the weights.

3.6.3 Distance Matrix Image

Although MDS can give us a nice visualization of the cases in 2D, much visual information is lost when moving from one set of weights to another. This is because the cases tend to move in seemingly unpredictable ways. The distance matrix image does not suffer from these problems. The main disadvantage, however, is that it can often be hard to see small intensity changes.

To form the distance matrix image, each entry becomes a block of pixels in an image. The color of each block is proportional to the distance. A common color map is a linear gradient from black to white. In this color map, the brighter the color, the larger the distance. Each row i in the image corresponds to $D_{i,j}$ for all columns j. As the weights change in the distance function, the color of the pixels will change.

To facilitate the qualitative meaning of the image, adjacent rows should correspond to cases in the same class. Grouping by class allows us to observe changes in the structural properties of the distance function. Blocks along the diagonal represent the same classes. Off-diagonal blocks represent distance between pairs of classes, and classes that are well separated will have bright colors for these blocks. Cohesive classes in which the cases are all close to each other will have dark colors in the blocks along the diagonal. We typically assume that classes are cohesive, but this may not be true in general.

3.6.4 Examples of Learned Distance Functions

We show two example case bases before and after changing the weights. For the 2D case base, we demonstrate all three visualization methods. For the 9D case base, we demonstrate only the last two methods. In both cases, the objective of learning the objective function is to improve classification performance. The objective function should make cases in the same class closer and cases in different classes further apart.

Example: 2D Random Case Base

Figure 3.3 shows about 20 cases each from two 2D Gaussian distributions with means $\mu_1 = (0.8, 1)$, $\mu_2 = (1, 1)$, and covariance $\Sigma_1 = \Sigma_2 = \begin{bmatrix} 0.006 & 0 \\ 0 & 0.04 \end{bmatrix}$. In Figs. 3.3, 3.4, and 3.5, the weights are

$$w_u = (0.5, 0.5)$$
$$w_r = (0.7, 0.3)$$

where w_r has unequal weights to simulate learning the distance function.

Voronoi Diagrams

Figure 3.3 shows Voronoi diagrams for the cases with equal and unequal weights. The diagram with unequal weights has been stretched vertically but compressed along the horizontal axis. It may be unclear why the cells are taller when the vertical weight is smaller. This is because a smaller weight decreases distance along that dimension. As a result, more cases are closer to the cases in the cells along that dimension. In contrast, horizontal distance has increased leaving fewer cases close in the cells.

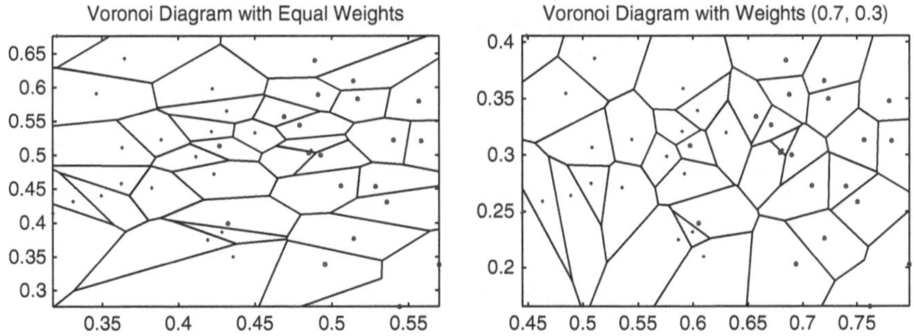

Fig. 3.3. Voronoi diagrams for (*left*) equal weights and (*right*) weights (0.7, 0.3). The scales are different in the figures because the algorithm that computes the Voronoi diagram only uses unweighted Euclidean distance

The Voronoi diagram indicates that the distance function meets our objectives. The separation between μ_1 and μ_2 is greatest along the horizontal component because $|\mu_1 - \mu_2| = (0.2, 0)$. The weights increase the horizontal distance causing the means to be further apart. The result is that the distance function can potentially improve classification performance.

Multidimensional Scaling

Figure 3.4 shows the result of multidimensional scaling with equal weights and unequal weights. Although the original case base was 2D, with unequal weights, the position of the cases and thus their distance has changed significantly compared to the distance with equal weights.

The figures show that with unequal weights, the grouping of the cases improves. As a whole, the classes appear to be further separated from each other in Fig. 3.4 (right). Within each class, the cases appear to be closer together.

Distance Matrix Image

Figure 3.5 shows the distance matrix image for the cases with equal and unequal weights. Since the cases are grouped by class, the lower-left and upper-right quadrants of the figure denote cases in the same class. With unequal weights, these regions are generally darker. This means that the distance between cases in the same class has decreased overall. The lower-right and upper-left quadrants of the figure denote the distance between cases in different classes. With unequal weights, these quadrants are generally brighter. This means that the distance between cases in different classes has increased overall.

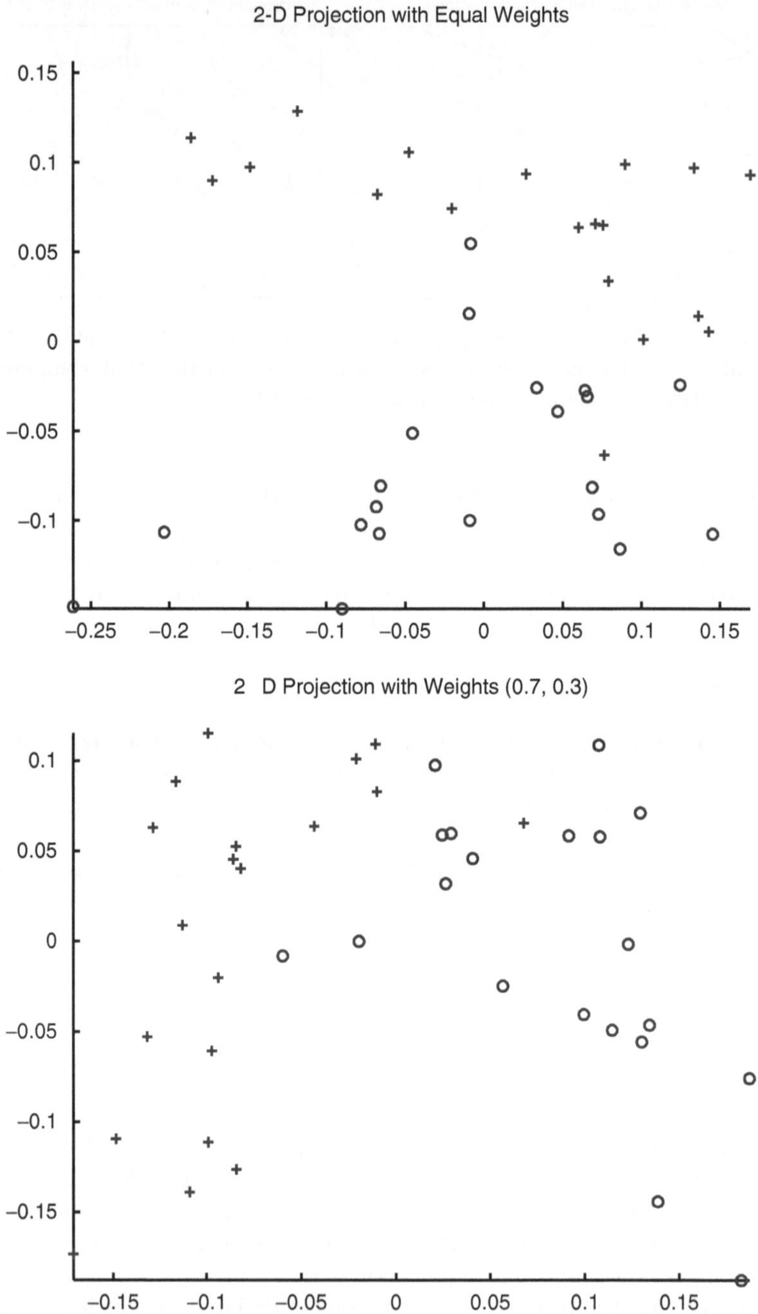

Fig. 3.4. Multidimensional scaling of the 2D case base with (*top*) equal weights in the original case base and (*bottom*) weights (0.7, 0.3)

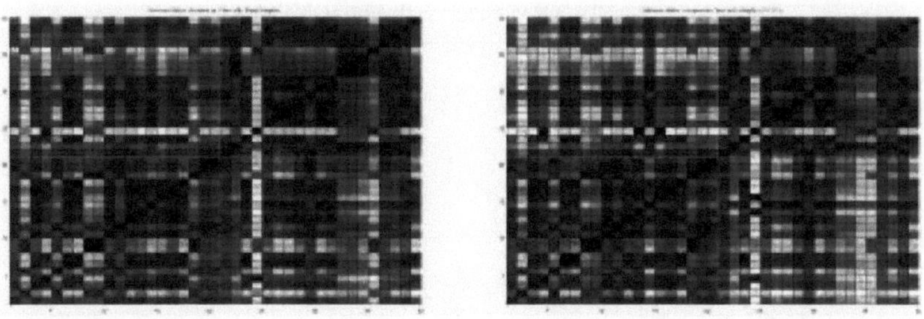

Fig. 3.5. Distance matrix images for the 2D case base with (*left*) equal weights and (*right*) weights (0.7, 0.3).

Example: Nine-Dimensional Case Base

With a 2D case base, the Voronoi diagrams were illustrative. Unfortunately, for more than three dimensions the Voronoi diagram is expensive to compute and difficult to display. For higher-dimensional case bases we use only the multidimensional scaling and the distance matrix images.

The case base has nine features with 214 cases from seven classes [4]. We demonstrate the effect of weights with two weight vectors:

$$w_u = \frac{1}{9}$$
$$w_r = (0.0646, 0.0936, 0.4283, 0.1275, 0.0466, 0.0340, 0.0826, 0.1121, 0.0109)$$

where w_u consists of uniform weights and w_r is the weight vector learned by Relief-F algorithm [30]. As discussed in Sect. 3.3, Relief is designed to find weights that increase the distance between cases of different classes and decrease the distance between cases of the same class.

Multidimensional Scaling

Figure 3.6 shows the multidimensional scaling of the case base with equal weights and with Relief weights. We see that the Relief algorithm works well on this case base. Unlike the 2D case, the cases have shifted significantly with the different weights. In general, the cases are further apart with the Relief weights because the scale of the figures is different. Cases in the same class are grouped together, which is desirable in classification problems. The clusters are better separated from clusters of cases belonging to another class. In general, the figure shows that cases are more tightly grouped within the same class and these groups are better separated from each other.

Distance Matrix Image

Figure 3.7 shows images of the distance matrix with equal weights and Relief weights. These figures show that the Relief weights increase the distance

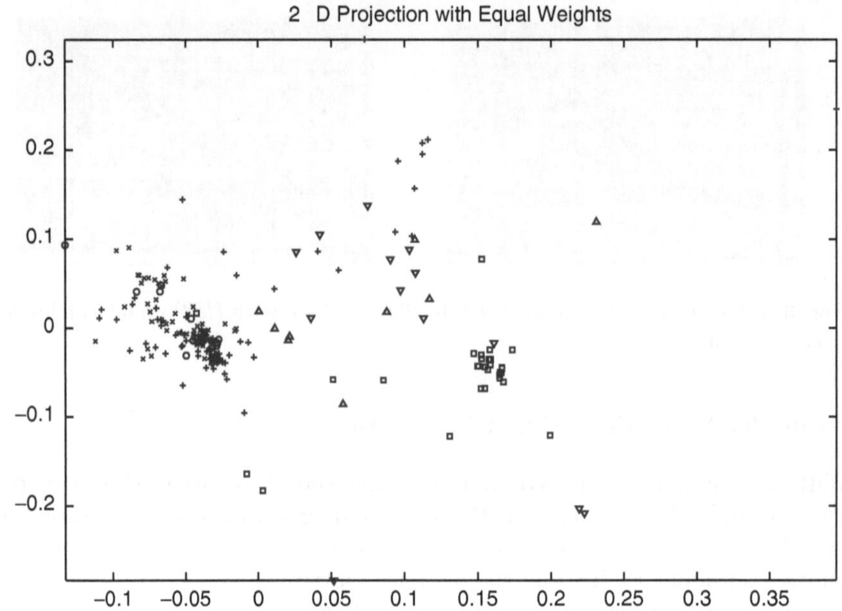

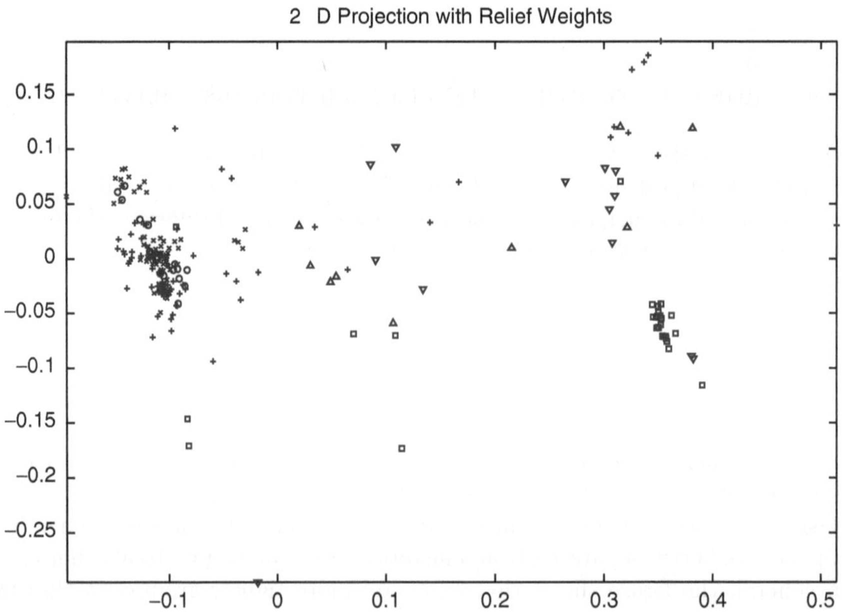

Fig. 3.6. Multidimensional scaling of the glass case base with (*top*) equal weights and (*bottom*) Relief weights [30]

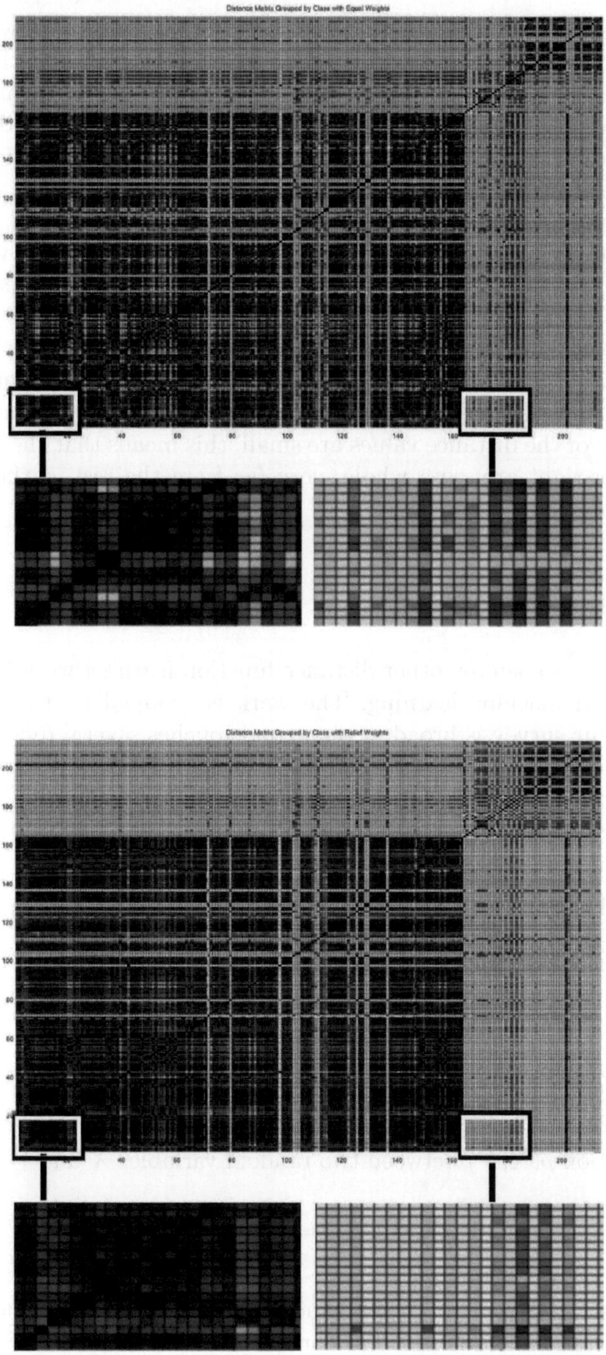

Fig. 3.7. Distance matrix images for the 9D case base with (*top*) equal weights and (*bottom*) Relief weights. The enlarged portions illustrate the difference between intraclass distances (*left*) and interclass distances (*right*)

between cases of different classes but decrease the distance between cases of the same class. The enlarged portions of the figure highlight the difference between the distance functions. With equal weights, the distance between cases in the same class decreases, causing the image to appear darker. With Relief weights, the distance between these cases and those of different classes increases, causing the image to appear brighter. Among the rest of the cases, the blocks along the diagonal appear darker with Relief weights. This means that cases in the same class are closer to each other with Relief weights. The off-diagonal blocks, particularly on the top and right, are brighter with Relief weights.

The distance matrix images also reveal some distance structure across the classes. If the classes were separated well, most of the distance matrix would be brightly colored. Only the regions along the diagonal would be dark. Since the majority of the distance values are small, this means that the three classes in the upper-right are, as a whole, very far from the rest of the cases. The Relief weights clarify this fact.

3.7 Related Work

In this section, we survey other distance function learning work in case-based reasoning and machine learning. The work is grouped by the method and objective. Our survey is broad in scope and touches several different fields of research, each placing different requirements on distance functions. Although we do not claim that our survey is complete, we hope it serves as a starting point for newcomers to the field.

3.7.1 Statistical Methods

Statistical methods make use of either simple statistical models or models of the probability distribution. The model is used to derive weights and usually depends on distributional assumptions about the case base.

Correlation

The correlation $\rho(X,Y)$ between two random variables X and Y is defined as

$$\rho(X,Y) = \frac{\Sigma_{X,Y}}{\sqrt{\sigma_X^2 \sigma_Y^2}}$$

where σ^2 is the variance of each random variable and Σ is the covariance between the random variables. The correlation has a range of $-1 \leq \rho(X,Y) \leq 1$. When $\rho=1$, the variables are said to be positively correlated, which means that as X increases, Y increases. When $\rho = -1$, the variables are negatively correlated, meaning that as X increases, Y increases. In distance function learning, we are interested in the magnitude of the correlation, which indicates the degree to which a feature is correlated (positive or negatively) with the target value. Both kinds of correlation mean that the feature should have a high weight.

Correlation is used for both selecting features and assigning weights. Yu and Liu select features that are highly correlated with the class labels but not correlated with each other [51]. This reduces redundant features because features that are highly correlated may be redundant. Doulamis and Doulamis set feature weights proportional to the correlation between the features values and a user-defined relevance measures, as in information retrieval [15]. Pan et al. find a highly correlated subset of features of a high-dimensional projection of the case base [34]. Projecting features into a high-dimensional space could introduce many irrelevant or redundant features. To eliminate redundant features, we compute a kernel matrix and perform an eigenvalue decomposition as in kernel PCA. The k largest eigenvalues correspond to the most important features. Weights for this subset of features are proportional to the correlation coefficient with respect to the target.

Variance

As we discussed in Sect. 3.5, many relevance feedback methods use the variance of features in the case base because it is easy to compute. It is especially useful in interactive case-based systems, because users can specify which cases are relevant. Kohlmaier et al. assign weights to a feature based on the degree to which it increases variance in the computed similarity values [29]. The variance in the similarity function is a good indicator of different cases. In addition to providing the single best response, a good case result should return a variety of different, but related, cases. We consider adding a single feature to the set of features. If this new feature increases the variance of the similarity function, it is believed to be a good starting point for obtaining good feedback from the users. If the variance is low with this feature, most of the cases are similar based on this feature, so it may not help the user find a good solution. The variance of similarity values can be used either alone or combined with weights. Results show that the method performs better than weights alone on several case-based retrieval tasks [40].

Expectation Maximization

Expectation maximization (EM) approaches iterative solve an optimization problem where the objective is to maximize the likelihood of the case base given the new model. In distance function learning, the model is the set of weights. The likelihood is the probability of finding the existing case base given the current weights. The EM algorithm has two basic steps. In the maximization step, we find those weights that best explain the case base. In the expectation step, we recompute the likelihood function, which changes when the weights change. The EM algorithm maximizes weights, computes the expected case base given the weights, and then repeats until the weights converge.

These methods are common in classification and clustering. In the maximum likelihood approach, we seek a set of weights that best explains the case base. If we have a set of pair-wise constraints, such as knowing that the two cases should not be considered similar, we can find weights that maximize the likelihood of separating cases [50]. Given the weights, we determine the degree to which the pair-wise constraints are violated and update our likelihood function. We can again find weights that maximize the likelihood. Huang et al. apply another EM-type algorithm to find weights that induce a good clustering, a maximum-likelihood clustering. Each cluster is modeled by a multivariate Gaussian distribution. Each case has a probability of being a member of each cluster. The weights are used to compute the membership probabilities for each case. The EM approach is to obtain a new clustering given the weights and then find the weights that improve the membership probabilities [24].

3.7.2 Changing the Set of Features

Often the original set of features is not adequate for the distance function. We would like to evaluate a feature, indicating its usefulness with respect to the class labels. When there are many features, some are irrelevant or redundant. Finding a suitable subset or subspace of the features, these features can improve the computational efficiency. When the features do not adequately express the target value, we can construct new features that are more expressive.

Feature Evaluation

In Sect. 3.3, we discussed the popular Relief method. Many other feature evaluation algorithms are commonly used to find weights for distance functions. Most of these were designed, like Relief, for classification problems and to compare the class distribution considering the different values of nominal features. Information gain measures the difference in entropy with respect to the class labels [16]. The case base is split into partitions with respect to a particular feature such that all cases in the partition have the same value for the feature. The entropy with respect to the class labels is calculated in each partition. The information gain is the difference between the initial entropy, without the partitions, and the average entropy after the partitioning.

As with the Relief algorithm, the output of a feature evaluation method can be used for weights in a distance function. In classification, the information gain increases weights for features that are similar to the class label. Features that are distributed like the class label have high weights as do those features with many values [14]. Text features, particularly in information retrieval, lend themselves to this method. A relevant word would only be in the set of relevant cases, so we would expect it to have high information gain. If the word is not relevant the class labels would tend to be distributed similarly in each split, so the information gain would be small [49].

Extending the concept of feature weights to individual feature values is also common in case-based reasoning literature. Such a local weight could, for example, place more emphasis on the difference between the values "red" and "blue" than on the difference between "red" and "pink". Nunez et al. compared several entropy-based weighting methods for local weighting [32]. They showed that the local weight approach can be useful. However, having more weights to control makes the learning process more difficult. As a result, more complex search methods such as genetic algorithms have to be used [25].

Decreasing the Number of Features

As we have seen, feature selection seeks an optimal subset of the initial set of features. In dimension reduction, we also seek a reduced set of features, but we consider combinations of features. As shown in our visualization cases in Sect. 3.6, multidimensional scaling can reduce a high-dimensional case base to a 2D case base that preserves clusters. There are three common methods for dimension reduction: principal component analysis, linear discriminant analysis, and multidimensional scaling.

Principal Component Analysis

The principal components of a case base are the directions (linear combination of features) that indicate the variance in the cases. They are the eigenvectors of the covariance matrix that correspond to the largest eigenvalues. The principal components become the new features. The covariance matrix allows to calculate the Mahalanobis distance, which projects the cases onto the inverse of the covariance matrix. The projected cases is "corrected" for the covariance such that an unweighted distance function can be used. The objective of these distance function learning algorithms is to find the best projection. For example, the projection should map the cases into well-separated classes [5,19].

Linear Discriminant Analysis

Rather than requiring a distance function to consist of good features, one can require that it lead to good classification results. Since predicting accuracy is computationally expensive, an alternative is to maximize the margin of separation between cases. Since the separation depends on the learned distance function, the problem is to find a distance that leads to the greatest separation between cases in different classes. An early algorithm for this purpose is Linear Discriminant Analysis (LDA). A local approach to LDA finds the discriminant among a small neighborhood of cases. The linear projection matrix that leads to the best discriminant is chosen for each case [22].

Multidimensional Scaling

From a geometric perspective, these covariance approaches reshape the cases to improve separation. The dimensionality of the projected cases is the same as the original cases. As we have seen in Figs. 3.4 and 3.6, projecting cases onto fewer dimensions can yield insight into the distances. Much work has shown that they are quantitatively effective as well. A desirable property of multidimensional scaling is that it should preserves distances in the projected space. Rather than preserving all the original distances, the projection can increase the distance between cases in different classes, leading to a better separation of classes in the lower-dimensional space [52].

Other projection methods utilize different aspects of relatedness such as knowledge of unwanted variance between cases of different classes [23] and the structure of local neighborhoods [20].

Increasing the Number of Features

Feature Construction

Often the original set of features for a case base are unable to represent the concept. For example, if the true target value is a nonlinear function of the features such as $x^2 + xy$, then it would be difficult for any weight vector to approach the function. A common solution is to construct a new, larger set of features. The similarity function and weights are then computed in this high-dimensional feature space. Examples include the set of polynomials of degree 2, yielding features such as $(x_1, x_2, x_1^2, x_2^2, x_1x_2)$. This feature construction method can introduce significant redundancy among the features. As a result, dimension reduction methods, like those discussed earlier, can be used. A specialized version of PCA, called kernel PCA, is ideally suited for this problem [34]. As the original number of features increases, feature construction like this tends to be very expensive, as $O(n^2)$ features have to be constructed. Another technique is to add new, derived features. An example is to find a set of weights for a subset of features and then use the weighted combination of feature values as a new feature [36]. The new feature can either replace or augment the existing set of features. It is particularly useful with image cases, when the individual low-level features are, by themselves, not good predictors of the class.

Kernels

In kernel-based machines, cases are represented as a vector of similarity values. Each case consists of its similarity to all other cases in the case base. A kernel function $k(x_1, x_2)$ defines the similarity between two cases x_1 and x_2. The kernel matrix is constructed from the original case base as follows. For a case base of size n, the kernel matrix has dimension $n \times n$. Each entry $K_{i,j}$ is the value of the kernel for two cases $K_{i,j} = k(x_i, x_j)$. Using the so-called kernel trick, any algorithm that operates on case bases with a dot product can be modified to use a kernel function. This kernelization allows many existing algorithms to be extended with kernels. Techniques that can be kernelized are the support vector machines, principal component analysis, and distance based algorithms [3, 41]. In kernel PCA, we compute a reduced-dimension version of the kernel matrix. The weights for this new feature vector indicate which cases contribute most to the overall covariance of the kernel matrix. As with traditional PCA, the kernel version improves the performance of algorithms when using the reduced set of features [34].

Recent work has established a relationship between kernels and distance functions [3, 41]. In particular, the work shows that good distance functions make for good kernels [3]. As a result, many researchers have applied distance function learning methods, like those discussed in this chapter, to learn kernel functions. As in distance function learning, feature evaluation methods like Relief can be used to find weights for kernel-based machines [9]. Weights can be learned for a parameterized kernel function in much the same way they are learned for distance functions [26]. Recent work shows that good distance function make for good kernels [3].

3.7.3 Extracting Weights from Learned Models

A common approach to assess the importance of features is to apply an algorithm that generates weights as part of the model. The weights are then extracted from the model and used as weights for the distance function or to select features.

Classifiers

A popular classifier that computes weights is the Logistic regression algorithm. This algorithm assumes that the probability of a case x belonging to a class y is

$$Pr(y \mid x, w) = \frac{1}{1 + e^{-w^T x}}$$

where w is a vector weight and includes a bias. The algorithm is trained using a gradient descent approach. The learned weights are then used for feature selection and weights in a distance function.

Arshadi and Jurisica have applied logistic regression to case-based classification of microarray cases [1]. Their objective was to select relevant features from a very high dimensional case base. They combine several classifiers in an ensemble. The classification approach is to retrieve several cases from the case base with the learned weights and then compute the majority class label. Their results show a significant improvement in accuracy. Wilson and Leake use this method to maintain the case base through clustering [48]. By clustering and then learning a set of weights, cases that are very distant from the prototype of the cluster are removed as irrelevant. Features with low weights are also removed as irrelevant. Their results showed that this lead to an improvement of classification accuracy.

Support vector machines also learn weights, classifying cases based on a separating hyperplane as follows:

$$f(x) = \text{sign}(w^T x + b)$$

where w is a weight vector and b is a bias. The learned hyperplane is typically one that separates cases into two classes by the largest possible margin. The decision function $f(x)$ is the distance between a case and the separating hyperplane, so this function can itself be used as a distance function. Alternatively, the weights are extracted and used as weights for a distance function [13].

Relevance Feedback

The relevance feedback methods like those we have previously discussed use relevant cases as the basis for weights. Their objective is to find weights that best separate relevant from nonrelevant cases. Applied both in text retrieval [8, 37, 39] and image retrieval [38], these methods are quite popular. We can also cluster the cases based on relevance and then find a distance function that separates the clusters [27]

3.7.4 Local Search Methods

Local search methods are popular for distance function learning. The most common of these are iterative methods such as hill-climbing. Here, an initial weight vector is updated until the objective function converges, e.g., to the peak of a hill in the objective function. Updates are computed either by the gradient of the objective function or with heuristics. Relief, as we discussed in Sect. 3.3, is a local search method.

Extensions to Relief

In the Relief algorithm, a weight is the degree to which a single feature can be used to predict the class label [28]. In practice this technique works well and has inspired several extensions. The most common extension, Relief-F, extends Relief from two classes to several classes, which allows for greater applicability and widespread use [30]. Rather than learning a single weight, we can find a pair of weights for each class: one for the nearest hit and miss cases [7]. Although widely used as a batch learning algorithm, a recent iterative version of Relief achieves comparable accuracy as an online learner and supports removing outliers [45].

Using Gradients

Gradient-based hill climbing methods, known as gradient descent methods, are the most common form of hill climbing algorithm. For distance function learning, these methods define an objective function in terms of weights and then compute its gradient. A common approach used in case-based retrieval is to find weights that compute an optimal ranking of the cases. For example, given a user's feedback of known ranks for a set of cases, weights can be learned that match the ranks. The error function is simply the squared difference between the known ranks and predicted ranks. This is a continuous, differentiable function, which lends itself to gradient-based methods [31, 43]. Coyle and Cunningham minimize the difference between a user's ranking of a set of retrieved cases versus those computed with uniform weights. The idea is to make the similarity with weights as different from uniform weights as possible. By saving these weights for individual users, a customized case base is created [11]. With a similarly objective function, Shiu et al. incorporate learned weights as a first step in their case base maintenance process [42]. Tsang et al. find weights that induce a good clustering to edit the case base. The objective is to improve cluster metrics such as intracluster distance (tightness of the cluster) [46]. The objective function for optimization minimizes the difference in the objective with learned weights vs. the set of uniform weights.

Nongradient Methods

In Sect. 3.4, we examined a nongradient approach that optimizes class purity [17]. Here the weights are changed along the single attribute that improves the purity the most. A subspace projection method, like hill climbing, updates weights along a predefined direction as long as the objective function improves [21]. This direction typically has components of several features.

3.7.5 Global Search Methods

Local search methods tend to converge to a point, known as a local optima. This local solution may not be the global, best solution. Global search methods are intended to find this global solution. Optimization methods are used when we know that there is only one optimal solution, typically because the objective function is convex. Most other search methods expand their search area with randomization.

Optimization

We can pose the distance function learning problem as one of the optimization. In general, we would like to find feature weights that minimize a global error function. Peleg and Meir find a subset of features that minimize the expected generalization error [35]. The objective is to minimize the margin cost for a feature. Features that can adequately separate the cases have low margin cost. Features that are poor separators are not useful for classification. Weinberger et al. extend the optimization problem in the Relief algorithm. For Relief we considered the distance between the nearest hit and miss case for each case in the training set. Keeping the hit and misses fixed during optimization, the objective is to minimize the distance between cases in the same class and maximize the distance between cases in different classes [47].

Randomized Search

Genetic algorithms can directly search for weights and can evolve expressions for the distance function. Stahl and Gabel find weights for specific feature values. Each individual is a similarity table containing weights for pairs of feature values. The fitness function is the accuracy of the ranks generated with the weights [44]. Jarmulak et al. select features by evolving a feature bit mask. The fitness function is cross-validation accuracy with classification cases [25]. Using genetic programming, feature indexes and arithmetic operators are added to a syntax tree forming an arithmetic expression. The fitness of the expression is its estimated prediction accuracy [12].

Because objective functions over a clustering can be expensive to compute, we seek search methods that do not evaluate many (very similar) solutions. We can view the search for weights as a transition from one clustering, induced by the original weights, to another, induced by the changed weights. If weights lead to a good cluster, these weights, and those that led to them, form a path of weights that lead to a good clustering. By remembering this path, consisting of which choices were good and bad, the search can focus on good paths while avoiding bad ones. If the path leads to a clustering that has already been seen, the search can quickly switch to a different path rather than repeating work already done. Conceptually, this is an application of reinforcement learning to search, in which the best choices are remembered and reused [2].

3.8 Summary

The objective of distance function learning for supervised similarity assessment is to find a distance function that groups together cases belonging to the same class, while separating cases of different classes. We introduced a framework for parameterized distance functions, which depends on a vector of weights. We then provided detailed descriptions of algorithms used in three different fields within case-based reasoning: case-based classification, case base maintenance, and case-based retrieval. We showed how to visualize the difference between good and bad distance functions for high-dimensional case bases. Finally, we surveyed recent work in the literature.

Distance function learning is particularly suited to applications with a large number of dimensions, when it is difficult for us to determine which features are the most important. In classification, the Relief algorithm finds features that separate cases from different classes. In case base maintenance, the inside–outside weight update concentrates cases of the same class into cohesive clusters. In case-based retrieval, relevance feedback adapts the distance function to a user's preference at run time.

Distance function learning is a very active research field, and it can benefit from a cross-fertilization of ideas from different fields. The algorithms discussed in this chapter originated in the fields of machine learning, data mining, and information retrieval. In the field of machine learning, recent work has established a relationship between kernels and distance functions. Distance function learning can be applied to obtain better kernels, and kernel methods can be used to derive good distance functions. This has been used in pattern classification. In data mining, distance function learning has found widespread use when users can specify an objective function to organize information. This suggests an interactive approach to unsupervised clustering in which users can explore a clustering by changing objective functions. In information retrieval, distance functions are tailored to individual users and their queries. As different modalities of information become available in addition to text (image, video, signals), distance function learning can be used to emphasize the relevant features with respect to a user's query. In all of these fields, distance function learning is the common thread helping us better assess the similarity between cases.

References

1. Niloofar Arshadi and Igor Jurisica. Maintaining case-based reasoning systems: A machine learning approach. In *Proc. 7th European Conf. on Adv. in Case-Based Reasoning*, LNAI 3155, pages 17–31, Madrid, Spain, September 2004.
2. Abraham Bagherjeiran, Christoph F. Eick, Chun-Sheng Chen, and Ricardo Vilalta. Adaptive clustering: Obtaining better clusters using feedback and past experience. In *Proc. 5th Int'l Conf. on Data Mining*, pages 565–568, Houston, TX, USA, November 2005. IEEE Computer Society.
3. Maria-Florina Balcan and Avrim Blum. On a theory of learning with similarity functions. In *Proc. 23rd Int'l Conf. on Machine Learning*, pages 73–80, 2006.
4. C.L. Blake and C.J. Merz. UCI repository of machine learning databases, 1998.
5. Leon Bobrowski and Magdalena Topczewska. Improving the k-NN classification with the Euclidean distance through linear data transformations. In Petra Perner, editor, *Adv. in Data Mining, Applications in Image Mining, Medicine and Biotechnology, Management and Environmental Control, and Telecommunications*, LNAI 3275, pages 23–32. Springer Verlag, 2004.
6. Ingwer Borg and Patrick Groenen. *Modern Multidimensional Scaling: Theory and Applications*. Springer, New York, 1997.
7. Derek Bridge and Alex Ferguson. Learning weight intervals for classification. In Diarmuid O'Donoghue, editor, *Proc. 12th Irish Conf. on Artif. Intell. & Cognitive Science*, pages 95–104, 2001.

8. Chris Buckley and Gerard Salton. Optimization of relevance feedback weights. In *Proc. 18th Int'l Conf. on Research and Development in Information Retrieval*, pages 351–357, New York, NY, USA, 1995. ACM Press.
9. Stephane Canu and Yves Grandvalet. Adaptive scaling for feature selection in SVMs. In *Adv. in Neural Information Processing Systems 15*, pages 553–560. MIT Press, 2002.
10. T. Cover and P. Hart. Nearest neighbor pattern classification. In *IEEE Trans. on Information Theory*, volume 13, pages 21–27, 1967.
11. Lorcan Coyle and Pádrain Cunningham. Improving recommendation ranking by learning personal feature weights. In *Proc. 7th European Conf. on Adv. in Case-Based Reasoning*, LNAI 3155, pages 560–572, Madrid, Spain, September 2004.
12. Ronan Cummins and Colm O'riordan. Evolving general term-weighting schemes for information retrieval: Tests on larger collections. *Artificial Intelligence Review*, 24(3–4):277–299, 2005.
13. Carlotta Domeniconi and D. Gunopulos. Adaptive nearest neighbor classification using support vector machines. In Thomas G. Dietterich, Suzanna Becker, and Zoubin Ghahramani, editors, *Adv. in Neural Information Processing Systems 14*, pages 665–672. MIT Press, 2001.
14. Carlotta Domeniconi, Jing Peng, and Dimitrios Gunopulos. An adaptive metric machine for pattern classification. In Todd K. Leen, Thomas G. Dietterich, and Volker Tresp, editors, *Adv. in Neural Information Processing Systems 13*, pages 458–464. MIT Press, 2000.
15. Anastasios D. Doulamis and Nikolaos D. Doulamis. A recursive optimal relevance feedback scheme for content based image retrieval. In *Proc. 2001 Int'l Conf. on Image Processing*, volume 2, pages 741–744, 2001.
16. Richard O. Duda, Peter E. Hart, and David G. Stork. *Pattern Classification*. John Wiley & Sons, New York, 2nd edition, 2001.
17. Christoph F. Eick, Alain Rouhana, Abraham Bagherjeiran, and Ricardo Vilalta. Using clustering to learn distance functions for supervised similarity assessment. *Int'l Scientific Journal of Engineering Applications of Artificial Intelligence*, 19(4):395–401, June 2006.
18. Christoph F. Eick and Nidal M. Zeidat. Using supervised clustering to enhance classifiers. In *Proc. 15th Int'l Symp. on Methodologies for Intelligent Systems*, pages 248–256, Saratoga Springs, NY, May 2005.
19. Amir Globerson and Sam Roweis. Metric learning by collapsing classes. In Y. Weiss, B. Schölkopf, and J. Platt, editors, *Adv. in Neural Information Processing Systems 18*, pages 451–458. MIT Press, Cambridge, MA, 2005.
20. Jacob Goldberger, Sam Roweis, Geoffrey Hinton, and Ruslan Salakhutdinov. Neighbourhood components analysis. In Lawrence K. Saul, Yair Weiss, and Léon Bottou, editors, *Adv. in Neural Information Processing Systems 17*, pages 513–520. MIT Press, Cambridge, MA, 2004.
21. M. Halkidi, D. Gunopulos, N. Kumar, M. Vazirgiannis, and C. Domeniconi. A framework for semi-supervised learning based on subjective and objective clustering criteria. In *Proc. 5th Int'l Conf. on Data Mining*, pages 637–640, 2005.
22. Trevor Hastie and Robert Tibshirani. Discriminant adaptive nearest neighbor classification and regression. In David S. Touretzky, Michael C. Mozer, and Michael E. Hasselmo, editors, *Adv. in Neural Information Processing Systems 8*, pages 409–415. The MIT Press, 1996.

23. Aharon Bar Hillel, Tomer Hertz, Noam Shental, and Daphna Weinshall. Learning distance functions using equivalence relations. In *Proc. 20th Int'l Conf. on Machine Learning*, pages 11–18, 2003.
24. Joshua Zhexue Huang, Michael K. Ng, Honqiang Rong, and Zichen Li. Automated variable weighting in k-means type clustering. *Trans. on Pattern Analysis and Machine Intelligence*, 27(5):657–668, May 2005.
25. Jacek Jarmulak, Susan Craw, and Ray Rowe. Genetic algorithms to optimise CBR retrieval. In *Proc. 5th European Workshop on Adv. in Case-Based Reasoning*, pages 136–147, Tento, Italy, September 2000.
26. Jaz Kandola, John Shawe-Taylor, and Nello Cristianini. Learning semantic similarity. In *Adv. in Neural Information Processing Systems 15*. MIT Press, 2002.
27. Deok-Hwan Kim and Chin-Wan Chung. QCluster: relevance feedback using adaptive clustering for content-based image retrieval. In *Proc. ACM Int'l Conf. on Management of Data*, pages 599–610, New York, NY, USA, 2003. ACM Press.
28. Kenji Kira and Larry A. Rendell. A practical approach to feature selection. In *Proc. 9th Int'l Conf. on Machine Learning*, pages 249–256, Aberdeen, Scotland, UK, 1992. Morgan Kaufmann.
29. Andreas Kohlmaier, Sascha Schmitt, and Ralph Bergmann. A similarity-based approach to attribute selection in user-adaptive sales dialogs. In *Proc. 4th Int'l Conf. on Case-Based Reasoning*, LNAI 2080, pages 306–320, Vancouver, BC, Canada, July 2001.
30. Igor Kononenko. Estimating attributes: Analysis and extensions of RELIEF. In *Proc. 7th European Conf. on Machine Learning*, pages 171–182, 1994.
31. David G. Lowe. Similarity metric learning for a variable-kernel classifier. Technical Report TR-93-43, 1993.
32. Héctor Núñez, Miquel Sànchez-Marrè, and Ulises Cortés. Improving similarity assessment with entropy-based local weighting. In *Proc. 5th Int'l Conf. on Case-Based Reasoning*, LNAI 3689, pages 377–391, Trondheim, Norway, June 2003.
33. Atsuyuki Okabe, Barry Boots, and Kokichi Sugihara. *Spatial Tessellations: Concepts and Applications of Voronoi Diagrams.* John Wiley & Sons, Hoboken, NJ, 1992.
34. Rong Pan, Qiang Yang, and Lei Li. Case retrieval using nonlinear feature-space transformation. In *Proc. 7th European Conf. on Adv. in Case-Based Reasoning*, LNAI 3155, pages 361–374, Madrid, Spain, September 2004.
35. Dori Peleg and Ron Meir. A feature selection algorithm based on the global minimization of a generalization error bound. In Lawrence K. Saul, Yair Weiss, and Léon Bottou, editors, *Adv. in Neural Information Processing Systems 17*, pages 1065–1072. MIT Press, Cambridge, MA, 2004.
36. Petra Perner. Why case-based reasoning is attractive for image interpretation. In *Proc. 4th Int'l Conf. on Case-Based Reasoning*, LNAI 2080, pages 27–43, Vancouver, BC, Canada, July 2001.
37. J.J. Rocchio. Relevance feedback in information retrieval. In Gerard Salton, editor, *The SMART Retrieval System–Experiments in Automatic Document Processing*, chapter 14, pages 313–323. Prentice Hall, 1971.
38. Yong Rui, Thomas S. Huang, and Sharad Mehrotra. Relevance feedback techniques in interactive content-based image retrieval. In *Storage and Retrieval for Image and Video Databases*, pages 25–36, San Jose, CA, January 1998.
39. Gerard Salton and Michael J. McGill. *Introduction to Modern Information Retrieval.* McGraw-Hill, New York, NY, USA, 1983.

40. Sascha Schmitt, Philipp Dopichaj, and Patricia Domínguez-Marín. Entropy-based vs. similarity-influenced: Attribute selection methods for dialogs tested on different electronic commerce domains. In *Proc. 6th European Conf. on Adv. in Case-Based Reasoning*, LNAI 2416, pages 380–394, Aberdeen, Scotland, UK, September 2002.
41. Bernhard Schölkopf. The kernel trick for distances. In Todd K. Leen, Thomas G. Dietterich, and Volker Tresp, editors, *Adv. in Neural Information Processing Systems 13*, pages 301–307. MIT Press, 2000.
42. Simon Chi Keung Shiu, Cai Hung Sun, Xi Zhao Wang, and Daniel So Yeung. Maintaining case-based reasoning systems using fuzzy decision trees. In *Proc. 5th European Workshop on Adv. in Case-Based Reasoning*, pages 285–296, Tento, Italy, September 2000.
43. Armin Stahl. Learning feature weights from case order feedback. In *Proc. 4th Int'l Conf. on Case-Based Reasoning*, LNAI 2080, pages 502–516, Vancouver, BC, Canada, July 2001.
44. Armin Stahl and Thomas Gabel. Using evolution programs to learn local similarity measures. In *Proc. 5th Int'l Conf. on Case-Based Reasoning*, LNAI 3689, pages 537–551, Trondheim, Norway, June 2003.
45. Yijun Sun and Jian Li. Iterative RELIEF for feature weighting. In *Proc. 23rd Int'l Conf. on Machine Learning*, pages 913–920, 2006.
46. Eric C.C. Tsang, Simon C.K. Shiu, X.Z. Wang, and Martin Lam. Clustering and classification of cases using learned global feature weights. 2001.
47. Kilian Weinberger, John Blitzer, and Lawrence Saul. Distance metric learning for large margin nearest neighbor classification. In Y. Weiss, B. Schölkopf, and J. Platt, editors, *Adv. in Neural Information Processing Systems 18*, pages 1473–1480. MIT Press, Cambridge, MA, 2005.
48. D. Wilson and D. Leake. Maintaining case-based reasoners: Dimensions and directions. *Computational Intelligence*, 17:196–212, 2001.
49. Nirmalie Wiratunga, Ivan Koychev, and Stewart Massie. Feature selection and generalisation for retrieval of textual cases. In *Proc. 7th European Conf. on Adv. in Case-Based Reasoning*, LNAI 3155, pages 806–820, Madrid, Spain, September 2004.
50. Liu Yang, Rong Jin, Rahul Sukthankar, and Yi Liu. An efficient algorithm for local distance metric learning. In *Proc. 21st Nat'l Conference on Artificial Intelligence*, 2006.
51. Lei Yu and Huan Liu. Feature selection for high-dimensional data: A fast correlation-based filter solution. In *Proc. 20th Int'l Conf. on Machine Learning*, pages 856–863, 2003.
52. Zhihua Zhang. Learning metrics via discriminant kernels and multidimensional scaling: Toward expected euclidean representation. In *Proc. 20th Int'l Conf. on Machine Learning*, pages 872–879, 2003.

4

Induction of Similarity Measures for Case Based Reasoning Through Separable Data Transformations*

L. Bobrowski[1,2] and M. Topczewska[1]

[1]Faculty of Computer Science, Bialystok Technical University
[2]Institute of Biocybernetics and Biomedical Engineering, PAS, Warsaw

Summary. Inducing similarity measures for the *case based reasoning* scheme through separable data transformations is considered in this chapter. Particular attention is paid to linear transformations of multidimensional data on visualising planes. Separable linear transformations are based both on solutions of eignevalue problems used in the *principal componet analysis* or in the *discriminant analysis* as well as on minimization of the convex and piecewise linear (CPL) criterion functions. The *perceptron* and the *differential* criterion functions belong among others to the CPL family. Such functions give possibility for flexible and efficient designing separable transformations of data sets.

4.1 Introduction

Decision support systems are often based on the case based reasoning (CBR) method [1]. An essential part of the CBR scheme is a search for such records in a database which are most similar to the case actually analysed [1]. Such a paradigm is also used in the nearest neighbours (K-NN) technique developed in the framework of the pattern recognition [2,3]. One of the central problems during implementation of the CBR or the K-NN scheme is the choice of a similarity measure or the distance function between the database records [4]. The quality of the decision support rules can be improved by adjusting the similarity measures or adequately tailoring the distance functions [5].

Here, we analyse possibilities of applying linear transformations of reference data sets for inducing similarity measures and diagnosis support rules from the learning sets. Particular attention is paid to linear transformations of multidimensional data on visualising planes. Designing linear transformation scheme that results from separability postulates is considered [5, 6]. The

*This work was partially supported by the grant W/WI/1/2005 from the Bialystok University of Technology, the KBN grant 3T11F01130 and by the grant 16/St/2007 from the Institute of Biocybernetics and Biomedical Engineering PAS.

L. Bobrowski and M. Topczewska: *Induction of Similarity Measures for Case Based Reasoning Through Separable Data Transformations*, Studies in Computational Intelligence (SCI) **73**, 127–148 (2008)
www.springerlink.com

separability postulates are reinforced through minimization of the convex and piecewise linear (CPL) criterion functions. The basis exchange algorithms, similar to linear programming, allow one to find a minimum of the CPL criterion functions efficiently, even in the case of large, multidimensional data sets [7].

4.2 Separabilty of Reference Sets

Let us assume that object descriptions stored in a database are represented as the so called feature vectors $\mathbf{x} = [\mathrm{x}_1, \ldots\ldots, \mathrm{x}_\mathrm{n}]^\mathrm{T}$, or as points in the n-dimensional feature space $\boldsymbol{F}[n]$ [3]. The components $\mathrm{x_i}$ of vectors $\mathbf{x}$ are numerical results of various examinations of a given object. The feature vectors $\mathbf{x}$ can be of the mixed, qualitative–quantitative type because their components can be both real numbers ($\mathrm{x_i} \in \mathrm{R}$) as well as binary ones ($\mathrm{x_i} \in \{0, 1\}$).

We assume that a database contains descriptions of m objects $\mathbf{x}_\mathrm{j}(k)$ ($j = 1, \ldots\ldots, m$) labelled in accordance with their class (*category*) $\omega_\mathrm{k}(k = 1, \ldots, K')$. The labelling of the feature vectors should be done in accordance with an additional knowledge about particular decision support problem. For example, a clinical database contains the descriptions of m patients $\mathbf{x}_\mathrm{j}(k)(j = 1, \ldots\ldots, m)$ labelled in accordance with their clinical diagnosis ω_k. The reference (learning) set $\mathrm{C_k}$ contains m_k labelled feature vectors $\mathbf{x}_\mathrm{j}(k)$ (*precedents*) related to the kth class ω_k.

$$\mathrm{C_k} = \{\mathbf{x}_\mathrm{j}(k)\} \quad (j \in I_\mathrm{k}) \tag{4.1}$$

where I_k is the set of indices j of m_k feature vectors $\mathbf{x}_\mathrm{j}(k)$ belonging to the class ω_k.

Definition 1. *The learning sets* $\mathrm{C_k}$ *(4.1) are separable in the feature space* $\boldsymbol{F}[n]$ *if they are disjoined in this space. It means that each of the feature vectors* $\mathbf{x}_\mathrm{j}$ *belongs to only one set* $\mathrm{C_k}$*:*

$$(\forall \boldsymbol{x}_j(k) \in C_k) \;\; and \;\; (\forall \boldsymbol{x}_{j'}(k') \in C_{k'},\; k \neq k') \quad \boldsymbol{x}_{j'}(k') \neq \boldsymbol{x}_j(k) \tag{4.2}$$

In accordance with Definition 1, the feature vectors $\mathbf{x}_\mathrm{j}(k)$ *and* $\mathbf{x}_{\mathrm{j}'}(k')$ *from different reference sets* $\mathrm{C_k}$ *and* $\mathrm{C_{k'}}$ *cannot be equal.*

We are also considering separation of the sets $\mathrm{C_k}$ (4.1) by the hyperplanes $H(\mathbf{w}_\mathrm{k}, \theta_\mathrm{k})$ in the feature space $\boldsymbol{F}[n]$

$$H(\boldsymbol{w}_k, \theta_k) = \{\boldsymbol{x} : \; \boldsymbol{w}_k^T \boldsymbol{x} = \theta_k\} \tag{4.3}$$

where $\mathbf{w}_\mathrm{k} = [\mathrm{w_{k1}}, \ldots, \mathrm{w_{kn}}]^\mathrm{T} \in \mathrm{R^n}$ is the weight vector, $\theta_\mathrm{k} \in \mathrm{R}^1$ is the threshold, and $(\mathbf{w}_\mathrm{k})^\mathrm{T}\mathbf{x}$ is the inner product.

The feature vector $\mathbf{x}$ is situated on the *positive side* of the hyperplane $H(\mathbf{w}_\mathrm{l}, \theta_\mathrm{l})$ if and only if $(\mathbf{w}_\mathrm{k})^\mathrm{T}\mathbf{x} > \theta_\mathrm{l}$. Similarly, the vector $\mathbf{x}$ is situated on the *negative side* of $H(\mathbf{w}_1, \theta_1)$ if and only if $(\mathbf{w}_\mathrm{k})^\mathrm{T}\mathbf{x} < \theta_1$.

Definition 2. *The reference sets are linearly separable if each of the sets* C_k *(4.1) can be fully separated from the sum of the remaining sets* $C_{k'}$ *by some hyperplane* $H(\mathbf{w}_{k'}\theta_k)$ *(4.4):*

$$(\forall k \in \{1,\ldots,K'\}) \quad (\exists \boldsymbol{w}_k, \theta_k)\,(\forall \boldsymbol{x}_j(k) \in C_k) \;\; \boldsymbol{w}_k^T \boldsymbol{x}_j(k) > \theta_k$$
$$and \quad (\forall \boldsymbol{x}_{j'}(k') \in C_{k'}) \;\; \boldsymbol{w}_k^T \boldsymbol{x}_{j'}(k') > \theta_k \qquad (4.4)$$

In accordance with the relation (4.4), all the vectors $\mathbf{x}_j(k)$ *belonging to the learning set* C_k *are situated on the positive side of the hyperplane* $H(\mathbf{w}_k, \theta_k)$ *(4.2) and all the feature vectors* $\mathbf{x}_{j'}(k')$ *from the remaining sets* $C_{k'}$ *are situated on the negative side of this hyperplane.*

It can be proved that the sets C_k *(4.1) are linearly separable if all the feature vectors* $\mathbf{x}_j(k)$ *are linearly independent (sufficient condition).*

4.3 Distance Functions Induced by Linear Transformations of the Feature Space $F[n]$

The nearest neighbours decision support rules are based on the distances $\delta(\mathbf{x}_0, \mathbf{x}_j(k))$ between the feature vector $\mathbf{x}_0$ of a new object and the labelled vectors $\mathbf{x}_j(k)$ from the reference sets C_k (4.1) [2]. Let us assume for a moment that m labelled feature vectors $\mathbf{x}_j(k)$ (4.1) are *ranked* $\{\mathbf{x}_{j(1)}, \mathbf{x}_{j(2)}, \ldots, \mathbf{x}_{j(n)}\}$ in respect to the distances $\delta(\mathbf{x}_0, \mathbf{x}_j(k))$ between the vectors $\mathbf{x}_0$ and $\mathbf{x}_j(k)$:

$$(\forall i \in \{1,\ldots,m-1) \quad \delta(\mathbf{x}_0, \mathbf{x}_{j(i)}) \leq \delta(\mathbf{x}_0, \mathbf{x}_{j(i+1)}) \qquad (4.5)$$

Let us define the *reference ball* $B_x(\mathbf{x}_0, K)$ which is centred in $\mathbf{x}_0$ and contains K first vectors $\mathbf{x}_{j(i)}(k)$:

$$B_x(\mathbf{x}_0, K) = \{\mathbf{x}_j(k) : \; \delta(\mathbf{x}_0, \mathbf{x}_{j(i)}) \leq \delta(\mathbf{x}_0, \mathbf{x}_{j(K)})\} \qquad (4.6)$$

In accordance with the K-nearest neighbours (*K-NN*) classification rule, the object $\mathbf{x}_0$ is allocated into this class ω_k ($k = 1,\ldots,K'$) where most of the labelled feature vectors $\mathbf{x}_j(k)$ from the ball $B_x(\mathbf{x}_0, K)$ (10) belong [2]:

$$\boldsymbol{if}\; (\forall l \subset \{1,\ldots,K'\})\; n_k \geq n_l \;\boldsymbol{then}\; \mathbf{x}_0 \in \omega_k \qquad (4.7)$$

where n_k is the number of the vectors $\mathbf{x}_j(k)$ from the set C_k (4.1) contained in the ball $B_x(\mathbf{x}_0, K)$ (6).

The decision rule similar to (4.11) is applied also in the case based reasoning scheme. It is assumed in this case that the reference ball $B_x(\mathbf{x}_0, K)$ contains such vectors $\mathbf{x}_{j(i)}(k)$ which are most *similar* to the vector $\mathbf{x}_0$.

The Euclidean distance $\delta_E(\mathbf{x}_0, \mathbf{x}_{j(i)})$ between the feature vectors $\mathbf{x}_0$ and $\mathbf{x}_{j(i)}$ is commonly used in the case based reasoning or in the nearest neighbours classification rule (8):

$$\delta_E{}^2(\mathbf{x}_0, \mathbf{x}_{j(i)}) = (\mathbf{x}_0 - \mathbf{x}_{j(i)})^T(\mathbf{x}_0 - \mathbf{x}_{j(i)}) \qquad (4.8)$$

A quality of the decision rule (4.11) based on the Euclidean distance δ_E $(\mathbf{x}_0,\ \mathbf{x}_{\mathrm{j(i)}})$ can be improved in some cases through modification of the distance function through transformations of the feature space $\boldsymbol{F}[n]$.

The Mahalanobis distance function $\delta_{\mathrm{M}}(\mathbf{x}_0, \mathbf{x}_{\mathrm{j}})$ in the feature space $\boldsymbol{X}$ is defined on the basis of the covariance matrix Σ [4]

$$\delta_M{}^2(\mathbf{x}_0, \mathbf{x}_{\mathrm{j(i)}}) = (\mathbf{x}_0 - \mathbf{x}_{\mathrm{j(i)}})^{\mathrm{T}}\Sigma^{-1}(\mathbf{x}_0 - \mathbf{x}_{\mathrm{j(i)}}) \tag{4.9}$$

The Mahalanobis distance function $\delta_{\mathrm{M}}{}^2(\mathbf{x}_0, \mathbf{x}_{\mathrm{j(i)}})$ takes into account the linear dependencies in the pairs of the features x_{k} and x_1. When the covariance matrix Σ is equal to the unit matrix $\boldsymbol{I}_{\mathrm{n}}$, then the Mahalanobis distance $\delta_{\mathrm{M}}{}^2(\mathbf{x}_0, \mathbf{x}_{\mathrm{j(i)}})$ is reduced to the Euclidean distance $\delta_{\mathrm{M}}{}^2(\mathbf{x}_0, \mathbf{x}_{\mathrm{j(i)}}))$ (8).

Let us consider the linear transformations of the feature vectors $\mathbf{x}_{\mathrm{j}}(k)$. Such transformations can be represented in the matrix form given below:

$$\mathbf{y}_{\mathrm{j}}(k) = \boldsymbol{W}^{\mathrm{T}}\mathbf{x}_{\mathrm{j}}(k) \quad (j = 1, \ldots.., m) \tag{4.10}$$

where $\boldsymbol{W} = [\mathbf{w}_1,\ \mathbf{w}_2, \ldots\ldots,\ \mathbf{w}_{\mathrm{n}'}]$ is the matrix of dimension $(n \times n')$ with $1 \leq n' \leq n$.

The relation (4.10) allows one to generate the transformed learning sets $\mathrm{C'}_{\mathrm{k}}$, where

$$\mathrm{C'}_{\mathrm{k}} = \{\mathbf{y}_{\mathrm{j}}(k)\} \quad (j \in I_{\mathrm{k}}) \tag{4.11}$$

Let us define the *induced distance function* $\delta_{\mathrm{I}}(\mathbf{x}_0,\ \mathbf{x}_{\mathrm{j(i)}})$ between the feature vectors $\mathbf{x}_0$ and $\mathbf{x}_{\mathrm{j(i)}}$ as the Euclidean distance function $\delta_{\mathrm{E}}(\mathbf{y}_0,\ \mathbf{y}_{\mathrm{j(i)}})$ (4.8) between adequate points $\mathbf{y}_0$ and $\mathbf{y}_{\mathrm{j(i)}}$ transformed in accordance with (4.10).

$$\begin{aligned}\delta_{\mathrm{I}}{}^2(\mathbf{x}_0, \mathbf{x}_{\mathrm{j(i)}}) &= \delta_{\mathrm{E}}{}^2(\mathbf{y}_0, \mathbf{y}_{\mathrm{j(i)}}) = (\mathbf{y}_0 - \mathbf{y}_{\mathrm{j(i)}})^{\mathrm{T}}(\mathbf{y}_0 - \mathbf{y}_{\mathrm{j(i)}}) \\ &= (\mathbf{x}_0 - \mathbf{x}_{\mathrm{j}}(k))^{\mathrm{T}}\boldsymbol{W}\boldsymbol{W}^{\mathrm{T}}(\mathbf{x}_0 - \mathbf{x}_{\mathrm{j}}(k))\end{aligned} \tag{4.12}$$

The *induced ball* $B_{\mathrm{I}}(\mathbf{x}_0, K)$ can be defined by using the distance function $\delta_{\mathrm{I}}(\mathbf{x}_0, \mathbf{x}_{\mathrm{j(i)}})$ (4.12).

$$\begin{aligned}B_{\mathrm{I}}(\mathbf{x}_0, K) &= \{\mathbf{x}_{\mathrm{j}}(k) : \delta_{\mathrm{E}}{}^2(\mathbf{y}_0, \mathbf{y}_{\mathrm{j(i)}}) \leq \delta_{\mathrm{E}}{}^2(\mathbf{y}_0, \mathbf{y}_{\mathrm{j(K)}})\} \\ &= \{\mathbf{x}_{\mathrm{j}}(k) : \delta_{\mathrm{I}}{}^2(\mathbf{x}_0, \mathbf{x}_{\mathrm{j(i)}}) \leq \delta_{\mathrm{I}}{}^2(\mathbf{x}_0, \mathbf{x}_{\mathrm{j(K)}})\}\end{aligned} \tag{4.13}$$

where points $\mathbf{x}_{\mathrm{j(i)}}$ are *ranked* $\{\mathbf{x}_{\mathrm{j(1)}},\ \mathbf{x}_{\mathrm{j(2)}}, \ldots\ldots..,\ \mathbf{x}_{\mathrm{j(n)}}\}$(4.5) in accordance with the induced distance function $\delta_{\mathrm{I}}(\mathbf{x}_0,\ \mathbf{x}_{\mathrm{j(i)}})$ (4.12).

The induced ball $B_{\mathrm{y}}(\mathbf{x}_0, K)$ contains such K feature vectors $\mathbf{x}_{\mathrm{j}}(k)$ which are the most similar to $\mathbf{x}_0$ in accordance with the distance function $\delta_{\mathrm{I}}(\mathbf{y}_0,\ \mathbf{y}_{\mathrm{j(i)}})$ (4.11).

The case based reasoning (*CBR*) or the nearest neighbours decision rules (4.7) can be based on the induced ball $B_{\mathrm{I}}(\mathbf{x}_0, K)$ (4.13):

If most of the labelled vectors $\mathbf{x}_{\mathrm{j}}(\mathrm{k})$ from the induced ball $\mathrm{B}_{\mathrm{I}}(\mathbf{x}_0, \mathrm{K})$ belongs to the class ω_{k}, **then** the object represented by $\mathbf{x}_0$ should be assigned to this class.

(4.14)

The performance of the above decision (*classification*) rule can be optimised through a special choice of the vectors $\mathbf{w}_i$ ($i = 1, \ldots.., n'$) in the transformation (4.10). A basic measure of the classification rule performance is the *error rate* – the fraction of new objects that are assigned to the wrong category [3].

4.4 Whitening of Reference Sets

An important role in classification is played by such linear transformations (4.9), which reduce correlation of the learning sets C_k (4.1) [2]. Such transformations can be built on the basis of the eigenvectors $\mathbf{k}_i$ and the eigenvalues λ_i of the covariance matrix Σ. Let us take into consideration the covariance matrix Σ_k estimated on the set C_k (4.1)

$$\Sigma_k = \sum_{j \in I_k} (x_j(k)\text{-}\mu_k)\,(x_j(k)\text{-}\mu_k)^T / (m_k - 1) \tag{4.15}$$

where μ_k is the mean vector in the set C_k

$$\mu_k = \sum_{j \in I_k} \mathbf{x}_j(k)/m_k \tag{4.16}$$

The eigenvalue problem with the covariance matrix Σ_k is formulated as the search for the eigenvectors $\mathbf{k}_i$ and the eigenvalues λ_i of the covariance matrix Σ_k. The eigenvectors $\mathbf{k}_i$ and the eigenvalues λ_i fulfil the below equation

$$\Sigma_k\ \mathbf{k}_i = \lambda_i\ \mathbf{k}_i \tag{4.17}$$

with an additional condition of the unit length

$$\mathbf{k}_i{}^T\mathbf{k}_i = 1 \tag{4.18}$$

The eigenvectors $\mathbf{k}_i$ and $\mathbf{k}_k$ corresponding to different eigenvalues λ_i and λ_k ($\lambda_i \neq \lambda_k$) are orthogonal

$$\mathbf{k}_i{}^T\mathbf{k}_k = 0 \tag{4.19}$$

Let us assume that the linear transformations (4.10) is defined by n' ($1 \leq n' \leq n$) orthogonal eigenvectors $\mathbf{k}_i$ with positive eigenvalues λ_i ($\lambda_i > 0$). Typically, the eigenvectors $\mathbf{k}_i$ and the greatest eigenvalues λ_i are taken into consideration. We are considering the linear transformation (4.10) with the columns of the matrix $\boldsymbol{W}$ formed by the vectors $\mathbf{k}_i/(\lambda_i)^{1/2}$

$$\boldsymbol{W}_k = [\mathbf{k}_1/(\lambda_1)^{1/2}, \ldots\ldots, \mathbf{k}_{n'}/(\lambda_{n'})^{1/2}] \tag{4.20}$$

The transformed vectors $\mathbf{y}_j(k)$ (4.9) form the set C'_k (4.11) with the mean vectors $\mu_{k'}$ (4.16). The correlation matrix $\Sigma_k{}'$ (15) are defined on the transformed vectors $\mathbf{y}_j(k)$ (9) from one set C'_k (4.11)

$$\begin{aligned} \Sigma_k{}' &= \sum_{j \in I_k}(\mathbf{y}_j(k) - \mu_k{}')(\mathbf{y}_j(k) - \mu_k{}')^T/(m_k - 1) \\ &= \boldsymbol{W}^T \sum_{j \in I_k}(\mathbf{x}_j(k) - \mu_k)(\mathbf{x}_j(k) - \mu_k)^T \boldsymbol{W}/(m_k - 1) \\ &= \boldsymbol{W}^T \Sigma_k \boldsymbol{W} = \boldsymbol{I}_{n' \times n'} \end{aligned} \tag{4.21}$$

where $\boldsymbol{I}_{n' \times n'}$ is the unit matrix of the dimension $(n' \times n')$.

It could be seen that the decision rule (4.13) with the Euclidean distance $\delta_E(\mathbf{y}_0, \mathbf{y}_{j(i)})$ (4.8) in the transformed space is equivalent to the decision rule (4.7) with the Mahalanobis distance functions $\delta_M{}^2(\mathbf{x}_0, \mathbf{x}_{j(i)})$ (4.9) in the feature space $\boldsymbol{F}[n]$, where the points $\mathbf{y}_0$ and $\mathbf{y}_{j(i)}$ are obtained through the transformation (4.10) with (4.20) of the points $\mathbf{x}_0$ and $\mathbf{x}_{j(i)}$, adequately.

In accordance with the equation (4.20), the transformation (4.10) decorrelates the set C'_k (4.11). The classification rule (4.14) based on the ball $B_I(\mathbf{x}_0, K)$ (4.13) which is induced by the transformation (4.10) gives the possibility to decrease the error rate [5]. Results of some experiments which support this statement are described in a farther part of the presented chapter. In these experiments the induced ball $B_I(\mathbf{x}_0, K)$ (13) has been defined on the basis of the Euclidean distance function $\delta_E{}^2(\mathbf{y}_0, \mathbf{y}_{j(i)})$ (4.8) in the transformed space. Generally, the decision rule (4.7) with the Euclidean distance $\delta_E{}^2(\mathbf{y}_0, \mathbf{y}_{j(i)})$ (4.8) can be matched in the best manner to data sets C'_k (4.11) with the unit correlation matrix $\Sigma_k{}'$.

Example 1. The numerical experiment has been performed on two-dimensional data sets C_k and C'_k (points on the plane). Data were generated from normal distributions with different covariance matrices and had different mean vectors. They belonged to three overlapping classes. The correlation coefficients were accordingly $\rho_1 = -0.9$, $\rho_2 = 0.2$ and $\rho_3 = -0.6$. There were 30 objects in every class. To check the differences between classification quality using whitening process the original data had been transformed. First transformation was the whitening based on the transformation matrix built using covariance matrix of the first class, analogically – on the second and on the third class covariance matrix. The last transformation has been performed using transformation matrix built using pooled estimate of the common covariance matrix (Fig. 4.1).

The results of classification errors for *K-NN* rule for the number of neighbours from $K = 1$ to $K = 10$ are shown in Table 4.1.

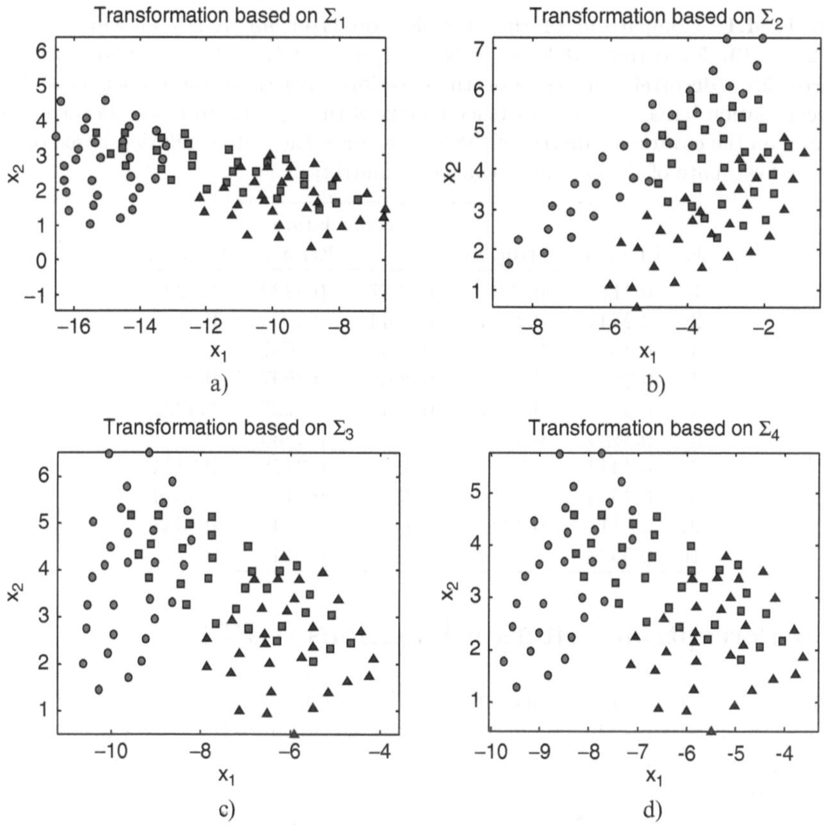

Fig. 4.1. Plots of transformed data (a) based on Σ_1, (b) based on Σ_2, (c) based on Σ_3 and (d) based on Σ_W

The mean value of the classification error for original data using *K-NN* rule was 43%. Adapting decorrelations we have achieved 33% as a mean of error_Σ_1, 39% (mean of error_Σ_2), 35% (mean of error_Σ_3) and 35% (mean of error_Σ_W). We can observe in the above results that the decorrelation of the learning sets C_k can improve the *K-NN* rule based on the Euclidean distance $\delta_E(\mathbf{x}_0, \mathbf{x}_j)$. The decorrelation of the learning sets C_k has entailed including the Mahalonobis distance $\delta_M(\mathbf{x}_0, \mathbf{x}_j)$ from these sets. With such interpretation, we can claim that the replacement of the Euclidean distance $\delta_E(\mathbf{x}_0, \mathbf{x}_j)$ by Mahalonobis distance $\delta_M(\mathbf{x}_0, \mathbf{x}_j)$ can lead to the improvement of the *K-NN* or the CBR rule. In the case of more than two classes using transformations based on single class C_k is preferred, because of the effect of conjoin different covariance matrices into one pooled estimate of the common covariance matrix. In our example the best classification quality we achieved for transformed data with transformation matrix built using covariance matrix for the second class Σ_2 (77%).

Table 4.1. Comparison of the classification error for *K-NN* classifiers ($K = 1, 2, \ldots, 10$) for correlated learning sets C_k (error_nd) and decorrelated sets C'_k (error_Σ_1 – decorrelation based on the covariance matrix of the C_1 set, error_Σ_2 – decorrelation based on the covariance matrix of the C_2 set, error_Σ_3 – decorrelation based on the covariance matrix of the C_3 set, error_Σ_W – decorrelation based on the pooled estimate of the common covariance matrix)

		Decorrelation			
K	Error_nd	Error_Σ_1	Error_Σ_2	Error_Σ_3	Error_Σ_W
1	0.4111	0.4667	0.5667	0.5222	0.5222
2	0.2444	0.2556	0.3444	0.3	0.3111
3	0.4	0.4	0.5333	0.4667	0.4778
4	0.3667	0.2	0.3667	0.2667	0.2667
5	0.5	0.3889	0.4444	0.4222	0.4222
6	0.3889	0.3111	0.3	0.2556	0.2
7	0.4444	0.3778	0.3889	0.3778	0.3444
8	0.4556	0.2778	0.2556	0.3111	0.2889
9	0.5111	0.4222	0.4	0.3444	0.3333
10	0.5333	0.2889	0.3111	0.2444	0.3111

4.5 Perceptron Criterion Functions (CPL)

The *perceptron* criterion function $\Phi(\mathbf{w}, \theta)$ originated from neural networks theory [3,9]. $\Psi(\mathbf{w}, \theta)$ is the convex and piecewise linear (CPL) criterion function. The designing transformation (4.10) can be based on the minimisation of the perceptron criterion function [6].

It is convenient to define the perceptron criterion function $\Phi(\mathbf{w}, \theta)$ by using the positive G^+ and the negative G^- sets of the feature vectors $\mathbf{x}_j$ (1).

$$G^+ = \{\mathbf{x}_j\} \quad (j \in J^+) \quad \textit{and} \quad G^- = \{\mathbf{x}_j\} \quad (j \in J^-) \tag{4.22}$$

Each element $\mathbf{x}_j$ of the set G^+ defines the positive penalty function $\upsilon_j^+(\mathbf{w}, \theta)$

$$\varphi_j^+(\mathbf{w}, \theta) = \begin{cases} 1 - \mathbf{w}^T\mathbf{x}_j + \theta & \textit{if} \quad \mathbf{w}^T\mathbf{x}_j - \theta \leq 1 \\ 0 & \textit{if} \quad \mathbf{w}^T\mathbf{x}_j - \theta > 1 \end{cases} \tag{4.23}$$

Similarly, each element $\mathbf{x}_j$ of the set G_l^- defines the negative penalty function $\upsilon_j^-(\mathbf{w}, \theta)$

$$\upsilon_j^-(\mathbf{w}, \theta) = \begin{cases} 1 + \mathbf{w}^T\mathbf{x}_j - \theta & \textit{if} \quad \mathbf{w}^T\mathbf{x}_j - \theta \geq -1 \\ 0 & \textit{if} \quad \mathbf{w}^T\mathbf{x}_j - \theta < -1 \end{cases} \tag{4.24}$$

where $\mathbf{w} = [w_1, \ldots, w_n]^T \in R^n$ is the weight vector and $\theta_k \in R^1$ is the threshold.

Both the penalty functions $\upsilon_j^+(\mathbf{w},\theta)$ and $\upsilon_j^-(\mathbf{w},\theta)$ are convex and piecewise linear. The penalty function $\upsilon_j^+(\mathbf{w},\theta)$ is aimed at placing the vector $\mathbf{x}_j(\mathbf{x}_j \in G^+)$ on the positive side of the hyperplane $H(\mathbf{w},\theta)$ (4.3). Similarly, the function $\upsilon_j^-(\mathbf{w},\theta)$ should insert the vector $\mathbf{x}_j(\mathbf{x}_j \in G^-)$ on the negative side of this hyperplane.

The perceptron criterion function $\Phi(\mathbf{w},\theta)$ is determined by the sets G^+ and G^- is the weighted sum of the penalty functions $\upsilon_j^+(\mathbf{w}, \theta)$ and $\upsilon_j^-(\mathbf{w}, \theta)$

$$\Phi(\mathbf{w},\theta) = \sum_{j \in J^+} \alpha_j^+ \varphi_j^+(\mathbf{w},\theta) + \sum_{j \in J^-} \alpha_j^- \varphi_j^-(\mathbf{w},\theta) \tag{4.25}$$

where α_j^+ $(\alpha_j^+ > 0)$ and $\alpha_j^-(\alpha_j^- > 0)$ are positive parameters (*prices*).

$$\Phi^* = \Phi(\mathbf{w}^*,\theta^*) = \min_{\mathbf{w},\theta} \Phi(\mathbf{w},\theta) \geq 0 \tag{4.26}$$

The basis exchange algorithms which are similar to the linear programming allow one to find the minimum of the criterion function $\Phi(\mathbf{w},\theta)$ efficiently even in the case of large, multidimensional data sets G^+ and G^- (4.22) [7].

It has been proved that the minimum value Φ^* of the perceptron criterion function $\Phi(\mathbf{w},\theta)$ (4.25) is equal to zero ($\Phi^* = 0$) if and only if the positive G^+ and the negative G^- sets (4.22) are linearly separable (4.4). In this case, all elements $\mathbf{x}_j$ of the set G^+ (4.22) are located on the positive side of the hyperplane $H(\mathbf{w}^*,\theta^*)$ (4.3) and all elements $\mathbf{x}_j$ of the set G^- are located on the negative side:

$$\begin{aligned} &(\forall \mathbf{x}_j \in G^+)(\mathbf{w}^*)^T\mathbf{x}_j > \theta_1^* \\ \textit{and } &(\forall \mathbf{x}_{j'} \in G^-)(\mathbf{w}^*)^T\mathbf{x}_{j'} < \theta_1^* \end{aligned} \tag{4.27}$$

If the sets G^+ and G^- (4.22) are not linearly separable (4.4), then the above relation is fulfilled not by all but by a majority of the elements $\mathbf{x}_j$ of these sets.

Minimization of the function $\Phi(\mathbf{w},\theta)$ (4.25) allows one to find optimal parameters $(\mathbf{w}^*,\theta^*)$ which can define the hyperplane $H(\mathbf{w}^*,\theta^*)$ (4.3), which separates relatively well two sets G^+ and G^- (4.22). The vector $\mathbf{w}_1{}^*$ can be used also as one of the columns of the transformation matrix $\boldsymbol{W}$ (4.10).

4.6 Four-Fields Diagnostic Maps of the System *Hepar*

The computer system *Hepar* aggregates the clinical database with tools for data exploration and diagnosis support [8]. The database of the system *Hepar* contains hepathological data. An essential part of the system is data visualisation module. For the purpose of data visualisation there are used linear transformations from multidimensional feature space $\boldsymbol{F}[n]$ on a plane. Such transformations allow for inducing the distance function $\delta_I{}^2(\mathbf{x}_0, \mathbf{x}_{j(i)})$ (4.12) based both on the Euclidean distance $\delta_{\mathrm{E}}{}^2(\mathbf{y}_0, \mathbf{y}_{\mathrm{j(i)}})$ (4.8) as well as on a subjective measures of similarity.

The parameters $\mathbf{w}^*$ and θ^* determining minimum (4.26) of the criterion function $\Psi(\mathbf{w}, \theta)$ (4.25) can be also used in definition of the affine transformation of the feature vectors $\mathbf{x}$ on a line (4.9):

$$y = (\mathbf{w}^*)^{\mathrm{T}}\mathbf{x} - \theta^* \tag{4.28}$$

where $\mathbf{w} = [\mathrm{w}_1, \ldots\ldots, \mathrm{w}_{\mathrm{n}}]^{\mathrm{T}}$ is the parameter vector which determines direction of the line.

Such transformations have been applied in the system *Hepar* for definition of the visualising planes. This system allows for designing pairs of special visualizing transformations (4.28), which result in the so called *diagnostic maps.* Two linearly independent transformations (4.28) give possibility to produce such a visualizing plane (*diagnostic map*), which relatively well separates four groups of patients. The diagnostic maps are used for inducing the similarity measure between feature vector of a new patient $\mathbf{x}_0$ and the vectors $\mathbf{x}_{\mathrm{j}}(k)$ from the reference sets C_{k} (4.1).

The affine transformation of the feature vectors $\mathbf{x}_{\mathrm{j}}$ (4.1) on a plane can be represented in a below manner

$$\mathbf{y}_{\mathrm{j}} = [\mathrm{y}_{\mathrm{j}1}, \mathrm{y}_{\mathrm{j}2}]^{\mathrm{T}} = [(\mathbf{w}_1{}^*)^{\mathrm{T}}\mathbf{x}_{\mathrm{j}} - \theta_1{}^*, (\mathbf{w}_2{}^*)^{\mathrm{T}}\mathbf{x}_{\mathrm{j}} - \theta_2{}^*]^{\mathrm{T}} \tag{4.29}$$

where $\mathbf{w}_{\mathrm{i}}{}^* = [\mathrm{w}_{\mathrm{i}1}, \ldots\ldots, \mathrm{w}_{\mathrm{in}}]^{\mathrm{T}}$ $(i = 1, 2)$ are the parameter vectors that span a plane.

The *scatterplots* or, in other words, the *maps* of data can be generated as a result of visualisation of the transformed points $\mathbf{y}_{\mathrm{j}}(k)$. If the vectors $\mathbf{w}_{\mathrm{i}}$ are orthogonal $((\mathbf{w}_1{}^*)^{\mathrm{T}}\ \mathbf{w}_2 = 0)$ and have the unit length $((\mathbf{w}_1{}^*)^{\mathrm{T}}\ \mathbf{w}_1{}^* = (\mathbf{w}_2{}^*)^{\mathrm{T}}\ \mathbf{w}_2{}^* = 0)$ then the transformations (4.2) describes the *projection* of the feature vectors $\mathbf{x}_{\mathrm{j}}(k)$ on the visualizing plane $P(\mathbf{w}_1{}^*,\ \mathbf{w}_2{}^*; \theta^*)$

$$P(\mathbf{w}_1, \mathbf{w}_2; \theta) = \{\mathbf{x} : \mathbf{x} = \alpha_1\mathbf{w}_1 + \alpha_2\mathbf{w}_2 + \theta,\ \text{where } \alpha_i \in R^1\} \tag{4.30}$$

where $\theta_{\mathrm{i}} = [\theta_1, \theta_2]^{\mathrm{T}}$

Example 2. Let us consider this example in order to explain the basic principles of the diagnostic map designing in the framework of the system *Hepar.* We have taken into consideration four learning sets C_{k} (4.1) extracted from the *Hepar* database [8]

$$\begin{array}{ll} C_9\text{–Hepatitis chronica activa} & \text{–91 patients} \\ C_{13}\text{–Steatosis hepatis} & \text{–67 patients} \\ C_{15}\text{–Hiperbilirubinemia functionalis} & \text{–56 patients} \\ C_{22}\text{–Cirrhosis hepatis billiaris primaria} & \text{–272 patients} \end{array} \tag{4.31}$$

Patients from these sets C_{k} have been described by the feature vectors $\mathbf{x}_{\mathrm{j}}(k)$ of dimensionality n equal to 106. The components x_{i} of the vectors $\mathbf{x}_{\mathrm{j}}(k)$

were numerical results of various diagnostic examinations of a given patient. Numerical results of both laboratory tests ($x_i \in R$) as well as patients symptoms ($x_i \in \{0,1\}$) have been taken as features x_i.

The maps can reflect an actual diagnostic hypothesis of a medical doctor (an user). The user declares which classes ω_k should be located into particular quarters of the map and which features (tests) x_i are to be used in the visualizing transformation or, in other words, used for hypothesis examination. The above map (Fig. 4.2) resulted from the affine transformation (4.29) of the 106 – dimensional feature vectors $\mathbf{x}_j(k)$ on a visualizing plane.

The affine transformation (4.29) of the feature vectors $\mathbf{x}_j$ on a visualising plane is determined by two pairs of parameters ($\mathbf{w}_1{}^*$, $\theta_1{}^*$) and ($\mathbf{w}_2{}^*$, $\theta_2{}^*$). These parameters have been induced from the sets (4.30) through minimisation of two perceptron criterion functions $\Phi_1(\mathbf{w},\theta)$ and $\Phi_2(\mathbf{w}, \theta)$ (4.25). Each function $\Phi_k(\mathbf{w}, \theta)$ was defined by their own pair of the sets $G_k{}^+$ and $G_k{}^-$ (22), where

$$G_1{}^+ = C_{13} \cup C_{15} \quad \mathit{and} \quad G_1{}^- = C_9 \cup C_{22} \tag{4.32}$$

$$G_2{}^+ = C_9 \cup C_{13} \quad \mathit{and} \quad G_2{}^- = C_{15} \cup C_{22} \tag{4.33}$$

If each pair of the sets $G_k{}^+$ and $G_k{}^-$ ($k = 1,\ 2$) is linearly separable (4.4), then the transformation (4.29) based on the above sets assures the exact placements of the learning sets C_k (4.30) in an adequate quarter of the diagnostic map (Fig. 4.2).

The transformation (4.29) defines the coordinates $\mathbf{y}_j(k) = [y_{j1}(k),\ y_{j2}(k)]$ on the map of particular feature vectors $\mathbf{x}_j(k)$. The vector of $\mathbf{x}_0$ of a new patient can be located on the map as $\mathbf{y}_0$ by using the transformation (4.29). As a result, the system can be used in the diagnosis support in accordance with the CNR or the *K-NN* schemes (4.7), which are based on the Euclidean distances $\delta_E(\mathbf{y}_0, \mathbf{y}_j[k])$ (4.8) between the transformed vectors $\mathbf{y}_0$ and $\mathbf{y}_j[k]$. It has been demonstrated experimentally that despite the significant reduction of the problem dimensionality (from $n = 106$ to $n' = 2$), the replacement of the distances $\delta_E(\mathbf{x}_0, \mathbf{x}_j[k])$ by the distances $\delta_E(\mathbf{y}_0, \mathbf{y}_j[k])$ induced through a diagnostic map gives possibility to reduce the error rate of the classification rules [8,9].

4.7 Fisher Linear Discriminant and Principal Components

Let us consider further linear transformations (4.10) of data sets C_k (4.1) from n-dimensional feature space $\boldsymbol{F}[n]$ onto line. Such problem is analysed in the *discriminant analysis*. Discriminant analysis seeks direction $\mathbf{w}$ that are efficient in separation on a line of two data sets C_1 and C_2 (4.1).

$$y = (\mathbf{w})^T \mathbf{x} \tag{4.34}$$

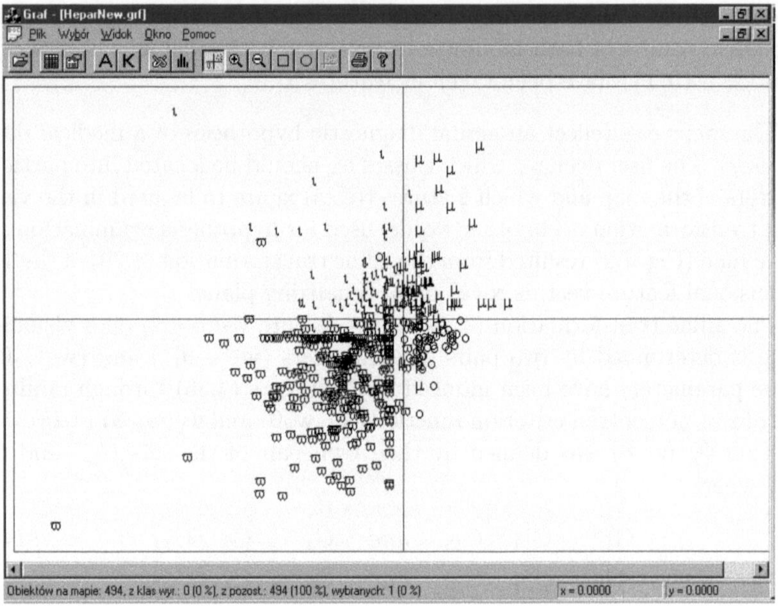

Fig. 4.2. The diagnostic map with the following structure: C_9, *the upper-left quarter*; C_{13}, *the upper-right quarter*; C_{15}, *the lower-right quarter*; C_{22}, *the lower-left quarter*

The direction vectors $\mathbf{w}$ which are used in the discriminant analysis have the unit length

$$\mathbf{w}^T\mathbf{w} = 1 \tag{4.35}$$

The linear transformation (4.34) with an additional condition (4.33) describes the projection $y_j = (\mathbf{w})^T\ \mathbf{x}_j$ of the corresponding vectors $\mathbf{x}_j$ onto a line in the direction of $\mathbf{w}$.

A fundamental role in the *discriminant analysis* is played by the Fisher's criterion function $J(\mathbf{w})$.

$$J(\mathbf{w}) = |\mu_1(\mathbf{w}) - \mu_2(\mathbf{w})|/(s_1(\mathbf{w})^2 + s_2(\mathbf{w})^2) \tag{4.36}$$

where $|\mu_1(\mathbf{w}) - \mu_2(\mathbf{w})|$ is the distance between the projected mean vectors μ_1 and μ_2 (4.16)

$$|\mu_1(\mathbf{w}) - \mu_2(\mathbf{w})| = |\mathbf{w}^T(\mu_1 - \mu_2)| \tag{4.37}$$

and $s_k(\mathbf{w})^2$ ($k = 1,\ 2$) is the *within-class scatter* (a measure of variance) of the projected points y_j from the set C_k.

$$s_k(\mathbf{w})^2 = \sum_{j \in I_k} (y_j(k) - \mu_k(\mathbf{w}))^2 \tag{4.38}$$

The below optimization problem is based on the Fisher's criterion function $J(\mathbf{w})$.

$$J(\mathbf{w}^*) = \underset{\mathrm{w}}{max}\, J(\mathbf{w}) \tag{4.39}$$

In accordance with the Fisher's criterion, the vector $\mathbf{w}^*$ that constitutes maximum of the function $J(\mathbf{w})$ (4.30) determines the best discriminant line. The optimal vector $\mathbf{w}^*$ determines large distance between the projected means μ_1 and μ_2 (4.31) relatively to some measure of the variance of the projected points $\mathrm{y_j}$.

The criterion function $J(\mathbf{w})$ can be represented in a matrix form. Let us define for this purpose the scatter matrices $\boldsymbol{S}_{\mathrm{k}}(k = 1, 2)$ and $\boldsymbol{S}_{\mathrm{W}}$

$$S_{\mathrm{k}} = \sum_{\mathrm{j} \in \mathrm{I_k}} (\mathbf{x}_{\mathrm{j}}(k) - \mu_{\mathrm{k}})(\mathbf{x}_{\mathrm{j}}(k) - \mu_{\mathrm{k}})^{\mathrm{T}} \tag{4.40}$$

and

$$S_{\mathrm{W}} = S_1 + S_2 \tag{4.41}$$

$\boldsymbol{S}_{\mathrm{W}}$ is called the *within-class scatter matrix*. The scatter $\mathrm{s_k}(\mathbf{w})^2$ (4.38) can be expressed as

$$\mathrm{s_k}(\mathbf{w})^2 = \sum_{j \in \mathrm{I_k}} (\mathbf{w}^{\mathrm{T}}\mathbf{x}_{\mathrm{j}}(k) - \mathbf{w}^{\mathrm{T}}\mu_{\mathrm{k}})^2 = \Sigma \mathbf{w}^{\mathrm{T}}(\mathbf{x}_{\mathrm{j}}(k) - \mu_{\mathrm{k}})(\mathbf{x}_{\mathrm{j}}(k) - \mu_{\mathrm{k}})^T \mathbf{w} = \mathbf{w}^{\mathrm{T}} S_{\mathrm{k}} \mathbf{w} \tag{4.42}$$

thus

$$\mathrm{s_1}(\mathbf{w})^2 + \mathrm{s_2}(\mathbf{w})^2 = \mathbf{w}^{\mathrm{T}} S_{\mathrm{W}} \mathbf{w} \tag{4.43}$$

Similarly,

$$(\mu_1(\mathbf{w}) - \mu_2(\mathbf{w}))^2 = (\mathbf{w}^{\mathrm{T}}\mu_1 - \mathbf{w}^{\mathrm{T}}\mu_2)^2 = \mathbf{w}^{\mathrm{T}}(\mu_1 - \mu_2)(\mu_1 - \mu_2)^{\mathrm{T}}\mathbf{w} = \mathbf{w}^{\mathrm{T}} S_B \mathbf{w} \tag{4.44}$$

where

$$S_{\mathrm{B}} = (\mu_1 - \mu_2)(\mu_1 - \mu_2)^{\mathrm{T}} \tag{4.45}$$

$\mathbf{S}_{\mathrm{B}}$ is called as the between-class scatter matrix.

The criterion function $J(\mathbf{w})$ (4.36) can be written as

$$J(\mathbf{w}) = \mathbf{w}^{\mathrm{T}} S_{\mathrm{B}} \mathbf{w} / \mathbf{w}^{\mathrm{T}} S_{\mathrm{W}} \mathbf{w} \tag{4.46}$$

The vector $\mathbf{w}^*$ that maximizes $J(\mathbf{w})$ must satisfy a generalized eigenvalue problem for some constant λ

$$S_{\mathrm{B}} \mathbf{w} = \lambda S_W \mathbf{w} \tag{4.47}$$

The vector $\mathbf{w}_{\mathrm{F}}$ that maximizes $J(\mathbf{w})$ is known as

$$\mathbf{w}_{\mathrm{F}} = {S_{\mathrm{W}}}^{-1}(\mu_1 - \mu_2) \tag{4.48}$$

Fishor's linoar discriminant $\mathrm{y} = (\mathbf{w}_{\mathrm{F}})^{\mathrm{T}}\mathbf{x}$ (4.34), which is determined by the optimal vector $\mathbf{w}_{\mathrm{F}}$, yields the maximum ratio of the between-class scatter matrix to the within-class scatter on the projecting line.

In the case of the probabilistic normal model, when the conditional densities $f(\mathbf{x}/\omega_1)$ and $f(\mathbf{x}/\omega_2)$ are multivariate normal distributions $N(\mu_1, \Sigma)$ and $N(\mu_2, \Sigma)$ with the same covariance matrix Σ, then the optimal (*Bayesian*) decision boundary is the hyperplane $H(\mathbf{w}_B, \theta_B)$ (4.3), where

$$(\mathbf{w}_B)^T\mathbf{x} = \theta_B \tag{4.49}$$

and $\mathbf{w}_B$ is determined by an equation similar to (4.48)

$$\mathbf{w}_B = \mathbf{\Sigma}^{-1}(\mu_1 - \mu_2) \tag{4.50}$$

In this case, the optimal decision rule has the following form

$$\begin{array}{lll} \textit{if} & \mathbf{w}_B{}^T\mathbf{x} > \theta_B & \textit{then } \mathbf{x} \text{ should be allocated into the class } \omega_1 \\ \textit{if} & \mathbf{w}_B{}^T\mathbf{x} < \theta_B & \textit{then } \mathbf{x} \text{ should be allocated into the class } \omega_2 \end{array} \tag{4.51}$$

The above considerations have been related to discrimination between only two classes ($K' = 2$). When the number of classes c is greater than 2 ($K' > 2$), then the generalization of Fisher's linear discrimination involves $K' - 1$ linear discriminant functions [3]. In this case, it is designed projection (4.10) from n-dimensional space to a $(K' - 1)$-dimensional space.

Discriminant analysis seeks a projection that best separates data in a last squares sense. In contrast, principal component analysis (PCA) or *Karhunen-Loeve transform* seeks a projection that best represents data in a last squares sense. PCA deals with dimensionality reduction through such linear transformations (4.10) from n-dimensional space to a n' - dimensional space which preserve variability in data as much as possible.

Principal component analysis is based on n' linear transformations $y_i = (\mathbf{k}_i)^T\mathbf{x}$ that are defined by the eigenvectors $\mathbf{k}_i = [k_{i1}, k_{i2}, \ldots\ldots\ldots, k_{in}]^T$ of the covariance matrix Σ_k (4.15).

$$y_i = (\mathbf{k}_i)^T\mathbf{x} = k_{i1}x_1 + k_{i2}\, x_2 + \ldots\ldots\ldots\ldots + k_{in}x_n \tag{4.52}$$

The covariance matrix Σ_k (4.15) of dimensionality $n \times n$ can have up to n normalised (4.18) eigenvectors $\mathbf{k}_i$ with positive eigenvalues λ_i. The eigenvectors $\mathbf{k}_i$ are ranked in accordance with the eigenvalues λ_i.

$$\mathbf{k}_1, .\mathbf{k}_2, \ldots\ldots\ldots\ldots, \mathbf{k}_n \tag{4.53}$$

where

$$\lambda_1 \geq \lambda_2 \geq \ldots\ldots\ldots\ldots \geq \lambda_n > 0$$

The first principal component $y_1 = (\mathbf{k}_1)^T\,\mathbf{x}$ is defined by the eigenvector $\mathbf{k}_1$ with the largest eigenvalue λ_1. The second principal component $y_1 = (\mathbf{k}_1)^T\,\mathbf{x}$ is defined by the second eigenvector $\mathbf{k}_2$ and so on. Principal components are defined by n' eigenvectors $\mathbf{k}_i$ $(1 \leq n' \leq n)$ with the largest eigenvalues λ_i. Often, there are just a few large eigenvalues λ_i and it can be assumed that remaining

$n - n'$ dimensions contain noise. Considerable dimensionality reduction can be achieved in such case through linear transformation of data.

The variance $\sigma_i{}^2$ of the ith principal component $y_i = (\mathbf{k}_i)^T\mathbf{x}$ can be estimated in the following manner (4.15 and 4.38)

$$\sigma_i{}^2 = \sum_{j \in I_k} (y_j(k) - \mu_k)^2/(m_k - 1) = \sum_{j \in I_k} ((\mathbf{k}_i)^T(\mathbf{y}_j(k) - \mu_k)^2/(m_k - 1) \quad (4.54)$$
$$= \sum_{j \in I_k} ((\mathbf{k}_i)^T(\mathbf{y}_j(k) - \mu_k)(\mathbf{y}_j(k) - \mu_k)^T\mathbf{k}_i/(m_k - 1) = (\mathbf{k}_i)^T\Sigma_k\mathbf{k}_i = \lambda_i$$

As it results from the above relation, the first principal component $y_1 = (\mathbf{k}_1)^T\ \mathbf{x}$ has the largest variance $\sigma_1{}^2$, and (4.53)

$$\sigma_1 \geq \sigma_2 \geq \ldots\ldots\ldots \geq \sigma_n > 0 \quad (4.55)$$

Example 3. Two eigenvectors $\mathbf{k}_i$ and $\mathbf{k}_{i'}$ with eigenvalues λ_i and $\lambda_{i'}$ of the kth covariance matrix Σ_k (4.15) can be used in the below visualising transformation (4.20 and 4.29), the data map determined by the kth set C_k (4.1).

$$\mathbf{y} = [y_1, y_2]^T = [(\mathbf{k}_i/(\lambda_i)^{1/2})^T(\mathbf{x} - \mu_k), (\mathbf{k}_{i'}/(\lambda_{i'})^{1/2})^T(\mathbf{x} - \mu_k)]^T \quad (4.56)$$

where μ_k is the mean vector (4.15) of the set C_k (4.1).

In accordance with the relation (4.55), all feature vectors $\mathbf{x}_j(k)$ (4.1) are transformed into the points $\mathbf{y}_j(k)$ on the visualising plane $P_k(\mathbf{k}_1, \mathbf{k}_2; \mathbf{0})$ (4.30) determined by the kth set C_k (4.1).

The mean value μ'_k (4.16) of the transformed points $\mathbf{y}_j(k)$ from the set C_k (4.1) is equal to zero.

$$\mu'_k = \sum_{j \in I_k} \mathbf{y}_j(k)/m_k = [0, 0]^T = \mathbf{0} \quad (4.57)$$

The covariance matrix Σ'_k (4.15) of the transformed points $\mathbf{y}_j(k)$ from the set C_k (4.1) is equal the unit matrix $\boldsymbol{I}_{2\times2}$.

$$\Sigma'_k - \sum_{j \in I_k} (\mathbf{y}_j(k) - \mu'_k)(\mathbf{y}_j(k) - \mu'_k)^T/(m_k \quad 1)$$
$$= \sum_{j \in I_k} \mathbf{y}_j(k)\mathbf{y}_j(k)^T/(m_k - 1) = (W'_k)^T\Sigma_k W'_k = I_{2x2} \quad (4.58)$$

where Σ_k is the covariance matrix (4.15) and the matrix $\boldsymbol{W}'_k$ has the following form (4.20)

$$W'_k = [\mathbf{k}_i/(\lambda_i)^{1/2}, \mathbf{k}_{i'}/(\lambda_{i'})^{1/2}] \quad (4.59)$$

As it results from the relation (4.50) the transformed features y_1 and y_2 (where $\mathbf{y} = [y_1, y_2]^T$ (4.48)) are uncorrelated. Each of these features y_i has the mean value equal to zero and the variance equal to one in the set C_k (4.1).

In accordance with the considerations given in the *Example* 1 it should be profitable to use the visualizing transformation (4.48) and the diagnostic map for an inducing of similarity measure for the CBR or *K-NN* decision rules (4.7) with the Euclidean distance $\delta_E{}^2(\mathbf{y}_0, \mathbf{y}_{j(i)})$ (4.8).

For this purpose, the decisionic rule (4.7) can be modified in the following manner.

$$\textit{if} \quad (\forall l \in \{1, \ldots, K'\}) n_k(k) \quad \geq \quad n_l(k) \quad \textit{then } \mathbf{x}_0 \quad \in \quad \omega_k \tag{4.60}$$

where $n_l(k)$ is the number of the vectors $\mathbf{y}_j(l)$ from the set C_l (4.1) contained in the ball $B_k(\mathbf{y}_0, K)$, and

$$B_k(\mathbf{y}_0, K) = \{\mathbf{y}_j : \delta(\mathbf{y}_0, \mathbf{y}_{j(i)}) \quad \leq \delta(\mathbf{y}_0, \mathbf{y}_{j(K)})\} \tag{4.61}$$

where points $\mathbf{y}_{j(i)}$ are *ranked* $\{\mathbf{y}_{j(1)}, \ \mathbf{y}_{j(2)}, \ldots\ldots.., \ \mathbf{y}_{j(n)}\}$ (4.5) in accordance with the Euclidean distance function $\delta_E(\mathbf{y}_0, \mathbf{y}_{j(i)})$ (4.8) on the plane P_k $(\mathbf{k}_1, \mathbf{k}_2; \mathbf{0})$ (4.30) determined by the kth set C_k (4.1).

The visualizing plane P_k $(\mathbf{k}_1, \mathbf{k}_2; \mathbf{0})$ (4.30) design in the above manner can be called as the *one-field diagnostic map.* Each of the maps P_k $(\mathbf{k}_1, \mathbf{k}_2; \mathbf{0})$ is centered on one of reference sets C_k (4.1).

4.8 Dipolar Separability Postulates

The linear transformations (4.10) can be defined on a variety of principles. Let us use for this purpose the concept of the mixed and clear dipoles formed by the feature vectors $\mathbf{x}_j(k)$ (4.1) [6].

Definition 3. *A pair of the feature vectors* $(\mathbf{x}_j(k), \mathbf{x}_{j'}(k'))$ $(\mathbf{x}_j(k) \neq \mathbf{x}_{j'}(k'),$ $j' > j)$ *constitutes a mixed dipole if and only if the vectors* $\mathbf{x}_j(k)$ *and* $\mathbf{x}_{j'}(k')$ *belong to different classes* ω_k $(k \neq k')$. *Similarly, a pair of different feature vectors from the same class* ω_k *constitutes a clear dipole* $(\mathbf{x}_j(k), \mathbf{x}_{j'}(k))$.

The dipoles $\{\mathbf{x}_j(k), \ \mathbf{x}_{j'}(k')\}$ of the length $\delta_x(j, j')$ are transformed by (4.10) into the dipoles $\{\mathbf{y}_j(k), \ \mathbf{y}_{j'}(k')\}$ – the pairs of the points $\mathbf{y}_j(k)$ and $\mathbf{y}_{j'}(k)$ situated in the Euclidean distance $\delta_y(j, j')$, where

$$\delta_x{}^2(j, j') = (\mathbf{x}_j(k) - \mathbf{x}_{j'}(k'))^T(\mathbf{x}_j(k) - \mathbf{x}_{j'}(k') \tag{4.62}$$

$$\begin{aligned} \delta_y{}^2(j, j') &= (\mathbf{y}_j(k) - \mathbf{y}_{j'}(k'))^T(\mathbf{y}_j(k) - \mathbf{y}_{j\prime}(k')) \\ &= (\mathbf{x}_j(k) - \mathbf{x}_{j'}(k'))^T WW^T(\mathbf{x}_j(k) - \mathbf{x}_{j\prime}(k')) \end{aligned} \tag{4.63}$$

We are interested in designing such transformations (4.10) which fulfil the following *separability inequalities*:

$$(\forall (j, j') \in I_c) \quad \delta_y{}^2(j, j') \leq \quad \rho_c{}^2(j, j') \tag{4.64}$$

$$(\forall (j, j') \in I_m) \quad \delta_y{}^2(j, j') \geq \quad \rho_m{}^2(j, j') \tag{4.65}$$

where I_c and I_m are the so called *control sets* (the sets of indices (j, j') of selected clear and mixed dipoles adequately, $\rho_c(j, j')$ and $\rho_m(j, j')$ are non-negative parameters (*margins*).

Separability postulate. *The linear transformation* (4.10) *should shorten the clear dipoles* $\{\mathbf{x}_j(k),\ \mathbf{x}_{j'}(k)\}$ *from the* <u>*control set*</u> I_c *to the length* $\delta_y{}^2(j, j')$ *less than* $\rho_c(j, j')$ (4.64) *and lengthen the mixed dipoles* $\{\mathbf{x}_j(k),\ \mathbf{x}_{j'}(k')\}$ *from the control set* I_m *to the length* $\delta_y{}^2(j, j')$ *more than* $\rho_m(j, j')$ (4.65).

The above separability postulate is aimed at designing such linear transformations (4.10) which enhance differences between categories ω_k. This postulate can be treated as an alternative which is complementary to the Fisher's criterion (4.39) used in discriminant analysis [3].

In the case of a linear transformation on the ith line $y = (\mathbf{w}_i)^T\mathbf{x}$, the separability inequalities (4.64) and (4.65) can be represented as the following sets of inequalities ($i = 1, 2, \ldots\ldots, n'$, with $1 \leq n' \leq n$):

$$(\forall (j, j') \in I_{ci}) - \rho_{ci}(j, j') < (\mathbf{w}_i)^T(\mathbf{x}_{j'}(k) - \mathbf{x}_j(k)) < \rho_{ci}(j, j') \tag{4.66}$$
$$(\forall (j, j') \in I_{mi}{}^+)\ (\mathbf{w}_i)^T(\mathbf{x}_{j'}(k) - \mathbf{x}_j(k')) > \rho_{mi}(j, j') \tag{4.67}$$
$$(\forall (j, j') \in I_{mi}{}^-)\ (\mathbf{w}_i)^T(\mathbf{x}_{j'}(k) - \mathbf{x}_j(k')) < -\rho_{mi}(j, j') \tag{4.68}$$

where $I_{mi}{}^+$ and $I_{mi}{}^-$ are disjoined subsets of the control set I_{mi} ($I_{mi}{}^+ \cap I_{mi}{}^- = \emptyset$ and $I_{mi}{}^+ \cup I_{mi}{}^- = I_{mi}$) of the mixed dipoles $\{\mathbf{x}_j(k), \mathbf{x}_{j'}(k')\}$, $\rho_{c_i}(j, j')$ and $\rho_{m_i}(j, j')$ are the clear and the mixed margins defined on the ith line.

Remark 1: If two orthonormal vectors $\mathbf{w}_1$ and $\mathbf{w}_2$ ($\mathbf{w}_i{}^T\mathbf{w}_i = 1,\ \mathbf{w}_i{}^T\mathbf{w}_k = 0$) fulfil the inequalities (separability postulate, (4.67), and (4.68)), then the inequalities (4.64) and (4.65) with the below parameters $\rho_c{}^2(j, j')$ and $\rho_m{}^2(j, j')$ are also fulfilled

$$(\forall (j, j') \in I_c)\ \ \rho_c{}^2(j, j') = \rho_{c1}{}^2(j, j') + \rho_{c2}{}^2(j, j') \tag{4.69}$$
$$(\forall (j, j') \in I_m)\ \ \rho_m{}^2(j, j') = \rho_{m1}{}^2(j, j') + \rho_{m2}{}^2(j, j') \tag{4.70}$$

4.9 Reinforcement of the Separability Postulates Through the Differential Criterion Function

The *differential* criterion function $\Psi(\mathbf{w})$ similar to the perceptron function $\Phi(\mathbf{w}, \theta)$ (25) can be used for the purpose of finding such vector of parameters $\mathbf{w}_i$, which fulfil in a best manner (fully or partly) the inequalities (4.64–4.67) [6, 9]. The criterion functions $\Psi(\mathbf{w})$ is a positive combination of the penalty functions $\pi_{jj'}{}^+(\mathbf{w})$, $\pi_{jj'}{}^-(\mathbf{w})$ and $\pi_{jj'}{}^0(\mathbf{w})$ defined on the *differential vectors* $\mathbf{r}_{jj'}$:

$$(\forall (j, j') \in I_c \cup I_m)\mathbf{r}_{jj'} = \mathbf{x}_{j'} - \mathbf{x}_j \tag{4.71}$$

whoro $\mathbf{x}_j(k) \neq \mathbf{x}_{j'}(k')$ and $j' > j$.

The *CPL* penalty functions $\pi_{jj'}{}^+(\mathbf{w})$, $\pi_{jj'}{}^-(\mathbf{w})$ and $\pi_{jj'}{}^0(\mathbf{w})$ is defined in a similar manner to $\upsilon_j{}^+(\mathbf{w},\ \theta)$ (4.23) and $\upsilon_j{}^-(\mathbf{w},\ \theta)$ (4.24)

$$(\forall(j,j') \in I_m{}^+) \qquad (4.72)$$

$$\psi_{jj'}{}^+(\mathbf{w}) = \begin{cases} \rho_m(j,j') - \mathbf{w}^T\mathbf{r}_{jj'} & \text{if} \quad \mathbf{w}^T\mathbf{r}_{jj'} \leq \rho_m(j,j') \\ 0 & \text{if} \quad \mathbf{w}^T\mathbf{r}_{jj'} > \rho_m(j,j') \end{cases}$$

where $\rho_m(j,j') = \rho_{m_i}(j,j')$ (4.67). The penalty functions $\pi_{jj'}{}^+(\mathbf{w})$ are aimed at reinforcement the inequalities (4.67).

$$(\forall(j,j') \in I_m{}^-) \qquad (4.73)$$

$$\psi_{jj'}{}^-(\mathbf{w}) = \begin{cases} \rho_{mi}(j,j') + \mathbf{w}^T\mathbf{r}_{jj'} & \text{if} \quad \mathbf{w}^T\mathbf{r}_{jj'} \geq -\rho_m(j,j') \\ 0 & \text{if} \quad \mathbf{w}^T\mathbf{r}_{jj'} < -\rho_m(j,j') \end{cases}$$

The penalty functions $\pi_{jj'}{}^-(\mathbf{w})$ are aimed at reinforcement the inequalities (4.68).

$$(\forall(j,j') \in I_c) \qquad (4.74)$$

$$\psi_{jj'}{}^0(\mathbf{w}) = \begin{cases} -\rho_c(j,j') - \mathbf{w}^T\mathbf{r}_{jj'} & \text{if} \quad \mathbf{w}^T\mathbf{r}_{jj'} \geq -\rho_c(j,j') \\ 0 & \text{if} \quad -\rho_c(j,j') < \mathbf{w}^T\mathbf{r}_{jj'} < \rho_c(j,j') \\ -\rho_c(j,j') + \mathbf{w}^T\mathbf{r}_{jj'} & \text{if} \quad \mathbf{w}^T\mathbf{r}_{jj'} \geq \rho_c(j,j') \end{cases}$$

where $\rho_c(j,j') = \rho_{c_i}(j,j')$ (separability postulate). The penalty functions $\pi_{jj'}{}^0(\mathbf{w})$ are aimed at reinforcement the inequalities (separability postulate).

The criterion function $\Psi(\mathbf{w})$ is the weighted sum of the above penalty functions

$$\Psi(\mathrm{w}) = \sum_{(j,j')\in I_m{}^+} \gamma_{jj'}\psi_{jj'}{}^+(\mathbf{w}) + \sum_{(j,j')\in I_m{}^-} \gamma_{jj'}\psi_{jj'}{}^-(\mathbf{w}) + \sum_{(j,j')\in I_c} \gamma_{jj'}\psi_{jj'}{}^0(\mathbf{w}) \qquad (4.75)$$

where $\gamma_{jj'}$ ($\gamma_{jj'} > 0$) are positive parameters (*prices*) related to particular dipoles $(\mathbf{x}_j(k), \mathbf{x}_{j'}(k'))$.

The criterion function $\Psi(\mathbf{w})$ belongs to the family of the convex and piecewise linear (CPL) criterion functions.

The function $\Psi(\mathbf{w})$ (4.72) can be specified as the criterion function $\Psi_i(\mathbf{w}_i)$ linked to the ith axis ($i = 1, \ldots, n'$) of the transformed space. The specification of the criterion function $\Psi_i(\mathbf{w}_i)$ to the ith axis is done through an adequate choice of the function parameters. The sets of dipoles I_{ci} (separability postulate), $I_{mi}{}^+$ (4.67) and $I_{mi}{}^-$ (4.68) and the sets of margins $\rho_{c_i}(j,j')$ (4.64) and $\rho_{m_i}(j,j')$ ((4.65), separability postulate) can be specified in a different manner for particular axis.

Minimization of the function $\Psi_i(\mathbf{w})$ allows one to find the parameters vector $\mathbf{w}_i{}^*$, which defines (4.6) the ith column of the transformation matrix $\boldsymbol{W}$ (4.10) or the ith axis of the ($i = 1, \ldots, n'$) of the transformed space.

$$\Psi_i{}^* = \Psi_i(\mathrm{w}_i{}^*) = \min_{\mathrm{w}} \Psi_i(\mathrm{w}) \geq 0 \qquad (4.76)$$

The transformation (4.10) defined by the optimal vectors $\mathbf{w}_i{}^*$ (4.76) will be called as the *dipolar* one. The basis exchange algorithms allow to find the minimal value $\Psi_i{}^*$ of the criterion function $\Psi_i(\mathbf{w})$ in an efficient manner [5]. It can be proved that the minimal value $\Psi_i{}^*$ is equal to zero ($\Psi_i{}^* = 0$) if and only if all the inequalities (separability postulate, (4.67), and (4.68)) can be fulfilled on some line $y = (\mathbf{w})^T\mathbf{x}$. In this case, all the inequalities (separability postulate, (4.67), and (4.68)) are fulfilled on the optimal line $\mathrm{y} = (\mathbf{w}_i{}^*)^T\mathbf{x}$.

Example 4. Let us examine such dipolar transformation (4.10) of the feature vectors $\mathbf{x}_j(k)$ on the visualising plane ($n' = 2$), which fulfil both the inequalities (4.64) and (4.65) or the separability postulate. There is a structural difference between the separability inequalities (4.64) and (4.65). All the inequalities (4.64) should be realised by both the axes $\mathbf{w}_1{}^*$ and $\mathbf{w}_2{}^*$ of the visualising plane. Realisation of each inequality (4.65) by only one axis $\mathbf{w}_1{}^*$ or $\mathbf{w}_2{}^*$ is sufficient for fulfilling of the separability postulate. In other words, if the length $\delta_y(j, j')$ (4.64) of the mixed dipole $(\mathbf{x}_j(k), \mathbf{x}_{j'}(k')\}$ along one axis $\mathbf{w}_i{}^*$ is greater than $\rho_m(j, j')$, then the length of this dipole on the plane is also greater then $\rho_m(j, j')$. The length $\delta_y(j, j')$ of the mixed dipole $\{\mathbf{x}_j(k), \mathbf{x}_{j'}(k')\}$ along the ith axis is greater then $\rho_m(j, j')$ (4.65) if and only if one of the inequalities (4.67) or (4.68) is fulfilled by the optimal vector $\mathbf{w}_i{}^*$. In a consequence, the indices (j, j') of such mixed dipoles which are sufficiently long on the first axis $\mathbf{w}_1{}^*$ cannot be considered on the second axis $\mathbf{w}_2{}^*$. In a result, the indices (j, j') from the set I_m could be divided along two axis of the visualising plane. Such division reduces the sets I_{m1} and I_{m2} of mixed dipoles $\{\mathbf{x}_j(k), \mathbf{x}_{j'}(k')\}$ considered on particular axis and, in result, increases chance for fulfilling all the inequalities (separability postulate, (4.67), and (4.68)) on the optimal line $\mathrm{y} = (\mathbf{w}_i{}^*)^T\mathbf{x}$.

To realise the inequality (4.64) for the clear dipole $\{\mathbf{x}_j(k), \mathbf{x}_{j'}(k)\}((j, j') \in I_c)$, both the axes $\mathbf{w}_1{}^*$ and $\mathbf{w}_2{}^*$ of the visualising plane should produce small enough lengths $|((\mathbf{w}_i{}^*)^T(\mathbf{x}_{j'}(k) - \mathbf{x}_j(k))|$ (separability postulate). If the vectors $\mathbf{w}_1{}^*$ and $\mathbf{w}_2{}^*$ are orthogonal and the first vector $\mathbf{w}_1{}^*$ produces the length $|(\mathbf{w}_1{}^*)^T(\mathbf{x}_{j'}(k) - \mathbf{x}_j(k))| \le \rho_c{}^2(j, j')$, then the second vector $\mathbf{w}_2{}^*$ should fulfill the below condition (4.64)

$$|(\mathrm{w}_2{}^*)^T(\mathbf{x}_{j'}(k) - \mathbf{x}_j(k))| \le \rho_c{}^2(j, j') - |(\mathbf{w}_1{}^*)^T(\mathbf{x}_j'(k) - \mathbf{x}_j(k))| \qquad (4.77)$$

To fulfill the separability postulate, the second vector $\mathbf{w}_2{}^*$ should produce such length $|(\mathbf{w}_2{}^*)^T(\mathbf{x}_{j'}(k) - \mathbf{x}_j(k))|$, which is small enough in accordance with (4.77).

The linear transformations (4.10) based on the dipolar model can be used in designing diagnostic maps. Such maps give possibility to enhance clusters of points $\mathbf{y}_j(k)$ on the visualizing plane. The number L of clusters (fields) on the diagnostic map can be equal or greater than the number K of the classes ω_k. The number L of clusters on the diagnostic map can be equal to the number of classes K, if all the feature vectors $\mathbf{x}_j(k)$ from each set C_k (4.1) are used in producing clear dipoles $\{\mathbf{x}_j(k), \mathbf{x}_{j'}(k)\}$. It means that the set I_{ck} (separability postulate) contains all clear dipoles based on the set C_k (4.1).

$$I_{ck} = \{(j, j') : (j' > j) \wedge (\mathbf{x}_j(k) \in C_k) \wedge (\mathbf{x}_{j'}(k) \in C_k)\} \tag{4.78}$$

and

$$I_c = I_{c1} \cup \ldots\ldots \cup \quad I_{ck} \tag{4.79}$$

In this case, the dipolar transformation similarly as the Fishers one is aimed at focussing all the transformed points $\mathbf{x}_j\ (k)$ from each set C_k (class ω_k) into one cluster on the map under condition of preserving the classes ω_k separability.

In some cases, given set C_k has its own internal structure and it could be profitable to enhance such structure by dividing this set into more than one cluster. For this purpose, the set I_{ck} (4.78) can be modified in a below manner:

$$I_{ck} = \{(j, j') : (j' > j) \wedge (\mathbf{x}_j(k) \in C_k) \wedge (\mathbf{x}_{j'}(k) \in C_k) \wedge (\delta_x(j, j') \leq \rho_0)\} \tag{4.80}$$

where $\delta_x(j, j')$ is the dipol length (4.62) and ρ_0 is a 'small' parameter.

As it results from the relation (4.80), the set I_{ck} contains the indices (j, j') of only such clear dipoles $\{\mathbf{x}_j\ (k), \mathbf{x}_{j'}(k)\}$ which are 'short'.

4.10 CPL Criterion Functions with Feature Costs

Data sets C_k (4.1) used in decision support systems are often multidimensional. Many features (attributes) x_i are used for description of particular objects $\mathbf{x}_j(k)$. A large part of these features x_i can be unimportant or redundant in decision support rules. Such features should be removed in accordance with one of feature selection procedures.

The feature selection procedure can be based on the CPL criterion functions with feature costs. Let us introduce for this purpose the modified perceptron criterion function $\Phi(\mathbf{w}, \theta)$ (4.25) and the modified differential criterion function $\Psi(\mathbf{w})$ (4.75). The modified perceptron function $\Phi_\lambda'(\mathbf{w}, \theta)$ can have the following form:

$$\begin{aligned} \Phi_{\lambda}'(\mathbf{w}, \theta) &= \Phi(\mathbf{w}, \theta) + \lambda \Sigma \gamma_i \phi_i(\mathbf{w}, \theta) \\ & \qquad\qquad\qquad i \in \{0, 1, \ldots., n\} \\ &= \Sigma \alpha_j \varphi_j^{+}(\mathbf{w}, \theta) + \Sigma \alpha_j \varphi_j^{-}(\mathbf{w}, \theta) + \lambda(\Sigma \gamma_i |w_i| + \gamma_0 \\ & \qquad j \in J^{+} \quad j \in J^{-} \qquad i \in \{1, \ldots., n\} \end{aligned} \tag{4.81}$$

where $\alpha_j \geq 0$, $\lambda \geq 0$, $\gamma_i > 0$, $\mathbf{w} = [w_1, \ldots., w_n]^T$ is the weight vector, the function $\Phi(\mathbf{w}, \theta)$ is defined by the formula (4.25), and the cost functions $\phi_i(\mathbf{w}, \theta)$ are equal to modulus $|w_i|$ of particular weights w_i.

Minimization of the CPL criterion function $\Phi_\lambda'(\mathbf{w}, \theta)$ (4.81) allows to find the optimal parameters $\mathbf{w}^*$ and θ^*.

$$\Phi_\lambda^* = \Phi_\lambda'(w^*, \theta^*) = \min_{w,\theta} \Phi_\lambda'(w, \theta) \quad \geq 0 \tag{4.82}$$

Minimum of the function $\Phi_\lambda'(\mathbf{w}, \theta)$ (4.81) can be found by using the basis exchange alghorithm.

It can be shown that in the case of linearly separable (4.4) sets G^+ and G^- (4.22), the minimal value $\Phi_\lambda{}^*$ of the criterion function $\Phi_\lambda'(\mathbf{w}, \theta)$ (4.81) with sufficient small value of the parameter $\lambda (0 < \lambda < \lambda_g)$ is equal to

$$\Phi_\lambda^* = \Phi_\lambda'(\mathbf{w}^*, \theta^*) = \lambda(\Sigma \quad \gamma_i |w_i^*| + \gamma_0 |\theta|) \quad > 0 \tag{4.83}$$

and (4.25)

$$\Phi(w^*, \theta^*) = 0 \tag{4.84}$$

The optimal parameters $\mathbf{w}^*$ and θ^* give a balance between an *increasing tendency* resulting from the penalty functions $\upsilon_j{}^+(\mathbf{w}, \theta)$ (4.23) and $\upsilon_j{}^-(\mathbf{w}, \theta)$ (4.24) and *decreasing tendency* resulting from the cost functions $\phi_i(\mathbf{w}, \theta)$ (4.81). An influence of the cost functions $\phi_i(\mathbf{w}, \theta)$ (4.81) decreases with the value of the parameter λ.

The feature selection rules can be based on the optimal parameters $\mathbf{w}^* = [w_1{}^*, \ldots, w_n{}^*]^T$ (4.82):

$$\textit{if} \quad w_i^* = 0 \quad \textit{then} \quad \textit{theith feature } x_i \textit{ can be neglected}, \tag{4.85}$$

or/

$$\textit{if} \quad |w_i{}^*| \quad < \quad \varepsilon \quad \textit{then} \quad \textit{theithfeature } x_i \textit{ can be neglected}, \tag{4.86}$$

where ϖ is a small parameter.

The differential criterion function $\Psi(\mathbf{w})$ (4.75) can be also modified by adding the cost functions $\phi_i(\mathbf{w}, \theta)$ in a manner similar to (4.81). The feature selection rules similar to (4.85) and (4.86) can be based on the modified differential function $\Psi(\mathbf{w})$ (4.75).

4.11 Concluding Remarks

Similarity measures for the *case based reasoning* scheme of decision support can be induced through separable data transformations. In particular, linear transformations of data sets corresponding to particular categories allow to reduce dimensionality of the data sets under the condition of preserving the categories separability. Separable linear transformations can be designed both through solutions of eignevalue problems used in the *principal componet analysis* or in the *discriminant analysis* [4] as well as through minimization of the convex and piecewise linear (CPL) criterion functions [9]. The *perceptron* and the *differential* criterion functions belong, among others, to the CPL family.

Functions from the CPL family give possibility for flexible modelling and solving many problems of exploratory data analysis [9]. In particular, the feature selection problem can be solved through minimization of the CPL criterion functions. The basis exchange algorithms, which are similar to the linear programming, allow one to find the minimum of the CPL criterion functions efficiently even in the case of large, multidimensional data sets [7].

References

1. P. Perner, *Data Mining on Multimedia Data*, Springer, Berlin 2002
2. K. Fukunaga: *Statistical Pattern Recognition*, Academic Press, Inc., San Diego, 1990
3. O.R. Duda, P.E. Hart: *Pattern Classification*, Sec.. Edition. J. Wiley, New York, 2001
4. R.A. Johnson, D.W. Wichern: *Applied Multivariate Statistical Analysis*, Prentice-Hall, Inc., Englewood Cliffs, New York, 1991
5. Bobrowski L., Topczewska M., Improving the *K-NN* classification with the Euclidean distance through linear data transformations, pp. 22–32 in: *ICDM* 2004, Eds. P. Perner et al., Lipsk, Germany, Springer Verlag 2004, *Springer Lecture Notes in Artificial Intelligence* 3275
6. L. Bobrowski, M. Topczewska: "Linear visualising transformations and convex, piecewise linear criterion functions", *Bioc. and Biom. Eng.*, pp. 69–78, Vol. 22, Nr.1, 2002
7. L. Bobrowski: "Design of piecewise linear classifiers from formal neurons by some basis exchange, *Pattern Recognition*, 24(9), pp. 863–870, 1991
8. L. Bobrowski, H. Wasyluk, Diagnosis supporting rules of the *Hepar* system, pp. 1309–1313 in: *MEDINFO* 2001, Eds: V. L. Petel, R. Rogers, R. Haux, IOS Press, Amsterdam 2001
9. Bobrowski L.: *Eksploracja danych oparta na wypukłych i odcinkowo-liniowych funkcjach kryterialnych* (*Data mining based on convex and piecewise linear (CPL) criterion functions)* (in Polish), Technical University Białystok, 2005

5

Graph Matching

X. Jiang[1] and H. Bunke[2]

[1] Department of Mathematics and Computer Science, University of Münster
Einsteinstrasse 62, D-48149 Münster, Germany
xjiang@math.uni-muenster.de

[2] Institute of Informatics and Applied Mathematics, University of Bern
Neubrückstrasse 10, CH-3012 Bern, Switzerland
bunke@iam.unibe.de

5.1 Introduction

Graphs are a powerful and universal tool widely used in information processing. Numerous methods for graph analysis have been developed. Examples include the detection of Hamiltonian cycles, shortest paths, vertex coloring, graph drawing, and so on [5]. In particular, graph representations are extremely useful in image processing and understanding, which is the complex process of mapping the initially numeric nature of an image (or images) into symbolic representations for subsequent semantic interpretation of the sensed world.

Case-based reasoning (CBR) provides us powerful strategies to fulfill the high demands on robustness, accuracy, and flexibility of image interpretation systems [53]. In CBR systems a concept is described by a case base and an associated similarity measure. Cases can be organized into a flat case base or in a hierarchical fashion. In a flat organization, we have to calculate similarity between the problem case and each case in memory. It is clear that this will take a considerable amount of time. To speed up the retrieval process, a more sophisticated, hierarchical organization of the case base is necessary. This organization should allow separating the set of similar cases from those cases not similar to the recent problem at the earliest stage of the retrieval process. In case base creation, maintenance, and retrieval, a central issue is that of case (object) similarity. In this chapter we are concerned with structural case representations [20, 54], which are common in computer vision and image interpretation [6], building design [31], timetabling [19], etc. Given such representations, we consider the problem of determining the equality or similarity of graphs, which is generally referred to as *graph matching*.

Standard concepts in *exact* graph matching include graph isomorphism and subgraph isomorphism. Two graphs are called *isomorphic* if they have identical structure. There exists a *subgraph isomorphism* between two graphs

X. Jiang and H. Bunke: *Graph Matching*, Studies in Computational Intelligence (SCI) **73**, 149–173 (2008)
www.springerlink.com

if one graph contains a subgraph that is isomorphic to the other. Subgraph isomorphism is useful to find out if a given object is part of another object or a collection of several objects. Although exact graph matching offers a rigorous way to describe structure equality in mathematical terms, it is generally only applicable to a restricted set of real-world problems. *Inexact*, or *error-tolerant*, graph matching methods, on the other hand, are able to cope with strong inner-class distortion, which is often present in real-world applications.

In this chapter we provide an overview of graph matching. We mainly concentrate on the fundamental concepts and some recent developments. The reader is referred to the recent survey [23] for a detailed discussion of the numerous graph matching algorithms and also the vast applications of graph matching. Other recent collections of papers on graph matching (and mining) can be found in [18, 24, 26, 29].

5.2 Basic Definitions and Notation

Attributed graphs with an unrestricted label alphabet is one of the most general ways to define graphs. It turns out that the definition given below is sufficiently flexible for a large variety of applications.

Definition 1 (Graph). *A graph is a 4-tuple $g = (V, E, \alpha, \beta)$, where*

- *V is the finite set of vertices*
- *$E \subseteq V \times V$ is the set of edges*
- *$\alpha : V \rightarrow L_V$ is a function assigning labels to the vertices*
- *$\beta : E \rightarrow L_E$ is a function assigning labels to the edges*

Edge (u, v) originates at node u and terminates at node v. The labeling function can be used to integrate information about nodes and edges into graphs by assigning attributes from L_V and L_E to nodes and edges, respectively. Usually, there are no constraints imposed on the label alphabets. In practical applications, however, label alphabets are often defined as vector spaces $\Re^k$ of a fixed dimension k or discrete sets of symbols $\{s_1, s_2, \ldots, s_k\}$. In principle, nodes and edges may also have other, arbitrarily complex labels. The notation $|g|$ will be used for the number of nodes of graph g.

The graph definition introduced above includes a number of special cases. To define undirected graphs, for instance, we require that $(v, u) \in E$ for every edge $(u, v) \in E$ such that $\beta(u, v) = \beta(v, u)$. In the case of nonattributed graphs, the label alphabets are defined by $L_V = L_E = \{\phi\}$, so that every node and edge gets assigned the null label ϕ.

For some applications, it is important to detect whether a smaller graph is present in a larger graph – for instance, if the larger graph represents an aggregation of objects and the smaller graph a specific object in the larger context. This intuitively leads to the formal definition of a subgraph.

Definition 2 (Subgraph). *Let $g_1 = (V_1, E_1, \alpha_1, \beta_1)$ and $g_2 = (V_2, E_2, \alpha_2, \beta_2)$ be graphs; g_2 is a* subgraph *of g_1, written as $g_2 \subseteq g_1$, if*

- $V_2 \subseteq V_1$
- $E_2 = E_1 \cap (V_2 \times V_2)$
- $\alpha_1(v) = \alpha_2(v)$ *for all* $v \in V_2$
- $\beta_1(e) = \beta_2(e)$ *for all* $e \in E_2$

Conversely, graph g_1 is called a supergraph of g_2 if g_2 is a subgraph of g_1. Sometimes, the second condition of this definition is replaced by $E_2 \subseteq E_1 \cap (V_2 \times V_2)$, and a subgraph fulfilling the more stringent condition given above is called an induced subgraph. The notion of subgraph can be used to approach more complex problems such as the largest common part of several graphs, which will be discussed below.

5.3 Exact Graph Matching

In exact graph matching, the objective is to determine whether or not the structure and labels, or part of the structure, of two graphs are identical.

Definition 3 (Graph isomorphism). *Let g_1 and g_2 be graphs. A* graph isomorphism *between g_1 and g_2 is a bijective mapping $f : V_1 \rightarrow V_2$ such that*

- $\alpha_1(v) = \alpha_2(f(v))$ *for all* $v \in V_1$
- *for any edge $e_1 = (u, v) \in E_1$ there exists an edge $e_2 = (f(u), f(v)) \in E_2$ such that $\beta_1(e_1) = \beta_2(e_2)$, and for any edge $e_2 = (u, v) \in E_2$ there exists an edge $e_1 = (f^{-1}(u), f^{-1}(v)) \in E_1$ such that $\beta_1(e_1) = \beta_2(e_2)$*

Two graphs g_1 and g_2 are called isomorphic if there exists a graph isomorphism between them.

From this definition we conclude that isomorphic graphs are identical in terms of structure and labels. To establish an isomorphism one has to map each node from the first graph to a node of the second graph such that the edge structure is preserved and the node and edge labels are consistent.

The graph isomorphism problem is of considerable practical importance and also of theoretical interest due to its relationship to the concept of NP-completeness. Despite intensive research for over three decades [30,55,58] still no efficient (polynomial-bound) algorithm for graph isomorphism is known. Neither has the conjecture been proved that no such algorithm can exist. While it is easy to determine equality of patterns in case of feature vectors or strings, the same computation is much more complex for graphs. Because the nodes and edges of a graph cannot be ordered in general, unlike the components of a feature vector or the symbols of a string, the problem of graph equality (graph isomorphism) is computationally very demanding. The most straightforward approach to checking the isomorphism of two graphs is to traverse a search tree considering all possible node-to-node correspondences [61].

The expansion of tree branches is continued until the edge structure implied by the node mapping does not correspond in both graphs. If nodes and edges are additionally endowed with labels, matching nodes and edges must also be consistent in terms of their labels. Reaching a leaf node of the search tree is equivalent to successfully mapping all nodes without violating the structure and label constraints and is therefore equivalent to having found a graph isomorphism. In general, the computational complexity of this procedure is exponential in the number of nodes of either graph.

By imposing certain restrictions on the underlying graphs, however, it is possible to derive algorithms of polynomial-time complexity. For instance, Luks [45] described a polynomially bounded method for the isomorphism detection of graphs with bounded valence. For the special case of trivalent graph isomorphism, it was shown in [45] that algorithms with a computational complexity of $O(n^4)$ exist. Low-order polynomial-time methods [35, 36, 64] are also known for planar graphs. Quadratic-time algorithms [37, 38] have been reported for ordered graphs, in which the edges incident to a vertex are uniquely ordered. Further special graph classes, for which the isomorphism problem is solvable in polynomial time, are trees [1], interval graphs [9], permutation graphs [22], chordal (6, 3) graphs [3], graphs with bounded genus [48], graphs with bounded treewidth [7], graphs with bounded eigenvalue multiplicity [2], and rooted directed path graphs [4].

Closely related to graph isomorphism is the problem to detect if a smaller graph is present in a larger graph. If graph isomorphism is regarded as a formal notion of graph equality, subgraph isomorphism can be seen as subgraph equality.

Definition 4 (Subgraph isomorphism). *Let $g_1 = (V_1, E_1, \alpha_1, \beta_1)$ and $g_2 = (V_2, E_2, \alpha_2, \beta_2)$ be graphs. An injective function $f : V_1 \rightarrow V_2$ is called a subgraph isomorphism from g_1 to g_2 if there exists a subgraph $g \subseteq g_2$ such that f is a graph isomorphism between g_1 and g.*

A subgraph isomorphism exists from g_1 to g_2 if the larger graph g_2 can be turned into a graph that is isomorphic to the smaller graph g_1 by removing some nodes and edges. Subgraph isomorphism can also be determined with the procedure outlined above for graph isomorphism [61]. It is known that subgraph isomorphism belongs to the class of NP-complete problems.

5.4 Inexact Graph Matching

In graph representations of real-world patterns, it is often the case that graphs from the same class differ in terms of structure and labels. Hence, graph matching systems need to take structural errors into account. In this section several variants of realizing this goal are discussed.

5.4.1 Graph Edit Distance

Graph edit distance offers an intuitive way to integrate error-tolerance into the graph matching process and is applicable to virtually all types of graphs. Originally, edit distance has been developed for string matching [62] and a considerable amount of variants and extensions to the edit distance have been proposed for strings and graphs. The key idea is to model structural variation by edit operations reflecting modifications in structure, such as the removal of a single node or the modification of an attribute attached to an edge. A standard set of edit operations consists of a node insertion, node deletion, node substitution, edge insertion, edge deletion, and edge substitution operation.

Definition 5 (Edit path). *Let* $g_1 = (V_1, E_1, \alpha_1, \beta_1)$ *and* $g_2 = (V_2, E_2, \alpha_2, \beta_2)$ *be graphs. Any bijective function* $f : \hat{V}_1 \rightarrow \hat{V}_2$*, where* $\hat{V}_1 \subseteq V_1$ *and* $\hat{V}_2 \subseteq V_2$*, is called an edit path from* g_1 *to* g_2*.*

We say that node $u \in \hat{V}_1$ is *substituted* by node $v \in \hat{V}_2$ if $f(u) = v$. If $\alpha_1(u) = \alpha_2(f(u))$ then the substitution is called an *identical* substitution. Otherwise, it is termed a *nonidentical* substitution. Furthermore, any node from $V_1 - \hat{V}_1$ is *deleted* from g_1, and any node from $V_2 - \hat{V}_2$ *inserted* in g_2 under f. We will use $\hat{g}_1$ and $\hat{g}_2$ to denote the subgraphs of g_1 and g_2 that are induced by the sets $\hat{V}_1$ and $\hat{V}_2$, respectively.

The mapping f *directly* implies an edit operation on each node in g_1 and g_2, i.e., nodes are substituted, deleted, or inserted as described above. Additionally, the mapping f *indirectly* implies edit operations on the edges of g_1 and g_2. If $f(u_1) = v_1$ and $f(u_2) = v_2$ and there exist edges $(u_1, u_2) \in E_1$ and $(v_1, v_2) \in E_2$ then edge (u_1, u_2) is substituted by (v_1, v_2) under f. If $\beta_1((u_1, u_2)) = \beta_2((v_1, v_2))$ then the edge substitution is called an *identical* substitution. Otherwise, it is termed a *nonidentical* substitution. If there exists no edge $(u_1, u_2) \in E_1$, but an edge $(v_1, v_2) \in E_2$, then edge (v_1, v_2) is inserted. Similarly, if $(u_1, u_2) \in E_1$ exists but no edge (v_1, v_2), then (u_1, u_2) is deleted under f. If a node u is deleted from g_1, then any edge incident to u is deleted, too. Similarly, if a node u' is inserted in g_2, then any edge incident to u' is inserted, too. Obviously, any edit path f can be understood as a set of edit operations (substitutions, deletions, and insertions of both nodes and edges) that transform a given graph g_1 into another graph g_2.

Example 1. A graphical representation of two graphs is given in Fig. 5.1. For those graphs, we have the following:

$L_V = \{X, Y, Z\}$; $L_E = \{a, b, c\}$

$V_1 = \{1, 2, 3\}$; $V_2 = \{4, 5, 6, 7\}$

$E_1 = \{(1,2), (1,3), (2,3)\}$; $E_2 = \{(4,5), (4,6), (4,7), (5,6), (5,7)\}$

α_1: $1 \mapsto X$, $2 \mapsto X$, $3 \mapsto Y$

α_2: $4 \mapsto X$, $5 \mapsto X$, $6 \mapsto Y$, $7 \mapsto Z$

β_1: $(1,2) \mapsto a$, $(1,3) \mapsto b$, $(2,3) \mapsto b$

β_2: $(4,5) \mapsto a$, $(4,6) \mapsto c$, $(4,7) \mapsto b$, $(5,6) \mapsto c$, $(5,7) \mapsto b$

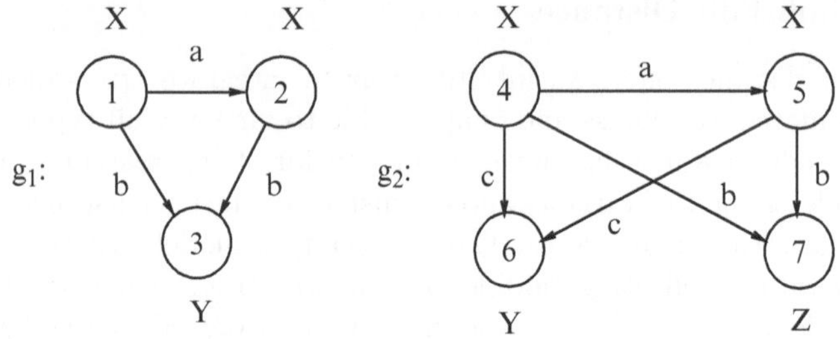

Fig. 5.1. An example of edit paths (see text)

Three examples of edit paths are the following:

- $f_1 : 1 \mapsto 4, 2 \mapsto 5, 3 \mapsto 7$ with $\hat{V_1} = \{1, 2, 3\}$ and $\hat{V_2} = \{4, 5, 7\}$
- $f_2 : 1 \mapsto 4, 2 \mapsto 5, 3 \mapsto 6$ with $\hat{V_1} = \{1, 2, 3\}$ and $\hat{V_2} = \{4, 5, 6\}$
- $f_3 : 1 \mapsto 4, 2 \mapsto 5$ with $\hat{V_1} = \{1, 2\}$ and $\hat{V_2} = \{4, 5\}$

Under f_1, nodes 1, 2, and 3 are substituted by nodes $4, 5$, and 7, respectively. Consequently, edges $(1, 2), (1, 3)$, and $(2, 3)$ are substituted by $(4, 5), (4, 7)$, and $(5, 7)$, respectively. The substitution of nodes 1 and 2 by 4 and 5 are identical substitutions that involve no label change; there are no label changes involved in the edge substitutions, either. The label Y of node 3 is substituted by Z of node 7, and node 6 together with its incident edges $(4, 6)$ and $(5, 6)$ are inserted in g_2. There are, of course, many other paths from g_1 to g_2.

With substitutions, deletions, and insertions for both nodes and edges at our disposal, any graph can be transformed into any other graph by iteratively applying edit operations. Consequently, the concept of graph editing can be used to define a dissimilarity measure on graphs. To quantify how strongly an edit operation modifies the structure of a graph, it is common to use an edit cost function that assigns a cost value to each edit operation. An edit operation associated with a low cost is assumed to only slightly alter the graph under consideration, while an edit operation with a high cost is assumed to strongly modify the graph. To obtain a cost function on edit paths, we simply accumulate individual edit operation costs of the edit path.

Definition 6. *The* cost *of an edit path* $f : \hat{V_1} \rightarrow \hat{V_2}$ *from a graph* $g_1 = (V_1, E_1, \alpha_1, \beta_1)$ *to a graph* $g_2 = (V_2, E_2, \alpha_2, \beta_2)$ *is given by*

$$c(f) = \textstyle\sum_{u \in \hat{V_1}} c_{ns}(u) \quad + \sum_{u \in V_1 - \hat{V_1}} c_{nd}(u) + \sum_{u \in V_2 - \hat{V_2}} c_{ni}(u)$$
$$+ \textstyle\sum_{e \in E_s} c_{es}(e) + \sum_{e \in E_d} c_{ed}(e) \qquad + \sum_{e \in E_i} c_{ei}(e),$$

where

- $c_{ns}(u)$ *is the cost of substituting node* $u \in \hat{V_1}$ *by* $f(u) \in \hat{V_2}$
- $c_{nd}(u)$ *is the cost of deleting node* $u \in V_1 - \hat{V_1}$ *from* g_1

- $c_{ni}(u)$ *is the cost of inserting node* $u \in V_2 - \hat{V}_2$ *in* g_2
- $c_{es}(e)$ *is the cost of substituting edge* e
- $c_{ed}(e)$ *is the cost of deleting edge* e
- $c_{ei}(e)$ *is the cost of inserting edge* e

and E_s, E_d, *and* E_i *are the sets of edges that are substituted, deleted, and inserted, respectively. All costs are nonnegative real numbers.*

Notice that the sets E_s, E_d, and E_i are implied by the mapping f. That is, if edge $e = (u_1, u_2) \in E_1$, $f(u_1) = v_1$, $f(u_2) = v_2$, and $(v_1, v_2) \notin E_2$, then $e \in E_d$. Similarly, if $(u_1, u_2) \notin E_1$, $f(u_1) = v_1$, $f(u_2) = v_2$, and $e = (v_1, v_2) \in E_2$, then $e \in E_i$. Likewise, if $e_1 = (u_1, u_2) \in E_1$, $f(u_1) = v_1$, $f(u_2) = v_2$, and $e_2 = (v_1, v_2) \in E_2$, then $e_1 \in E_s$. Because an edge is deleted (inserted) whenever one or both of its incident nodes are deleted (inserted), we furthermore observe that (1) $e \in E_d$ if $e \in (V_1 \times V_1) - (\hat{V}_1 \times \hat{V}_1)$ and (2) $e \in E_i$ if $e \in (V_2 \times V_2) - (\hat{V}_2 \times \hat{V}_2)$.

The problem of measuring the dissimilarity of two graphs is then equivalent to the problem of finding the edit path that models the structural difference of two graphs in the least costly way. Consequently, the graph edit distance of two graphs is defined by the minimum cost edit path from the first to the second graph.

Definition 7 (Graph edit distance). *Let* $g_1 = (V_1, E_1, \alpha_1, \beta_1)$ *and* $g_2 = (V_2, E_2, \alpha_2, \beta_2)$ *be graphs and let* $c(f)$ *denote the cost of edit path* f. *The edit distance of* g_1 *and* g_2 *can be defined by*

$$d(g_1, g_2) = \min_{\text{all edit paths } f \text{ from } g_1 \text{ to } g_2} c(f)$$

If the two graphs under consideration are very similar in terms of structure and labels, it can be assumed that only minor edit operations are required to transform the first into the second graph, which results in a low-cost optimal edit path. In this case, the resulting edit distance will be small. Conversely, if the two graphs differ significantly, every edit path will necessarily include strong modifications and hence result in high costs.

In the following it is assumed, for the purpose of simplicity, that the costs $c_{nd}(x)$, $c_{ni}(x)$, and $c_{ns}(x)$ do not depend on node x; neither do $c_{ed}(e)$, $c_{ei}(e)$, and $c_{es}(e)$ depend on edge e. In other words, $c_{nd}(x)$, $c_{ni}(x)$, and $c_{ns}(x)$ will be the same for any node x, and $c_{ed}(e)$, $c_{ei}(e)$, and $c_{es}(e)$ will be the same for any edge e. Hence, the notation $c_{nd}(x) = c_{nd}$, $c_{ni}(x) = c_{ni}, \ldots, c_{es}(e) = c_{es}$ will be used and a cost function is given by the 6-tuple $C = (c_{nd}, c_{ni}, c_{ns}, c_{ed}, c_{ei}, c_{es})$. Unless otherwise stated, it is assumed that the cost of an identical node or edge substitution is zero, while the cost of any other edit operation is greater than zero.

Example 2. Consider the uniform cost function $C = (c_{nd}, c_{ni}, c_{ns}, c_{ed}, c_{ei}, c_{es}) = (1, 1, 1, 1, 1, 1)$. Then the edit path f_1 given in Example 1 has cost $c(f_1) = 4$

(one node label substitution, one node insertion, and two edge insertions). It can be easily verified that there is no other edit path from g_1 to g_2 that has a smaller cost under C. For example, $c(f_2) = 5$ (two edge label substitutions, one node insertion, and two edge insertions) and $c(f_3) = 9$ (one node and two edge deletions, two node and four edge insertions). However, if we change the cost function and consider $C' = (1, 1, 3, 1, 1, 1)$ then $c(f_1) = 6$, $c(f_2)$=5, and $c(f_3) = 9$. Thus, f_1 is no longer optimal under C' and it can be easily verified that f_2 has in fact the smallest cost among all possible paths from g_1 to g_2. If we consider a third cost function $C'' = (1, 1, 7, 1, 1, 7)$ then $c(f_1) = 10$, $c(f_2) = 17$, and $c(f_3) = 9$. Under this cost function, f_3 is optimal.

If a cost function satisfies the conditions of positive definiteness and symmetry as well as the triangle inequality at the level of single edit operations, the resulting edit distance is known to be a metric [10]. This fact legitimates the use of the term distance in graph edit distance.

In practical applications, graph edit distance is usually accomplished by means of heuristic A*-based tree search procedures [10]. Because of the exponential nature of the problem, such procedures are normally limited to rather small graphs. Recently, however, several methods to speed up edit distance computation have been proposed. In [8, 59] the authors proposed to optimize local rather than global criteria, which results in a suboptimal procedure. Other suboptimal techniques were introduced in [51, 56]. In Justice and Hero [42] a linear programming method for computing the edit distance of graphs with unlabeled edges is proposed. This method can be used to derive lower and upper edit distance bounds in polynomial time.

5.4.2 Graph Distance Functions Based on mcs

The definition of subgraph isomorphism naturally leads us to the formal definition of the largest common part of two graphs.

Definition 8 (Maximum common subgraph). *Let $g_1 = (V_1, E_1, \alpha_1, \beta_1)$ and $g_2 = (V_2, E_2, \alpha_2, \beta_2)$ be graphs. A common subgraph of g_1 and g_2, $cs(g_1, g_2)$, is a graph $g = (V, E, \alpha, \beta)$ such that there exist subgraph isomorphisms from g to g_1 and from g to g_2. We call g a maximum common subgraph of g_1 and g_2, $mcs(g_1, g_2)$, if there exists no other common subgraph of g_1 and g_2 that has more nodes than g.*

A maximum common subgraph of two graphs represents the maximal part of both graphs that is identical in terms of structure and labels. Note that, in general, the maximum common subgraph is not uniquely defined, that is, there may be more than one common subgraph with a maximal number of nodes. A standard approach to computing maximum common subgraphs is based on solving the maximum clique problem in an association graph [44,46]. The association graph of two graphs represents the whole set of possible node-to-node mappings that preserve the edge structure of both graphs. Finding a maximum clique in the association graph, that is, a fully connected maximal subgraph, is equivalent to finding a maximum common subgraph. In Bunke et al. [16] the reader can find a comparison of algorithms for maximum common subgraph computation on randomly connected graphs.

Graph dissimilarity measures can be derived from the maximum common subgraph of two graphs. Intuitively speaking, the larger a maximum common subgraph of two graphs is, the more similar are the two graphs. This observation leads to graph dissimilarity measures that are able to cope with structural errors. Bunke and Shearer [12] have introduced such a distance measure:

$$d_{MCS}(g_1, g_2) = 1 - \frac{|mcs(g_1, g_2)|}{\max\{|g_1|, |g_2|\}} \tag{5.1}$$

where $|\ldots|$ indicates the size of a graph (usually taken to be the number of nodes). Note that, whereas the maximum common subgraph of two graphs is not uniquely defined, the d_{MCS} distance is. If two graphs are isomorphic, their d_{MCS} distance is 0; if two graphs have no part in common, their d_{MCS} distance is 1. The d_{MCS} distance accounts for a certain amount of tolerance towards errors, as two graphs need not be completely identical for a successful match. However, a small d_{MCS} distance, and hence a high graph similarity, can only be obtained if large portions of both graphs are isomorphic. It has been shown that d_{MCS} is a metric and produces a value in $[0, 1]$.

A second distance measure which has been proposed by Wallis et al. [63], based on the idea of graph union, is

$$d_{WGU}(g_1, g_2) = 1 - \frac{|mcs(g_1, g_2)|}{|g_1| + |g_2| - |mcs(g_1, g_2)|}$$

By "graph union" it is meant that the denominator represents the size of the union of the two graphs in the set-theoretic sense. This distance measure behaves similarly to d_{MCS}. The motivation of using graph union in the denominator is to allow for changes in the smaller graph to exert some influence over the distance measure, which does not happen with d_{MCS}. This measure was also demonstrated to be a metric and creates distance values in $[0, 1]$.

A similar distance measure [11] which is not normalized to the interval $[0, 1]$ is

$$d_{UGU}(g_1, g_2) = |g_1| + |g_2| - 2 \cdot |mcs(g_1, g_2)|$$

Fernandez and Valiente [27] have proposed a distance measure based on both the maximum common subgraph and the minimum common supergraph

$$d_{MMCS}(g_1, g_2) = |MCS(g_1, g_2)| - |mcs(g_1, g_2)|$$

where $MCS(g_1, g_2)$ is the minimum common supergraph of graphs g_1 and g_2, which is the complimentary idea of minimum common subgraph.

Definition 9 (Minimum common supergraph). *Let $g_1 = (V_1, E_1, \alpha_1, \beta_1)$ and $g_2 = (V_2, E_2, \alpha_2, \beta_2)$ be graphs. A common supergraph of g_1 and g_2,*

$CS(g_1, g_2)$, is a graph $g = (V, E, \alpha, \beta)$ such that there exist subgraph isomorphisms from g_1 to g and from g_2 to g. We call g a minimum common supergraph of g_1 and g_2, $MCS(g_1, g_2)$, if there exists no other common supergraph of g_1 and g_2 that has less nodes than g.

The concept that drives the distance measure above is that the maximum common subgraph provides a "lower bound" on the similarity of two graphs, while the minimum supergraph is an "upper bound." If two graphs are identical, then both their maximum common subgraph and minimum common supergraph are the same as the original graphs and $|g_1| = |g_2| = |MCS(g_1, g_2)| = |mcs(g_1, g_2)|$, which leads to $d_{MMCS}(g_1, g_2) = 0$. As the graphs become more dissimilar, the size of the maximum common subgraph decreases, while the size of the minimum supergraph increases. This in turn leads to increasing values of $d_{MMCS}(g_1, g_2)$. For two graphs with no maximum common subgraph, the distance will become $|MCS(g_1, g_2)| = |g_1| + |g_2|$. The distance $d_{MMCS}(g_1, g_2)$ has also been shown to be a metric, but it does not produce values normalized to the interval $[0, 1]$, unlike d_{MCS} or d_{WGU}. We can also create a version of this distance measure which is normalized to $[0, 1]$ as follows:

$$d_{MMCSN}(g_1, g_2) = 1 - \frac{|mcs(g_1, g_2)|}{|MCS(g_1, g_2)|}$$

Note that if the conditions holds that $|MCS(g_1, g_2)| = |g_1| + |g_2| - |mcs(g_1, g_2)|$, then d_{UGU} and d_{MMCS} are identical. The same is true for d_{WGU} and d_{MMCSN}.

5.4.3 Relaxation Approaches

We use a matching matrix M to indicate the compatibility of nodes in the two graphs being matched. If the ith row and jth column element M_{ij} is 1, then node i in graph g_1 is matched with node j in graph g_2; otherwise there is no match and $M_{ij} = 0$. Constraints can be imposed on M so that each row has exactly one 1 and no column has more than one 1. Such a representation and the algorithms applied to it for determining graph matching are straightforward; however, they can require generating all the permutations of possible node matchings over the matrix.

To improve time complexity, we can instead attempt to approximate the optimal solution by finding good suboptimal solutions. A method that is sometimes used to achieve this for graph matching problems is called *relaxation* (or more specifically, *discrete relaxation*). Put simply, discrete relaxation is a method of transforming a discrete representation (such as the matrix M used for graph matching) into a continuous representation. Thus, we can transform a discrete optimization problem into a continuous one. Compared to the typical state-space search approaches to graph matching, relaxation is a nonlinear optimization approach. Gold and Rangdarajan [32] applied relaxation to the graph matching problem. They have posed the problem of attributed graph matching in terms of an optimization problem:

$$E = -\frac{1}{2}\sum_{a=1}^{|V_1|}\sum_{i=1}^{|V_2|}\sum_{b=1}^{|V_1|}\sum_{j=1}^{|V_2|} M_{ai}M_{bj}\sum_{r=1}^{R} C_{aibj}^{(2,r)} + \alpha\sum_{a=1}^{|V_1|}\sum_{i=1}^{|V_2|} M_{ai}\sum_{r=1}^{S} C_{ai}^{(1,s)}$$

Here M is the matching matrix as before, R is the number of edge types, S is the number of node types, α is a weighting factor, and the C's are compatibility measures between the edges of the two graphs. The goal is then to minimize the objective function given above. In [32] the authors use the graduated assignment algorithm to find an M which minimizes E.

Medasani et al. [47] gave a procedure based on fuzzy assignments and relaxation similar to the method just described. The objective function for this approach is

$$J(M, C) = \sum_{i=1}^{|V_1|+1}\sum_{j=1}^{|V_2|+1} M_{ij}^2 f(C_{ij}) + \eta\sum_{i=1}^{|V_1|+1}\sum_{j=1}^{|V_2|+1} M_{ij}(1 - M_{ij})$$

where M is now a fuzzy membership matrix ($0 \leq M_{ij} \leq 1$) that relates the degree of match between nodes, C is a compatibility matrix between nodes (rather than edges as above), η is a control parameter, and

$$f(C_{ij}) = e^{-\beta C_{ij}}$$

The summations in the objective function are under the constraint that $(i, j) \neq (|V_1| + 1, |V_2| + 1)$; the extra nodes in the graphs are dummy nodes. The authors then go on to derive the necessary update equations for M and C in order to minimize $J(M, C)$ and propose an algorithm which updates these matrices in an alternating fashion.

5.4.4 Probabilistic Approaches

In this section we summarize the probabilistic approach proposed by Wilson and Hancock [66]. We attempt to match a data graph g_D and a stored model graph g_M, both being attributed graphs. In Wilson and Hancock [66] an attributed graph is defined to be one $g = (V, E, A)$, where A is a set of attributes associated with each node, $A = x_v^y,\ \forall v \in V$.

The attributes in the data graph are to be matched to those in the model graph, such that the matched nodes have the same or similar attributes. Edges may also have associated attributes, but they are not considered in this approach. Next, we have the concept of super-clique of a node. A *super-clique* [66] of a node i in graph $g = (V, E, A)$ is defined as $C_i = i \cup \{j|(j, i) \in E\}$. In other words, the super-clique of a node i is the set of nodes which contain i and all nodes connected to it by edges. The goal is then to match all super-cliques in the data graph with super-cliques in the model graph.

The set of all possible matches between super-clique C_i in the data graph g_D and super-cliques in the model graph g_M is called a dictionary and denoted Θ_i. To cope with size differences between the data and model super-cliques dummy (or null) nodes ϕ are allowed to be inserted into S_j so that both graphs have the same number of nodes. The function matching a node in C_i to a node in S is $f : V_D \rightarrow V_M \cup \phi$. The probability of matching errors (a node in g_D is matched to the wrong node in g_M) is denoted P_e and the probability of structural errors (a node in g_D is matched to a dummy node in g_M) is denoted P_ϕ. Given these definitions, some assumptions, and through application of Bayes' rule and other probability theoretic constructions, Wilson and Hancock arrive at a mathematical description for the probability of a super-clique matching between two graphs (denoted Γ_j for super-clique C_j):

$$P(\Gamma_j) = \frac{K_{C_j}}{|\Theta_j|} \sum_{S_j \in \Theta_j} \exp\{-(k_e H(\Gamma_j, S_i) + k_\phi[\psi(\Gamma_j, S_i) + \Psi(\Gamma_j)])\}$$

where

$$K_{C_j} = [(1 - P_e)(1 - P_\phi)]^{|C_j|}$$
$$k_e = \ln \frac{1 - P_e}{P_e}$$
$$k_\phi = \frac{(1 - P_e)(1 - P_\phi)}{P_\phi}$$

$H(\Gamma_j, S_i)$ is the Hamming distance between the super-clique of g_D under the mapping f and the super-clique of g_M, $\psi(\Gamma_j, S_i) = |C_j| - |S_i|$ (i.e., the number of null nodes inserted into S_i), and $\Psi(\Gamma_j)$ is the number of nodes in C_j which are mapped onto null nodes in S_i. The deviation of $P(\Gamma_j)$ is beyond the scope of this chapter, but the equation contains three parts which are fairly straightforward. The part associated with K_{C_j} is the probability of no error occurring. The part associated with k_e is concerned with the probability of matching error occurring. Finally, the part associated with k_ϕ deals with the probability of structural errors occurring. For an in-depth derivation of these equations, the reader is referred to [66].

The authors then go on to derive rules which can be applied to update the matching function f under three different methods (null-labeling, constraint filtering, and graph edit operations). The methods use update rules of the form

$$f(u) = \arg \max_{v \in V_M} \frac{P(u, v | x_u^D, x_v^M)}{P(u, v)} \sum_{j \in C_u} P(\Gamma_j)$$

Here $P(u, v)$ indicates the prior probability that node u in g_D corresponds to node v in g_M, while the other probability in the numerator is the conditional *a posteriori* probability, given the corresponding attributes related with the nodes.

An advantage of this framework is that it can be applied in many situations. For example, an extension of the work [65] deals with multiple graph matching through computations of fuzzy consistency matrices. Finch et al. [28] developed an energy function for graph matching based on the probabilistic framework of this section. A method using this approach for the fitness function in a genetic search for graph matching is described in [25]. A similar probabilistic framework for hierarchical graphs is given in [67]. Myers et al. [49] modified the approach described here to include graph edit distance; the new method achieves better complexity by removing the need to inset null nodes in the model graph.

5.4.5 Distance Preservation Approach

In [21] Chartrand et al. describe an approach for graph distance calculation based on preserving the distance between nodes. The idea comes from the fact that when two graphs are isomorphic, the distance (meaning in this context the number of edges traversed) between every pair of nodes are identical in both graphs. Given a graph $g = (V, E)$, the distance between two nodes $x, y \in V$, denoted $d_g(x, y)$, is defined as the minimum number of edges that need to be traversed when traveling from x to y [21]. Further, the ϕ-distance [21] between two graphs g_1 an g_2, denoted $d_\phi(g_1, g_2)$, is defined as

$$d_\phi(g_1, g_2) = \sum_{\forall x \forall y \in V_1} |d_{g_1}(x, y) - d_{g_2}(x, y)|$$

where ϕ is a one-to-one mapping (but not necessarily an isomorphism) between g_1 and g_2.

If ϕ is an isomorphism, then $d_\phi(g_1, g_2) = 0$; if g_1 and g_2 are not isomorphic, then $d_\phi(g_1, g_2) > 0$. This leads to a definition of distance between two graphs as

$$d(g_1, g_2) = \min_{\forall \phi} d_\phi(g_1, g_2)$$

In [21] the authors also go on to show that we can make some other, less expensive calculations if the graphs meet certain requirements. For example, if g_1 and g_2 are connected graphs with an equal number of nodes, then we can determine the lower bound on their distance by

$$d(g_1, g_2) \geq |td(g_1) - td(g_2)|$$

where

$$td(g) = \sum_{\forall u, v \in V} d(u, v)$$

or, in other words, the sum of distances between all pairs of nodes in graph. Further theoretical contributions related to this approach can be found in [21].

5.5 Theoretical Foundations of Graph Matching

The graph distance measure according to (5.1) is based on the maximum common subgraph of two graphs. Obviously, it can be regarded an alternative to graph edit distance. Indeed, it was recently shown that there is a direct relation between graph edit distance and maximum common subgraph in the sense that graph edit distance and maximum common subgraph computation are equivalent to each other under a certain cost function [11]. In [11] the following cost function was considered:

$$\begin{aligned}
c_{ns}(x) &= \left\{ \begin{array}{l} 0, \text{ if } \alpha_1(x) = \alpha_2(f(x)) \\ \infty, \text{ otherwise} \end{array} \right\} \text{for any } x \in \hat{V}_1, \\
c_{nd}(x) &= 1 \text{ for any } x \in V_1 - \hat{V}_1, \\
c_{ni}(x) &= 1 \text{ for any } x \in V_2 - \hat{V}_2, \\
c_{es}(e) &= \left\{ \begin{array}{l} 0, \text{ if } \beta_1((x,y)) = \beta_2((f(x), f(y))) \\ \infty, \text{ otherwise} \end{array} \right\} \\
&\quad \text{for any } e = (x,y) \in \hat{V}_1 \times \hat{V}_1, \\
c_{ed}(e) &= 0 \text{ for any } e = (x,y) \in (V_1 \times V_1) - (\hat{V}_1 \times \hat{V}_1), \\
c_{ei}(e) &= 0 \text{ for any } e = (x,y) \in (V_2 \times V_2) - (\hat{V}_2 \times \hat{V}_2).
\end{aligned} \tag{5.2}$$

Under this cost function, any node deletion and insertion has a cost equal to one. Identical node and edge substitutions have zero cost, while substitutions involving different labels have infinity cost. The insertion or deletion of an edge incident to a node that is inserted or deleted, respectively, has no cost. Intuitively speaking, it is assumed that the cost of a node deletion (insertion) includes the cost of deleting (inserting) the incident edges. As for any two graphs $g_1 = (V_1, E_1, \alpha_1, \beta_1)$ and $g_2 = (V_2, E_2, \alpha_2, \beta_2)$ there is always an edit path f with cost $c(f) = |V_1| + |V_2|$ (corresponding to the case where all nodes together with their incident edges are deleted from g_1, and all nodes with their incident edges are inserted in g_2), any edit operation with infinity cost will never need to be considered when looking for an optimal edit path. Thus, we may think of edit operations with infinity cost as nonadmissible. In other words, under the given cost function we can restrict our attention on edit paths involving only insertions, deletions, and identical node and edge substitutions, but no nonidentical substitutions. For example, for the edit path f_3 discussed in Example 1, we have $c(f_3) = 3$ under the considered cost function. Obviously both f_1 and f_2 have infinity cost.

It was shown in [11] that under this cost function the following equation holds true for any two graphs g_1 and g_2, and a maximum common subgraph g of g_1 and g_2 (this maximum common subgraph may be empty):

$$d(g_1, g_2) = |g_1| + |g_2| - 2|g| \tag{5.3}$$

Obviously, this equation establishes a relation between the size $|g|$ of the maximum common subgraph of two graphs g_1 and g_2, and their edit distance $d(g_1, g_2)$. Thus, given one of the two quantities and the size of g_1 and g_2, we can immediately calculate the other. It was furthermore shown in [11] that the mapping $f : \hat{V}_1 \rightarrow \hat{V}_2$, defining an optimal edit path according to Def. 7, represents a maximum common subgraph of g_1 and g_2, i.e., f is a graph isomorphism between $\hat{g_1}$, the graph induced by $\hat{V}_1$, and $\hat{g_2}$, the graph induced by $\hat{V}_2$, and there are no larger subgraphs in g_1 and g_2, respectively, that are isomorphic to each other.

This theoretical result has an interesting practical consequence, namely, any algorithm for graph edit distance computation can be applied for maximum common subgraph computation if it is run under the cost function given in (5.2). Conversely, any algorithm that computes the maximum common subgraph of two graphs can be used for graph edit distance computation under cost function (5.2), using (5.3). A similar relation between string edit distance and longest common subsequence has been known for long [60].

The results derived in [11] were recently shown to hold not only for the cost function given in (5.2), but for a whole class consisting of infinitely many cost functions. In [13] cost functions C with $c_{ns} = c_{es} = 0$ for identical substitutions and

$$c_{nd} + c_{ni} < c_{ns} \text{ and } c_{nd} + c_{ni} < c_{es} \tag{5.4}$$

are considered. (Note that (5.2) is a special case of this class.) It is shown that for this whole class of cost functions the minimum cost mapping $f : \hat{V}_1 \rightarrow \hat{V}_2$ represents a maximum common subgraph of g_1 and g_2 and, conversely, any maximum common subgraph represents a minimum cost mapping in the sense of Def. 7. Intuitively speaking, the conditions in (5.4) imply that a node deletion together with a node insertion will be always preferred over a node or an edge substitution because of a smaller cost. This means that all nodes and edges in g_1 that cannot be mapped to a node or an edge with an identical label in g_2 will be deleted from g_1. Similarly, all nodes and edges in g_2 that are not part of the mapping f (i.e., that do not have a corresponding node or edge with identical label, respectively) will be inserted. What remains for the mapping f is exactly the maximum common subgraph of g_1 and g_2. An example is the edit path f_3 in Example 1. It is optimal under the cost function $C'' = (1, 1, 7, 1, 1, 7)$ as explained in Example 2. As a matter of fact, f_3 corresponds to the maximum common subgraph of g_1 and g_2 in Fig. 1, and cost function C'' satisfies conditions (5.4).

The equivalence of maximum common subgraph and graph edit distance computation shown in [13] is based on the assumption $c_{ei}(e) = c_{ed}(e) = 0$ for any edge e from $(V_1 \times V_1) - (\hat{V}_1 \times \hat{V}_1)$ and $(V_2 \times V_2) - (\hat{V}_2 \times \hat{V}_2)$, respectively, see (5.2). Thus, no individual costs for the deletion of edges from $(V_1 \times V_1) - (\hat{V}_1 \times \hat{V}_1)$ and no individual costs for the insertion of edges in $(V_2 \times V_2) - (\hat{V}_2 \times \hat{V}_2)$ are taken into regard. The reason is that these operations are automatically implied by the deletion of nodes from $(V_1 - \hat{V}_1)$ and the insertion of nodes in $(V_2 - \hat{V}_2)$, respectively. Thus, it is assumed that their costs are included in the costs of the corresponding node deletions and insertions. In other words, the cost of a node deletion (insertion) includes not only the cost of deleting (inserting) a node, but also the deletion (insertion) of the edges that connect it to the other nodes of the graph. This assumption may be justified in many applications.

The equivalence of graph edit distance and maximum common subgraph shown in [13] yields additional insight on the measure $d_{MCS}(g_1, g_2)$ of (5.1). Although no *explicit* costs of graph edit operations are needed to compute $d_{MCS}(g_1, g_2)$, there are, nevertheless, costs involved in an *implicit* manner, because the quantity $|mcs(g_1, g_2)|$ in (5.1) is equivalent to the graph edit distance $d(g_1, g_2)$ in the sense of (5.3), assuming a cost function satisfying (5.4). In other words, whenever we compute the maximum common subgraph of two graphs we may consider this as a graph edit distance computation under an arbitrary cost function belonging to the class studied in [13]. From this point of view, the measure defined in (5.1) may be regarded an advantage over conventional graph edit distance computation because it is robust against changing the costs of the underlying graph edit operations over a fairly wide range.

Another important result shown in [13] is the existence of classes of cost functions that always result in the same optimal mapping $f : \hat{V}_1 \rightarrow \hat{V}_2$ for any two given graphs g_1 and g_2. Intuitively speaking, if we consider two cost functions C and C', where C' is a scaled version of C, i.e., $c'_{nd} = \alpha c_{nd}, c'_{ni} = \alpha c_{ni}, \ldots, c'_{ei} = \alpha c_{ei}$ for some $\alpha > 0$, then we expect that any edit path f that is optimal under C is also optimal under C' for any two given graphs g_1 and g_2. Just the absolute cost of the two optimal edit paths would differ by a factor α. In [13] it was shown that any optimal edit path under a cost function C is optimal under another cost function C' not only if C' is a scaled version of C, but for a much larger class of cost functions C'. If the conditions

$$(c_{ni} + c_{nd})/c_{ns} = (c'_{ni} + c'_{nd})/c'_{ns} \qquad \text{and} \tag{5.5}$$

$$c_{es}/c_{ns} = c'_{es}/c'_{ns} \tag{5.6}$$

for cost functions C and C' are satisfied then any edit path f is optimal under C if and only if it is optimal under C' for any two given graphs g_1 and g_2. Furthermore, there is a relation between the values $c(f)$ obtained under two different cost functions that is similar to (5.3). Given the edit distance under cost function C we can analytically compute the edit distance under C' using just the parameters of C and C' and the size of the two graphs under consideration. Hence, given an algorithm that was designed for a particular cost function C, we can use the same algorithm for any other cost function C' for which (5.5) and (5.6) are satisfied. The existence of similar classes of cost functions for string edit distance has been discovered recently Rice et al. [57].

As discussed above, maximum common subgraph computation is a special case of graph edit distance under a particular class of cost functions. It was furthermore shown in [13] that also graph isomorphism and subgraph isomorphism are special cases of edit paths. If we define $c_{nd} = c_{ni} = c_{ns} = c_{ed} = c_{ei} = c_{es} = \infty$ then an edit path f between g_1 and g_2 with $c(f) < \infty$ exists if and only if there exists a graph isomorphism between g_1 and g_2. Clearly, any such graph isomorphism f is optimal and $c(f) = 0$. Similarly, if

$$\begin{aligned}
&c_{nd} = c_{ns} = \infty,\\
&0 \leq c_{ni} < \infty,\\
&c_{es}(e) = c_{ei}(e) = c_{ed}(e) = \infty \text{ if } e \in \hat{V}_1 \times \hat{V}_1,\\
&c_{ed}(e) = 0 \text{ if } e \in (V_1 \times V_1) - (\hat{V}_1 \times \hat{V}_1),\\
&c_{ei}(e) = 0 \text{ if } e \in (V_2 \times V_2) - (\hat{V}_2 \times \hat{V}_2),
\end{aligned}$$

then an optimal edit path f with $c(f) < \infty$ between g_1 and g_2 exists if and only if there exists a subgraph isomorphism from g_1 to g_2. Any optimal edit path f is in fact a subgraph isomorphism and $c(f) = (|g_2| - |g_1|)c_{ni}$.

5.6 Some Recent Developments

In this section we discuss some of the recent developments, in particular, automatic learning of graph edit distance, median graph, and weighted mean of two graphs.

5.6.1 Learning Edit Costs

One of the major difficulties in the application of edit distance in graph matching is the definition of adequate edit costs. The edit costs essentially govern how the structural matching is performed. For some graph representations, it may be crucial whether a node is missing or not, while for other representations, the connecting edges are more important than the nodes. The question of how to define edit costs can therefore only be addressed in the context of an application-specific graph representation.

In the case of labels from n-dimensional space of real numbers, a simple edit cost model that has been used often is based on the distance of labels. The idea is to assign edit costs to substitutions that are proportional to the Euclidean distance of the two labels. Substituting an edge by another edge with the same label therefore does not involve any costs. For nonidentical labels, the further the two labels differ from each other, the higher will be the corresponding substitution cost. Insertions and deletions are often assigned constant costs in this model. The advantage of this simple model is that only a few parameters are involved and the edit costs are defined in a very intuitive way. However, it turns out that for some applications this model is not sufficiently flexible. For instance, it does not take into account that some label components may be more relevant than others. Also, the absolute values of the labels are not evaluated, but only the distance of labels, which means that all regions of the label space are equally weighted in terms of edit costs.

Recently, automatic approaches [50, 52] have been proposed to learn the edit costs. In [50] an approach based on self-organizing maps (SOM) is proposed. SOMs [43] are two-layer neural networks consisting of an input layer and a competitive layer. The role of the input layer is to forward input patterns to the competitive layer. The competitive neurons are arranged in a homogeneous grid structure such that every neuron is connected to its neighbors. The idea is that the competitive layer reflects the space of input elements, that is, every competitive neuron corresponds to an element of the input space. For pattern representations in terms of feature vectors, this is usually accomplished by assigning weight vectors of the same dimension as the input space to neurons. Upon feeding an input pattern into the network, neurons of the competitive layer compete for the position of the input element. To this end, the weight of the closest competitive neurons and their neighbors are adapted so as to shift them towards the position of the input element. The longer this procedure is carried out, the more the competitive neurons tend to migrate to areas where many input elements are present. That is, the competitive layer can be regarded as a model of the pattern space, where the neuronal density reflects the pattern density. This unsupervised learning procedure is called self-organization as it does not rely on predefined classes of input elements.

SOMs can be used to model the distribution of edit operations. The idea is to use a SOM to represent the label space. A sample set of edit operations is derived from pairs of graphs from the same class in a manner that is equivalent to the probabilistic model. The self-organization process turns the initially regular grid of the competitive layer into a deformed grid. From the deformed grid, we obtain a distance measure for substitution costs and a density estimation for insertion and deletion costs. In the case of substitutions, the SOM is trained with pairs of labels, where one label belongs to the source node (or edge) and one to the target node (or edge). The competitive layer of the SOM is then adapted so as to draw the two regions corresponding to the source and the target label closer to each other. The edit cost of node and edge substitutions is then defined proportional to the distance in the trained SOM. That is, instead of measuring label distance by the Euclidean distance, we measure the deformed distance in the corresponding SOM. In the case of insertions and deletions, the SOM is adapted so as to draw neurons closer to the inserted or deleted label. The edit cost of insertions and deletions is then defined according to the competitive neural density at the respective position. That is, the more neurons at a certain position in the competitive layer are, the lower is the respective insertion or deletion cost. The self-organizing training procedure for substitutions, insertions, and deletions hence results in lower costs for those pairs of graphs that are in the training set and belong to the same class.

5.6.2 Median Graph

The concept of median graph does not give us an indication of graph similarity, but is useful in summarizing a group of graphs. This is among others needed in applications such as clustering, where we have to represent a group of graphs by some representative exemplar graph.

Given a set of graphs $S = \{g_1, g_2, \ldots, g_n\}$ defined over labels from L_V and L_E and a distance function $d()$ that measures the dissimilarity of two graphs, the concept of median graph is given as follows [39]:

Definition 10. *Let U be the set of all graphs that can be constructed using labels from L_V and L_E. The generalized median graph $\overline{g}$ and the set median graph $\hat{g}$ of $S \subset U$ are defined by*

$$\overline{g} = \arg\min_{g \in U} \sum_{i=1}^{n} d(g, g_i)$$

and

$$\hat{g} = \arg\min_{g \in S} \sum_{i=1}^{n} d(g, g_i)$$

respectively.

Both the generalized median and the set median graph minimize the sum of distances to all input graphs and the only difference lies in the graph space where the median is searched for. The generalized median is the more general concept and therefore usually a better representation of the given patterns than the set median. Notice that $\overline{g}$ is usually not a member of S. In general several generalized median graphs and several set median graphs may exist. However, this is usually not a drawback in practice since any such graph may serve equally well as a representative of the given set.

Conceptually, searching for the set median graph of n input graphs is an easy task since it suffices to compute $\frac{1}{2}n(n-1)$ pairwise graph distances. Due to the high expense of graph distance computation, however, it is often desired to determine the set median graph more efficiently than under the naive approach. Some suggested methods for this purpose can be found in [41].

Determining the generalized median graph is computationally more complex. On the one hand, the computation time is clearly exponential in the size of the input graphs. On the other hand, it is also exponential in terms of the number of input graphs. The reason for this behavior is that already for the special case of strings, the required time is exponential in the number of input strings [34]. As a consequence, we are generally forced to resort to approximate solutions that can be found in reasonable time. In [39] a genetic solution is proposed for this purpose. A comparison of the genetic algorithm against combinatorial search is described in [14].

When approximative approaches are involved, the question of accuracy of the approximative generalized median graph $\tilde{g}$ arises. In [40] a lower bound is proposed to answer this question. An approximate computation method gives us a solution $\tilde{g}$ such that

$$\mathrm{SOD}(\tilde{g}) = \sum_{p \in S} d(\tilde{g}, p) \geq \sum_{p \in S} d(\overline{g}, p) = \mathrm{SOD}(\overline{g})$$

where SOD stands for sum of distances and $\overline{g}$ represents the (unknown) true generalized median graph. The quality of $\tilde{g}$ can be measured by the difference $\mathrm{SOD}(\tilde{g}) - \mathrm{SOD}(\overline{g})$. Since $\overline{g}$ and $\mathrm{SOD}(\overline{g})$ are unknown in general, we resort to a lower bound $\Gamma \leq \mathrm{SOD}(\overline{g})$ and measure the quality of $\tilde{g}$ by $\mathrm{SOD}(\tilde{g}) - \Gamma$. Note that the relationship

$$0 \leq \Gamma \leq SOD(\overline{g}) \leq SOD(\tilde{g})$$

holds. Obviously, $\Gamma = 0$ is a trivial, and also useless, lower bound. We thus require Γ to be as close to $\mathrm{SOD}(\overline{g})$ as possible.

The distance function $d(p, q)$ is assumed to be a metric. The generalized median graph $\overline{g}$ is characterized by

$$\begin{aligned}
&\text{minimize } \mathrm{SOD}(\overline{g}) = d(\overline{g}, g_1) + d(\overline{g}, g_2) + \cdots + d(\overline{g}, g_n) \text{ subject to} \\
&\forall i, j \in \{1, 2, \ldots, n\}, i \neq j, \begin{cases} d(\overline{g}, g_i) + d(\overline{g}, g_j) \geq d(g_i, g_j) \\ d(\overline{g}, g_i) + d(g_i, g_j) \geq d(\overline{g}, g_j) \\ d(\overline{g}, g_j) + d(g_i, g_j) \geq d(\overline{g}, g_i) \end{cases} \\
&\forall i \in \{1, 2, \ldots, n\}, \; d(\overline{g}, g_i) \geq 0
\end{aligned}$$

Note that the constraints except the last set of inequalities are derived from the triangular inequality of the metric $d(p, q)$. By defining n variables x_i, $i = 1, 2, \ldots, n$, we replace $d(\overline{g}, g_i)$ by x_i and obtain the linear program LP:

$$\begin{aligned}
&\text{minimize } x_1 + x_2 + \cdots + x_n \text{ subject to} \\
&\forall i, j \in \{1, 2, \ldots, n\}, i \neq j, \begin{cases} x_i + x_j \geq d(g_i, g_j) \\ x_i + d(g_i, g_j) \geq x_j \\ x_j + d(g_i, g_j) \geq x_i \end{cases} \\
&\forall i \in \{1, 2, \ldots, n\}, \; x_i \geq 0
\end{aligned}$$

If we denote the solution of LP by Γ, then the true generalized median $\overline{g}$ satisfies $\Gamma \leq \mathrm{SOD}(\overline{g})$, i.e., Γ is a lower bound for $\mathrm{SOD}(\overline{g})$.

5.6.3 Weighted Mean of Two Graphs

If we consider two points x and y in the k-dimensional real space $\Re^k$, their weighted mean can be defined as a point z such that

$$z = (1 - \gamma) \cdot x + \gamma \cdot y, \quad 0 \leq \gamma \leq 1$$

Clearly, if $\gamma = \frac{1}{2}$ then z is the (normal) mean of x and y. If z is as defined above, then $z - x = \gamma \cdot (y - x)$ and $y - z = (1 - \gamma) \cdot (y - x)$. In other words, z is a point on the line segment in n dimensions that connects x and y, and the distance between z and both x and y is controlled via the parameter γ.

Conceptually, the same idea can be easily transformed into other domains, such as strings [17] and graphs [15]. According to Bunke and Günter [15] the weighted mean of two graphs g_1 and g_2 is a graph g such that

Fig. 5.2. Example of a series of weighted means of two graphs

$$d(g, g_1) = \gamma \cdot d(g_1, g_2)$$

and

$$d(g, g_2) = (1 - \gamma) \cdot d(g_1, g_2)$$

where $0 < \gamma < 1$.

An algorithm for finding the weighted mean of two graphs is given in [15]. It is an extension of the method for weighted mean of strings [17] and involves finding a subset of edit operations (given the lowest edit cost between two graphs) for the given γ in order to determine the weighted mean graph. Figure 5.2 shows an example of a series of weighted means of two graphs. In some sense the concept of weighted mean graph gives us a tool of "morphing" one graph into another graph.

If $\gamma = 0.5$, then we obtain the mean of two graphs g_1 and g_2 [33] and

$$d(g_1, g) = d(g, g_2), \quad d(g_1, g_2) = d(g, g_1) + d(g, g_2)$$

holds. In other words, the mean graph g is equidistant from both g_1 and g_2. Clearly, the mean will depend on the distance function chosen. There may be more than one graph satisfying these conditions; it is also possible that no (exact) mean graph exists for a given pair of graphs.

5.7 Conclusions

Graph matching has successfully been applied to various problems in image processing and understanding. In the case of exact graph matching, the graph extraction process is assumed to be structurally flawless, i.e., the conversion of image data of a single class into graphs always results in identical structures or substructures. Otherwise, graph isomorphism or subgraph isomorphism detection are rather unsuitable, which seriously restricts the applicability of graph isomorphism algorithms. The main advantages of isomorphism algorithms are their mathematically stringent formulation and the existence of well-known procedures to derive optimal solutions.

Error-tolerant methods, sometimes also referred to as inexact or error-correcting methods, are characterized by their ability to cope with errors, or noncorresponding parts, in structure and labels of graphs. Hence, in order for two graphs to be positively matched, they need not be identical at all, but only similar. The notion of graph similarity depends on the error-tolerant matching method that is to be applied.

In this chapter we have given an overview of both exact and inexact graph matching. The emphasis has been the fundamental concepts and the recent developments concerned with the automatic learning of edit costs, median graph, and weighted mean of two graphs. Particularly, the concept of inexact graph matching provides us a means of measuring the similarity of graphs and thus lays the foundation of using the versatile and flexible tool of graphs in case-based reasoning in dealing with images.

References

1. Aho AV *et al.* (1974) The design and analysis of computer algorithms. Reading: Addison-Wesley
2. Babai L *et al.* (1982) Isomorphism of graphs with bounded eigenvalue multiplicity. Proc. of 14th ACM Symposium on Theory of Computing, pp. 310–324
3. Babel L (1995) Isomorphism of chordal (6, 3) graphs. Computing 54:303–316
4. Babel L *et al.* (1996) The isomorphism problem for directed path graphs and for rooted directed path graphs. Journal of Algorithms 21:542–564
5. Balakrishnau VK (1997) Theory and Problems of Graph Theory. McGraw-Hill
6. Bischof WF, Caelli T (2001) Learning spatio-temporal relational structures. Applied Artificial Intelligence 15:707–722
7. Bodlaender HL (1990) Polynomial algorithms for graph isomorphism and chromatic index on partial k-tress. Journal of Algorithms 11:631–643
8. Boeres MC, Ribeiro CC, Bloch I (2004) A randomized heuristic for scene recognition by graph matching. In Ribeiro CC, Martins SL (Eds.), Proc. of 3rd Workshop on Efficient and Experimental Algorithms, LNCS 3059, pp. 100–113, Springer
9. Booth KS, Lueker GS (1979) A linear-time algorithm for deciding interval graph isomorphism. JACM 26:183–195
10. Bunke H, Allermann G (1983) Inexact graph matching for structural pattern recognition. Pattern Recognition Letters 1:245–253
11. Bunke H (1997) On a relation between graph edit distance and maximum common subgraph. Pattern Recognition Letters 18:689–694
12. Bunke H, Shearer K (1998) A graph distance metric based on the maximal common subgraph. Pattern Recognition Letters 19:255–259
13. Bunke H (1999) Error correcting graph matching: On the influence of the underlying cost function. IEEE Trans. on Pattern Analysis and Machine Intelligence 21:917–922
14. Bunke H, Münger A, Jiang X (1999) Combinatorial search versus genetic algorithms: a case study based on the generalized mean graph problem. Pattern Recognition Letters 20:1271–1277
15. Bunke H, Günter S (2001) Weighted mean of a pair of graphs. Computing 67:209–224

16. Bunke H, Foggia P, Guidobaldi C, Sansone C, Vento M (2002) A comparison of algorithms for maximum common subgraph on randomly conntected graphs. In Caelli T, Amin A, Duin R, Kamel M, de Ridder D (Eds.), Structural, Syntactic, and Statistical Pattern Recognition, pp. 123–132, Springer
17. Bunke H, Jiang X, Kandel A (2002) On the weighted mean of a pair of strings. Pattern Analysis and Applications 5:23–30
18. Bunke H, Caelli T, Eds. (2004) Special Issue on Graph Matching in Pattern Recognition and Machine Vision, in International Journal of Pattern Recognition and Artificial Intelligence 18(3)
19. Burke EK, MacCarthy B, Petrovic S, Qu R (2001) Case-based reasoning in course timetabling: An attributed graph approach. In Aha DW, Watson I (Eds.), Case-Based Reasoning Research and Development, LNAI 2080, pp. 90–103, Springer
20. Champin PA, Solnon C (2003) Measuring the similarity of labeled graphs. In Ashley KD, Bridge DG (Eds.), Case-Based Reasoning Research and Development, Proc. of 5th Int. Conf. on Case-Based Reasoning, LNCS 2689, pp. 80–95, Springer
21. Chartrand G, Kubicki G, Schultz M (1998) Graph similarity and distance in graphs. Aequationes Mathematicae 55:129–145
22. Colbourn GJ (1981) On testing isomorphism of permutation graphs. Networks 11:13–21
23. Conte D, Foggia P, Sansone C, Vento M (2004) Thirty years of graph matching in pattern recognition. International Journal of Pattern Recognition and Artificial Intelligence 18:265–298
24. Cook DJ, Holder LB, Eds. (2007) Mining Graph Data. Wiley Interscience
25. Cross ADJ, Wilson RC, Hancock ER (1997) Inexact graph matching using genetic search. Pattern Recognition 30:953–970
26. Dickinson SJ, Pelillo M, Zabih R, Eds. (2001) Special Section on Graph Algorithms and Computer Vision, in IEEE Trans. on Pattern Analysis and Machine Intelligence 23(10)
27. Fernandez ML, Valiente G (2001) A graph distance metric combining maximum common subgraph and minimum common supergraph. Pattern Recognition Letters 22:753–758
28. Finch AM, Wilson RC, Hancock ER (1998) An energy function and continuous edit process for graph matching. Neural Computation 10:1873–1894
29. Foggia P, Sansone C, Vento M, Eds. (2003) Special Issue on Graph-based Representations in Pattern Recognition, in Pattern Recognition Letters 24(8)
30. Gati G (1979) Further annotated bibliography on the isomorphism disease. Journal of Graph Theory 3:95–109
31. Gebhardt F, Voss A, Gräther W, Schmidt-Belz B (1997) Reasoning with Complex Cases. Kluwer
32. Gold S, Rangarajan A (1996) A graduated assignment algorithm for graph matching. IEEE Trans. on Pattern Analysis and Machine Intelligence 18: 377–388
33. Günter S, Bunke H (2002) Self-organizing map for clustering in the graph domain. Pattern Recognition Letters 23:401–417
34. de la Higuera C, Casacuberta F (2000) Topology of strings: Median string is NP-complete. Theoretical Computer Science 230:39–48
35. Hopcroft JE, Tarjan RE (1973) A $VlogV$ algorithm for isomorphism of triconnected planar graphs. Journal of Computer and System Sciences 7:323–331

36. Hopcroft JE, Wong JK (1974) Linear time algorithm for isomorphism of planar graphs. Proc. of 6th Annual ACM Symposium on Theory of Computing, pp. 172–184
37. Jiang X, Bunke H (1998) On the coding of ordered graphs. Computing 61:23-38
38. Jiang X, Bunke H (1999) Optimal quadratic-time isomorphism of ordered graphs. Pattern Recognition, 32:1273-1283
39. Jiang X, Münger A, Bunke H (2001) On median graphs: properties, algorithms, and applications. IEEE Trans. on Pattern Analysis and Machine Intelligence 23:1144–1151
40. Jiang X, Bunke H (2002) Optimal lower bound for generalized median problems in metric space. In Caelli T, Amin A, Duin E, Kamel M, de Ridder D (Eds.), Structural, Syntactic, and Statistical Pattern Recognition, pp. 143–151, Springer
41. Juan A, Vidal E (1998) Fast median search in metric spaces. In Amin A, Dori D (Eds.), Advances in Pattern Recognition, pp. 905–912, Springer
42. Justice D, Hero A (2006) A binary linear programming formulation of the graph edit distance. IEEE Trans. on Pattern Analysis and Machine Intelligence 28:1200–1214
43. Kohonen T (1995) Self-Organizing Maps. Springer
44. Levi, G (1972) A note on the derivation of maximal common subgraphs of two directed or undirected graphs. Calcolo 9:341–354
45. Luks EM (1982) Isomorphism of graphs of bounded valence can be tested in polynomial time. Journal of Computer and System Science 25:42–65
46. McGregor JJ (1982) Backtrack search algorithms and the maximal common subgraph problem. Software Practice and Experience 12:23–34
47. Medasani S, Krishnapuram R, Choi YS (2001) Graph matching by relaxation of fuzzy assignments. IEEE Trans. on Fuzzy Systems 9:173–182
48. Miller GL (1980) Isomorphism testing for graphs with bounded genus. Proc. of 12th ACM Symposium on Theory of Computing, pp. 225–235
49. Myers R, Wilson RC, Hancock ER (2000) Bayesian graph edit distance. IEEE Trans. on Pattern Analysis and Machine Intelligence 22:628–635
50. Neuhaus M, Bunke H (2005) Self-organizing maps for learning the edit costs in graph matching. IEEE Trans. on Systems, Man, and Cybernetics - Part B: Cybernetics 35:503–514
51. Neuhaus M, Riesen K, Bunke H (2006) Fast suboptimal algorithms for the computation of graph edit distance. Proc. Joint IAPR Workshops on Structural and Syntactic Pattern Recognition and Statistical Techniques in Pattern Recognition, Hong Kong, 163–172
52. Neuhaus M, Bunke H (2007) Automatic learning of cost functions for graph edit distance. Information Sciences 177:239–247
53. Perner P, Holt A, Richter M (2005) Image processing in case-based reasoning. The Knowledge Engineering Review 20:311–314
54. Perner P (2006) Case-base maintenance by conceptual clustering of graphs. Engineering Applications of Artificial Intelligence 19:381–393
55. Read RC, Corneil DG (1977) The graph isomorphism disease. Journal of Graph Theory 1:339–363
56. Riesen K, Neuhaus M, Bunke H (2007) Bipartite graph matching for computing the edit distance of graphs. Proc. of 6th Int. Workshop on Graph-based Representations, Alicante, Spain
57. Rice S, Bunke H, Nartker T (1997) Classes of cost functions for string matching. Algorithmica 18:271–280

58. Schöning U (1988) Graph isomorphism is in the low hierarchy. Journal of Computer and System Sciences 37:312–323
59. Sorlin S, Solnon S (2005) Reactive tabu search for measuring graph similarity. In Brun L, Vento M (Eds,), Proc. of 5th Int. Workshop on Graph-based Representations in Pattern Recognition, LNCS 3434, pp. 172–182. Springer
60. Stephen GA (1994) String Searching Algorithms. World Scientific, Publ. Co.
61. Ullman, JR (1976) An algorithm for subgraph isomorphism. Journal of the Association for Computing Machinery 23:31–42
62. Wagner RA, Fischer MJ (1974) The string-to-string correction problem. JACM 21:168–173
63. Wallis WD, Shoubridge P, Kraetzl M, Ray D (2001) Graph distances using graph union. Pattern Recognition Letters 22:701–704
64. Weinberg L (1966) A simple and efficient algorithm for determining isomorphism of planar triply connected graphs. IEEE Trans. on Circuit Theory 13:142–148.
65. Williams ML, Wilson RC, Hancock ER (1997) Multiple graph matching with Bayesian inference. Pattern Recognition Letters 18:1275–1281
66. Wilson RC, Hancock ER (1997) Structural matching by discrete relaxation. IEEE Trans. on Pattern Analysis and Machine Intelligence 19:634–647
67. Wilson RC, Hancock ER (1999) Graph matching with hierarchical discrete relaxation. Pattern Recognition Letters 20:1041–1052

6

Memory Structures and Organization in Case-Based Reasoning

I. Bichindaritz

Institute of Technology, University of Washington 1900 Commerce street, 98402 Tacoma USA

Summary. Case-based reasoning (CBR) methodology stems from research on building computational memories capable of analogical reasoning, and require for that purpose specific composition and organization. This main task in CBR has triggered very significant research work and findings, which are summarized and analyzed in this article. In particular, since memory structures and organization rely on declarative knowledge and knowledge representation paradigms, a strong link is set forth in this article between CBR and data mining for the purpose of mining for memory structures and organization. Indeed the richness of data mining methods and algorithms applied to CBR memory building, as presented in this chapter, mirrors the importance of learning memory components and organization mechanisms such as indexing. The article proceeds through an analysis of this link between data mining and CBR, then through an historical perspective referring to the theory of the dynamic memory, and finally develops the two main types of learning related to CBR memories, namely mining for memory structures and mining for memory organization.

6.1 Introduction

Case-based reasoning (CBR) systems have tight connections with machine learning and data mining. They have been tagged by machine learning researchers as *lazy* learners because they defer the decision of how to generalize beyond the training set until a target new case is encountered [37], by opposition to most other learners, tagged as *eager*. Even though a large part of the inductive inferences are definitely performed at *Retrieval* time in CBR [3], mostly through sophisticated similarity evaluation, most CBR systems also perform inductive inferences at *Retain* time. There is a long tradition within this research community to study what is a memory and what its components and organization should be. Indeed CBR methodology focuses more on the memory part of its intelligent systems [51] than any other artificial intelligence (AI) methodology, and this often entails learning declarative memory structures and organization. Therefore this article proposes to focus on studying

I. Bichindaritz: *Memory Structures and Organization in Case-Based Reasoning*, Studies in Computational Intelligence (SCI) **73**, 175–194 (2008)
www.springerlink.com

CBR from the *memory* standpoint more than the inference standpoint, which opens a different, and complementary, perspective on the lazy/eager learning comparison. For CBR, this antagonism can be more adequately called the lazy/eager retrieval dilemma or pondering when it is more appropriate to learn: at *Retrieval* time or at *Retain* time [1]. The lazy learners are also the eager retrievers, and the eager learners are the lazy retrievers. Both though are full-fledged CBR systems. More precisely, the approach taken here has an analogy in the medical domain. Instead of studying the patient – CBR is here the object of our study – from the physiological standpoint, we will adopt the anatomical perspective, and attempt to answer this question: what structures and organization constitute the anatomy of a CBR system memory? We will see that a variety of data mining tasks and methods are performed in CBR, and that this richness reinforces the well known fact that CBR systems are indeed powerful data mining systems. The second section explains the relationship between CBR and data mining, and the motivation behind mining in CBR. The third section revisits data mining in early CBR systems. The fourth section concentrates on mining for memory structures, and the fifth on mining for memory organization. It is followed by a discussion on CBR and data mining, and by the conclusion.

6.2 Data Mining and Case-based Reasoning

Data mining is the analysis of observational data sets to find unsuspected relationships and to summarize the data in novel ways that are both understandable and useful to the data owner [22]. CBR systems are generally classified as data mining systems simply because they answer this definition. From a set of data – called cases in CBR – they perform one of the classical data mining tasks such as prediction for instance, which gives the case base a competency beyond what the data provide. In this chapter, we will focus more on another aspect, namely what data mining tasks and methods are used in CBR and what is their result in the CBR memory.

First of all, since data mining emerged in the nineties from scaling up machine learning algorithms to large datasets, let us review what machine learning authors have been saying about CBR. Machine learning authors consider case-based reasoning systems as either analogical reasoning systems [11, 15, 33, 55] or instance based learners [37]. Michalski presents the analogical inference, at the basis of case-based retrieval, as a dynamic induction performed during the matching process [33]. Mitchell refers to CBR as a kind of instance based learner [37]. This author labels these systems as *lazy* learners because they defer the decision about how to generalize beyond the training data until each new query instance is encountered. He also praises CBR systems for not committing to a global approximation once and for all during the training phase of machine learning, but for being able to generalize specifically for each target case, therefore, to fit its approximation bias, or induction bias, to the case at hand. He points here to the drawback of overgeneralization that is well known for eager learners, to which instance-based learners are exempt [37].

These authors focus their analysis on the inferential aspects of learning in case-based reasoning [2, 17, 25]. Historically CBR systems have evolved from the early work of Schank in the theory of the dynamic memory [51], where this author proposes to design intelligent systems primarily by modeling their memory. Ever since Schank's precursory work on natural language understanding, one of the main goals of case-based reasoning has been to unify as much as possible memory and inferences for the performance of intelligent tasks. Therefore, focusing on studying how case-based reasoning systems learn, or mine, their memory structures and organization can prove at least as fruitful as studying and classifying them from an inference standpoint.

From a memory standpoint, learning in CBR consists in the creation, update, and organization of the structures and organization in memory. It is often referred to as case base maintenance [54, 58]. In the general cycle of CBR, learning takes place within the general reasoning cycle – see Aamodt and Plaza [1] for this classical cycle. It completely serves the reasoning, and therefore one of its characteristics is that it is an *incremental* type of mining. It is possible to fix it after a certain point, though, in certain types of applications, but it is not a tradition in CBR: *learning is an emergent behavior from normal functioning* [26]. Ideally, CBR systems start reasoning from an empty memory, and their reasoning capabilities stem from their progressive learning from the cases they process. The decision to stop learning because the system is judged competent enough is not taken from definitive criteria. It is the consequence of individual decisions made about each case, to keep it or not in memory depending upon its potential contribution to the system. Thus often the decisions about each case, each structure in memory, allow the system to evolve progressively toward states as different as ongoing learning, in novice mode, and its termination, in expert mode. If reasoning, and thus learning, are directed from the memory, learning answers to a process of prediction of the conditions of cases recall (or retrieval). As the theory of the dynamic memory showed, recall and learning are closely linked [51]. Learning in case-based reasoning answers a disposition of the system to anticipate future situations: *the memory is directed toward the future.* The anticipation deals both with avoiding situations having caused a problem, and with reinforcing the performance in success situations.

More precisely, learning in case-based reasoning takes the following forms:

1. *Adding a case to the memory*: it is at the heart of CBR systems, traditionally one of the main phases in the reasoning cycle, and the last one: *Retain* [1]. It is the most primitive learning kind, also called learning by consolidation or rote learning
2. *Explaining*: the ability of a system to find explanations for its successes and failures, and by generalization the ability to anticipate

3. *Choosing the indices*: it consists in anticipating *Retrieval*, the first reasoning step
4. *Learning memory structures*: these may be learnt by generalization from the cases or be provided from the start to hold the indices, for example. These learnt memory structures can play additional roles, such as facilitating the reuse or the retrieval
5. *Organizing the memory*: the memory comprises a network of cases, given memory structures, and learnt memory structures, organized in efficient ways. Flat and hierarchical memories have been traditionally described
6. *Refining cases*: cases may be updated, refined based upon the CBR result
7. *Refining knowledge*: the knowledge at the basis of the case-based reasoning can be refined, such as modifying the similarity measure (weight learning) or situation assessment refinement

6.3 Data Mining in Early CBR Systems

Roger Schank pioneered the methodology that would become CBR from a cognitive background [51]. Based on cognitive science research, he proposed a model of dynamic memory [51] capable of evolving the events it encounters and of learning both from successes and failures. The theory of the dynamic memory presents memory structures and organization that were later implemented in some of the first CBR systems. Their principles have been followed up to current CBR sytems.

6.3.1 The Theory of the Dynamic Memory

Memory structures are of two types, domain dependent and domain independent. The domain dependent structures are called *scripts*, defined as generalized standardized episodes. All the other structures are either organizational structures or generalized structures. The organizational structures are *generalized scenes, MOPs* (memorization organization packets), and *meta-MOPs*. Generalizations of these are, respectively, *universal scenes, u-MOPs*, and *universal u-MOPs* [51] (see Fig. 6.1). The domain independent structures are *TOPs* (thematic organization packets).

6.3.2 Generalization Based Memory

The memory of IPP (integrated partial parser) is a generalization-based memory [27,28]. IPP is a natural language understanding system working from textual information from news about international terrorism. Text understanding is presented in [28] as memory directed, but accomplishes a case-based type of search through the memory for specific events linked with the text in entry,

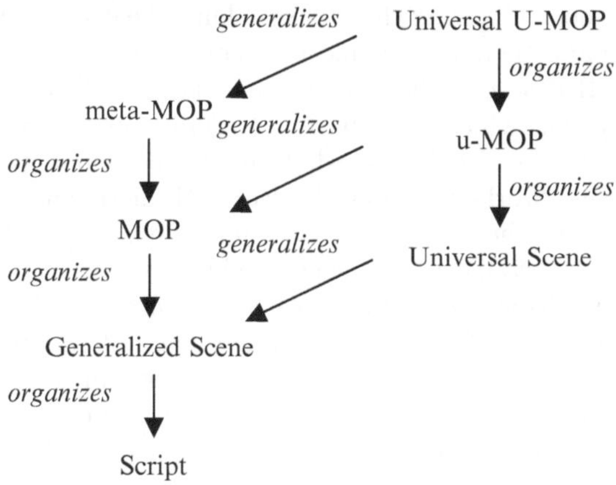

Fig. 6.1. The two dimensions of the dynamic memory and its domain dependent structures

giving it a meaning. IPP is a system implementing closely the theory of the dynamic memory.

The memory structures are on the one hand the initial structures, and on the other hand the structures learnt by the system from documents. The initial structures reproduce the stereotypical aspects of the situations: these are the S-MOPs (simple MOPs) and the AUs (action units). The S-MOPs or Simple MOPs describe abstract situations such as acts of extortion and attack. The AUs or Action Units represent concrete events such as shootings or hostage liberations. The AUs are components of the S-MOPs. The learnt structures are the spec-MOPs, containing the traits common to the structures indexed under them. These common traits are conjunctions of triplets <attribute1, attribute2, value>, such as, for example:

(TARGET NATION DOMINICAN-REPUBLIC)
(TARGET TYPE GOVERNMENT)
(LOCATION COUNTRY COLUMBIA)
(METHOD AU OCCUPATION)...

In this last example, the method used is represented by an AU: Occupation. At the highest level, the memory is a set of S-MOPs, under which are indexed some spec-MOPs forming a network. Each spec-MOP contains a discrimination network being a set of indexes to the events (AUs and role values associated with them) close to this spec-MOP. Moreover, these indexes take as values the differences between the events and the spec-MOP from which they depend.

The memory is organized as a generalization based memory, which means that the S-MOPs are the most general structures, and that the degree of generalization decreases with the deepness from the root of the structures traversed in memory following the indexes. Thus the events not having structures indexed under them are the most specific structures.

Learning in IPP follows the same mechanism as in GBM (generalization-based memory) [30], which is a concept learning system particularly interesting because it can also be linked, from its methodology, to case-based reasoning systems. Works by Lebowitz deal with hierarchical clustering as a way of implementing the theory of the dynamic memory [27]. GBM's memory is composed of GEN-NODES (for generalized nodes), also called concepts and instances. The GEN-NODES, similar to MOPs, are built by factoring common traits in the form of <attribute, value> pairs of the instances indexed underneath them. Indexing is similar to that performed in IPP. Attribute values are qualitative. Moreover a discrimination network, called D-NET, is associated to each GEN-NODE, like in IPP. The D-NET indexes the instances depending upon it by the traits differing from the norm carried by the GEN-NODE. Learning in such a system is particularly flexible. Through the bias of a counter carried by a node, each of these traits can be confirmed, by incrementation, or infirmed, by decrementation, during the search through the memory for a new instance presented to the system. When a minimum threshold is reached, a system parameter, the trait is removed from the concept, and the concept is removed when it does not comprise any trait anymore. Inversely, the system constantly searches for new concepts to create. The system memory is a dynamic memory containing both instances and concepts. It was used in two systems, UNIMEM [31] and RESEARCHER [29].

Some issues for the system have been the dependency of learnt concepts upon the order of presentation of the instances. To remedy it, Lebowitz proposed to postpone as much as possible the formation of new concepts [32]. When a new instance cannot be incorporated to a concept, because of a trait for which the counter is not high, the system prefers to wait before rejecting this concept that the concept evolution permits to definitively incorporate the instance or not.

6.4 Mining for Memory Structures

Memory structures in CBR are not only cases. A case is defined as a contextualized piece of knowledge representing an experience that teaches a lesson fundamental to achieving the goals of a reasoner [26]. For many systems, cases are represented as truthfully as possible from the application domain. Additionally, data mining methods have been applied to cases themselves, features, and generalized cases. These techniques can be applied concurrently to the same problem or selectively. If the trend is now to use them selectively, probably in the near future CBR systems will use these methods more and more concurrently.

6.4.1 Case Mining

Case mining refers to the process of mining potentially large data sets for cases [60]. Researchers have often noticed that cases simply do not exist in electronic format, that databases do not contain well-defined cases, and that the cases need to be created before CBR can be applied. Another option is to start CBR with an empty case base. When large databases are available, preprocessing these to learn cases for future CBR permits to capitalize on the experience dormant in these databases. Yang and Cheng propose to learn cases by linking several database tables [60]. *Clustering* and *support vector machines (SVM)* techniques permit to mine for cases in [60].

6.4.2 Feature Mining

Feature mining refers to the process of mining data sets for features. Many CBR systems select the features for their cases and/or generalize them. Wiratunga et al. notice that transforming textual documents into cases requires dimension reduction and/or feature selection [59], and show that this preprocessing improves the classification in terms of CBR accuracy and efficiency. These authors induce a kind of decision tree called boosted *decision stumps* because they comprise only one level, in order to select features, and *induce rules* to generalize the features. In biomedical domains, in particular when data vary continuously, the need to abstract features from streams of data is particularly prevalent. Recent, and notable, examples include Montani et al., who reduce their cases time series dimensions through *discrete Fourier transform* [39], approach adopted by other authors for time series [42]. Niloofar and Jurisica propose an original method for generalizing features. Here the generalization is an abstraction that reduces the number of features stored in a case [41]. Applied to the bioinformatics domain of micro arrays, the system uses both *clustering* techniques to group the cases into clusters containing similar cases, and *feature selection* techniques. The goal in their system is to abstract cases in a domain where there are many attributes, and few samples, where the pitfall is the famous "curse of dimensionality." The clustering method chosen is *spectral clustering* and the feature selection technique is *logistic regression*. Applying these methods to the case base improved the case-based reasoning along several dimensions, among which improved accuracy, less error, and less undecided cases (those for which there is a tie in the similarity score) [41].

6.4.3 Generalized Case Mining

Generalized case mining refers to the process of mining databases for generalized and/or abstract cases. Generalized cases are named in varied ways, such as prototypical cases, abstract cases, prototypes, stereotypes, templates, classes, categories, concepts, and scripts – to name the main ones [36]. Although all these terms refer to slightly different concepts, they represent structures that have been abstracted or generalized from real cases either by the CBR system or by an expert. When these prototypical cases are provided by a domain expert, this is a knowledge acquisition task [7, 8]. More frequently they are learnt from actual cases. In CBR, prototypical cases are often learnt to structure the memory. Therefore most of the prototypical cases presented here will also be listed in the section on structured memories.

Many authors mine for *prototypes*, and simply refer to *induction* for learning these. CHROMA [4] uses induction to learn prototypes corresponding to general cases, which each contain a pair <*situation*, *plan*>, where the situation is an object whose slots have several values possible – values are elements of a set. Bellazzi et al. organize their memory around prototypes [10]. The prototypes can either have been acquired from an expert or induced from a large case base. Schmidt and Gierl point that prototypes are an essential knowledge structure to fill the gap between general knowledge and cases in medical domains [53]. The main purpose of this prototype learning step is to guide the retrieval process and to decrease the amount of storage by erasing redundant cases. A generalization step becomes necessary to learn the knowledge contained in stored cases. They use several threshold parameters to adjust their prototypes, such as the number of cases the prototype is filled with, and the minimum frequency of each contraindication for the antibiotic therapy domain [53].

Others specifically refer to *generalization*, so that their prototypes correspond to generalized cases. An example of system inducing prototypes by generalization is a computer aided medical diagnosis system interpreting electromyography for neuropathy diagnosis [34]. The first prototypes are learnt from the expert by supervised learning, then the prototypes are automatically updated by the system by generalizing from cases. Prototypes can fusion if one is more general than the other ones, or new prototypes can be added to the memory. Malek proposes to use a *neural network* to learn the prototypes in memory for a classification task, such as diagnosis [35]. A similar connectionist approach is proposed by [50]. Portinale and Torasso [47] in ADAPTER organize their memory through E-MOPs [26] learnt by generalization from cases for diagnostic problem-solving. E-MOPs carry the common characteristics of the cases they index, in a discrimination network of features used as indices to retrieve cases. Mougouie and Bergmann [40] present a method for learning generalized cases. This method, called the Topkis-Veinott method, provides a solution to the computation of similarity for generalized cases over an n-dimensional Real values vector. Maximini et al. [36] have studied the different structures induced from cases in CBR systems. They point out that several different terms exist, such as generalized case, prototype, schema, script, and abstract case. The same terms do not always correspond to the same type of entity. They define three types of cases. A point case is what we refer to as a real case. The values of all its attributes are known. A generalized case is an arbitrary subspace of the attribute space.

There are two forms: the attribute independent generalized case, in which some attributes have been generalized (interval of values) or are unknown, and the attribute dependent generalized case, which cannot be defined from independent subsets of their attributes.

Yet other authors refer to *abstraction* for learning abstract cases. Branting proposes case abstractions for its memory of route maps [16]. The abstract cases, which also contain abstract solutions, provide an accurate index to less abstract cases and solutions. Perner [44] learns prototypes by abstracting cases as well.

Finally, many authors learn *concepts* through *conceptual clustering.* MNAOMIA [12–14] learns concepts and trends from cases through *conceptual clustering* similar to GBM [30] (see Fig. 6.2). Perner learns a hierarchy of classes by *hierarchical conceptual clustering*, where the concepts represent clusters of prototypes [44].

Dìaz-Agudo and Gonzàlez-Calero use *formal concept analysis* (FCA) – a mathematical method from data analysis – as another induction method for extracting knowledge from case bases, in the form of *concepts* [18]. The concepts learnt comprise some cases, and have both an intent– the set of attributes shared by these cases represented by a concept. Retrieval step is a classification in a concept hierarchy, as specified in the FCA methodology, which provides such algorithms. The concepts can be seen as an alternate form of indexing structure. The authors point to one notable advantage of this method, during adaptation. The FCA structure induces dependencies among the attributes that guide the adaptation process [19].

6.5 Mining for Memory Organization

Efficiency at case retrieval time is conditioned by a judicious memory organization. Two main classes of memory are presented here: unstructured – or flat – memories, and structured memories.

6.5.1 Flat Memories

Flat memories are memories in which all cases are organized at the same level. Retrieval in such memories processes all the cases in memory. Classical nearest neighbor (NN) retrieval is a method of choice for retrieval in flat memories. Flat memories can also contain prototypes, but in this case the prototypical cases do not serve as indexing structures for the cases. They can simply replace a cluster of similar cases that has been deleted from the case base during case base maintenance activity. They can also have been acquired from experts. Flat memories are the memories of predilection of NN retrieval methods [3].

Among these are so called memory-based systems, such as ANON [43]. Although capable of case-based reasoning, they have also their own characteristics. The memory of memory-based systems is completely directed by the

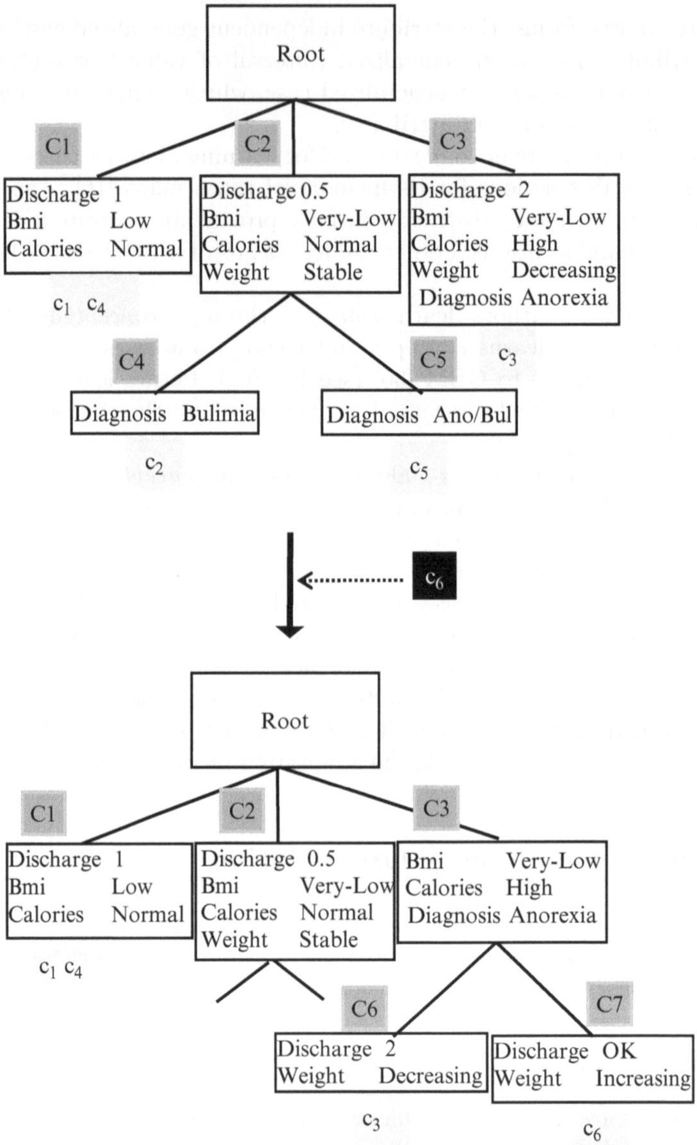

Fig. 6.2. Conceptual clustering example in MNAOMIA: a new case c6 being CBR processed creates two new concepts C6 and C7, and concept C3 is abstracted by losing two features

inferences. Implemented on parallel machines, indexation is replaced by the attribution of labels to the different cases, corresponding to the traditional indices of case-based reasoning. Feature extraction and the search through the memory correspond to the same inferences. There is generally a compromise between the importance of the inferences during the features extraction to constitute the indices, and during the search through the memory. Two approaches are possible: maximizing the inferences during the extraction of the features to constitute the indices, and maximizing the inferences during the search through the memory. The features serving as indices in both approaches are by the way different: having a semantic connotation in the former and a syntactic one in the latter. ANON proposes to integrate both by using first the features with syntactic load to perform a preselection, then the features with semantic load are extracted on the retained cases, powered by processors working in parallel. These memories are called active [43] because they must perform an important inferential effort during the search through the memory, to compensate for the absence of a learnt structuring.

6.5.2 Structured Memories

Among the different structured organizations, the accumulation of generalizations or abstractions facilitates the evaluation of the situation and allows a control of indexation.

Like GBM and related early systems, many CBR memories are organized *hierarchically* through *generalization* or *abstraction*. SWALE [24, 52] is a case-based explanation system. An explanation is a causal chain. Its cases are generalized explanations. When a new case must be integrated to the memory, it constructs a schema generalized from this case and the retrieved case used for reasoning. This constructed schema, called an explanation schema, is similar to a MOP. The authors refer to a generalization process for learning this schema. SWALE's memory is organized hierarchically, and the most abstract explanation schemas are located near the root, while the least abstract ones are positioned near the leaves, made up of the explanation cases. CANDIDE [9] is a system for language acquisition from similar cases. Based on a conceptual clustering algorithm [46], its memory is organized in categories, induced by abstraction of the common elements of two cases. Nevertheless, learning goes through three phases, the first one being totally supervised (an expert provides a set of cases and a generalization hierarchy). The second phase interacts with the expert, helping him by extracting similar memorized cases. The third phase is then an unsupervised recognition phase. This form of learning is close to knowledge acquisition. Similarly, AQUA [49] builds categories by induction. But the traits chosen for the generalization are selected in function of their pertinence. As previously, pertinence is evaluated by the construction of a causal explanation. The generalization is constrained by explanations: it is an explanation based generalization [38], and is a form of axiomatic learning. Branting's case abstractions assist case indexing, matching, and adaptation [16]. The abstract solutions contain the most important aspects of the less abstract solutions. Matching is less expensive because new cases are compared with the abstract cases, which contain many less features than the specific cases and also are less numerous. The adaptation effort is less costly too because many nonpertinent features need not be adapted. The system has been applied to route finding. This author evaluates the comparative performance of ground-level CBR, heuristic search (A*), REFINEMENT (hierarchical problem solving), and S_{CBR} (Stratified case-based reasoning) [16]. Like S_{CBR}, REFINEMENT is a form of hierarchical problem solving, in which a solution at one level of the hierarchy guides the search at a lower level in the hierarchy. The difference between REFINEMENT and S_{CBR} is that S_{CBR} starts by matching with the most specific case in the hierarchy, not systematically the most abstract ones only. The match in this particular domain is the match between a given abstract start position and a given abstract goal position. In both methods, the search starts from the root of the hierarchy and proceeds top-down. S_{CBR} proved to be an improvement over ground-level CBR and heuristic search in terms of number of levels of abstraction, the size of the case library, and the resemblance among cases. S_{CBR} was also an improvement over REFINEMENT in the same dimensions when the number of levels in the hierarchy reached at least three [16]. Some hierarchies are nonrefinable, which means that a solution may show at an abstract level, but not be valid at a more specific level. In this type of hierarchy, S_{CBR} outperforms REFINEMENT particularly in nonrefinable hierarchies.

Structured memories, dynamic, present the advantage of being declarative. The important learning efforts in declarative learning are materialized in the structures and the dynamic organization of their memories. Perner learns a hierarchy of classes by *hierarchical conceptual clustering*, where the concepts are clusters of prototypes [44]. She notes the advantages of this method: a more compact case base, and more robust (error-tolerant). The same author explains that an important aspect of case base maintenance – beyond the classic trio addition, removal and revision of cases – is learning the memory organization as well as the prototypes in memory [45]. Case-based organization is based on approximate graph subsomption. The nodes in the graph can be represented by a prototype. MNAOMIA [14] proposes to use *incremental concept learning* [20,21], which is a form of hierarchical clustering, to organize the memory. Concepts are composed of pairs of <*attribute*, *value*> common to all the cases indexed under these concepts. This system integrates highly data mining with CBR because it reuses the learnt structures to answer higher level tasks such as generating hypotheses for clinical research (see Fig. 6.3), as a side effect of CBR for clinical diagnosis and treatment decision support. Therefore this system illustrates that by learning memory structures in the form of concepts, the classical CBR classification task improves, and at the same time the system extracts what it has learnt, thus adding a knowledge discovery dimension to the classification tasks performed.

Another notable class of systems is composed of those who perform *decision tree induction* [48,56] to organize their memory. INRECA [5] project studied how to integrate CBR and decision tree induction. They propose to preprocess the case base by an induction tree, namely a decision tree. The system is based on similar approach in KATE and PATDEX from the authors.

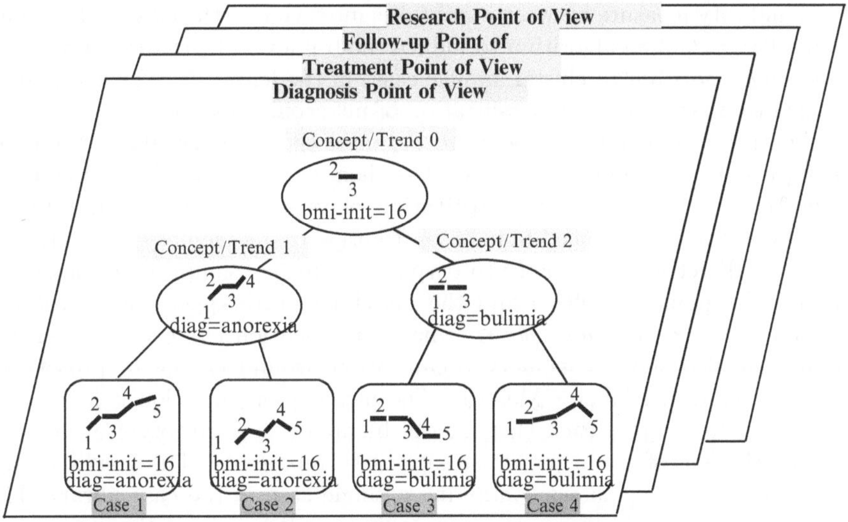

Fig. 6.3. Hierarchical memory organization in MNAOMIA: concepts are learnt during CBR for diagnosis, treatment, and/or follow-up, and can be reused by research task

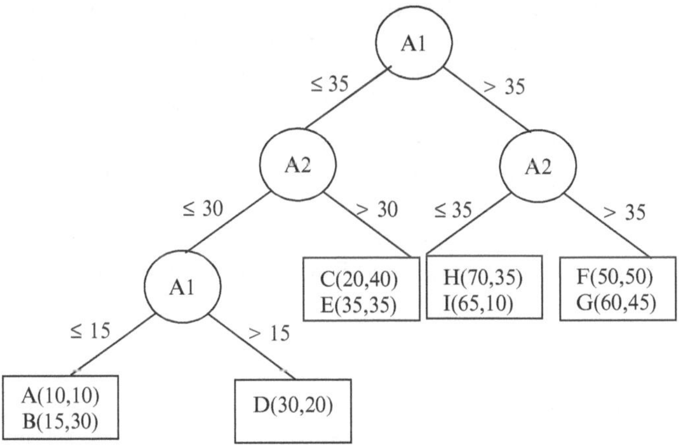

Fig. 6.4. Tree memory organization in INRECA using k-d trees

The decision tree partitions the case base around nodes composed of a single attribute and two branches per node, splitting the values on each branch in the median, based on the interquartile distance. Later refined into an *INRECA tree* [6] (see Fig. 6.4), which is a hybrid between a decision tree and a k-d tree, this method allows both similarity based retrieval and decision tree retrieval, is incremental, and speeds up the retrieval. The structures are a set of classes, each class carrying a rule to determine whether a case belongs to it or not. Each condition in the rule is sufficient and concerns a single attribute.

The similarity measure between cases takes into account the classes. Jarmulak uses a *tree induction* algorithm to induce the top level of the structure, and a *clustering* algorithm to cluster similar cases in the leaves [23]. This system is applied to imaging for the classification of ultrasonic B-scans.

Another important method is to organize the memory like a hierarchy of objects, by *subsomption*. Retrieval is then a classification in a hierarchy of objects, and functions by substitution of values in slots. CHROMA [4] uses its prototypes, induced from cases, to organize its memory. The retrieval step of CBR retrieves relevant prototypes by using subsomption in the object oriented language NOOS to find the matching prototypes. The prototypes contain a pair <*situation*, *plan*> where the situation is an object. Bellazzi et al. [10] also show a memory organization around classes of prototypes in the domain of Diabetes Mellitus. The memory organization is a tree-like structured taxonomy: each class in the hierarchy is a prototypical description of the set of problems or situations it represents. The leaves in the taxonomy are basic classes containing a single case in the case library. The authors stress that in many domains, some knowledge about the structuring of the domain is available, or can be induced. This is the case in object oriented approaches both for database management and programming languages. Every knowledge-based methodology derived from frames and semantic nets rely on that type of knowledge. In such a hierarchy, the retrieval step is two folded: first a classification in a hierarchy of objects, in this system a Bayesian classification, followed by a NN technique on the cases in the classes selected by the first step. This method is called PBR (pivoting based retrieval). An evaluation shows that PBR retrieves cases linearly with the size of the case base, in comparison with the NN technique, which grows quadratically with the number of cases [10].

Many systems use personalized memory organizations structured around several layers or *networks*. Malek and Rialle in the domain of neuropathy diagnosis construct a memory of prototypical cases that is reused in the retrieval phase. The memory structure has two levels: the upper level contains prototypes, each of them representing a group of cases; the lower level contains analyzed patient cases organized into groups of similar cases [34]. A small memory of prototypes learnt by generalization decreases the retrieval time in comparison with a large memory of cases. Malek uses a neural network to learn the prototypes in memory for a classification task, such as diagnosis [35]. Here the memory is organized in three layers: an input layer containing one unit for each attribute, a hidden layer containing the prototypes, and an output layer containing one unit for each class.

Another type of memory organization is the *formal concept lattice*. Dìaz-Agudo and Gonzàlez-Calero organize through formal concept analysis (FCA) the case base around *Galois lattices* [18]. Retrieval step is a classification in a concept hierarchy, as specified in the FCA methodology, which provides such algorithms. The concepts can be seen as an alternate form of indexing structure.

Yet other authors take advantage of the *B-tree structure* implementing databases. West and McDonald propose a method using database SQL query language to retrieve cases over a large case base stored in a database [57]. The method makes implicit use of the optimized B-tree structure underlying the relational databases implementation for fast retrieval, and computes the similarity measure during retrieval. The look up time for retrieving from a case library of any size is constant – and low – and the inclusion of the similarity assessment varies less than linearly [57].

6.6 Discussion

CBR systems make efficient use of most data mining tasks defined for descriptive modeling. We can list among the main ones encountered cluster analysis, rule induction, hierarchical cluster analysis, and decision tree induction. The motivations for performing an incremental type of data mining during CBR are several folds, and their efficiency has been measured to validate the approach. The main motivations are the following:

- Increase efficiency of retrieval mostly, but also of reuse, revise, and retain steps
- Increase robustness, tolerance to noise
- Increase accuracy of reasoning
- Improve storage needs
- Follow a cognitive model
- Add a synthetic task such as generating new research hypotheses as a side effect of normal CBR functioning

The memory organization maps directly into the retrieval method used. For example, object-oriented taxonomies will retrieve cases by subsomption mechanism and not by NN retrieval as in flat memories. Generalized cases and the like are used both as indexing structures and organizational structures. We can see here a direct mapping with the theory of the dynamic memory, which constantly influences the CBR approach. The general idea is that the learnt memory structures and organizations condition what inferences will be performed and how. This is a major difference with database approaches, which concentrate only on retrieval, and also with data mining approaches, which concentrate only on the structures learnt, and not on how they will be used. The ideal CBR memory is one which at the same time speeds up the retrieval step and improves the accuracy and robustness of the task performed by the reasoner, and particularly the reuse performed, influencing positively both the retrieval, the reuse, and the other steps. Researchers do not want to settle for a faster retrieval at the expense of less accuracy due to an overgeneralization. And they succeed at it.

6.7 Conclusion

Data mining in CBR consists mainly in incremental mining for memory structures and organization with the goal to improve performance of retrieval, reuse, revise, and retain steps. Memory structures mining comprises case mining, feature mining, and generalized case mining. CBR memories are rich in a variety of generalized structures such as concepts, prototypes, and abstract cases. These structures can be organized in flat memories or in structured memories, among hierarchies, conceptual hierarchies, decision trees, object-oriented taxonomies, formal concept lattices, and B-trees. Researchers are aiming at the ideal memory as described in the theory of the dynamic memory, which follows a cognitive model, while also improves performance and accuracy in retrieve, reuse, revise, and retain steps. Many have succeeded in showing that their memories indeed both decrease retrieval time and increase accuracy of reasoning. This demanding goal is what motivates the constant search for novel mining methods specific for CBR, and that cannot be met by methodologies that simply do not share the same goals. The variety of approaches as well as the specific and complex purpose lead to thinking that there is still space for future models and theories of CBR memories.

References

1. Aamodt A, Plaza E (1994) Case-Based Reasoning: Foundational Issues, Methodologies Variations, and Systems Approaches. AI Communications, IOS Press, Vol. 7: 1:39–59
2. Aha DW (1991) Case-Based Learning Algorithms. In: Bareiss R (ed) Proceedings of a Workshop on case-based reasoning (DARPA), Washington, D.C. Morgan Kaufmann, San Mateo, California, pp 147–157
3. Aha DW (1997) Lazy Learning. Artificial Intelligence Review 11:7–10
4. Armengol E, Plaza E (1994) Integrating induction in a case-based reasoner. In: Keane M, Haton JP, Manago M (eds) Proceedings of EWCBR 94. Acknosoft Press, Paris, pp 243–251
5. Auriol E, Manago M, Althoff KD, Wess S, Dittrich S (1994) Integrating Induction and Case-Based Reasoning: Methodological Approach and First Evaluations. In: Keane M, Haton JP, Manago M (eds) Proceedings of EWCBR 94. Acknosoft Press, Paris, pp 145–155
6. Auriol E, Wess S, Manago M, Althoff KD, Traphöner R (1995) INRECA: A Seamless Integrated System Based on Inductive Inference and Case-Based Reasoning. In: Veloso M, Aamodt A (eds) Proceedings of ICCBR 95. Springer-Verlag, Lecture Notes in Artificial Intelligence 1010, Berlin, Heidelberg, New York, pp 371–380
7. Bareiss R (1989a) Exemplar-Based Knowledge Acquisition. Academic Press, Inc., San Diego, CA
8. Bareiss R (1989b) The Experimental Evaluation of a Case-Based Learning Apprentice. In: Hammond KJ (ed) Proceedings of a Workshop on case-based reasoning (DARPA). Morgan Kaufmann, San Mateo, CA, pp 162–166

9. Beck HW (1991) Language Acquisition from Cases. In: Bareiss R (ed) Proceedings of a Workshop on case-based reasoning (DARPA), Washington, D.C. Morgan Kaufmann, San Mateo, California, pp 159–169
10. Bellazzi R, Montani S, Portinale L (1998) Retrieval in a Prototype-Based Case Library: A Case Study in Diabetes Therapy Revision. In: Smyth B, Cunningham P (eds) Proceedings of ECCBR 98. Springer-Verlag, Lecture Notes in Artificial Intelligence 1488, Berlin, Heidelberg, New York, pp 64–75
11. Bichindaritz I (1990) Alexia: un système de résolution de problèmes par analogie utilisant une mémoire des exemples méta-indexée par un modèle causal appliqué à la détermination de l'étiologie de l'H.T.A. Rapport de stage de DEA, EHEI, Université René Descartes-Paris V
12. Bichindaritz I (1994) A case-based assistant for clinical psychiatry expertise. In: Proceedings 18th Symposium on Computer Applications in Medical Care. AMIA, Washington DC, pp 673–677
13. Bichindaritz I (1995a) Case-Based Reasoning and Conceptual Clustering: For a Co-operative Approach. In: Watson I, Fahrir M (eds) Advances in Case-Based Reasoning. Springer-Verlag, Lecture Notes in Artificial Intelligence 1020, Berlin, Heidelberg, New York, pp 91–106
14. Bichindaritz I (1995b) A case-based reasoner adaptive to several cognitive tasks. In: Veloso M., Aamodt A (eds) Proceedings of ICCBR 95. Springer-Verlag, Lecture Notes in Artificial Intelligence 1010, Berlin, Heidelberg, New York, pp 391–400
15. Bichindaritz I, Séroussi B (1992) Contraindre l'analogie par la causalité. Technique et sciences informatiques Volume 11, n 4: 69–98
16. Branting KL (1997) Stratified Case-Based Reasoning in Non-Refinable Abstraction Hierarchies. In: Leake D, Plaza E (eds) Proceedings of ICCBR 97. Springer-Verlag, Lecture Notes in Artificial Intelligence 1266, Berlin, Heidelberg, New York, pp 519–530
17. Cain T, Pazzani MJ, Silverstein G (1991) Using Domain Knowledge to Influence Similarity Judgment. In: Bareiss R (ed) Proceedings of a Workshop on case-based reasoning (DARPA), Washington, D.C. Morgan Kaufmann, San Mateo, California, pp 191–202
18. Dìaz-Agudo B, Gonzàlez-Calero P (2001) Classification Based Retrieval Using Formal Concept Analysis. In: Aha DW, Watson I (eds) Proceedings of ICCBR 01. Springer-Verlag, Lecture Notes in Artificial Intelligence 2080, Berlin, Heidelberg, New York, pp 173–188
19. Dìaz-Agudo B, Gervàz P, Gonzàlez-Calero P (2003) Adaptation Guided Retrieval Based on Formal Concept Analysis. In: Ashley K, Bridge DG (eds) Proceedings of ICCBR 03. Springer-Verlag, Lecture Notes in Artificial Intelligence 2689, Berlin, Heidelberg, New York, pp 131–145
20. Fisher D (1987) Knowledge acquisition via incremental conceptual clustering. Machine Learning 2:139–172
21. Gennari JH, Langley P, Fisher D (1989) Models of Incremental Concept Formation. Artificial Intelligence 40:11–61
22. Hand D, Mannila H, Smyth P (2001) Principles of Data Mining. The MIT Press, Cambridge, Massachusetts
23. Jarmulak J (1998) Case-Based Classification of Ultrasonic B-Scans: Case-Base Organization and Case Retrieval. In: Smyth B, Cunningham P (eds) Proceedings of ECCBR 98. Springer-Verlag, Lecture Notes in Artificial Intelligence 1488, Berlin, Heidelberg, New York, pp 100–111

24. Kass A, Leake D, Owens C (1986) SWALE, A Program that Explains. In: Schank RC (ed) Explanation Patterns. Understanding Mechanically and Creatively. Laurence Erlbaum Associates, Publishers, Hillsdale, New Jersey, pp 232–256
25. King J, Bareiss R (1989) Similarity Assessment and Case-Based Reasoning. In: Hammond KJ (ed) Proceedings of a Workshop on case-based reasoning (DARPA), Pensacola Beach, Florida. Morgan Kaufmann, San Mateo, CA, pp 67–71
26. Kolodner JL (1993) Case-Based Reasoning. Morgan Kaufmann Publishers, San Mateo, California
27. Lebowitz M (1982) IPP. In: Schank RC (ed) Dynamic memory: A theory of reminding and learning in computers and people. Cambridge University Press, Cambridge, MA, pp 197–207
28. Lebowitz M (1983) Generalization From Natural Language Text. Cognitive Science 7:1–40
29. Lebowitz M (1985) Researcher: An Experimental Intelligent Information System. In: Proceedings IJCAI 85, pp 858–862
30. Lebowitz M (1986) Concept Learning in a Rich Input Domain: Generalization-Based Memory. In: Michalski RS, Carbonell JG, Mitchell TM (eds) Machine Learning: An Artificial Intelligence Approach, Vol 2. Morgan Kaufmann, Los Altos, CA
31. Lebowitz M (1987) Experiments with Incremental Concept Formation: UNIMEM. Machine Learning 2:103–138
32. Lebowitz M (1988) Deferred Commitment in UNIMEM: Waiting to learn. In: Proceedings 5th Machine Learning Conference, Ann Arbor, Michigan. pp 80–86
33. Michalski RS (1993) Toward a Unified Theory of Learning. In: Buchanan BG, Wilkins DC (eds) Readings in knowledge acquisition and learning, automating the construction and improvement of expert systems. Morgan Kaufmann Publishers, San Mateo, California, pp 7–38
34. Malek M, Rialle V (1994) A Case-Based Reasoning System Applied to Neuropathy Diagnosis. In: (Keane M, HAton JP, Manago M (eds) Proceedings of EWCBR 94. Acknosoft Press, Paris, pp 329–336
35. Malek M (1995) A Connectionist Indexing Approach for CBR Systems. In: Veloso M, Aamodt A (eds) Proceedings of ICCBR 95. Springer-Verlag, Lecture Notes in Artificial Intelligence 1010, Berlin, Heidelberg, New York, pp 520–527
36. Maximini K, Maximini R, Bergmann R (2003) An Investigation of Generalized Cases. In: Ashley KD, Bridge DG (eds) Proceedings of ICCBR 03. Springer-Verlag, Lecture Notes in Artificial Intelligence 2689, Berlin, Heidelberg, New York, pp 261–275
37. Mitchell TM (1997) Machine Learning. Mc Graw Hill, Boston, Massachusetts
38. Mitchell TM, Keller RM, Kedar-Cabelli S (1986) Explanation-based generalization: A unifying view. Machine Learning 1(1): 47–80
39. Montani S, Portinale L, Bellazzi R, Leornardi G (2004) RHENE: A Case Retrieval System for Hemodialysis Cases with Dynamically Monitored Parameters. In: Funk P, Gonzàlez Calero P (eds) Proceedings of ECCBR 04. Springer-Verlag, Lecture Notes in Artificial Intelligence 3155, Berlin, Heidelberg, New York, pp 659–672

40. Mougouie B, Bergmann R (2002) Similarity Assessment for Generalized Cases by Optimization Methods. In: Craw S, Preece A (eds) Proceedings of EWCBR 02. Springer-Verlag, Lecture Notes in Artificial Intelligence 2416, Berlin, Heidelberg, New York, pp 249–263
41. Niloofar A, Jurisica I (2004) Maintaining Case-Based Reasoning Systems: A Machine Learning Approach. In: Funk P, Gonzàlez Calero P (eds) Proceedings of ECCBR 04. Springer-Verlag, Lecture Notes in Artificial Intelligence 3155, Berlin, Heidelberg, New York, pp 17–31
42. Nilsson M, Funk P (2004) A Case-Based Classification of Respiratory sinus Arrhythmia. In: Funk P, Gonzàlez Calero P (eds) Proceedings of ECCBR 04. Springer-Verlag, Lecture Notes in Artificial Intelligence 3155, Berlin, Heidelberg, New York, pp 673–685
43. Owens C (1993) Integrating Feature Extraction and Memory Search. In: Kolodner JL (ed) Case-Based Learning. Kluwer Academic Publishers, Boston, pp 117–145
44. Perner P (1998) Different Learning Strategies in a Case-Based Reasoning System for Image Interpretation. In: Smyth B, Cunningham P (eds) Proceedings of ECCBR 98. Springer-Verlag, Lecture Notes in Artificial Intelligence 1488, Berlin, Heidelberg, New York, pp 251–261
45. Perner P (2003) Incremenetal Learning of Retrieval Knowledge in a Case-Based Reasoning System. In: Ashley KD, Bridge DG (eds) Proceedings of ICCBR 03. Springer-Verlag, Lecture Notes in Artificial Intelligence 2689, Berlin, Heidelberg, New York, pp 422–436
46. BW, Bareiss R, Holte RC (1990) Concept Learning and Heuristic Classification in Weak-Theory Domains. Artificial Intelligence, 45:229–263
47. Portinale L, Torasso P (1995) ADAPTER: An Integrated Diagnostic System Combining Case-Based and Abductive Reasoning. In: Veloso M, Aamodt A (eds) Proceedings of ICCBR 95. Springer-Verlag, Lecture Notes in Artificial Intelligence 1010, Berlin, Heidelberg, New York, pp 277–288
48. Quinlan JR (1986) Induction of decision trees. Machine Learning 1(1):81–106
49. Ram A (1993) Indexing, Elaboration and Refinement: Incremental Learning of Explanatory Cases. In: Kolodner JL (ed) Case-Based Learning. Kluwer Academic Publishers, Boston, pp 7–54
50. Reategui E, Campbell JA, Borghetti S (1995) Using a Neural Network to Learn General Knowledge in a Case-Based System. In: Veloso M, Aamodt A (eds) Proceedings of ICCBR 95. Springer-Verlag, Lecture Notes in Artificial Intelligence 1010, Berlin, Heidelberg, New York, pp 528–537
51. Schank RC (1982) Dynamic memory. A theory of reminding and learning in computers and people. Camdridge University Press, Cambridge
52. Schank RC, Leake DB (1989) Creativity and Learning in a Case-Based Explainer. Articifial Intelligence 40:353–385
53. Schmidt R, Gierl L (1998) Experiences with Prototype Designs and Retrieval Methods in Medical Case-Based Reasoning Systems. In: Smyth B, Cunningham P (eds) Proceedings of ECCBR 98. Springer-Verlag, Lecture Notes in Artificial Intelligence 1488, Berlin, Heidelberg, New York, pp 370–381
54. Smyth B, McKenna E (1999) Building Compact Competent Case-Bases. In: Proceedings of ICCBR 99. Springer-Verlag, Berlin, Heidelberg, New York, pp 329–342

55. Veloso MM, Carbonell JG (1993) Derivational Analogy in PRODIGY: Automating Case Acquisition, Storage and Utilization. In: Kolodner JL (ed) Case-Based Learning. Kluwer Academic Publishers, Norwell, MA, pp 55–84
56. Utgoff PE (1989) Incremental induction of decision trees. Machine Learning, 4, 2:161–186
57. West GM, McDonald JR (2003) An SQL-Based Approach to Similarity Assessment within a Relational Database. In: Ashley K, Bridge DG (eds) Proceedings of ICCBR 03. Springer-Verlag, Lecture Notes in Artificial Intelligence 2689, Berlin, Heidelberg, New York, pp 610–621
58. Wilson DC, Leake DB (2001) Mainting Case-based Reasoners: Dimensions and Directions. Computational Intelligence Journal, Vo. 17, No. 2:196–213
59. Wiratunga N, Koychev I, Massie S (2004) Feature Selection and Generalisation for Retrieval of Textual Cases. In: Funk P, Gonzàlez Calero P (eds) Proceedings of ECCBR 04. Springer-Verlag, Lecture Notes in Artificial Intelligence 3155, Berlin, Heidelberg, New York, pp 806–820
60. Yang Q, Cheng H (2003) Case Mining from Large Databases. In: Ashley K, Bridge DG (eds) Proceedings of ICCBR 03. Springer-Verlag, Lecture Notes in Artificial Intelligence 2689, Berlin, Heidelberg, New York, pp 691–702

7

Learning a Statistical Model for Performance Prediction in Case-Based Reasoning

B. Bhanu and R. Wang

Center for Research in Intelligent Systems, University of California at Riverside, California, 92521, USA

Summary. This chapter is concentrated with the performance characterization of a case-based reasoning (CBR) system. Based on the match score and nonmatch score computed from the cases in the case library, we develop a statistical model for prediction. We estimate the size of a subset of cases, called gallery size, that can generate the optimal error estimate and its confidence on a large population (relative to the size of the gallery). The statistical model is based on a generalized two-dimensional prediction model that combines a hypergeometric probability distribution model with a binomial model explicitly and considers the data distortion problem in large populations. Learning is incorporated in the prediction process in order to find the optimal small gallery size and to improve the prediction performance. During the prediction, the expectation–maximization (EM) algorithm is used to learn the match score and the nonmatch score distributions that are represented as mixture of Gaussians. By learning, the optimal size of small gallery is determined and at the same time the upper bound and the lower bound for the prediction on large populations are obtained. Results are shown using a real-world database with the increasing size of the case library.

7.1 Introduction

Case-based approaches are characterized by how the learner represents what it has learned so far, as well as the analogical methods which are used to transfer the learned experience [1]. In CBR, "past" experiences are stored in memory as cases and are used to solve a new problem case. Given a problem to be solved, the case-based method retrieves from the memory the solution to a similar problem encountered in the past, adapts the previous solution to the current problem, and stores the new problem-solution packet as another case in the memory. The major concerns with CBR are the selection of the indexing scheme to organize cases in the memory, the method for choosing the most relevant cases at reasoning time, the adaptation heuristics to modify pervious cases to fit the current problem, and maintenance of the case library.

B. Bhanu and R. Wang: *Learning a Statistical Model for Performance Prediction in Case-Based Reasoning*, Studies in Computational Intelligence (SCI) **73**, 195–212 (2008)
www.springerlink.com

An important problem for a CBR system is the prediction of its performance as the case library grows. In this chapter, we present a statistical model for performance characterization.

Recognition/classification systems can classify images, signals, or other types of measurements into a number of classes. In a simplistic way, we can view a CBR system as a model-based recognition/classification system [2,3], which stores a set of models in its case library and classifies the incoming cases by performing match with the database models. In this process, the case library will be continuously growing as more cases/models become part of the case base. It is like an incremental learning system for object recognition, which not only involves traditional recognition/classification of complete/occluded objects but also new model acquisition and refinement of existing model [3,4]. This chapter provides a method for the life-time problem [5] of a CBR system. We define the following terms in the concept of CBR: gallery/training set, probe/testing set, populations, algorithm/system, data, recognition/classification.

We define the following terms in the context of CBR: The gallery set and the probe set are the training set and the testing set, respectively. The algorithm/system is the set of recognition/classification programs. The population is the data that the system under consideration may encounter in its lifetime.

Since the recognition performance of an algorithm/system is usually based on limited data, it is difficult to estimate this performance for additional data: the limited test data may, after all, not accurately represent a larger population. Before we can evaluate and predict the performance of a recognition algorithm/system on large populations, we need to answer some fundamental questions. When we use a small gallery to estimate the algorithm/system performance on large populations how can we find the optimal size of the small gallery and how accurate is the estimation? Since the prediction is based on the selected recognition algorithm/system, we can give the confidence interval for the performance estimation on a large population [6]. The confidence interval can describe the uncertainty associated with the estimation. This gives an interval within which the true algorithm/system performance for the large population is expected to fall, along with the probability that it is expected to fall there [7].

Given limited data we can use the Bayesian parameter estimation or nonparametric estimation methods to estimate the data distribution. The expectation–maximization (EM) algorithm, one of the parameter estimation methods, assumes that the underlying distribution is known. It is an iterative method to estimate the mixture parameters by maximum likelihood techniques. The parzen window and the K-nearest-neighbor are nonparametric estimation methods that are used to estimate the data distribution in case the underlying distribution is unknown. These two methods converge the distribution to the unknown distribution. The parzen window method estimates the density, while the K-nearest-neighbor method determines the K closest neighbors [8].

In this chapter, we use a generalized prediction model that combines a hypergeometric probability distribution model with a binomial model. This prediction model takes into account distortion that may occur in large populations. It also provides performance measurements as a function of rank, large population size, number of distorted images, and similarity score (match and nonmatch score) distributions. While we use the EM algorithm to estimate the match score and the nonmatch score distributions, we introduce learning to feed back similarity scores (match scores and nonmatch scores) to increase the small gallery size. In this way, we can find the optimal size of the small gallery to predict the large population performance. Meanwhile, we provide the upper and the lower bounds for the prediction performance of a large population. In this chapter, we use two different statistical methods – Chernoff's inequality and Chebychev's inequality – to obtain the relationship between the small gallery size and the confidence interval given a margin of error. The small gallery size for prediction that we get from the learning process is smaller than the size determined by statistical methods.

The paper is organized as follows. Related work and contributions are presented in Sect. 7.2. The details of the technical approach are given in Sect. 7.3. It includes the integrated model, the procedure of learning for similarity score distributions in the prediction, and the statistical methods to find the optimal sample size. Experimental results are shown in Sect. 7.4. The integrated model with learning is tested on the NIST-4 fingerprint database. Conclusions are given in Sect. 7.5.

7.2 Related Work and Contributions

7.2.1 Related Work

Until now the prediction models are mostly based on the feature space or similarity scores [9]. The statistical approaches are used by many researchers to estimate the recognition system performance. Wayman [10] and Daugman [11] develop a binomial model that uses the nonmatch score distribution. This model underestimates recognition performance for large populations. Phillips et al. [12] develop a moment model, which uses both the match score and the nonmatch score distributions. Since all the similarity scores are sampled independently, the probability of error is increased and the prediction results underestimate the identification performance. Wang and Bhanu [9] present a binomial model to predict the large fingerprint database recognition performance based on a small gallery. They present their early work on a generalized two-dimensional model, which integrates a hypergeometric probability distribution explicitly with a binomial distribution [13]. This work considers the distortion caused by sensor noise, feature uncertainty, feature occlusion, and feature clutter. However, there is no learning to determine the optimal small gallery size and the bounds on performance. Johnson et al. [14] improve the moment model by using a multiple nonmatch score set. They average match scores on the whole gallery. For each match score they count the number of times that a nonmatch score is larger than the match score, leading to an error. In this chapter, they assume that the distribution of the match score is uniform. Grother and Phillips [15] introduce the joint density function of the match score and the nonmatch score to estimate both the open-set and the closed-set identification performance. The closed-set identification is the identification for which all potential users are enrolled in the system. The open-set identification is the identification for which some potential users are not enrolled in the system. Since the joint density is generally impractical to estimate, they assume that the match score and nonmatch score are independent and their distributions are the same for large populations. They use the Monte Carlo sampling method to linearly interpolate the match score and the nonmatch score look-up tables.

Providing the upper and lower bounds for the prediction performance is another important topic in the recognition performance prediction. Lindenbaum [16] proposes a probabilistic method to derive bounds on the number of features required to achieve recognition with a certain degree of confidence. This method considered object similarity, bounded uncertainty, and occlusion. A similar approach presented in [17] can be used to analyze object recognition with uncertainty, similarity, and clutter. Boshra and Bhanu [18, 19] present a method to predict upper and lower bounds on the performance prediction. They predict performance by considering feature uncertainty, occlusion, clutter, and similarity simultaneously. In their method performance is predicted in two steps: compute the similarity between each pair of models; use the similarity information along with the statistical model to determine upper and lower bounds for the object recognition performance. Guyon et al. [1] propose guaranteed estimators to determine the test size for the independent identical distribution recognition error and the correlated recognition error, along with the assumption of the underlying probability distribution.

7.2.2 Contributions

In this chapter we address the problems associated with the prediction of performance on large populations and optimal small gallery size. The specific contributions are the following:

1. We use a generalized prediction model that combines a hypergeometric probability distribution model explicitly with a binomial model which takes into account distortions that may occur in large populations. Our distortion model includes feature uncertainty, feature occlusion, and feature clutter. In the prediction model, we use the EM algorithm to estimate similarity score (match score and nonmatch score) distributions and find the number of components of the distributions automatically.

2. We find the optimal size of a small gallery by an iterative learning process [20]. We use the Chernoff inequality and the Chebychev inequality to determine the small gallery size in theory which is related to the margin of error and the confidence interval. We find the upper bound and a good lower bound for predicting recognition performance on a large population.
3. The results are shown on a large data set (NIST-4) of fingerprint images [21]. We show the prediction results with the increasing size of the database.

7.3 Technical Approach

We are given two sets of data: a gallery set and a probe set. The gallery is a set of models saved in the database. The probe is a set of queries for the database. A large population is the unknown data set whose recognition performance needs to be estimated. Based on the given gallery and probe set we would like to estimate the recognition performance on large populations.

7.3.1 Methodology for Determining the Small Gallery Size

Figure 7.1 provides the detailed diagram for the implementation of our proposed approach to get the optimal small gallery size. For a given recognition system whose database size or the number of classes is N, we randomly pick n images from the database N to be our small gallery. By an authentication process, we can get a set of match scores and nonmatch scores for this small gallery. Then, we use the EM algorithm [8] to estimate distributions of the match score and the nonmatch score. Assume that the match score and nonmatch score distributions are Gaussian mixtures. Let $ms(x)$ represents the match score distribution and $ns(x)$ represents the nonmatch score distribution. We have

$$ms(x) = \sum_{i=1}^{m} \alpha_i ms_i(x) \tag{7.1}$$

and

$$ns(x) = \sum_{j=1}^{n} \beta_j ns_j(x) \tag{7.2}$$

where m and n are the number of components, α_i and β_j are the component proportions, $\sum_{i=1}^{m} \alpha_i = 1$, and $\sum_{j=1}^{n} \beta_j = 1$. We have $ms_i \sim N(\mu_{si}, \sigma_{si}^2)$, $ns_j \sim N(\mu_{nj}, \sigma_{nj}^2)$, where μ_s, μ_n, σ_s^2, and σ_n^2 are the mean and variance for the match score distribution and the nonmatch score distribution, respectively.

Based on these distributions, we use our prediction model, which combines a hypergeometric probability distribution model with a binomial model

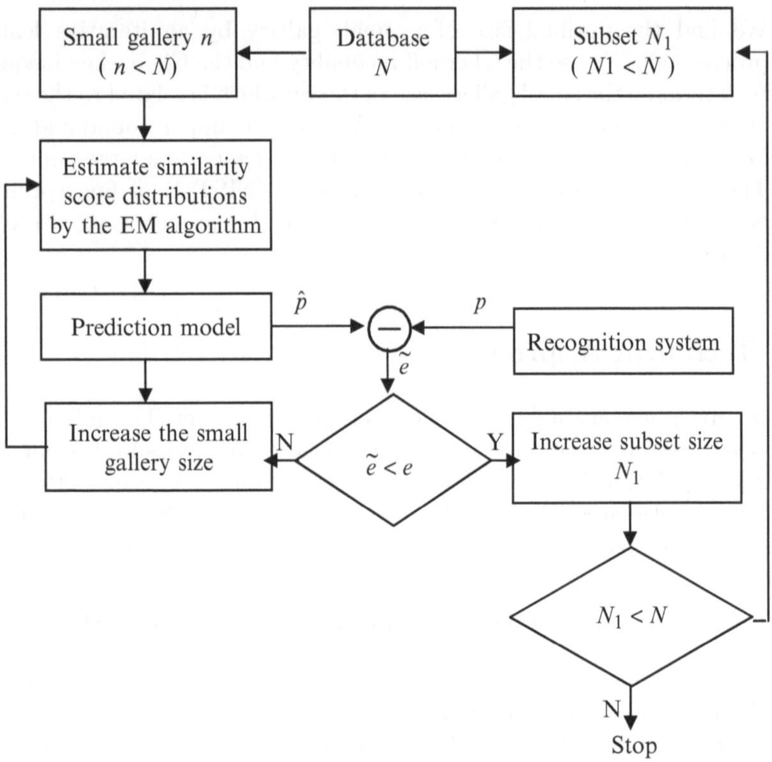

Fig. 7.1. Conceptual prediction model

to estimate the recognition system performance for a large population N_1, a subset of database N ($N_1 < N$). We assume the prediction performance on N_1 is $\hat{p}$. From the recognition system we can obtain the match score and the nonmatch score for N_1. Then, compute the actual recognition performance p for N_1. $\tilde{e}$ is the error between the predicted performance and the actual performance, $\tilde{e} = |\hat{p} - p|$. The margin of error e is the maximum specified error acceptable by the recognition system. If $\tilde{e}$ is larger than the margin of error e then we increase the small gallery size n and feed back match scores and nonmatch scores to the EM algorithm to estimate the similarity score distributions again. Otherwise, we increase N_1, the size of the large population, and repeat this process until the N_1 has increased to N. We use the Chernoff and Chebychev inequalities to find the relationship between the small gallery size and the prediction confidence interval given a margin of error. The small gallery size which we get from the inequalities is used to validate the learned optimal small gallery size. We will explain each part of the diagram in detail in this section.

7.3.2 Prediction Model

Usually a recognition system consists of three stages: image acquisition, feature extraction, and matching. Distortion often occurs in these stages and is caused by sensor noise, feature uncertainty, feature occlusion, and feature clutter. The effects of sensor and image noise are reflected in the feature uncertainty. Our two-dimensional prediction model considers the distortion problem which conforms to reality. Assume we have two kinds of different quality biometric images, group #1 and group #2. Group #1 is a set of biometric images without distortion. Group #2 is a set of biometric images with distortion. Let the size of these two groups be n_1 pairs and n_2 pairs, respectively. We randomly pick n pairs of images from group #1 and group #2 to be our small gallery. Then, the number of pairs of distorted images y which are chosen from group #2 follow the hypergeometric distribution

$$f(y) = \frac{C_{n-y}^{n_1} C_y^{n_2}}{C_n^{n_1+n_2}} \tag{7.3}$$

where $n_1 + n_2$ is the total number of images in these two groups and $n - y$ is the number of images chosen from group #1.

These n pairs of images are our small gallery. We split them into the gallery and the probe set. For each image in the probe set, we compute the match score and the nonmatch score with images in the gallery. Then, we have one match score and $n-1$ nonmatch scores for this image. We assume that the match score and the nonmatch score are independent. With all these similarity scores we can use the EM algorithm to estimate the match score and the nonmatch score distributions.

From the above discussion, we know that the match score and nonmatch score distributions depend not only on the similarity scores but also on the number of images with distortion. Let $ms(x|y)$ and $ns(x|y)$ represent the distributions of match scores and nonmatch scores given the number of distorted images. If the similarity score is higher then the object are more similar. The error occurs when a given match score is smaller than the nonmatch score corresponding to the same image. For a given number of distorted images, the probability that the nonmatch score is greater than or equal to the match score x is $NS(x)$, where

$$NS(x) = \int_x^{\infty} \sum_{y=0}^{n} ns(t|y) f(y) dt \tag{7.4}$$

Thus, the probability that the nonmatch score is smaller than the match score is $1 - NS(x)$.

If the size of the large population is N, then for the jth image we can have one match score and $N-1$ nonmatch scores. We rank the match score and the nonmatch score in the descending order. For a given number of images with distortion, the probability that the match score x is at rank r is given by the binomial probability distribution

$$C_{r-1}^{N-1}(1 - NS(x))^{N-r}(NS(x))^{r-1} \tag{7.5}$$

Integrating over all the match scores, for a given number of images with distortion, the probability that the match score is at rank r can be written as

$$\int_{-\infty}^{\infty} C_{r-1}^{N-1}(1 - NS(x))^{N-r}(NS(x))^{r-1} ms(x|y)dx \tag{7.6}$$

By summing over the images chosen from group #2, the probability that the match score is at rank r can be written as

$$\int_{-\infty}^{\infty} C_{r-1}^{N-1}(1 - NS(x))^{N-r}(NS(x))^{r-1} \sum_{y=0}^{n} ms(x|y)f(y)dx \tag{7.7}$$

In theory, a match score can be any value within $(-\infty, \infty)$. The probability that the match score is within rank r is

$$P(N, r) = \sum_{i=1}^{r} \int_{-\infty}^{\infty} C_{i-1}^{N-1}(1 - NS(x))^{N-i}(NS(x))^{i-1} \sum_{y=0}^{n} ms(x|y)f(y)dx \tag{7.8}$$

Given that the correct match takes place above a threshold t, the probability that the match score is within rank r becomes

$$P(N, r, t) = \sum_{i=1}^{r} \int_{t}^{\infty} C_{i-1}^{N-1}(1 - NS(x))^{N-i}(NS(x))^{i-1} \sum_{y=0}^{n} ms(x|y)f(y)dx \tag{7.9}$$

When rank $r = 1$ the prediction model with threshold t becomes

$$P(N, 1, t) = \int_{t}^{\infty} (1 - NS(x))^{N-1} \sum_{y=0}^{n} ms(x|y)f(y)dx \tag{7.10}$$

In this model, we make two assumptions: match scores and nonmatch scores are independent and large populations have distortion with model of feature uncertainty, occlusion, and clutter. We use a small gallery to estimate distributions of $ms(x|y)$ and $ns(x|y)$.

7.3.3 Estimation of The Small Gallery Size Based on Statistical Inequalities

In the following, we discuss the relationship between the prediction confidence interval and the size of the small gallery which could be used to validate the optimal small gallery size that we obtain by the learning process. We use limited data to estimate a large population recognition performance. Therefore, the prediction value may or may not be very accurate. This question can be mathematically expressed as

$$P\{|(p-\hat{p})| > e\} \leq (1-\alpha) \tag{7.11}$$

where $\hat{p}$ is the predicted performance for the recognition system which can be obtained from our prediction model, p is the actual performance of the recognition system, e is the margin of error for the system, and α is the confidence interval. Then, inequality (7.9) can be written as

$$P\{p > \hat{p} + e\} \leq (1-\alpha) \tag{7.12}$$

or

$$P\{p < \hat{p} - e\} \leq (1-\alpha) \tag{7.13}$$

Here, we consider inequality (7.12) since inequality (7.13) can be solved by the same procedure as inequality (7.12).

We assume that a recognition system recognizes (authentication) individuals with the probability $P\{X_i = 1\} = p$ and $P\{X_i = 0\} = 1 - p$, where $X_i = 1$ means an individual with a given object X_i is recognized correctly, $X_i = 0$ means the opposite, $0 \leq p \leq 1$. According to the Chernoff inequality [22], let $X_1, X_2, \ldots, X_n$ be independent random variables. We define the random variable

$$X = \frac{1}{n}\sum_{i=1}^{n} X_i \tag{7.14}$$

For any $t \geq 0$ we have

$$P\{X \geq E(X) + \frac{t}{n}\} \leq e^{-\frac{2t^2}{n}} \tag{7.15}$$

where $E(X)$ is the mean of X. Comparing with inequality (7.12), we can get

$$1 - \alpha = e^{-\frac{2t^2}{n}} \tag{7.16}$$

So,

$$t = \sqrt{-\frac{n\ln(1-\alpha)}{2}} \tag{7.17}$$

Thus, equation (7.15) becomes

$$P\{X \geq E(X) + \sqrt{-\frac{\ln(1-\alpha)}{2n}}\} \leq 1 - \alpha \tag{7.18}$$

From inequality (7.12), we know that

$$e = \sqrt{-\frac{\ln(1-\alpha)}{2n}} \tag{7.19}$$

Thus, we get

$$n = -\frac{\ln(1-\alpha)}{2e^2} \tag{7.20}$$

Equation (7.20) is the relationship between the small gallery size and the confidence interval under the given margin of error for the system with the underlying distribution.

In the above we assume that a recognition system can recognize object with a certain distribution. If we do not know the underlying distribution of the recognition system, then we can use the Chebychev inequality [22] which is distribution independent. Assume $X_1, X_2, \ldots, X_n$ are independent random variables. We define X as

$$X = \frac{1}{n}\sum_{i=1}^{n} X_i \tag{7.21}$$

For any $\varepsilon \geq 0$, we have

$$P\{|X - E(X)| \geq \varepsilon\} \leq \frac{\sigma^2}{n\varepsilon^2} \tag{7.22}$$

where σ^2 is the variance of X. Comparing with (7.12), we have

$$1 - \alpha = \frac{\sigma^2}{n\varepsilon^2} \tag{7.23}$$

From the above equation, we obtain

$$\varepsilon = \frac{\sigma}{\sqrt{n(1-\alpha)}} \tag{7.24}$$

From (7.22), (7.23), and (7.24) we have

$$P\{X \geq E(X) + \frac{\sigma}{\sqrt{2n(1-\alpha)}}\} \leq (1-\alpha) \tag{7.25}$$

Then

$$e = \frac{\sigma}{\sqrt{2n(1-\alpha)}} \tag{7.26}$$

So we have,

$$n = \frac{\sigma^2}{2(1-\alpha)e^2} \tag{7.27}$$

From equation (7.27), we obtain the relationship between the small gallery size and the confidence interval under the given margin of error for the system without the assumption of the underlying distribution. It is known that the Chernoff inequality is much tighter than the Chebychev inequality and the Chebychev inequality is distribution independent.

In the above we provide a statistical estimation of the small gallery size. Meanwhile, in our approach presented in Sect. 7.3.1, we learn the similarity score distribution to find the optimal size of the small gallery. The small gallery size which we get from the statistics can be used as a guide for learning. Under the assumptions that the randomly chosen small galleries can represent the distributions of similarity scores for other galleries of the same size, we use different small galleries with the learned optimal size to predict large population performance. We randomly choose several small galleries of the optimal size to predict the large population performance. Then, we obtain the maximum and minimum prediction performance on the large population. In this way, we can provide the upper bound and the good lower bound for performance prediction on large populations.

7.4 Experimental Results

In all the experiments, we use fingerprints from the *NIST Special Database 4* (NIST-4). It consists of 2,000 pairs of fingerprints. Each of the fingerprints is labeled with an ID number preceded by an "f" or an "s," which represents different impressions of the same fingerprint. The images are collected by scanning inked fingerprints from paper. The resolution of the fingerprint image is 500 DPI and the size of the image is 480 × 512 pixels.

7.4.1 Prediction Model

Distorted Data

Since large populations will have distortions which may not be presented in the small gallery, we simulate the distortion in our prediction model to estimate the recognition performance based on small galleries. The minutiae features used for the fingerprint recognition can be expressed as $f = (x, y, c, d)$, where x and y are the locations of a minutiae, c is the class of the minutiae which represents whether the minutiae is endpoint (0) or bifurcation (1), and d is the direction of the minutiae. We define the amount of the minutiae distortion for a fingerprint as $g\%$. In this chapter, we choose $g = 5\%$. Assume the number of minutiae is num_j. Usually one pair of fingerprints has a different number of minutiae so $j = 1, 2, \ldots, 4000$. We apply the distortion model [13] to these 2,000 pairs of fingerprints as follows:

(a) *Uncertainty*: Uniformly choose $U = 5\% \times num_j$ minutiae features out of num_j features and replace each $f_i = (x, y, c, d)$ with f'_i chosen uniformly from the set $\{(x', y', c', d')\}$, where $(x', y') \in 4NEIGHBOR(x, y)$, $c' = 1 - c, d' = d \pm 3°, i = 1, 2, \ldots, U$.

(b) *Occlusion*: Uniformly choose $O = 5\% \times num_j$ minutiae features out of num_j features and remove these minutiae.

(c) *Clutter*: Add $C = 5\% \times num_j$ additional minutiae, where each minutiae is generated by picking a feature uniformly at random from the clutter region. Here we choose the clutter region as $CR = \{(x, y, c, d),\ 50 \leq x \leq 450,\ 60 \leq y \leq 480,\ c = \{0, 1, 2, 3, 4\}, 10° \leq d \leq 350°\}$.

In our experiments we use the uniform distribution as the uncertainty *PDF* and the clutter *PDF*. The number of features with uncertainty, occlusion, and clutter is the same. We use the algorithm provided in [23] to extract minutiae and algorithm [21] for matching.

Verification

We use the algorithm provided in [23] to extract minutiae. Suppose there are M and Q minutiae in the gallery and query fingerprints, respectively. Δ_{m} and Δ_{q} are potential corresponding triangles. We assume $F(s, \theta, t_x, t_y)$ is the transformation between the query and gallery fingerprints, where s is a scale parameter, θ is a rotation parameter, t_x and t_y are translation parameters. If these parameters are within limits [21], then we apply this transformation as the transformation between potential corresponding triangles Δ_{m} and Δ_{q}. The details of how to estimate the transformation parameters are explained in [21]. Based on the transformation $F(s, \theta, t_x, t_y)$, we compute the distance d by using equation (7.28),

$$d = \underset{i}{\arg\min} \left\{ \left| \hat{F}\left(\begin{bmatrix} x_{j,1} \\ x_{j,2} \end{bmatrix} \right) - \begin{bmatrix} y_{i,1} \\ y_{i,2} \end{bmatrix} \right| \right\} \tag{7.28}$$

where $(x_{j,1}, x_{j,2})$ and $(y_{i,1}, y_{i,2})$ are two sets of minutiae in the gallery and query fingerprints, $j = 1, 2, \ldots M$ and $i = 1, 2, \ldots Q$. If d is smaller than a threshold, then we can say that $(x_{j,1}, x_{j,2})$ and $(y_{i,1}, y_{i,2})$ are the corresponding points. If the number of corresponding points is greater than a threshold [21], then we define Δ_{m} and Δ_{q} as the corresponding triangles between the template and the query fingerprints. The final match score is the number of corresponding triangles between the query and template fingerprints.

Prediction Results

We randomly choose 50 pairs of fingerprints from two kinds of fingerprint pairs (with and without distortion) as our small gallery following a hypergeometric distribution. For this small gallery, we get 50 match scores and 2,450 nonmatch scores. After we obtain these similarity scores we use the EM algorithm to estimate the match score distribution and the nonmatch score distribution. The EM algorithm can find the number of components automatically [24] and for each component the EM algorithm finds its mean, variance, and weight. In this chapter, the similarity scores are the number of matched triangles between two fingerprints, the match scores are positive integers and the nonmatch

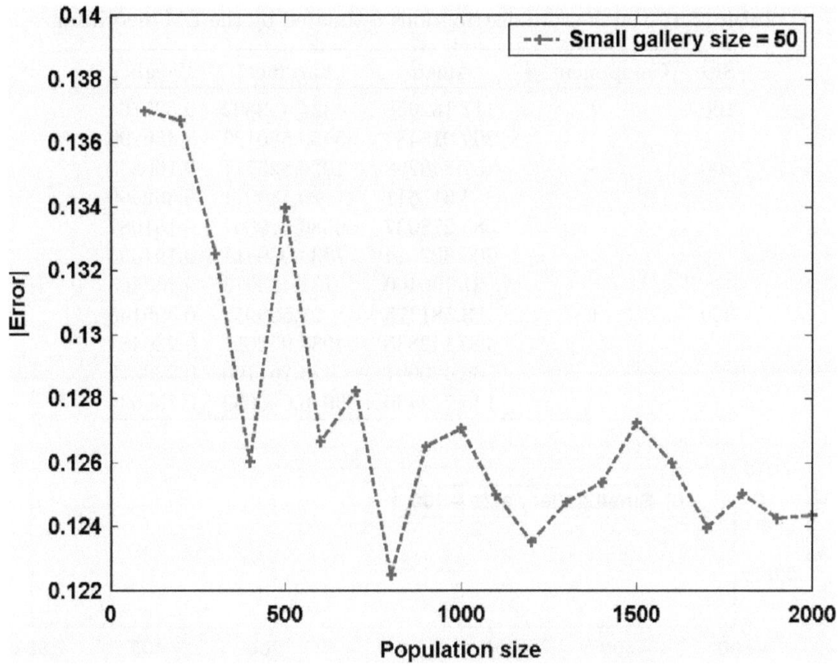

Fig. 7.2. Absolute error between the prediction and the actual performance when the small gallery size is 50

scores are close to 0. By applying the prediction model, we can estimate the fingerprints recognition performance on 2,000 pairs of fingerprints based on these 50 pairs of fingerprints. We repeat the experiment seven times and average the results to obtain the prediction performance which is shown in Fig. 7.2. Here, we choose the subset size $N_1 = 100$ and the margin of error $e = 0.06$. From this curve, we can see that for the large population size 100 the error between the prediction performance and the actual performance is 0.137, which is larger than the margin of error.

Now, we apply learning to the prediction process. We increase the small gallery size to $n = 100$. We feed back the match score and the nonmatch score from the randomly selected 100 pairs of fingerprint and repeat this process seven times. When the large population size is 100, the absolute error between the prediction performance and the actual performance is 0.135, which is greater than the margin of error 0.06. So, we increase the small gallery size to $n = 200$ and repeat the same process seven times. The absolute error is 0.09 when the large population size is 100. Then, we increase the small gallery size to $n = 300$ and repeat the same process seven times. The absolute error is 0.042 when the large population size is 100. We increase the large population size in steps of 100 until the large population size $N = 2,000$. For these three small galleries, most of the nonmatch scores are 0. Table 7.1 shows the

Table 7.1. Match score distributions estimated by the EM algorithm

Size	Component #	Mean	Variance	Weight
100	2	17.152658	334.452802	0.535764
		299.015489	55459.580193	0.450296
200	5	57.348298	1026.825771	0.160830
		3.615611	20.189071	0.362406
		585.278037	66686.529667	0.151087
		206.327514	7334.980411	0.191394
		27.106400	131.423073	0.133465
300	4	3.581775	21.569950	0.395165
		420.142835	64933.952657	0.236481
		35.420091	423.267100	0.228275
		143.774430	3016.000039	0.139634

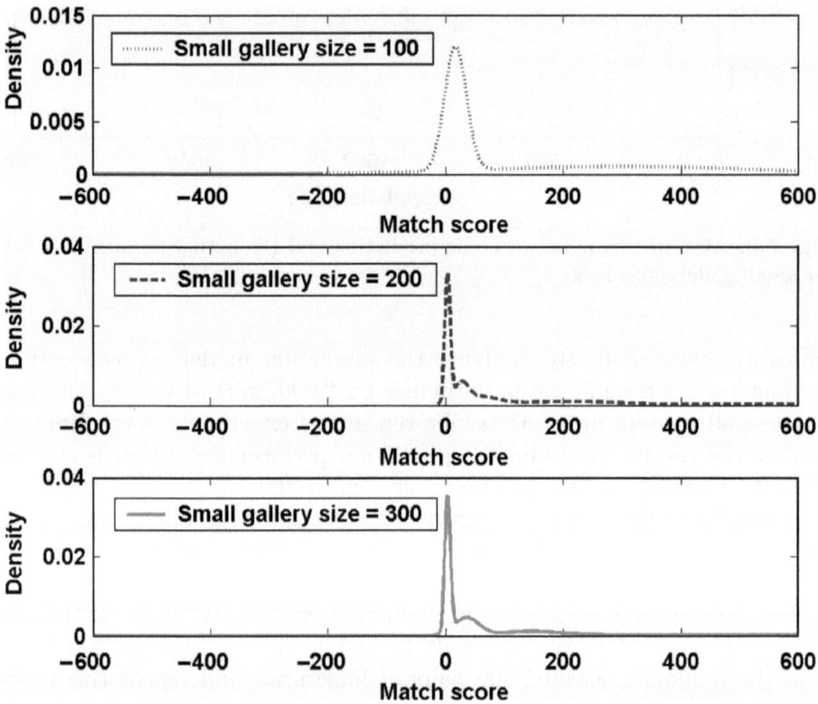

Fig. 7.3. Match score distributions for different small gallery sizes

estimation of the match score distributions with different small gallery sizes. The distributions are represented by the Gaussian mixture model. For each component we have its mean, covariance, and weight. Figure 7.3 shows the match score distribution curves on different small gallery sizes. For each small

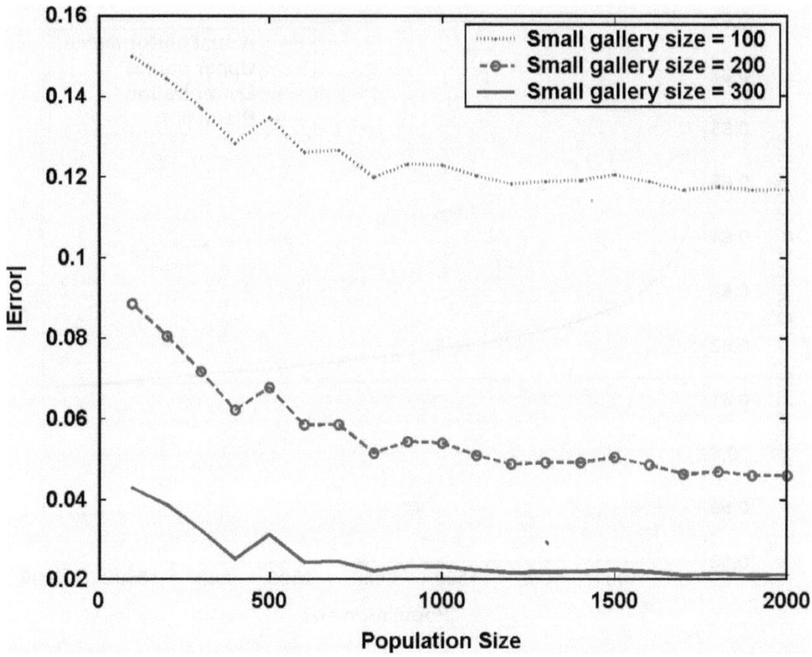

Fig. 7.4. Absolute error between the predicted and actual performance for different small gallery sizes

gallery size we provide two figures with different range of match score (X-axis) so that the distribution can be more closely examined. Figure 7.4 shows the absolute error between the prediction and the actual performance decreases when the gallery size increases. When the small gallery size is $n = 300$, the absolute error for the large population is smaller than the margin of error 0.06. At this point we can stop learning the small gallery size.

We use different small galleries with the learned optimal size to predict large population performance. Then, we select the maximum and the minimum prediction performance as our upper bound and lower bound for the performance prediction on the large population. Figure 7.5 gives the upper bound and lower bound on the prediction of large population performance when the small gallery size $n = 300$. Since we have 2,000 pairs of fingerprints, the actual recognition performance for the distorted images is shown in Fig. 7.5. Beyond this population size we can give the bounds for the prediction. From Fig. 7.5 it can be seen that the actual performance is within the upper bound and lower bound except when the population size is very small. Our experiments show that when the small gallery size is $n = 300$ the prediction error is less than 0.05.

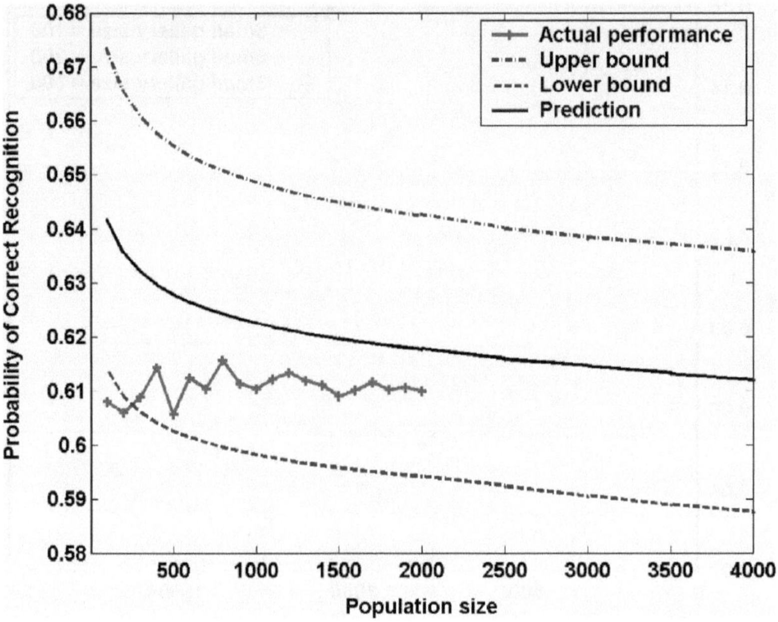

Fig. 7.5. The upper bound and lower bound on the large population when the small gallery size is 300. Note that the upper bound and lower bound are within 5%.

Table 7.2. Values of the confidence interval, the margin of error, and the small gallery size for Chernoff inequality and Chebychev inequality ($\sigma^2 = 1$)

$1 - \alpha$	0.05	0.05	0.1	0.1	0.15	0.15
e	0.06	0.04	0.06	0.04	0.06	0.04
n(Chernoff)	417	937	320	720	264	593
n(Chebychev)	2,778	6,250	1,389	3,125	926	2,083

7.4.2 Estimation of The Small Gallery Size Based on Statistical Inequalities

Table 7.2 shows different small gallery sizes given different confidence intervals and margins of error for Chernoff inequality and Chebychev inequality ($\sigma^2 = 1$). From the table we ascertain that the Chernoff inequality is much tighter than the Chebychev inequality. We compare our learning small gallery size with the Chernoff inequality. When the confidence interval $\alpha = 95\%$ and margin of error $e = 0.06$ then the small gallery size $n = 417$. From our experiment for the same margin of error the small gallery size is 300 and the confidence interval is $\alpha = 95\%$. Note that statistical methods give us a loose estimate of the small gallery size. Based on our recognition system we find a more accurate small gallery size by learning.

7.5 Conclusions

We focused on the problem of performance characterization of a simplified CBR system. In particular, we addressed the following questions: what is the optimal size of the small gallery that can give good error estimation and what is the confidence in the estimation? We use a generalized prediction model that combines a hypergeometric probability distribution model with a binomial model, taking into account distortion in large populations. We incorporate learning in the prediction process to find the optimal small gallery size and provide the upper and lower bounds for the performance prediction on large populations. The Chernoff inequality and the Chebychev inequality are used as a guide to obtain the small gallery size and the confidence interval given a margin of error. Experimental results show that the small gallery size obtained from the statistical methods are loose compared to the size provided by the proposed learning method. We believe that the methodology and results of this research will be useful for a wide range of applications of CBR in signal processing, image processing, computer vision, and pattern recognition.

References

1. Kolodner J (1993) Case-based Reasoning. Morgan Kaufmann Publisher.
2. Perner P, Perner H, Jänichen S (2006) Recognition of airborne fungi spores in digital microscopic images. J Arificial Intelligence in Medicine AIM, Special Issue on CBR, vol. 36, no. 2, pp. 137–157.
3. Perner P, Jänichen S (2004) Case acquisition and case mining for case-based object recognition. In: Peter Funk, Pedro A. González Calero (Eds.), Advances in Case-Based Reasoning, ECCBR2004, Springer Verlag 2004, vol. 3155, pp. 616–629.
4. Ming J, Bhanu B (1997) ORACLE: An integrated learning approach for object recognition. J Pattern Recognition and Artificial Intelligence, vol. 11, no. 6, pp. 961–990.
5. Minor M, Hanft A (1999) Cases with a life-cycle. In: Proc. Intl. Conf. on Case-Based Reasoning Workshops, 1999, pp. 3–8.
6. Guyon I, Makhoul J (1998) What size test set gives good error rate estimates? J IEEE Trans. on Pattern Analysis and Machine Intelligence, vol. 20, no. 1, pp. 52–64.
7. Mitchell TM (1997) Machine Learning. McGraw Hill.
8. Duda RO, Hart PE, Stork DG (2000) Pattern Classification. Wiley-Interscience Publication.
9. Wang R, Bhanu B (2007) Predicting fingerprint biometric performance from a small gallery. J Pattern Recognition Letters, vol. 28, pp. 40–48.
10. Wayman JL (1999) Error-rate equations for the general biometric system. J IEEE Robotics & Automation Magazine, vol. 6, issue 1, pp. 35–48.
11. Daugman J (2003) The importance of being random: statistical principles of iris recognition. J Pattern Recognition, vol. 36, no. 2, pp. 279–291.
12. Phillips PJ, Grother P, Micheals RJ, Blackburn DM, Tabassi E, and Bone M (2003) Face recognition vendor test 2002, Evaluation Report.

13. Wang R, Bhanu B, Chen H (2005) An integrated prediction model for biometrics. In: Proc. Audio- and Video-based Biometric Person Authentication, New York, pp. 355–364.
14. Johnson AY, Sun J, Boick AF (2003) Using similarity scores from a small gallery to estimate recognition performance for large galleries. In: Proc. IEEE Int. Workshop on Analysis and Modeling of Faces and Gestures, pp. 100–103.
15. Grother P, Phillips PJ (2004) Models of large population recognition performance. In: Proc. IEEE Conf. on Computer Vision and Pattern Recognition, vol. 2, pp. 68–75.
16. Lindenbaum M (1995) Bounds on shape recognition performance. J Pattern Analysis and Machine Intelligence, vol. 17, no. 7, pp. 666–680.
17. Lindenbaum M (1997) An integrated model for evaluating the amount of data required for reliable recognition. J Pattern Analysis and Machine Intelligence, vol. 19, no. 11, pp. 1251–1264.
18. Boshra M, Bhanu B (2000) Predicting performance of object recognition. J Pattern Analysis and Machine Intelligence, vol. 22, no. 9, pp. 956–969.
19. Boshra M, Bhanu B (2001) Predicting an upper bound on SAR ATR performance. J IEEE Trans. on Aerospace and Electronic Systems, vol. 37, no. 3, pp. 876–888.
20. Wang R, Bhanu B (2005) Learning models for predicting recognition performance. In: Proc. IEEE International Conf. on Computer Vision, vol. 2, pp. 1613–1618.
21. Tan X, Bhanu B (2002) Robust fingerprint identification. In: Proc. IEEE Int. Conf. on Image Processing, vol. 1, pp. 277–280.
22. Mood AM, Graybill FA, Boes DC (1974) Introduction to the Theory of Statistics. McGraw,Hill.
23. Bhanu B, Boshra M, Tan X (2000) Logical templates for feature extraction in fingerprint images. In: Proc. Int. Conf. on Pattern Recognition, vol. 3, pp. 850–854.
24. Figueiredo MAT, Jain AK (2002) Supervised learning of finite mixture models. J IEEE Trans. on Pattern Analysis and Machine Intelligence, vol. 24, no. 3, pp. 381–396.

8

A CBR Agent for Monitoring the Carbon Dioxide Exchange Rate from Satellite Images

J.M. Corchado, J. Aiken, and J. Bajo

Departamento Informática y Automática
Universidad de Salamanca
Plaza de la Merced s/n, 37008, Salamanca, Spain
corchado@usal.es
Centre for Air–Sea Interactions and fluxes
Plymouth Marine Laboratory, Prospect Place, Plymouth, PL1 3 DH, UK
jai@mail.pml.ac.uk
Universidad Pontificia de Salamanca
Compañía 5, 37002, Salamanca, Spain
jbajope@upsa.es

Summary. This work presents a multiagent system for evaluating automatically the interaction that exists between the atmosphere and the ocean surface. monitoring and evaluating within the ocean carbon dioxide exchange process is a function requiring working with a great amount of data: satellite images and in situ Vessel's data. The system presented in this work focuses on Ambient Intelligence (AmI) technologies since the vision of AmI assumes seamless, unobtrusive, and often invisible but also controllable interactions between humans and technology. The work presents the construction of an open multiagent architecture which, based on the use of deliberative agents incorporating Case-Based Reasoning (CBR) systems, offers a distributed model for such an interaction. This work also presents an analysis and design methodology that facilitates the implementation of CBR agent-based distributed artificial intelligent systems. Moreover, the architecture takes into account the fact that the working environment is dynamic and therefore it requires autonomous models that evolve overtime. In order to resolve this problem an intelligent environment has been developed, based on the use of CBR agents, which are capable of handling several goals, constructing plans from the data obtained through satellite images and research Vessels, acquiring knowledge, and of adapting to environmental changes, are incorporated. The artificial intelligence system has been successfully tested in the North Atlantic ocean, and the results obtained will be presented within this work.

8.1 Introduction

Ambient intelligent environments are characterized by their ubiquity, transparency, and intelligence [2]. The agents and multiagent systems (MASs) have become increasingly relevant for developing distributed and dynamic

J.M. Corchado et al.: *A CBR Agent for Monitoring the Carbon Dioxide Exchange Rate from Satellite Images*, Studies in Computational Intelligence (SCI) **73**, 213–246 (2008)
www.springerlink.com

intelligent environments. Agents and MAS have become increasingly relevant for developing applications in the internet, personalized user interfaces, oceanography, control systems, or robotic environments. Agents can be characterized through their capacities in areas such as autonomy, reactivity, proactivity, social abilities, reasoning, learning, and mobility. These capacities which can be modelled in various ways, using different methodologies [47], make the agents and MAS highly suited to intelligent environments. One of the possibilities to model the reasoning capacity is to use Case-Based Reasoning (CBR). In this work we present a distributed architecture whose main characteristic is the use of Case-Based Reasoning–Beliefs Desires Intentions (CBR–BDI) agents [13], which will be named in this chapter as CBR agents. These agents are capable of learning, from their initial knowledge and by interacting autonomously with their environment and with users of the system, adapting themselves accordingly. Both, the developed multiagent architecture and the Modelling agent are described in detail. The development of ambient intelligent systems is normally complicated due it novel connotations, this work presents a practical way for analyzing and designing MAS at the same time of describing the distributed ambient intelligent system developed.

The mission of the intelligent environment presented in this work, is to globally monitor the interaction between the ocean surface and the atmosphere, facilitating the work of oceanographers. Initially, the system is being used in order to evaluate and predict the amount of carbon dioxide (CO_2) absorbed or expelled by the ocean in the North Atlantic [6, 7, 15]. The main purpose of this work is to obtain an architecture that enables the construction of open, distributed, and dynamic systems capable of growing in dimension and of adapting their knowledge according to different changes that take place in their environment. There are many different architectures for constructing deliberative agents and many of them are based on the BDI model. In the BDI model, the internal structure of an agent and its capacity to choose is based on mental aptitudes. This has the advantage that it uses a natural model (human) and a high level of abstraction. The BDI model uses the agent' Beliefs as informational aptitudes, its Desires as motivational aptitudes and its Intentions as deliberative aptitudes. The method proposed in [5, 12, 13] facilitates the incorporation of CBR systems as a deliberative mechanism within BDI agents, allowing them to learn and adapt themselves, and lending them a greater level of autonomy than pure BDI architecture [25]. Moreover, this proposal differs from others [10, 22, 31, 36, 46] in that it proposes direct mapping between the concept of the agents and their implementation. CBR systems are also highly suited to some of the tasks in the study of carbon dioxide exchange between the ocean and the atmosphere, such as the interpretation of satellite images [40].

One of the major problems in the development of an architecture based on MAS is that there are currently no clear standards or well developed methodologies for defining the steps of analysis and design that need to be taken in order to define an intelligent environment. There are at present a number of methodologies: Gaia [48], AUML [8, 34, 35], MAS-CommonKADS [26],

MaSE [18], ZEUS [33], MESSAGE [21]. The problem with these methodologies is that they are generally not fully developed and present a number of limitations. For this study, we have decided to opt for a combination of elements from Gaia and Agent Unified Modelling Language (AUML) for our MAS. Gaia is a simple methodology that allows us to carry out a preliminary analysis and design with which to confront the problem at a general level. The great advantage is that we can carry out a rapid, broad study but problems arise when the design is at its completion because there tends to be an overly high level of abstraction. AUML, on the other hand, offers mechanisms which allow us to obtain a design that is sufficiently precise and able to pass directly to the implementation stage, but has the disadvantage of being too precise and detailed for the preliminary stages. Our goal is to take advantage of both methodologies by carrying out a preliminary analysis and design with Gaia and later on to carry out the appropriate changes by using a detailed AUML design. In this way we are able to obtain both a generalized vision of the problem in terms of organization, and a detailed MAS description which helps enormously in the development of such a research project.

In order to implement the BDI agents, various tools are used. One interesting tool is Jadex [41], which incorporates the BDI architecture into Jade agents [9]. In this sense, Jadex agents work with concepts of beliefs, goals, and plans, all of which become objects which can be created and manipulated within the agent. The beliefs represent any type of Java object and are stored in the beliefs' database. The goals represent specific motivations that influence the behaviour of the agent. The plans are procedures written in Java which are executed in order to reach the goals. Jadex has the advantage to allow the programmer to introduce his own deliberative planning mechanisms. In our case, this mechanism will be a CBR system. In addition, it offers all the advantages of Jade and allows the use of Jade and Jadex agents within the same MAS. The MAS incorporates "lightweight" agents that can live in mobile devices, such as phones, personal digital assistants (PDAs), etc. These agents make it possible for a oceanographer to interact with the MAS in a very simple way, downloading and installing a personal agent in his mobile phone or PDA.

In Sect. 8.2, we will explain the various relationships that can be established between CBR and BDI concepts. In Sect. 8.3 we will describe the oceanic/atmospheric problem that has led to most of this research. In Sect. 8.4, the MAS developed will be described, paying special attention to the CBR agents. Finally, some preliminary results and the conclusions will be presented.

8.2 CBR–BDI Agents

Ambient Intelligence has been widely studied and different artificial intelligence techniques have been applied. The application of agents and MAS facilitates taking advantage of the agent capabilities, such as mobility, proactivity, or social abilities, as well as the possibility of solving problems in a distributed way. Agents, in the context of an intelligent environment, must be able to respond to events, take the initiative according to their goals, communicate with other agents, interact with users, and make use of past experiences to find the best ways to achieve goals. There are many architectures for constructing deliberative agents and many of them are based on the BDI model [27,28]. In the BDI model, the internal structure of an agent and its capacity to choose, is based on mental aptitudes: agent behaviour is composed of beliefs, desires, and intentions [42]. The beliefs represent its information state, what the agent knows about itself and its environment. The desires are its motivation state, what the agent is trying to achieve. And the intentions represent the agent's deliberative states. Intentions are sequences of actions; they can be identified as plans. A BDI architecture has the advantage that it is intuitive and relatively simple to identify the process of decision-making and how to perform it. Furthermore, the notions of belief, desire, and intention are easy to understand. On the other hand, its main drawback lies in finding a mechanism that permits its efficient implementation.

CBR is a type of reasoning based on the use of past experiences [28]. The purpose of CBR systems is to solve new problems by adapting solutions that have been used to solve similar problems in the past. The fundamental concept when working with CBR is the concept of case. A case can be defined as a past experience, and is composed of three elements: A problem description which describes the initial problem, a solution which provides the sequence of actions carried out in order to solve the problem, and the final state which describes the state achieved once the solution was applied. A CBR system manages cases (past experiences) to solve new problems. The way in which cases are managed is known as the CBR cycle.

The deliberative agents, proposed in the framework of this investigation, use this concept to gain autonomy and improve their problem-solving capabilities. The method proposed in [13] facilitates the incorporation of CBR systems as a deliberative mechanism within BDI agents, allowing them to learn and adapt themselves, lending them a greater level of autonomy than pure BDI architecture [12]. Accordingly, CBR agents implemented using CBR systems could reason autonomously and therefore adapt themselves to environmental changes. The CBR system is completely integrated within the agents' architecture. The CBR–BDI agents incorporate a "formalism" which is easy to implement, in which the reasoning process is based on the concept of intention. Intentions can be seen as cases, which have to be retrieved, reused, revised, and retained. This makes the model unique in its conception and reasoning capacities. The structure of the CBR system has been designed around the concept of a case. A direct relationship between CBR systems and BDI agents can also be established if the problems are defined in the form of states and actions.

Case: <Problem, Solution, Result>	**BDI agent**
Problem: initial_state	Belief: state
Solution: sequence of <action, [intermediate_state]>	Intention: sequence of <action>
Result: final_state	Desire: set of <final_state>

The relationship between CBR systems and BDI agents can be established by implementing cases as beliefs, intentions, and desires which lead to the resolution of the problem. As described in [7, 16], in a CBR–BDI agent, each state is considered as a belief; the objective to be reached may also be a belief. The intentions are plans of actions that the agent has to carry out in order to achieve its objectives [11], so an intention is an ordered set of actions; each change from state to state is made after carrying out an action (the agent remembers the action carried out in the past, when it was in a specified state, and the subsequent result). A desire will be any of the final states reached in the past (if the agent has to deal with a situation, which is similar to a past one, it will try to achieve a similar result to that previously obtained).

8.3 Air–Sea Interaction Problem

One of the factors of greatest concern in climactic behaviour is the quantity of carbon dioxide present in the atmosphere. Carbon dioxide is one of the greenhouse gases that helps to make the earth's temperature habitable, so long it is maintains certain levels [43]. Traditionally, it has been considered that the main system regulating carbon dioxide in the atmosphere is the photosynthesis and respiration of plants. However, thanks to teledetection techniques, it has been shown that the ocean plays a highly important role in the regulation of carbon quantities, the full significance of which still needs to be determined [44]. Current technology allows us to obtain data and make calculations that were unthinkable some time ago. This data gives us an insight into the original source and the decrease in carbon dioxide as well as its causes [30], which allows us to make predictions on the behaviour of carbon dioxide in the future.

The need to quantify the carbon dioxide valence, and the exchange rate between the oceanic water surface and the atmosphere, has motivated us to develop the distributed system, presented here, that incorporates CBR agents capable of estimating such values using accumulated knowledge and updated information. The CBR agents receive data from satellites, oceanographic databases, oceanic, and commercial Vessels. The CBR system incorporated within the BDI agents allows the agents to optimize tasks such as the interpretation of images using various strategies [39]. The information received is composed of satellite images of the ocean surface, wind direction and strength, and other parameters such as water temperature, salinity, and fluorescence as can be

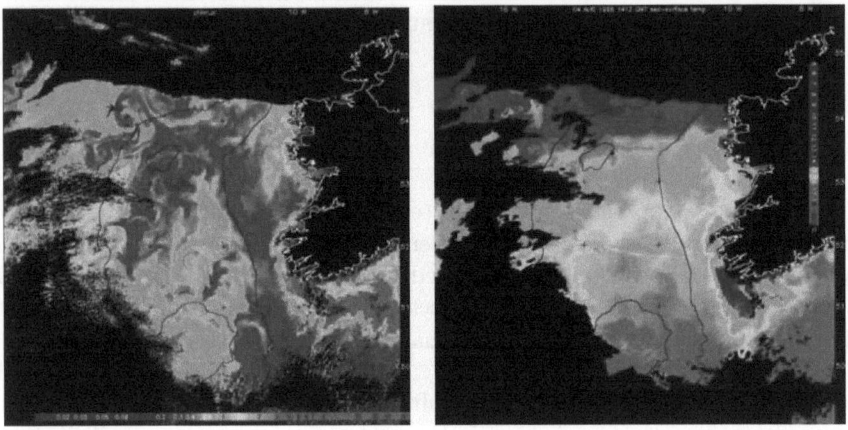

Fig. 8.1. Satellite colour pictures

seen in Fig. 8.1. An improvement over the monitoring and forecasting methods presented in [5, 6, 15] has been incorporated to the modelling CBR agents presented in this chapter.

The parameters obtained from the satellite images and which have most influence within our models are: temperature of the water and air, salinity of the water, wind strength, wind direction, and biological parameters such as chlorophyll. These parameters allow us to calculate the variables that define our models, such as the velocity of gas transfer, solubility, or the differentiation between partial pressures on the atmosphere and sea surface (a case structure is shown in Table 8.1). The majority of CO2 is dissolved in the sea water because of phytoplankton or accumulates at the bottom of the ocean in the form of organic material. The phytoplankton present in deep areas of the ocean is taken to the surface by surges or surface appearances that are no more than large upwards movements of cold water that bring nutrients to the sea surface. The principal cause of there surges are the winds. The way to detect them through satellites is to study the images captured with sensors that are sensitive to longitudes of thermal infrared waves (capable of detecting the sea surface temperature (SST)) and to identify the cold waters. Another possible way to detect them is to monitor the activity of the chlorophyll through sensors within the spectrum range found between blue and green which are associated with the presence of phytoplankton. In order to obtain the satellite images that contain information about these parameters it is necessary to use different sensors. The Earth Observation satellites that have been used to obtain images in the Northern Atlantic are NOAA, Orbview-2, and above all, the ENVISAT satellite of the European Space Agency. Below we shall briefly describe the sensor used in each one of these and the software for the digital processing of the images.

The thermal sensors allow us to measure the surface temperature of the sea. The NOAA satellites are equipped with the Advanced Very High-Resolution Radiometer (AVHRR) sensor that is capable of detecting electromagnetic energy reflected by objects present on the earth within five spectrum ranges (three bands in the visible and two in the thermal range). It has a receiving cycle of 12 hours with which it is possible to obtain up to six images per day at a resolution of 1 km^2. In order to determine the SST, the NOAA uses a multichannel algorithm for the water surface [29]. The ENVISAT satellite has an Avanced Along-Track Scaning Radiometer (AATSR) sensor with which it is capable of exploring the ocean surface at various infrared and visible frequencies in order to measure the exact temperature. Specifically, the temperature of the sea surface can be calculated with an accuracy of 0.3°C. [45]

There are also sensors that allow us to measure the concentration of chlorophyll. The Earth Observation satellite Orbview-2 uses a Sea-Viewing Wide Field-of-view Sensor (SeaWIFS) [29, 49], which is capable of giving images with information on eight bands or ranges of the electromagnetic spectrum. Of these eight bands, four around the blue–green are used for the detection of chlorophyll. In order to calculate these quantities, the Ocean Chlorophyll 4-band OCTS is used, included in the SeaWIFS Data Analysis System (SeaDAS) software developed by NASA [38]. The ENVISAT satellite has a Medium Resolution Image Spectrometer (MERIS) with which it is possible to take images of the planet surface and the clouds, capturing the light of the visible areas, and the infrared of the electromagnetic spectrum. In this way it is capable of knowing the exact colour of the ocean surface and coastal areas, from which it is possible to reflect the biological activity, to monitor cloud cover and to detect the vapour of the invisible water into the atmosphere. [3, 4, 45]

The processing of the images obtained may vary depending on the sensor that has taken them [19, 20]. The processing of the images is carried out at the CAXIS centre at the Plymouth Marine Laboratory (PML). The processing of the thermal images is carried out initially by taking a reading of the images in their original format as they were received. Then a calibration and a radiometric correction is made in order to reduce the atmospheric effects and a reference is made to a known cartography base. The next step is to mask the clouds and the land in order to eliminate distortions. Lastly, the SST is calculated applying a suitable algorithm. In order to process images of the chlorophyll concentration a reading is made of the images and decoded when necessary. Meteorological and ozone files are requested by the software. The clouds and land are masked and the chlorophyll image is calculated. Lastly, reference is made to a known cartographic base and compositions and midpoint images are made, which can take some days.

The MAS presented is aimed at modelling the flux of carbon dioxide exchanged between the atmosphere and the ocean surface. The oceans contain approximately 50 times more carbon dioxide in dissolved forms than the atmosphere, while the land biosphere including the biota and soil carbon contains about three times as much carbon (in carbon dioxide form) as the atmosphere [44]. The carbon dioxide concentration in the atmosphere is governed primarily by the exchange of carbon dioxide with these two dynamic reservoirs. Since the beginning of the industrial era, about 2,000 billion tons of carbon have been released into the atmosphere as carbon dioxide from various industrial sources including fossil fuel combustion and cement production. This amount, which is about 35% of the total amount of carbon in the preindustrial level, corresponds to approximately 590 billion tons as carbon. At present, atmospheric carbon dioxide content is increasing at an annual rate of about 3 billion tons which corresponds to one-half of the annual emission rate of approximately 6 billion tons from fossil fuel combustion. Whether the missing carbon dioxide is mainly absorbed by the oceans or by the land and their ecosystems has been debated extensively over the past decade.

It is important, therefore, to fully understand the nature of the physical, chemical, and biological processes which govern the oceanic sink/source conditions for atmospheric carbon dioxide [30, 44]. Satellite-borne instruments provide high-precision, high-resolution data on atmosphere, ocean boundary layer properties and ocean biogeochemical variables, daily, globally, and in the long term. All these new sources of information have changed our approach to oceanography and the data generated needs to be fully exploited. Wind stress, wave breaking, and the damping of turbulence and ripples by surface slicks, all affect the air–sea exchange of carbon dioxide. These processes are closely linked to the "roughness" of the sea surface, which can be measured by satellite radars and microwave radiometers. Sea surface roughness consists of a hierarchy of smaller waves upon larger waves. Different sensors give subtly different measurements of this roughness.

Our final aim is to model both the open ocean and shelf seas and it is believed that by assimilating Earth Observation (EO) data into artificial intelligence models these problems may be solved. Earth Observation data (both for assimilation and for validation) are vital for the successful development of reliable models that can describe the complex physical and biogeochemical interactions involved in marine carbon cycling. Satellite information is vital for the construction of oceanographic models, and in this case, to produce estimates of air–sea fluxes of carbon dioxide with much higher spatial and temporal resolution, using artificial intelligence models than can be achieved realistically by direct in situ sampling of upper ocean carbon dioxide. To handle all the potentially useful data to create daily models in a reasonable time and with a reasonable cost, it is necessary to use automated distributed systems capable of incorporating new knowledge. Our proposal is presented in Sect. 8.4.

8.4 Air–Sea Interaction Multiagent System

The option chosen to define an appropriate analysis and design methodology for the problem to be resolved is one that combines Gaia [48] and Agent-UML (AUML) [8, 34, 35], in an attempt to take advantage of both. Through Gaia it is possible to make an analysis of the problem using organizational criteria and a later design. After applying Gaia, the result consists of a design at the elevated abstraction level. At this point the Gaia design is transformed so that Agent-UML techniques can be applied. Figure 8.2 illustrates the paths followed in order to obtain the different models used. It shows how Gaia is used initially in order to obtain an analysis and high-level design and then Agent-UML is used in order to obtain a detailed, low-level design.

8.4.1 Gaia Analysis and Design

Gaia is a methodology for analysis and design in agent-based systems. It is very general and therefore applicable to a very wide range of MAS. It also allows the user to have a wide knowledge of the MAS both at an organizational (social) level and at a detailed level for each agent [48]. Through the Gaia analysis, two models are obtained: the role model and the interaction model. We analyse a problem in terms of organization, first by analyzing the different roles that our system could play. Studying the requirements of the problem we have come to the conclusion that we need six roles: a STORING role, for obtaining data that should be permanently available and stored in a database; a PROCESSING role, for transforming the images from the satellite into cases; a DATACAPTURING role for obtaining the data from the Vessel; a CONSTRUCTAPARTIALCO_2MODEL role, for generating a model; an OBTAINCO_2EXCHANGE role for calculating the rate of CO2 exchange

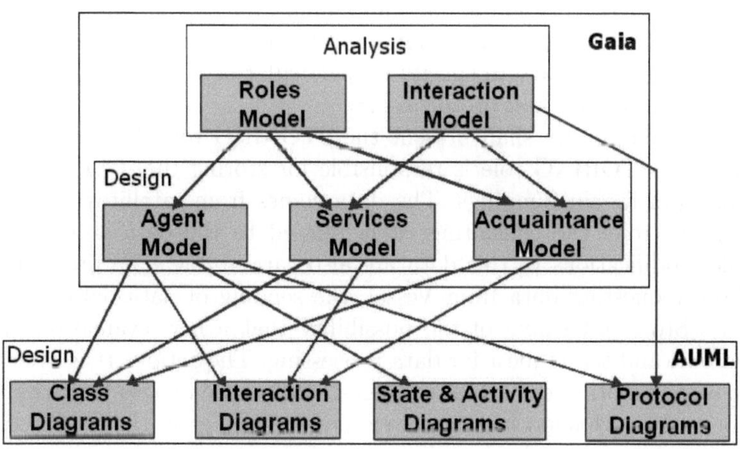

Fig. 8.2. Methodology followed

Role Schema: STORING (S)

Description:
Stores information about the environment i.e. raw data from satellites.
Besides allows consults on stored data.

Protocols and Activities:
StoringSatelliteData, StoringVesselData, RequestInsituData, ConsultCaData,
SendCaData, RequestProcessData, ChangeParamsStore, RequestEvaluation,
RequestCasesData

Permisions:
Reads: On-line Envisal Data, On-line Vessel Data

Changes: CaStore BD

Generates

Responsabilities:
Liveness:
STORESAT: (StoringSatelliteData)^{w}.RequestProcessData
STOREVESSELDATA: [RequestInsituData].StoringVesselData
.[RequestEvaluation]
CONSULTCA: ConsultCaData.[RequestCasesData].SendCaData
MODIFYCA: ChangeParamsStore
Safety: Succesful connection with satellite established.
Succesful connection with Vessel established.
Succesful connection with CaStore BD established.

Fig. 8.3. Gaia role model for the STORING role

using the data from the model; an AUTOEVALUATION role for assessing the model by contrasting the results offered by the model with the real data obtained by the sensors on the boat; and finally, a PROCESSINGINFORMATION role for allowing the user to interact with the system. For each of these roles, it will be necessary to specify its particular attributes: responsibilities, permission, activities, and protocols [48].

As an example, we shall present the STORING role: In Fig. 8.3 we can see how the STORING role is responsible for storing the data fed into the systems from its surroundings. The data comes from satellites or ships. As part of this role, the consulting tasks related to the data stored and the possible modifications to this data are also carried out. The protocols used are those requesting data from Vessel, the sending of data obtained to the database Store, informing of the possibility that a new evaluation may be carried out, and the request for data processing. The actions that are carried out consist of storing the satellite images, storing the images from the Vessel sensors, making changes in the storage parameters, or in the database data, when necessary. The role must have permission to access the data and the Vessel data via satellite. In addition, it must have permission for reading and writing over the Store database where the data received is stored. Its liveness responsibilities are as follows: STORESAT which continually stores the data received via satellite. When new data is received, it is stored in "raw" format and then a request is made to the PROCESSING role to carry out a data processing action. STOREVESSELDATA is responsible for storing the data that is received from the Vessels. In order for the data to arrive, there exist two possibilities, the data is either requested from the boat, or the boat sends them under its own initiative. In the first case, the sequence is to carry out a data request and store the data. In the second case, the sequence is to store the data, and the appropriate role is informed of the possibility to carry out an autoassessment of the current model with the new data. CONSULTCA carries out consultations concerning the Store database. MODIFYCA makes it possible to carry out modifications both on the parameters of the storage of the database Store, and on the data stored there. Lastly, the safety responsibilities that the STORING role has are those which can establish a valid connection either with the satellite and the Vessel, or with the database.

Once the role model has been obtained, the Gaia analysis is completed with the interaction model. The interaction model shows us the dependences and relationships between the roles. For each interaction between two roles there is a protocol. For our MAS, we have decided to use the following interaction protocols: ObtainVesselData is formed by protocols between the STORING role and the DATACAPTURING role, whereby the first protocol requests the data in situ from the Vessel (for a specific date) and the second protocol ensures that it is given. ObtainConstructData is an interaction through which the CPCM (CONSTRUCTAPARTIALCO_2MODEL) role wishes to construct a new model and in order to do so, requests new cases from the PROCESSING role. The PROCESSING role responds with the requested information. ObtainInsituData allows the AE (AUTOEVALUATION) role to obtain current data in situ aboard the Vessel. To do this, it is necessary to make a request to the STORING role to carry out a consultation of the Store database. In case the data requested is not available, a request will be made to the DATACAPTURING role to obtain it. ObtainStExchange is used by the PI (PROCESSINGINFORMATION) role to obtain the rate of exchange of CO_2 that is produced when applying the current model. OCE (OBTAINCO_2EXCHANGE) consults the model database and calculates the exchange. ObtainNewModelSuper allows the PI role to request the creation of a new model. In order to do this, it makes a request from CPCM, which will in turn have to consult if there are new cases available within the case database. ObtainNewModelAuto allows the AE role to request the creation of a new model in case the current one is considered inadequate. ObtainNewModelStoring is the protocol that is executed when the PROCESSING role informs the CPCM role that new images have arrived from the satellite and have been transformed. With the new data, a new model can be created. ObtainStModel enables the PI role to consult the information associated with a certain model. ObtainStore allows the PI role to consult the data stored in the Store database. In order to do this, it makes a request to the STORING role. ObtainVessel allows the PI role to consult the data stored in the EPROM memory of the Vessel. In order to do this it makes a request to DATACAPTURING role. ObtainEvaluationSuper is the protocol with which PI requests an evaluation of a model. To do this, it makes a request to the AE role. The AE role needs to know the current data in situ in order to make the evaluation. ObtainEvalautionDC enables the DATACAPTURING role to inform the AE role that the Vessel has carried out a new data collection. This implies that AE is able to carry out an evaluation of the current model. If the evaluation is not satisfactory, a request is made to generate a new model. The DATACAPTURING role does not have any direct communication with the AE role so it must make the request to the STORING role, which acts as an intermediary. Activate/Deactivate Sensors allow the PI role to activate or deactivate the Vessel sensors. In order to do this, it is necessary for it to communicate with the DATACAPTURING role. Delete EPROM allows the PI role to delete all the data from the Vessel's EPROM memory. ChangeStore enables the PI role to modify the storage parameters of the Store data. It communicates with the STORING role that carries out the modifications. ChangeCase allows the PI role to modify the storage parameters of the case memory store. To do this it needs to communicate with the PROCESSING role.

Figure 8.4 illustrates the interaction ObtainVesselData. It shows that the interaction uses two different protocols. In each protocol, a textual description indicates the type of interaction (RequestInsituData and SendInsituData), the role that initiates the interaction (STORING in the first one, and DATACAPTURING in the second one), and the role to which it is directed (DATACAPTURING in the first one and STORING in the second one), and a description of the process that is carried out during the interaction. Moreover, the entry information given by the role that initiates the interaction (InsituDataFromAGivenDate in the second protocol) and the exit information given by the role to which the interaction is directed (VesselInsituData in the second protocol) is also shown.

Once the analysis has been finalized, the Gaia design is carried out. Traditional techniques of software engineering are not followed in terms of detailing the analysis to the extent that a direct implementation can be made. Instead, the level of abstraction is reduced so that traditional techniques can be applied. In the design process three models are considered: agent model, services model, and acquaintance model [48]. The agent model shows the types of agents that are going to appear in the system, as well as the number of instances for each agent type that can be executed within the execution time.

Using the role models as a base, we have decided to use five types of agents: Store, Vessel, Modelling, User, and SuperUser. As Fig. 8.5 illustrates, each agent is responsible for carrying out some particular roles. For example, Store agent is responsible for carrying out STORING and PROCESSING

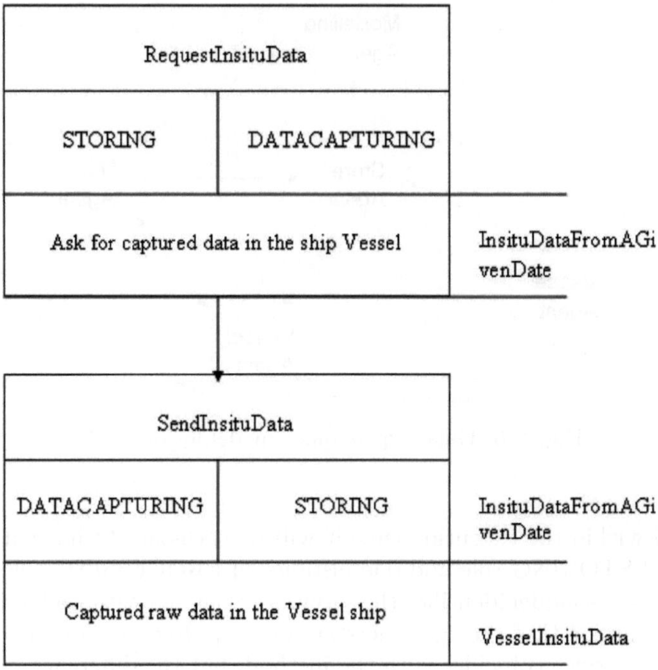

Fig. 8.4. Protocols for the ObtainVesselData interaction

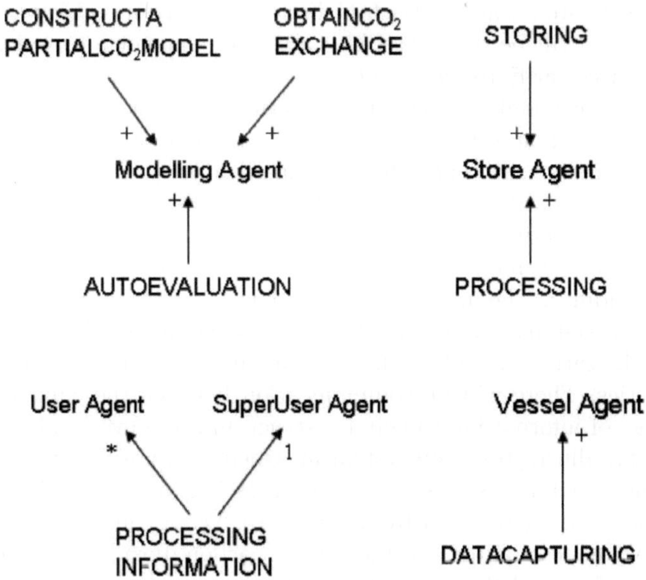

Fig. 8.5. Gaia agents model for our MAS

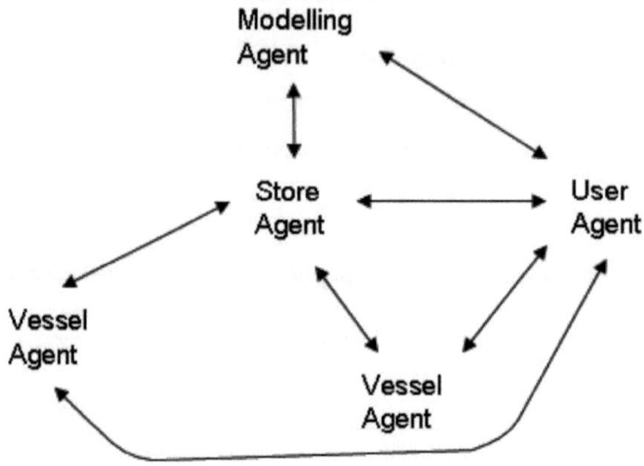

Fig. 8.6. Gaia acquaintance model for our MAS

roles, and within the execution time it will be necessary to have at least one instance of STORING role and one instance of PROCESSING role.

The services model identifies the services associated with each role, being a service a function that the agent needs to develop. In object oriented methodologies, the service coincides with the method, but the difference here is that they would not be available for other agents in the way methods were for other objects. A service will serve as a block of simple, individual, and coherent activities that create an agent. Each Gaia activity corresponds to a service, in other words, it will be developed into a service but not all services need to correspond necessarily to an activity.

Lastly, the acquaintance model defines the communication links that exist between the different types of agents. As can be seen in Fig. 8.6, the messages – or formats – are not defined, even when they are sent, but are only responsible for indicating the communication paths. At this point, it may be of interest to introduce a diagram that shows the architecture of our system, indicating the number of agents, the relationships between them and their surroundings. The aim of this project is to construct a MAS composed of various subsystems. Each one of these subsystems will be responsible for modelling the carbon dioxide exchange in an area of the ocean with particular characteristics. There will be communication between the subsystems, with an exchange of information to help construct and improve local models.

Figure 8.7 illustrates a subsystem in which is it possible to observe how a Modelling agent is responsible for the creation and evaluation of models in terms of the data received from the Store, Vessel, and User agents. This model allows us to monitor and predict the carbon dioxide exchange between the ocean surface and the atmosphere. The Store agent processes the images from the satellite and transforms them for use by the system. Each Vessel agent is installed in a ship and collects information in situ that allows us to evaluate the models created by the Modelling agent. The User agent can

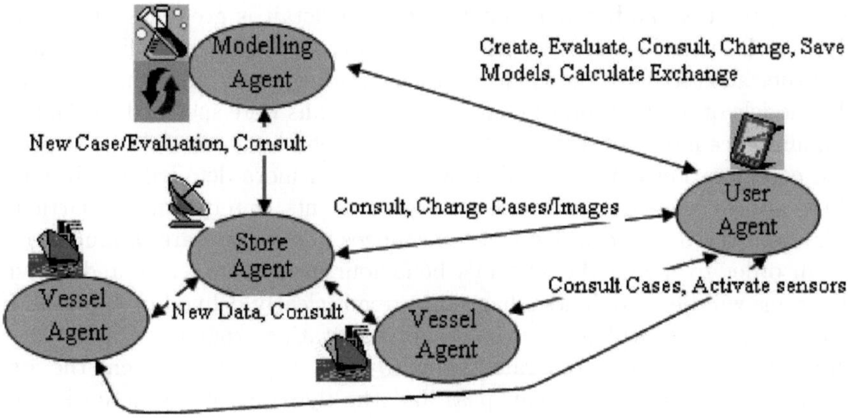

Fig. 8.7. Diagram of the architecture of our MAS

interact with any of the other agents. Figure 8.7 shows how the agents interact with each other and with their surroundings. From the oceanographic point of view, in order to resolve the problem that confronts us, the ocean has been divided into a series of zones. In each of these zones there will be a Modelling agent, a Store agent, and various Vessel agents.

8.4.2 Detailed Agent-UML Design

As well as presenting the proposal for the Agent-UML design established for the problem, it is necessary to establish a relationship between the Gaia methodology used during the analysis and the Agent-UML methodology. Moreover, its use needs to be justified. In contrast to Gaia, Agent-UML works at a highly detailed level, perhaps too high in its initial stages for large scale problems, as it is our case. We propose to use the low-level analysis made by Gaia and develop a design with Agent-UML, at a low level, but with sufficient detail to proceed with the implementation.

There are three concepts that diverge slightly between the meaning from the Gaia methodology and the Agent-UML methodology. Firstly, in Agent-UML a role is considered the result of social restrictions and individual behaviours and refers to the organization. Specifically, it makes reference to the behaviour of an agent within a society. One agent can play many roles in a MAS and may change role during its execution. Secondly, a service is defined within Agent-UML as the activity which an agent can develop and distribute among other agents. Lastly, a capability describes what the agent is capable of doing under certain particular conditions. Due to the existing differences in the definition that Gaia and Agent-UML provide of roles, services, and capacities, it is necessary to adapt the Gaia design to the Agent-UML standard. As far as the roles are concerned, we have divided those of Gaia into more specific Agent-UML roles. The services of Gaia are divided into Agent-UML services and capabilities, and, given that the level of detail is greater, it is possible that it will be necessary to consider some new service or capability, or divide and specialize some of the Gaia services. Another important change refers to the models of interaction. In Gaia the interactions were specified through the acquaintance model of services [48]. Now they will be used to obtain sequence and collaboration diagrams which will be much more detailed and in which there appear messages exchanged between agents (interpreting a particular role), and the order in which these exchanges of messages are produced.

In order to model the system's behaviour even more, state and activity diagrams will be used. These diagrams are not clearly defined in Agent-UML, and therefore we will have to make an adaptation from the Unified Modeling Language (UML) diagram types, so that they can represent the state through which our agents can pass and the dynamic of the activities that are produced within our system. After carrying out the appropriate changes, we begin the design of the Agent-UML by obtaining the class diagrams. The specifications that will be followed are those of the Foundation for Intelligent Physical Agents (FIPA) for the modelling of class diagrams for agents using Agent-UML [8]. We obtain a class diagram for one of the most prominent agents, the Modelling agent, the CBR agent.

Figure 8.8 shows the class diagram for the Modelling agent. The Modelling agent develops six capabilities and offers four services to other agents. The Jacobean Sensitivity Matrix (JSM) capability offers a mechanism to retrieve the beliefs that can be used to solve a given problem in a given situation.

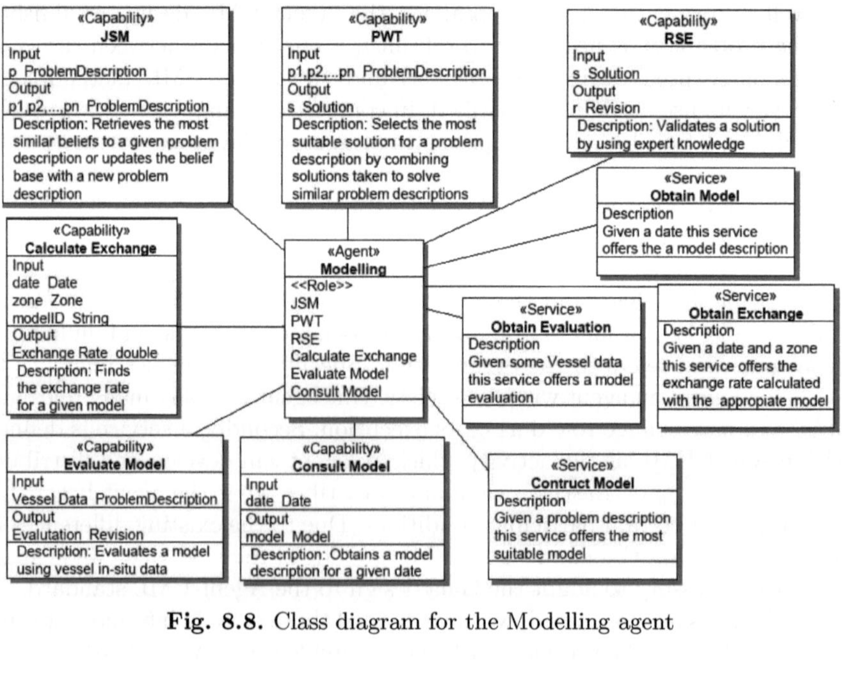

Fig. 8.8. Class diagram for the Modelling agent

These beliefs are given in the form of problem descriptions. Pondered Weigh Technique (PWT) is the capability of the Modelling agent through which it can calculate the most suitable solution for a given problem. The solution is calculated taking into consideration the beliefs retrieved by the JSM capability. The Revision Simulated Equation (RSM) capability enables the agent to compare the model obtained in the PWT capability with other oceanographic models and in situ data supplied from vessels. Calculate Exchange is the capability that allows it to calculate the rate of exchange of carbon dioxide that a specific ocean zone produces at a given time using that particular model. Evaluate Model allows the agent to evaluate the goodness of a model, in other words, it can measure the efficiency of a model by comparing the results that have been given with the results from the Vessel's sensors. Consult Model allows it to carry out consultations of the model database. As far as the services offered by the agent are concerned, we have: Obtain Exchange through which an agent can request a calculation of the rate of exchange of carbon dioxide that is produced in a given ocean zone at a given date, by using the specific model indicated. Obtain Model allows an agent to request the Modelling agent for data from a model that was being used at a given date. Construct Model offers the possibility to attend to construction requests from the models. Lastly, Obtain Evaluation offers any agent the possibility to request an evaluation for a particular model in a particular ocean zone at a given date based on the real data obtained from the Vessel sensors.

The Agent-UML design is completed by offering interaction diagrams which show the interaction between the MAS agents as well as the different roles that can be taken up by the different agents and the interactions between these roles. It is habitual to use a collaboration diagram, although a sequence diagram, which would be equivalent to the collaboration diagram, can also be used, [34]. We can differentiate ten different interactions.

Figure 8.9 shows the interactions between the Modelling and Store agents when a new problem descriptor or case is stored. When new satellite data is received, the Store Sat Data role of the Store agent is in charged of storing the data in the right format with the help of the Transform Im-Cases role. The image is digitally processed in order to obtain the corresponding problem description data. Finally, the Store agent moves into the Store Cases role to store the new problem description data. The Store agent establishes a communication process with the Modelling agent and transfers the new problem description information. The Modelling agent processes the new problem description, and may, creates a new model. The Modelling agent executes the Jacobean Sensitivity Matrix role, in which the agent consults the beliefs base in order to obtain the most similar beliefs (problem descriptions) to the initial problem sent by the Store agent. Once the Modelling agent has retrieved the beliefs, it changes to execute the Pondered Weigh Technique role. Now, the agent calculates the most suitable solution for the initial problem case provided by the Store agent. The most suitable solution is calculated using the cases retrieved by the Jacobean Sensitivity Matrix role as it is shown in

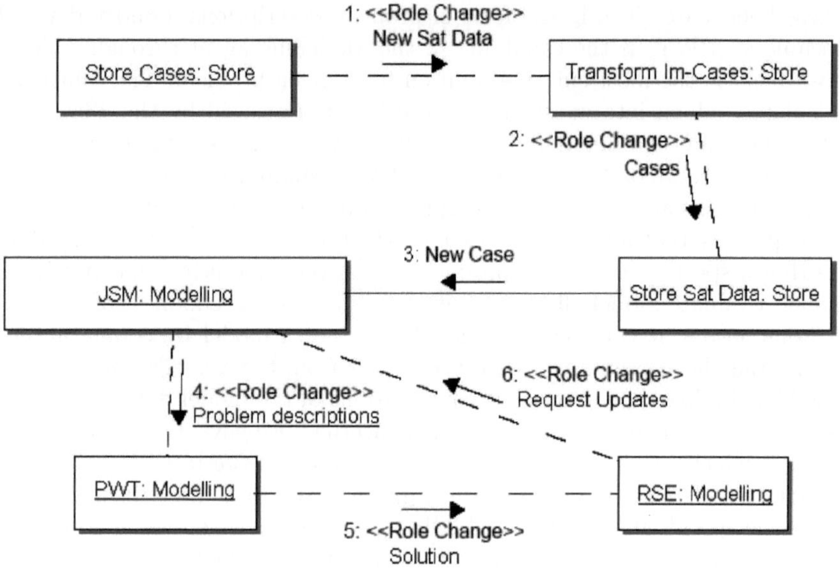

Fig. 8.9. Collaboration diagram corresponding to the interactions that occurs because of the arrival of a new satellite image to the Store agent.

Sect. 8.4.3. A model is created using the most suitable solution, and the RSE role is executed, which is in charge of the revision of the model. Finally, the RSE role transmits the result of the revision by changing back to the role Jacobean Sensitivity Matrix. Now the Modelling agent (with the Jacobean Sensitivity Matrix role) retains the problem description and the knowledge obtained after all this process. A new model may have been created or modified.

To finish the Agent-UML design, state, and activity diagrams are created to model the behaviour of the agents. We use UML state diagrams [37], which we have designed for the Store and Modelling agents. Figure 8.10 shows the state diagram for the Store agent. We believe that the Store agent can be found in three possible states: A state in which the agent is awaiting requests; a state in which the agent is modifying stored data; and thirdly, a state in which the agent carries out operations to store data. Some of these operations include obtaining particular parameters that characterize an image. An additional state allows it to be ready to receive new requests when it has no tasks to be carried out. Finally, to finish with the Agent-UML design, in terms of the activity diagrams, we focus on the activities that will be developed within the CBR cycle.

Once the design is complete, we go on to the implementation, using the Jadex tool, a tool that incorporates the BDI model within Jade agents and tool. With Jadex, the Modelling agents are built while the rest of the agents will be Jade. The communication mechanisms are the same as in Jade (Agent Communication

Language (ACL) is used) [9, 41]. The use of Jadex means that it is necessary to use Object Query Language (OQL) consulting language.

8.4.3 The CBR–BDI Modelling Agent

Once the architecture proposed has been studied, it would seem a good idea to deepen the Modelling agents – in the form of a deliberative agent that uses a CBR mechanism. This agent will have two principal functions. The first one is to generate models which are capable of predicting the atmospheric/oceanic interaction in a particular area of the ocean in advance. The second one is to permit the use of such models. In Fig. 8.10, we can see that a Modelling agent possesses two principal states: one to generate the forecasting models and the other to permit the use of the models. Moreover, the reasoning cycle is one of the activities carried out by the Modelling agent. We can see how the reasoning cycle of a CBR system is included among the activities, composed of stages of retrieval, reuse, revise, and retain. Also, an additional stage that introduces expert's knowledge is used. This reasoning cycle must correspond to the sequential execution of some of the agent roles.

The Modelling agent presents a deliberative architecture, based on the BDI model [12]. In this model, the internal structure and capabilities of the agents are based on mental aptitudes, using beliefs, desires, and intentions. This method facilitates the incorporation of CBR systems [1] as a deliberative mechanism within BDI agents, facilitating learning, and adaptation

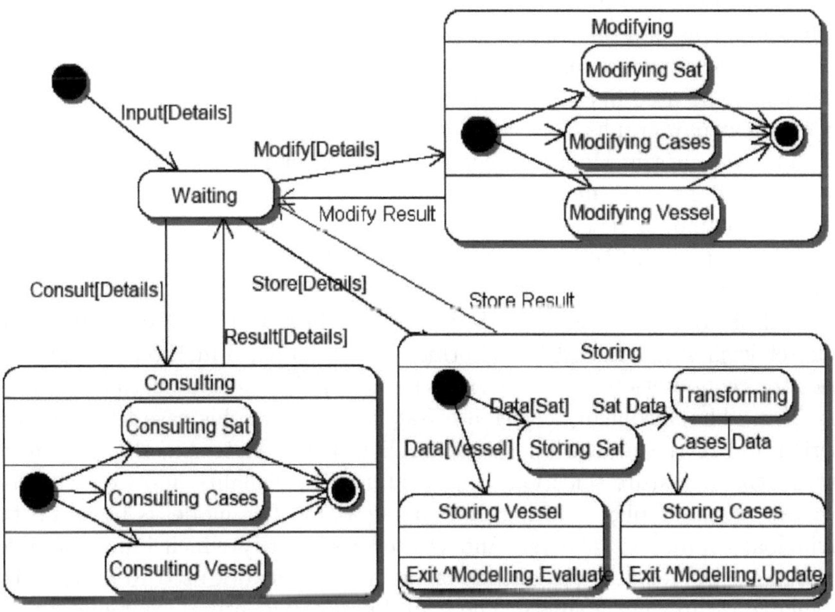

Fig. 8.10. State diagram for the Store agent

and providing a greater degree of autonomy than pure BDI architecture. To introduce a CBR motor into a BDI agent it is necessary to represent the cases used in a CBR system by means of beliefs, desires, and intentions, and implement a CBR cycle. A case is a past experience composed of three elements: an initial state or problem description that is represented as a belief; a final state that is represented as a set of goals; and the sequence of actions that makes it possible to evolve from an initial state to a final state. This sequence of actions is represented as intentions or plans. CBR consists of four sequential stages: retrieve stage to recover the most similar past experiences to the current one; reuse stage to combine the retrieved solutions in order to obtain a new optimal solution; revise stage to evaluate the obtained solution; and retain stage to learn from the new experience.

Figure 8.11 shows the internal structure of a CBR agent. Problem description (initial state) and solution (situation when final state is achieved) are represented as beliefs, the final state as a goal (or set of goals), and the sequences of actions as plans. The CBR cycle is implemented through goals and plans. When the goal corresponding to one of the stages is triggered, different plans (algorithms) can be executed concurrently to achieve the goal. Each plan can trigger new subgoals and, consequently, cause the execution of new plans.

Deliberative CBR agents, like Modelling agent, are able to incorporate other reasoning mechanisms that can coexist together with the CBR. Modelling is an autonomous agent that can survive in dynamic environment. However, is possible to incorporate communication mechanisms that allow it to be easily integrated into a MAS and work coordinately with other agents to solve problems in a distributed way.

The CBR motor is divided into four sequential stages and different algorithms can be used in each one. The reasoning structure is presented in detail in the next paragraphs.

The roles of the Modelling agent are shown in Fig. 8.8. The agent carries out roles to generate models such as Jacobean Sensitivity Matrix, Pondered Weigh Technique, RSE, and other roles that allow it to operate with the models calculated, like Forecast Exchange Rate, Evaluate Model, or Consult model. The roles used to carry out the stages of the CBR cycle are now described. Jacobean Sensitivity Matrix: This role is in charge of carrying out the retrieval stage. In order to do this it needs to use a method that guarantees the recuperation of cases whose characteristics are similar to the current problem. The Jacobean Sensitivity Matrix is used in this case for data clustering and retrieval [32]. The Jacobean Sensitivity Matrix method is a novel approach for feature selection. It can be used to visualize and extract information from complex, and highly dynamic data. The model is based on the principal component analysis and is used to identify which input variables have more influence in the output of the neural network used to perform the principal component analysis. The neural network identifies the beliefs stored

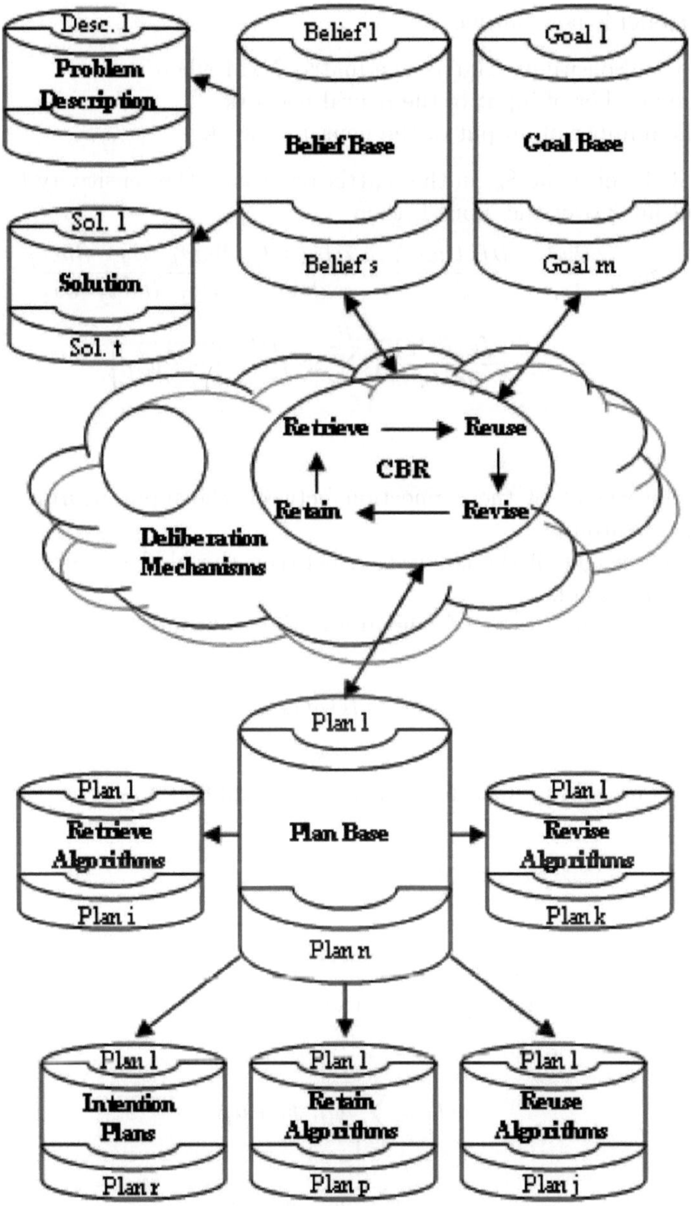

Fig. 8.11. CBR-agent (CBR–BDI) internal structure

by the agent that can be more useful to solve a given problem. The mathematical model is now outlined.

If Jacobean Sensitivity Matrix is a matrix NxM where
N: is the number of input of the neural network.
M: is the number of output of the neural network.

And if the element S_{ki} in the matrix represents the sensitivity (influence) of the output k over the input I, then

$$S_{ki} = \frac{\partial y_k}{\partial x_i} = \frac{\partial f_k\,(net_k)}{\partial x_i} = \frac{\partial f_k\,(net_k)}{\partial net_k}\frac{\partial net_k}{\partial y_j}\frac{\partial y_j}{\partial net_j}\frac{\partial net_j}{\partial x_i}$$
$$= \frac{\partial f_k\,(net_k)}{\partial net_k}\left(\sum_{j=1}^{H} w_{kj}\frac{\partial f_j\,(net_j)}{\partial net_j} w_{ji}\right) \tag{8.1}$$

where

w_{ji}: is the weight of the connection between the input neuron i and the hidden neuron j.
w_{kj}: is the weight of the connection between the hidden neuron j and the output neuron k.
y_k: is the Output obtained for neuron k of the output layer.

Then

$$y_k = f_k(net_k) \tag{8.2}$$

where

y_j: is the Output obtained for neuron j of the hidden layer.

Then

$$y_j = f_j(net_j) \tag{8.3}$$

where

x_i: is the Input for neuron i.
f_h: is the activation function in neuron h.

then

$$net_j = \sum_{i=1}^{N} w_{ji}\; x_i + \theta_j \tag{8.4}$$

$$net_k = \sum_{j=1}^{H} w_{kj}\; y_j + \theta_k \tag{8.5}$$

where

H: is the number of neurons in the hidden layer.
θ_j: is the value of threshold of neuron j of the hidden layer.
θ_k: is the value of threshold of neuron k of the output layer.

Pondered Weigh Technique: The reuse is carried out using the cases selected during the retrieval stage. The cases are pondered [17] and the bigger weight is given to the one that more resembles the current problem in the following way:

$$p^* = \frac{1}{\sum_{r=1}^{Z} e^{-|a-r|}} \sum_{r=1}^{Z} e^{-|a-r|} p^r \tag{8.6}$$

where:

p^*: is the solution prediction.
Z: is the number of retained cases from the base of beliefs.
a: is the measure of minimum similarity between the retained cases from the base of beliefs and the current case.
p^r: is the retained prediction rth from the base of beliefs.
r: is the measure of similarity between the retained cases rth from the base of beliefs and the current case.

RSE: During the revision stage an equation (F) is used to validate the proposed solution p^*.

$$F = kso(pCO_2SW - pCO_2AIR) \tag{8.7}$$

Where:

F: is the flux of CO_2.
k: is the gas transfer velocity.

Then

$$k = (-5,204Lat + 0,729Long + 2562,765)/3600 \tag{8.8}$$

Where:

Lat: is the Latitude.
Long: is the Longitude.
so: is the Solubility.
then it is verified that:

$$so = e^{\left(\frac{93,4517}{100tk} - 60,2409 + 23,3585 \log(100tk) + s(0,023517 - 0,023656 \bullet 100tk + 0,0047036 \bullet 1002tk)\right)} \tag{8.9}$$

$$\mathrm{tk} = 273,15 + \mathrm{t} \tag{8.10}$$

Where:

t: is the Temperature.
s: is the Salinity

$$pCO_2 = A + BLong + CLat + DSST + EYear \tag{8.11}$$

Table 8.1. Case attributes

Case field	Measurement
DATE	Date (dd/mm/yyyy)
LAT	Latitude (decimal degrees)
LONG	Longitude (decimal degrees)
SST	Temperature (°C)
S	Salinity (unitless)
WS	Wind strength ($m\,s^{-1}$)
WD	Wind direction (unitless)
Fluo_calibrated	Fluorescence calibrated with chlorophyll
SW pCO_2	Surface partial pressure of CO_2 (microatmospheres)
Air pCO_2	Air partial pressure of CO_2 (microatmospheres)
Flux of CO_2	CO_2 exchange flux (Moles m^{-2})

Table 8.2. Months\Coefficients values

Months\Coefficients	A	B	C	D	E
Feb	−2488	−0,42	4,98	−12,23	1,38
May	−7642	−0,9	−1,74	−20,77	4,14
Jun	−4873	−0,85	1,3	−15,64	2,66
Jul	−7013	−0,025	3,66	−7,07	3,64
Aug	−3160	−0,69	0,84	−11,31	1,8
Sep	−1297	0,43	−4,19	−17,06	1,05
Oct	83	−0,81	4,81	−10,92	0,076
Nov	747	0,2	−0,73	−17,3	−0,062
Dec	−4306	0,38	−0,22	−17,13	2,45

Where SST is the temperature of the marine surface or air as it corresponds to pCO_2SW or pCO_2AIR. The coefficients of the equation depend on the month, as shown in Table 8.2.

During the revision, the agent compares the obtained F value with the predicted one, and if the prediction differs in less than 10%, the case is stored on the base of beliefs. As it has been shown the CBR agents use a CBR system, at a low level of implementation, which is the reason for using cases. One case for the CBR consists of a problem (initial situation and a number of goals) and the plans to resolve it. For oceanic/atmospheric interaction, we define the problem in terms of the attributes shown in Table 8.1:

Table 8.1 shows the description of a case: DATE, LAT, LONG, SST, S, WS, WD, Fluo_calibrated, SW pCO_2, and Air pCO_2. Flux of CO_2 is the value to be identified.

As mentioned in Sect. 8.2 there is a correspondence between cases and BDI agents. To use a deliberative BDI model that utilizes a CBR mechanism, it is necessary to transform the case representation by the CBR system into a BDI formalisms. The BDI model deals with:

1. Beliefs, which represent the state of the problem, with certain knowledge about the surroundings and the agent itself. In our problem we shall use as belief the attributes DATE, LAT, LONG, SST, S, WS, WD, Fluo_calibrated, SW pCO_2, and Air pCO_2. A beliefs base will be used in which each belief is a ProblemDescription type and contains all the attributes mentioned in Table 8.1.
2. Desires, that represent those final states to which the agent wishes to arrive or reach. In this case, it deals with three goals:
   - Predict the flux of carbon dioxide exchanged between the sea surface and the atmosphere, using a window of two or three weeks.
   - Calculate the best parameters to use in order to improve the prediction for different window sizes.
   - Calculate the most suitable prediction window in relation to a maximum % error allowed.

   An agent stores all the goals in a similar way to the beliefs.
3. Intention, that represents the sequence of actions that should be followed in order to reach the final state or goal. This new attribute is introduced into the case description. The sequence of actions to be carried out is generally formed by the stages of the reasoning cycle and the different algorithms executed in each one of those stages. In general an agent will have available various predefined plans or intentions that could be called up and modified at the execution time. The selection of plans is made through the CBR agent, Jacobean Sensitivity Matrix, Pondered Weigh Technique, and Revise Simulated Equation mechanisms.

The tools offered by the Jadex platform [41] have been used for the storing and use of beliefs, desires, and intentions or plans. In this way, we have been able to construct a deliberative BDI agent capable of reasoning through the use of a CBR mechanism. The agent manages cases and carries out CBR cycles.

8.4.4 Communication Agents

To complete the proposal for the MAS, we outline the types of communication that the system agents employ. The Jadex tool has been used to carry out the implementation of the system. This tool is an extension of Jade and, among other features, which uses a standard for communications in accordance with the FIPA [23]. In this way, both ontologies and languages used are those proposed by the FIPA. Jade uses the ACL defined by the FIPA. The agents send and receive java objects that represent ACL messages in accordance with a series of protocols. The majority of protocols appear in the libraries offered by Jade [9]. Furthermore, the FIPA – Semantic Language (FIPA-SL) contents language is used.

The messages used have a syntax whereby the agent is instructed to send the message to the receiver of that message. It also indicates the content

```
{Request
        :sender User
        :receiver Modelling
        :content
                (RequestDataModel 10 100 10 100 27-11-
2005)
        :in-reply-to
        :reply-with
        :language FIPA-SL0
        :ontology RequestDataModel LONMin LONMax LATMin
LATMax DATE
}
```

Fig. 8.12. Example of the message used in the multiagent system. The User agent makes a request to the Modelling agent asking about the models generated for the maximum and minimum longitude and latitude coordinates the 27 November 2005.

of the message, information to identify the message, name of the language in which the content of the message is written, and the ontologies that define the meaning of the vocabulary used. Figure 8.12 shows an example of a message used by the Modelling agent to request the Storing agent for information on the cases stored in the case base. The language of contents used is FIPA Semantic Language. The ontology defines the vocabulary that is used within the message and the content makes reference to the request for a search to be carried out of the cases and the specific parameters of the search.

The types of messages used in the MAS proposed in this work: request, agree, refuse, cancel, inform, query-if, subscribe, propose, reject-proposal, accept-proposal, failure, and not-understood. As far as the protocols are concerned, the three used are defined by the FIPA: FIPA-request protocol, FIPA-query protocol, and FIPA-ContractNet protocol [23].

8.5 Results and Conclusions

The application of Artificial Intelligence techniques [1] is extremely useful in a field like oceanography and specifically in the study of the carbon dioxide exchange between the ocean surface and the atmosphere. The intelligent environment that has been developed allows oceanographers to maintain a seamless, unobtrusive, and often invisible but also controllable interaction with the available technology. The use of agents and MAS as the fundamental base for creating an intelligent environment is highly suited because of the characteristics of the agents themselves. Their mobility, proactivity, autonomy, social capabilities, reasoning, and capacity for learning all makes the MAS and transparent intermediate layer between the user of the system and the underlying technology. A user can access the system rapidly and efficiently using their personal agent. This agent is able to sit within a "light" device and can communicate through wireless technology with the other agents of the system. This allows oceanographers to be independent and unbound by location. Figure 8.13 shows the interface of a User agent accessed via a PDA. It is possible to see how the user can access the Modelling, Store, or Vessel agent. The appropriate Store or Vessel can be selected through a simple interface that only presents the necessary information and avoids showing too many elements on the screen. The oceanographers themselves can decide that amount of elements that they wish to see. Figure 8.13b presents the options that can be executed by the Modelling agent: To request the creation of a new model, for which it will be necessary to enter the appropriate parameters; predict the level of exchange in a particular zone of the ocean; make an

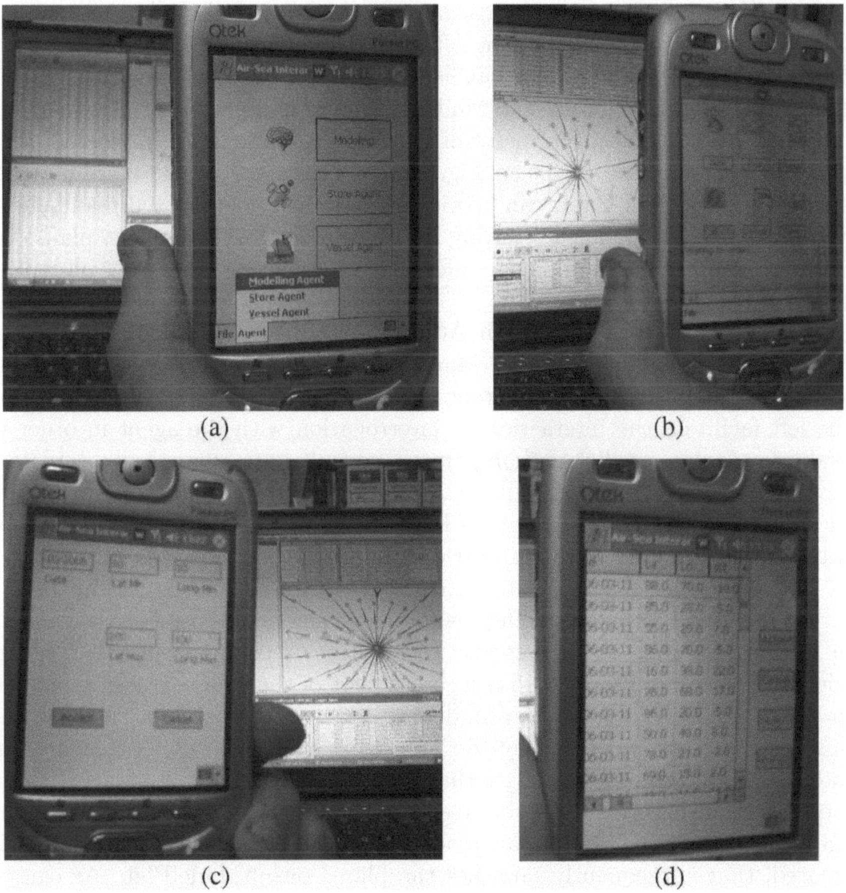

(a) (b) (c) (d)

Fig. 8.13. User agent screen shot. (**a**) Menu accessing subsystem agents. (**b**) User options for interacting with the Modelling agent from a light device (**c**) Inquiry about cases from a Store agent. (**d**) Result obtained from the inquiry made in (**c**) inquiry about the models stored; evaluate a model by entering real data; or saving the corresponding data to the models that are being currently used.

The User agent offers similar menus to allow the user to interact with the Vessel and Store agents. Figure 8.13c shows the parameters that an oceanographer needs to enter when he wishes to make and inquiry the cases stored by the Store agent for the zone of the Atlantic Ocean situated between 10° and 100° latitude and 10° and 100° longitude on the 11 March 2006. After the inquiry is made, the cases are shown to the user in a table as can be seen in Fig. 8.13d. The user can select each one of these cases and modify the parameters.

However, the agents do not represent a mere software that interacts between the user and the technology, but also has the capacity to make decisions and act for themselves in a distributed way, in order to respond and adapt to the changes that are produced within the environment and within its own internal knowledge structure. In this chapter we have described the construction of a MAS whose main component is the Modelling agent, a deliberative agent based on the BDI model [12] that uses CBR [1] as its reasoning mechanism. As can be seen in Fig. 8.14, the Modelling agent handles beliefs, desires, and intention from a conceptual point of view and cases from an implementation point of view. A case (a file in Fig. 8.14) is composed of the attributes described in Table 8.1. Cases can be viewed, modified, and deleted manually or automatically by the agent (during its revision stage). The agent plans (intentions) can be generated using different strategies since the agent integrates different algorithms.

Figure 8.14 shows the North Atlantic exchange rate calculating by the Modelling agent, during the 11 March 2006. The screen shot also presents the algorithms used in the different stages of the CBR cycle. The menus on the left facilitate the interaction or interrogation with the agent in order to evaluate models, predict exchange rates, consult data, change data create a new model or save the current model data. Figure 8.14 presents a view of the Modelling agent. These agents have their own interface and can also be accessed via the User or Super user agents.

The Modelling agent is fully integrated within the MAS. As can be observed in Fig. 8.15, the Modelling agent creates new goals based on the changes in its internal state or in response to messages received from other agents within the system. Figure 8.15 is a screen shot of the Jadex Tracer agent [41] in which the behaviour of the Modelling agent is represented. In it we can observe the goals generated by the Modelling agent, the plans that are put into operation, and the messages that it receives. For example, for a Request message inquiring about a model received from the Gui agent, the Modelling agent executes the update_plan model, from which a consult_model goal is created, that subsequently launches the plan ConsultModelPlan. As can be seen, the agent is capable of handling various goals and plans at the same time.

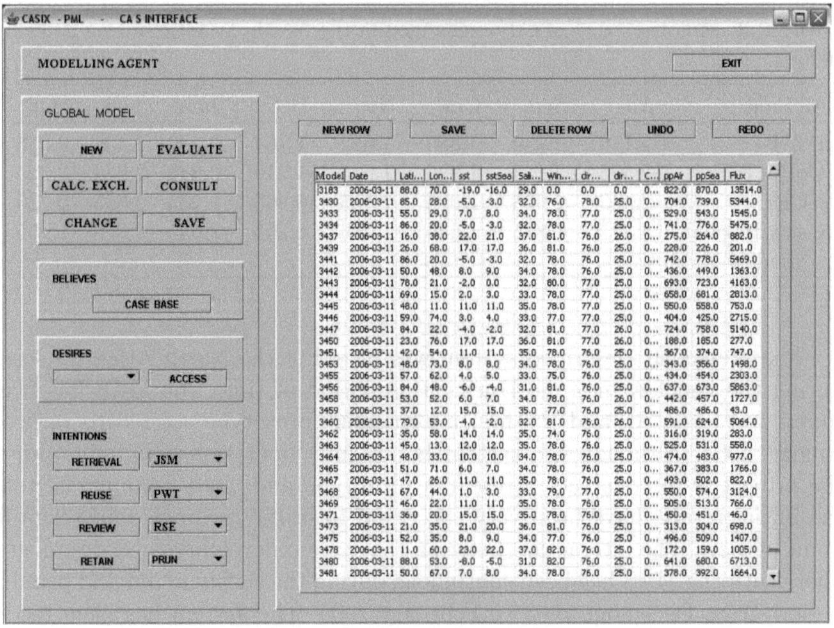

Fig. 8.14. Modelling agent User interface

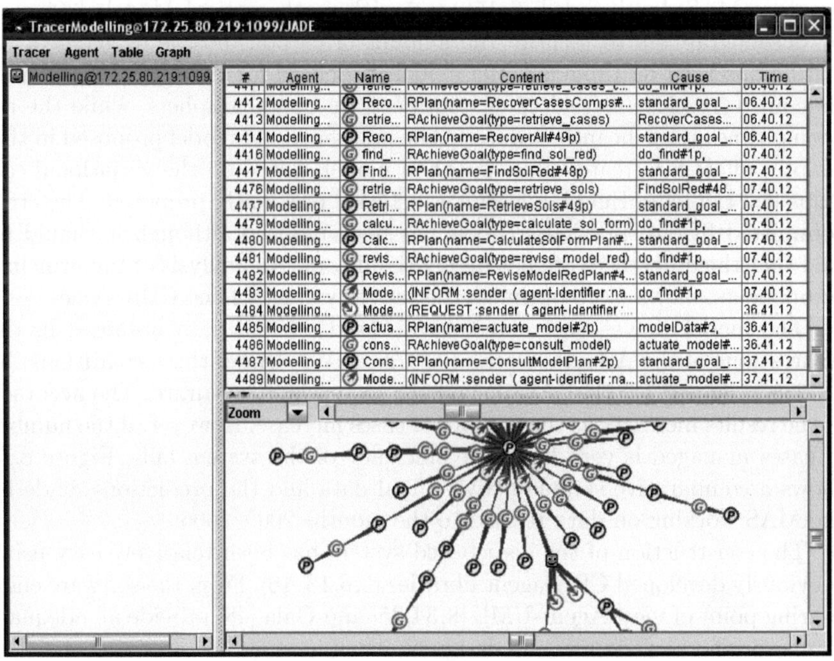

Fig. 8.15. Behaviour of the Modelling agent observed through a Jadex Tracer agent

Table 8.3. Mmol $m^{-2}d^{-1}$ of $C0_2$ exchanged in the North Atlantic during 2005 and 2006

Method	Aug.	Sep.	Oct.	Nov.	Dec.	Jan.	Feb.	Mar.
PWT	−2.119	−2.230	−1.885	−1.622	−1.164	0.495	2.435	2.235
CoHel IBR	−2.325	−2.126	−1.926	−1.625	−1.210	0.845	2.634	2.325
VCBP	−2.453	−2.965	−2.036	−1.155	0.965	−0.235	2.555	2.725
Casix manual models	−4.317	−1.875	−1.655	−1.233	0.205	2.035	3.978	1.955

The previously described system was tested in the North Atlantic Ocean during the last few months. During this period of time, the MAS has been tuned and updated, and the first autonomous prototype started to work In August 2004. Although the system is not fully operational and the aim of the project is to construct a research prototype and not a commercial tool, the initial results have been very successful from the technical and scientific point of view. The interaction between the system developers and oceanographers with the MAS has been continuous during the construction and pruning period, from December 2003 to February 2005. The system has been tested from September 2005 to March 2006 and the results have been very accurate. Table 8.3 presents the results obtained with the MAS proposed in this work, previous MAS developed [5, 6, 15], and with mathematical Models [30] used by oceanographers to identify the amount of carbon dioxide exchanged. As can be observed in Table 8.3, the models proposed for the MAS offer results that are very close to real values obtained by oceanographers, while the response time is significantly reduced. The mathematical model proposed in this chapter offers far greater precision than models based on the variational calculus [6,14,16] and Hebbian learning [5,15,24] previously proposed. The error committed by two previous models has been reduced although it should be said that the differences are not highly significant. An analysis of the principal components allows us to optimize the recovery stage of the CBR cycle.

The models have constructed cases based on real data obtained in the Azores zone of the Atlantic Ocean (± 37N, 25W). Under these conditions the models proposed for the MAS have been increasingly accurate. The accuracy of the results increases as the number of cases increase. However, if the number of cases managed is very high, the efficiency of the system falls. Figure 8.16 shows a comparative sample between real data and the predictions made by the MAS working on data related to the months 2005–2006.

The construction of the distributed system has been relatively easy using previously developed CBR-agent libraries [5,6,13–16]. From the software engineering point of view Agent-UML [8,34,35] and Gaia [48] provide an adequate framework for the analysis and design of distributed agent-based systems. The formalism defined in [25] facilitates the straight mapping between the agent definition and the CBR construction. Although the proposed system requires

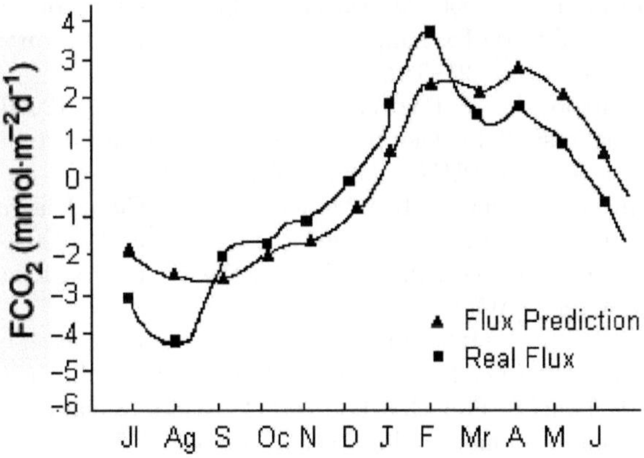

Fig. 8.16. CO_2 real flux and flux prediction

further improvements and more work, the initial results are very promising. The generated open framework facilitates the incorporation of new agents using different modelling techniques and learning strategies so further experiments will allow us to compare these initial results with the ones obtained by other techniques.

Acknowledgements

This work has been supported by the Spanish MCYT TIC2003-07369-C02-02 project and the PML's CASIX project.

References

1. Aamodt A, Plaza E (1994) Case-Based Reasoning: foundational Issues, Methodological Variations, and System Approaches, AICOM. Vol. 7., March, pp 39–59.
2. Aarts E (2004) Ambient intelligence, Third International Conference, AH 2004, Eindhoven, The Netherlands. LNCS Springer Verlag. ISBN: 3-540-22895-0, p. 1.
3. Aiken J., Fishwick J.R., Lavender S.J., Barlow R., Moore G.F., Sessions H., Bernard S., Ras J. & Hardman-Mountford N.J. (Submitted). Validation of MERIS reflectance and chlorophyll during the BENCAL cruise October, 2002: preliminary validation and new products for phytoplankton functional types and photosynthetic parameters. International Journal of Remote Sensing.
4. Aiken J., Hardman-Mountford N., Ufermann S., Woolf D., Challenor P., Robinson I. and the CASIX team, (2005). Exploiting ENVISAT-ERS data for deriving air-sea fluxes of CO2, Proceedings of the Envisat Symposium, Salzburg, ESA publication, in press.
5. Bajo J, Corchado JM (2005) Evaluation and monitoring of the air-sea interaction using a CBR-Agents approach. ICCBR 2005. Lecture Notes in Computer Science 3620, pp 50–62.

6. Bajo J, Corchado JM (2006) Multiagent architecture for monitoring the North-Atlantic carbon dioxide Exchange rate. CAEPIA 2005. Lecture Notes in Computer Science 4177, pp 321–330.
7. Bajo J, Corchado JM, de Paz Y, de Paz JF, Martín Q. (2006) A multiagent recommending system for shopping centres. Proceedings of the workshop on Recommender Systems (ECAI 2006). Vol 1, pp 92–06.
8. Bauer B, Huget MP (2003) FIPA Modeling: Agent Class Diagrams.
9. Bellifime F, Poggi A, Rimasa G (2001) JADE: a FIPA2000 compliant agent development environement. Proceedings of the 5th international conference on autonomous agents (ACM).
10. Bergmann R, Wilke W (1996) On the role of abstraction in case-based reasoning. Lecture Notes in Artificial Intelligence, 1186, pp. 28–43. Springer Verlag.
11. Bratman ME, Israel D, Pollack ME (1988) Plans and resource-bounded practical reasoning. Computational Intelligence, 4., pp. 349–355.
12. Bratman ME (1987) Intentions, Plans and Practical Reason. Harvard University Press, Cambridge, M.A.
13. Corchado JM, Laza R (2003) Constructing Deliberative Agents with Case-based Reasoning Technology, International Journal of Intelligent Systems. Vol 18, No. 12, December. pp. 1227–1241.
14. Corchado JM, Lees B (2001) A Hybrid Case-based Model for Forecasting. Applied Artificial Intelligence. Vol 15, no. 2, pp. 105–127.
15. Corchado JM, Aiken J, Corchado E, Lefevre N, Smyth T (2004) Quantifying the Ocean's CO2 Budget with a CoHeL-IBR System. 7th European Conference on Case-based Reasoning, Lecture Notes in Computer Science, Lecture Notes in Artificial Intelligence 3155, Springer Verlag. pp. 533–546.
16. Corchado J M, Pavón J, Corchado E, Castillo LF (2005) Development of CBR-BDI Agents: A Tourist Guide Application. 7th European Conference on Case-based Reasoning 2004. Lecture Notes in Artificial Intelligence 3155, Springer Verlag. pp. 547–559.
17. De Paz Santana Y (2005) Mixture Weibull distribution using artificial neural networks with csnsurated data PHD thesis, Chapter 3.
18. DeLoach S (2001) Anlysis and Design using MaSE and AgentTool. Proceedings of the 12th Midwest Artificial Intelligence and Cognitive Science Conference (MAICS).
19. Dransfeld S., Tatnall A.R., Robinson I.S. and Mobley C.D. (2004). A comparison of Multi-layer Perceptron and multilinear regression algorithms for the inversion of synthetic ocean colour spectra. Int. J. Remote Sens. 25 (21): 4829–4834.
20. Dransfeld S., Tatnall A.R., Robinson I.S. and Mobley C.D. (2005). Prioritizing ocean colour channels by neural network input reflectance perturbation. Int. J. Remote Sens. 26 (5): 1043–1048.
21. EURESCOM (2001) MESSAGE: Methodology for engineering systems of software agents. Technical report P907-TI1, EURESCOM.
22. Feret MP, Glasgow JI (1994) Explanation-Aided Diagnosis for Complex Devices, Proceedings of the 12th National Conference an Artificial Intelligence, (AAAI-94), Seattle, USA, August 94.
23. FIPA Foundation for Intelligent Physical Agents. http://www.fipa.org.
24. Fyfe C, Corchado ES (2002) Maximum Likelihood Hebbian Rules. European Symposium on Artificial Neural Networks. 2002.

25. Glez-Bedia M, Corchado JM, Corchado ES, Fyfe C (2002) Analytical Model for Constructing Deliberative Agents, Engineering Intelligent Systems, Vol 3, pp. 173–185.
26. Iglesias C, Garijo M, Gonzalez JC, Velasco JR (1998) Analysis and Design using MAS-CommonKADS. Inteligent Agents IV LNAI Volume 1365 Springer Verlag.
27. Kinny D, Georgeff M (1991). Commitment and effectiveness of situated agents. In: Proceedings of the Twelfth International Joint Conference on Artificial Intelligence (IJCAI'91), Sydney, Australia, pp. 82–88.
28. Kolodner J (1993) *Case-based reasoning.* Morgan Kaufmann.
29. Lavender, S.J., Pinkerton, M.H., Moore, G.F., Aiken, J and Blondeau-Patissier, D. (2005). Modification to the Atmospheric Correction of SeaWiFS Ocean Colour Images over Turbid Waters. Continental Shelf Research, 25, 539–555.
30. Lefevre N, Aiken J, Rutllant J, Daneri G, Lavender S, Smyth T (2002) Observations of pCO2 in the coastal upwelling off Chile: Sapatial and temporal extrapolation using satellite data. Journal of Geophysical research. Vol. 107, no. 0.
31. Martín FJ, Plaza E, Arcos JL (1999) Knowledge and experience reuse through communications among competent (peer) agents. International Journal of Software Engineering and Knowledge Engineering, Vol. 9, No. 3, pp. 319–341.
32. Montaño Moreno JJ, Palmer Pol A (2002) Artificial Neural Networks, opening the black box. Metodología de las Ciencias del Comportamiento 4(1) 77–93.
33. Nwana HS, Ndumu DT, Lee LC, Collins JC (1999) ZEUS: A Toolkit for Building Distributed Multi-Agent Systems. Applied Artificial Intelligence Jounal, vol 1, n°13, pp. 129–185.
34. Odell J, Levy R, Nodine M (2004) FIPA Modeling TC: Agent Class Superstructure Metamodel. FIPA meeting and interim work.
35. Odell J, Huget MP (2003) FIPA Modeling: Interaction Diagrams.
36. Olivia C, Chang CF, Enguix CF, Ghose AK (1999) Case-Based BDI Agents: An Effective Approach for Intelligent Search on the World Wide Web, AAAI Spring Symposium on Intelligent Agents, 22–24.
37. OMG Unified Modelling Language Specification. Version 1.3. http://www.omg.org.
38. O'Reilly J.E., S. Maritorena, B.G. Mitchell, D.A. Siegel, K.L. Carder, S.A. Garver, M. Kahru, C. McClain, 1998. Ocean color chlorophyll algorithms for SeaWiFS. Journal of Geophysical Research-Oceans, **103**(11), 24937–24953
39. Perner P (1998) Different Learning Strategies in a Case-Based Reasoning System for Image Interpretation. Lecture Notes in Computer Science 1488, pp 251–261.
40. Perner P (2001) Why Case-Based Reasoning Is Attractive for Image Interpretation. Lecture Notes in Computer Science 2080, pp 27–43.
41. Pokahr A, Braubach L, Lamersdorf W (2003) Jadex: Implementing a BDI-Infrastructure for JADE Agents, in: EXP - In Search of Innovation (Special Issue on JADE), Vol 3, Nr. 3, Telecom Italia Lab, Turin, Italy, September 2003, pp. 76–85. Jadex
42. Rao AS, Georgeff MP (1995) BDI Agents: From Theory to Practice. First International Conference on Multi-Agent Systems (ICMAS-95). San Franciso, USA.
43. Sarmiento JL, Dender M (1994) Carbon biogeochemistry and climate change. Photosynthesis Research, Vol. 39, pp. 209–234.
44. Takahashi T, Olafsson J, Goddard JG, Chipman DW, Sutherland SC (1993) Seasonal Variation of CO2 and nutrients in the High-latitude surface oceans: a comparative study. Global biochemical Cycles. Vol. 7, No. 4. pp. 843–878.

45. The Beam Project. http://www.brockmann-consult.de/beam/.
46. Wendler J, Lenz M (1998) CBR for Dynamic Situation Assessment in an Agent-Oriented Setting. Proc. AAAI-98 Workshop on CBR Integrations. Madison (USA).
47. Wooldridge M, Jennings NR (1995) Agent Theories, Architectures, and Languages: a Survey. In: Wooldridge and Jennings, editors, Intelligent Agents, Springer-Verlag, pp. 1–22.
48. Wooldridge M, Jennings NR, Kinny D (2000) The Gaia Methodology for Agent-Oriented Analysis and Design. Journal of Autonomous Agents and Multi-Agent Systems, 3 (3). pp. 285–312.
49. Zeichen M.M., Robinson I.S. (2004). Detection and monitoring of algal blooms using SeaWiFS imagery (vol 25, pg 1389, 2004). Int. J. Remote Sens. 25 (9): 1797–1798.

9

Extracting Knowledge from Sensor Signals for Case-Based Reasoning with Longitudinal Time Series Data

P. Funk and N. Xiong

Department of Computer Science and Electronics
Mälardalen University
SE-72123 Västerås, Sweden
{peter.funk, ning.xiong@mdh.se}

Summary. In many industrial and medical diagnosis problems it is essential to investigate time series measurements collected to recognize existing or potential faults/diseases. Today this is usually done manually by humans. However the lengthy and complex nature of signals in practice often makes it a tedious and hard task to analyze and interpret available data properly even by experts with rich experiences. The incorporation of intelligent data analysis method such as case-based reasoning is showing strong benefit in offering decision support to technicians and clinicians for more reliable and efficient judgments.

This chapter addresses a general framework enabling more compact and efficient representation of practical time series cases capturing the most important characteristics while ignoring irrelevant trivialities. Our aim is to extract a set of qualitative, interpretable features from original, and usually real-valued time series data. These features should on one hand convey significant information to human experts enabling potential discoveries/findings and on the other hand facilitate much simplified case indexing and similarity matching in case-based reasoning. The road map to achieve this goal consists of two subsequent stages. In the first stage it is tasked to transform the time series of real numbers into a symbolic series by temporal abstraction or symbolic approximation. A few different methods are available at this stage and they are introduced in this chapter. Then in the second stage we use knowledge discovery method to identify key sequences from the transformed symbolic series in terms of their cooccurrences with certain classes. Such key sequences are valuable in providing concise and important features to characterize dynamic properties of the original time series signals. Four alternative ways to index time series cases using discovered key sequences are discussed in this chapter.

9.1 Introduction

Case-based reasoning (CBR) [1] has been widely recognized as a powerful learning methodology for circumstances where generalized domain knowledge is not available or hard to obtain. Based on the tenet that similar problems

P. Funk and N. Xiong: *Extracting Knowledge from Sensor Signals for Case-Based Reasoning with Longitudinal Time Series Data*, Studies in Computational Intelligence (SCI) **73**, 247–284 (2008)
www.springerlink.com

have similar solutions, CBR attempts to solve new problems by retrieving previous similar cases for which solutions are already known. Usually condition parts of cases are represented as vectors of selected attribute values when making similarity matching between a query case and previous ones in the case base. Proper case index capturing truly relevant features has shown to be one of the crucial factors for the success of case retrieval.

Tackling time series cases is attaining increasing importance in applying case-based reasoning to various real-world problems. As long as processes in the underlying domain are inherently dynamic, cases should be constructed to reflect the phenomena that were evolving overtime rather than be depicted as snapshots at a given time instant. Unlike static cases described by time independent attributes, time series cases contain profiles of time-varying variables wherein pieces of data are associated with a times tamp and are valid only for a specific interval in the case duration. Temporal aspect of time series data has to be taken into account in the tasks of case indexing and case retrieval. Abstraction and representation of temporal knowledge in CBR systems were discussed in [7, 22, 42].

Signal analysis techniques have been applied to extract relevant features from time series signals such as sequential sensor readings. The most common methods used in applications are discrete Fourier transform (DFT) and wavelet analysis, see [9,35,36,51]. Both aims to capture significant characteristics of original signals by providing frequency related information. However, as noted in [11], such traditional analytical tools are only competent on signals with relatively simple dynamics, they fail to characterize patterns of more complex dynamics such as bifurcations and chaotic oscillations.

Another concern with signal analysis is the large number of coefficients produced during signal transformation such that feature selection is entailed to reduce the number of inputs to build similarity measures for case matching and retrieval. The issue of dimensionality reduction becomes particularly critical when measurements are gathered within a very long time span. Longitudinal time series signals are prevalent in circumstances such as patient monitoring for medical health care or condition-based maintenance of industrial equipments, where subjects monitored are expected to possibly change their behavior patterns during the long period of observation. Later in Sect. 9.2, we shall show that, using traditional signal processing methods, it is hard to acquire a moderate number of features as concise representation of original signals while retaining their time-varying properties.

This chapter suggests a general framework fostering compact and efficient representation of lengthy time series cases capturing important temporal behaviors while ignoring irrelevant trivialities. The aim is to extract a set of qualitative, interpretable features from original and usually real-valued time series data. These features should on one hand convey significant information to human experts enabling potential discoveries/findings and on the other hand facilitate much simplified case indexing and similarity matching in case-based reasoning. The road map to achieve this goal consists of two subsequent stages. In the first stage the time series of real numbers is transformed into a symbolic series by temporal abstraction or symbolic approximation. A few different methods are available to be utilized at this stage. Then, in the second stage, we use knowledge discovery method to identify key sequences from the transformed symbolic series in terms of their cooccurrences with certain classes. Such key sequences are valuable in providing important features to characterize dynamic properties of sensor signals, thus leading to concise index of longitudinal time series as well as reduced input dimensionality of similarity measures.

The remainder of this chapter is organized as follows. Section 9.2 gives a brief overview and outlines our general framework to handle longitudinal time series for efficient and compact case index. Approaches to transforming series of sensor measurements into symbolic ones are introduced in Sect. 9.3, followed by a knowledge discovery method presented in Sect. 9.4 to identify key sequences from symbolic series transformed. Subsequently, in Sect. 9.5, we discuss the utilities of key sequences discovered in case-based reasoning, e.g., case indexing, measures for similarity. Relevance to related works is discussed in Sect. 9.6. Finally Sect. 9.7 ends this chapter with concluding remarks.

9.2 Classification Based on Sensor Signals

Categories of subjects can be recognized by observing their relevant variables during their operation. Using sensor technology it is possible to measure the values of such variables and also record the profiles of their evolution with the time. We can then process and analyze the collected sensor recordings to find out hidden symptoms. These symptoms give us basis to reason about the class the subject belongs to or make prediction about a potential failure, that it is likely to emerge in the near future. A general road map for this purpose is illustrated in Fig. 9.1, which includes signal filtering, feature extraction, and pattern classifier as its functional components.

Signal filtering is used to purify original sensor readings by removing noises contained in the signals such that more reliable classification results will be

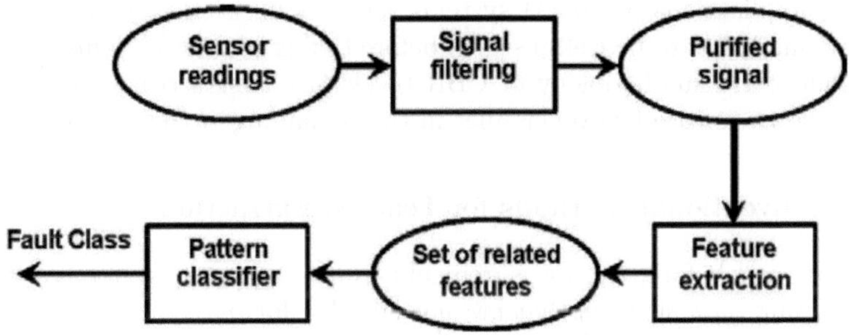

Fig. 9.1. Classification based on sensor signals

warranted. Usually there are two kinds of noises included in the received signals; one is measurement noise due to intrinsic imprecision of sensors and the other is external noise caused by disturbance from surroundings and which is added to the sensor data recorded. Signal recovery from external background noise has been achieved by applying signal processing methods like wavelet analysis and time domain averaging [29,30]. Reduction of measurement errors is outside the scope of this chapter, but interested readers can refer to sensor fusion systems in which Bayesian-based filtering approaches such as Kalman filtering [4] and particle filtering [15] merit to be used to acquire more accurate estimates of states for subjects.

Feature extraction is purported to identify characteristics of sensor signals as useful symptoms for further analysis. This stage is crucial in many industrial and medical domains where the process in consideration is dynamic such that measured variables generally change with the time. This means that it is not possible to depict sensor observations with static single values. Instead we need to identify a collection of features to characterize the evolution of a time-varying variable. The set of extracted features is desired to have a moderate size to reduce the input dimensionality for the pattern classifier (Fig. 9.1). On the other hand, features extracted also ought to be adequate to accommodate temporal information or transitional patterns of signals to be analyzed.

Regarding pattern classifier a number of methods might be considered. Expert systems were developed in support of gathering, representing, and utilizing human expert knowledge for problem solving but they suffer from the knowledge acquisition bottleneck. Regression functions distinguish objects by defining linear boundaries between classes using a moderate number of attributes as function variables. For problems with nonlinear boundaries artificial neural networks would be a suitable approach because they are capable of realizing arbitrary nonlinear mappings between input and output units. Nevertheless the functions of neural networks are rather like a black box, they hardly provide any reasons and arguments for decisions recommended by them. Comparatively, CBR is more transparent by making decisions according to similar cases retrieved such that human users are given reference information to understand, verify, and occasionally also modify the suggested results. The explanatory issue is quite important in many medical and industrial applications where AI systems serve as decision support and every decision made has to be well justified before taking into effect. This motivates us to adopt the methodology of CBR to classify time series signals and we narrow down to case-based classifier in the remaining of this chapter.

9.2.1 Conventional Methods for Feature Extraction

As mentioned before, the measurements from a dynamic system constitute time-varying data streams that are not suitable for immediate usage. Hence we need to "dig out" representative features hidden in the signals prior to classification. The features extracted are delivered to the case-based classifier as an index of the query case. Currently features extracted with traditional methods fall into two categories, namely statistical features and frequency-based features.

Statistical features are extracted from the profile of signal values with respect to calculated statistics as overall generalization. Typical features of this kind can be peak value, start time, overshoot, rising time, mean value, integral, standard deviation, etc. In practice what features to derive for case indexing is commonly ad hoc and domain dependent. An example of using statistical features for case-based circuit diagnosis was illustrated in [39]. However extracting statistical features has a weakness of converting dynamic data streams into static values, thus losing information of temporal relation between data.

Frequency-based features characterize sensor signals by groups of quantities related to a diversity of frequencies. As numerous signal transforms are available to yield frequency spectra, we seem to have more solid basis for extracting features based on frequency than for deriving features based on statistics. The two most common signal transform methods to this end are Fourier transform and wavelet analysis. We shall introduce them briefly in the following and also indicate their limitations facing longitudinal signals with substantial variations.

The Discrete Fourier Transform (DFT)

The DFT transforms a series of sampled values from a signal into spectral information about the signal. Let $x^*(t) = x(nT) = x(n)$ be a sampled function taking samples at times t = 0, T, ..., nT, ..., (N − 1)T, and T is the sampling period (the time between two consecutive samples). The components of DFT for the sampled signal x*(t) are given by a complex summation [47] as follows

$$X(k) = \sum_{n=0}^{N-1} x(nT) \exp\left[-jknT2\pi/(NT)\right] = \sum_{n=0}^{N-1} x(n) \exp\left[-jkn\Omega_0\right] \qquad (9.1)$$

where

$$k = 0, 1, 2, N-1 \qquad \text{and} \qquad \Omega_0 = 2\pi/N$$

If we substitute $\exp\left[-jkn\Omega_0\right]$ in (9.1) with Euler identity, the DFT components can be equivalently written as

$$X(k) = \sum_{n=0}^{N-1} x(n)\cos(kn\Omega_0) - j\sum_{n=0}^{N-1} x(n)\sin(kn\Omega_0) \qquad (9.2)$$

Note that X(k) shows the signal characteristics at the frequency k/(N-1)T. Thus the frequency spacing between two neighboring components in the DFT spectrum is

$$\nabla f = \frac{1}{(N-1)T} Hz \tag{9.3}$$

In general, the DFT component X(k) is a complex consisting of a real and an imaginary part. We adopt the magnitude of X(k)

$$|X(k)| = \sqrt{\left(\sum_{n=0}^{N-1} x(n)\cos(kn\Omega_0)\right)^2 + \left(\sum_{n=0}^{N-1} x(n)\sin(kn\Omega_0)\right)^2} \tag{9.4}$$

as the feature value corresponding to the frequency k/(N-1)T when doing Fourier transforms for feature extraction.

Two limitations of using DFT for feature extraction have to be pointed out here. First, it is clear that, for every k = 0, 1, ...N − 1, there is a DFT term X(k) such that the number of features equals the number of sampled data N. This would lead to an explosion in the number of features extracted when dealing with longitudinal signals common in practice. The sampling period of a signal must be kept below an upper bound according the Shanon theorem in order to avoid distortion of the original signal.

The second limitation with DFT is that the features extracted from the magnitudes of X(k) cannot guarantee to distinguish different orders of patterns within a signal. To show this point more clearly, let us consider two sampled signals with six sampling periods as follows:

$$x_1^*(t) = \Big[\underbrace{3,\ 3,\ 3,}_{\text{mode } A}\ |\ \underbrace{5,\ 5,\ 5}_{\text{mode } B}\Big]$$

$$x_2^*(t) = \Big[\underbrace{5,\ 5,\ 5,}_{\text{mode } B}\ |\ \underbrace{3,\ 3,\ 3}_{\text{mode } A}\Big]$$

Obviously the two signals differ only in the order of appearances of values (x_1^* takes the value of 3 in the first three sampling instances followed by 5 later, whereas x_2^* appears in the opposite order). The DFT terms of these two signals are calculated according to (9.2) with the results given by

$$X_1(0) = 24, \quad X_1(1) = -2.000 + 3.4641j, \quad X_1(2) = 0$$
$$X_1(3) = -2, \quad X_1(4) = 0, \quad X_1(5) = -2.000 - 3.4641j$$

$$X_2(0) = 24, \quad X_2(1) = 2.000 - 3.4641j, \quad X_2(2) = 0$$
$$X_2(3) = 2, \quad X_2(4) = 0, \quad X_2(5) = 2.000 + 3.4641j$$

It is clearly seen from the above that the magnitudes of $X_1(k)$ and $X_2(k)$ are identical for any k = 0, 1, 2, ..., 5. As a consequence the signals x_1^* and x_2^* cannot be distinguished by using the features extracted by DFT. This example, although simple, implies a potential problem for signals with varying modes in the whole duration, because temporal information of mode A followed by B or mode B followed by A might be completely indifferent to the DFT features.

Wavelet Transform

Wavelet transform (WT) is a relatively recently introduced signal analysis method [5, 19], which aims to provide a means of analyzing local behavior of signals. In this sense it is fundamentally different from global transforms such as the Fourier transform. The basic principle underlying WT is to represent a signal, x(t), of interest as a weighted sum of wavelets and scaling function by

$$x(t) = A_1\varphi(t) + A_2\psi(t) + \sum_{\substack{n\in +Z \\ m\in Z}} A_{n,m}\psi(2^{-n}t - m) \tag{9.5}$$

where $\psi(t)$ is the mother wavelet function and $\varphi(t)$ denotes the scaling function. Principally any function with positive and negative areas canceling out can be adopted as a wavelet. In other words the only condition imposed on a wavelet function is that it satisfies

$$\int_{-\infty}^{\infty} \psi(t)dt = 0 \tag{9.6}$$

In practice a very frequently used wavelet function is the Haar function which is defined as

$$\psi(t) = \begin{cases} 1 & if \quad 0 \leq t < 0.5 \\ -1 & if \quad 0.5 \leq t < 1 \\ 0 & \quad otherwise \end{cases} \tag{9.7}$$

Dilations and translations of the mother wavelet function (9.7) create child wavelets functions as expressed by

$$\psi_{s,l}(t) = 2^{-\frac{s}{2}}\psi(2^{-s}t - l) \tag{9.8}$$

where parameters s and l are integers according to which the mother wavelet function $\psi(t)$ is scaled and dilated. The child wavelets constitute an orthonormal basis of the Haar system. Using this orthonormal basis, time series x can now be formulated as a linear combination of the Haar wavelets:

$$x = x^0 + \sum_{s=1}^{\log_2 N} \sum_{l=0}^{\frac{N}{2^s}-1} c_{s,l}\Psi_{s,l}(t) \tag{9.9}$$

Here N is a power of 2 representing the number of data points in the time series. By x^0 we denote the coarsest approximation of the signal. The coefficients $c_{s,l}$ are considered as features obtained from wavelet transform. The WT features can be derived by a procedure of averaging and differencing applied on a finite signal, as illustrated in the following example.

Assume a finite time series x = [2, 5, 8, 9, 7, 4, −1, 1]. We now want to express the signal into the form of (9.9) using Haar basis. This can be achieved with three steps below.

Step 1: Perform averaging and differencing at the level corresponding to s = 1 such that

$$x = [2+5, 8+9, 7+4, -1+1, 2-5, 8-9, 7-4, -1-1]/\sqrt{2} = [7, 17, 11, 0, -3, -1, 3, -2]\sqrt{2}$$

Here we obtain the WT features in the highest frequency subband

$$WF(1) = [c_{1,0}, c_{1,1}, c_{1,2}, c_{1,3}] = [-3, -1, 3, -2]/\sqrt{2}$$

, which reflects the changing rates of the signal within every two sampling periods in the time dimension.

Step 2: Perform averaging and differencing at the level corresponding to s=2 such that

$$x = \left[\frac{7+17}{\sqrt{2}}, \frac{11+0}{\sqrt{2}}, \frac{7-17}{\sqrt{2}}, \frac{11-0}{\sqrt{2}}, -3, -1, 3, -2\right] \Big/ \sqrt{2} = \left[\frac{24}{\sqrt{2}}, \frac{11}{\sqrt{2}}, \frac{-10}{\sqrt{2}}, \frac{11}{\sqrt{2}}, -3, -1, 3, -2\right] \Big/ \sqrt{2}$$

Here we obtain the WT features in the medium frequency subband

$$WF(2) = [c_{2,0}, c_{2,1}] = \left[\frac{-10}{\sqrt{2}}, \frac{11}{\sqrt{2}}\right] \Big/ \sqrt{2} = [-5.00, 5.50]$$

, which reflects the changing rates of the signal within every four sampling periods in the time dimension.

Step 3: Perform averaging and differencing at the level corresponding to s=3 such that

$$\mathrm{x} = \left[\frac{24+11}{(\sqrt{2})^2}, \frac{24-11}{(\sqrt{2})^2}, \frac{-10}{\sqrt{2}}, \frac{11}{\sqrt{2}}, -3, -1, 3, -2\right] \Big/ \sqrt{2} = \left[\frac{35}{2}, \frac{13}{2}, \frac{-10}{\sqrt{2}}, \frac{11}{\sqrt{2}}, -3, -1, 3, -2\right]$$

$$\Big/ \sqrt{2} \approx [12.4, 4.60, -5.00, 5.50, -2.12, -0.707, 2.12, -1.41]$$

Here we obtain the WT features in the lowest frequency subband

$$WF(3) = [c_{3,0}] = \left[\frac{13}{2}\right] \Big/ \sqrt{2} = [4.60]$$

, which reflects the changing rate of the signal in the whole duration of eight sampling periods.

With the above example we understand that the WT coefficients can be divided into different frequency subbands, each of which reflects how fast the signal increases or decreases its values in the corresponding frequency. The total number of coefficients is equal to $N/2 + N/4 + \Lambda + 1 = N - 1$, which is almost the same as the length of the time series N. In order to get a reduced number of features for case indexing, common practices so far are to choose a dominant coefficient as representative of the subband [36] or to derive statistic values from each frequency subband [50]. Such methods work well with relatively short time series exhibiting simple dynamics. However, considering that WT coefficients themselves constitute a dynamic time series in a frequency subband, how to extract complete and compact information to characterize lengthy, time-varying subbands is still an unresolved issue.

9.2.2 The Proposal of Hybridizing Symbolization and Knowledge Discovery

As was noted in Sect. 9.2.1, traditional methods for feature extraction suffer from some drawbacks, such as undesired large number of features as well as the risk of loss of temporal relationship, when they are applied to complex, longitudinal series of measurements. The reason for this can be attributed to the primary representation of time series based on which features are derived. The data streams utilized are data rich but poor in information content. They only record measurements at every sampling point whereas contain no generalized descriptions of how the data in series evolve with the time. Pure signal processing and mathematical manipulations do not suffice to ensure the derivation of concise and complete dynamic information from primary sampling point-based data records.

The solution we propose in this chapter is to convert the sampling point-based representation of the time series into an interval-based representation. An interval consists of a set of consecutive sampling points and thus encompasses multiple sampling periods in the time dimension. Then data within an interval have to be generalized and aggregated into one symbolic value; the symbolization is conducted via discretization of the range for possible values of the signal. By doing this, the primary time series is now transformed into a symbolic series associated with intervals. Symbolization of primary numerical (usually real valued) time series signals brings the following merits:

1. Symbolic series are shorter in length and more intensive in information content (every symbol is considered as a generalization of the signal behavior in the associated interval), while much of the important temporal information is still retained.
2. Symbolic series facilitate higher computational efficiency; require less computational resource and memory space.

3. Symbolic data are more robust, less sensitive to measurement noises, and also enhance human understanding.
4. With symbolic time series data, it is relatively easier to apply data mining and knowledge discovery methods, algorithms, and tools to find novel, interesting knowledge, and patterns [12], which would in turn help better indexing and characterization of time series cases.

After a numerical signal has been converted into symbolic series, we have to focus on transitions of symbols in it rather than single symbolic values for interpreting and characterizing the series. This is supported by the fact that behaviors in dynamic processes are reflected from transitional patterns over time and occurrences of certain sequences are believed to be significant evidences to identify properties of sequential records. For instance, in medical domains, sequence of symptoms of patients are crucial for diagnosis by physicians, and frequently conditional changes with patients are more important than their static states within single time intervals. Deciding key sequences for case characterization is domain dependent. We need knowledge acquisition and discovery to find knowledge about key sequences when it is not known in advance.

The process of knowledge discovery for key sequences is highlighted in Fig. 9.2. It first entails converting the original database of numerical time series into a database of symbolic ones by means of symbolization. Subsequently the symbolic database with classified series is delivered as input to the knowledge discovery module, which then searches for qualified sequences in the space of all possible sequences. All competent sequences are to be picked up into the library of key sequences as final results.

Once the knowledge about key sequences has been made available, they are utilized as reference to capture important contents in a time series of query. This is shown in Fig. 9.3. The symbolic series transformed from the numerical one is checked thoroughly to detect any occurrences of key sequences stored in

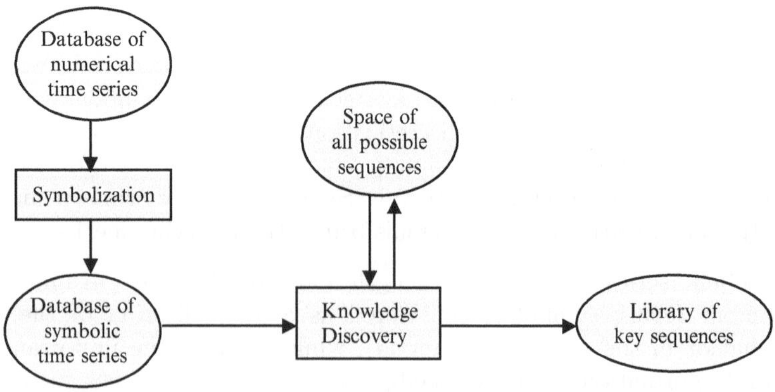

Fig. 9.2. Knowledge discovery to find key sequences

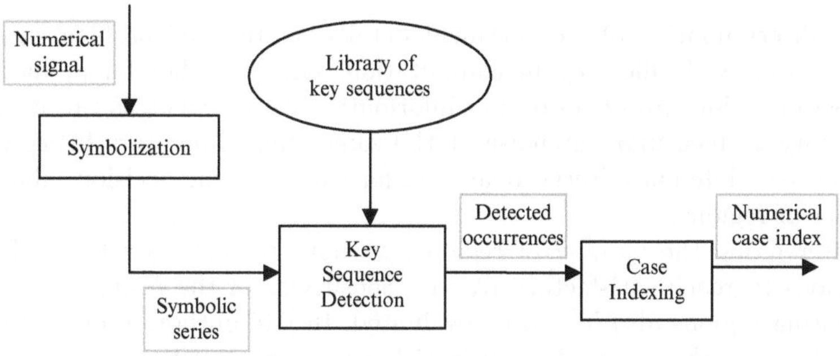

Fig. 9.3. Detection of key sequences in a time series

the key sequence library. Then the information derived about whether a key sequence has occurred and with how many times is made use of for building a numerical case index. This case index is concise since it only considers appearances of key sequences while ignoring other trivial randomness. Various ways to construct such case indexes will be addressed with details in Sect. 9.5.

At this point we can summarize the road map of our suggested solution as

$$Numericaltimeseries \longrightarrow Symbolictimeseries \longrightarrow Numericalcaseindex$$

Symbolization is performed as an intermediate stage to create symbolic expressions of cases which are more abstract and information intensive to facilitate knowledge discovery. Key sequences found by knowledge discovery capture the most significant transitions in a signal to identify its property. Finally case indexing in terms of key sequences creates compact descriptions of cases by quantifying influences of key sequences occurrences into numerical values.

9.3 Transformation into Symbolic Time Series

As noted before, symbolization of original numerical time series is a prestep for performing knowledge discovery to find key sequences. Fortunately many previous studies have shown that it is feasible to do so. Various approaches to making such transformation have been reported in the literature and they will be briefly outlined in this section.

9.3.1 Defining Symbols

Defining appropriate symbols is the first task to conduct during symbolization of signals. It involves partitioning the range of possible values of measurements into a set of regions. Each region corresponds to a specific symbol and each measurement value is thus uniquely mapped into the symbol of the region in which it falls in. The number of regions (symbols) reflects the level of resolution for the information that is retained. A low number of regions implies coarse discrimination of measurement details yet reduced problem space as well as improved efficiency in computation. On the other hand, increased number of regions preserves deeper information details whereas causes higher sensitivity to measurement noise at the same time. There are hence trade-offs between different criteria to account for when making decisions about the number of regions.

After fixing the number of regions, we have to select locations of these partitions to reach satisfactory results. Sensitivity of the results to the way of locating regions also has to be evaluated. In [10] authors proposed a theoretical approach to choosing optimal locations of partitions for noise-free, deterministic processes. However, selecting theoretically optimal partitions is hardly possible for practical sensor measurements. The reason is that the sensor data are expected to be produced from practical, uncertain processes with hidden dynamics and unknown characteristics of noise, such that a universally strict optimization method does not exist. In practice problem dependent ad hoc techniques are widely employed to determine suitable ways to partition sensor measurements.

In some cases partitioning can be conducted according to the context of the problem provided that the underlying physics betrays a natural choice for granulation. This means the situations where systems in consideration involve dynamics with natural borderlines dividing system states into distinct physical areas. For example, in neurobiological and chemical systems, there is often an excitability threshold above which oscillations will be activated [8,24]. Natural partitioning based on problem context gives a means of accommodating physical knowledge and makes meaningful results easy for human understanding.

In most of other cases traditional methods are to use data mean, midpoint, or median, equal-sized regions, or regions with equal probability to divide the whole range of sensor measurements. In [48] binary symbols corresponding to regions separated by sample median were adopted for reconstruction of dynamics of nonlinear models in light of existing heavy noise. Equal-sized partitions were developed by [18] in dealing with EEG signals. Kim and his colleagues [27] used combinations of sample mean and standard deviations to define regions when analyzing heart-rate dynamics. Finally symbols with equal probability were addressed in [14] by dividing the whole data range into regions in which observation values have identical likelihood to fall.

9.3.2 Symbolic Approximation

The method of symbolic approximation was proposed in [43] with the aim to convert a primary real-valued time series into a condensed symbolic sequence of much shorter length. In doing this the whole duration of the signal is divided into equally sized intervals, i.e., each interval encompasses the same amount of sampling periods. The data in each interval is averaged into a mean value, thus creating a sequence of real numbers summarizing signal behaviors in consecutive time intervals. We term this sequence as PAA (piecewise aggregate approximation) representation of the original signal. Then the PAA sequence is further transformed into a symbolic form by mapping the real numbers in it into corresponding symbols.

To be more concrete, PAA is performed to convert a real-valued time series (primary sensor signal) $C = [c_1, c_2, \ldots, c_n]$ into a shorter sequence $\tilde{C} = [\tilde{c}_1, \tilde{c}_2, \ldots, \tilde{c}_w]$ $(w < n)$. The element $\tilde{c}_i$ in $\tilde{C}$ represents the mean value of the data collected during the ith interval, thus it is given by

$$\tilde{c}_i = \frac{w}{n} \sum_{j=\frac{n}{w}(i-1)+1}^{\frac{n}{w}i} c_j \tag{9.10}$$

where w is the total number of time intervals. PAA creates a substantial data reduction by aggregating signal values within intervals into single values. The PAA sequence, as an intermediate representation, is visualized in Fig. 9.4 where the original signal is approximated by pieces of horizontal segments reflecting the signal's average levels during respective time intervals.

After the PAA sequence $\tilde{C}$ is obtained, it has to be transformed into a symbolic sequence $\hat{C} = [\hat{c}_1, \hat{c}_2, \ldots, \hat{c}_w]$ via symbolization. This entails discretization of the range of signal values into a set of nonoverlapping regions. According to [43] it is desirable to define regions (symbols) with equiprobability [3, 31]. The equal probability of symbols demands that the ordered list of breakpoints $B = \beta_1, \beta_2, \ldots, \beta_{r-1}$ separating regions be defined in such a way that

$$\int_{\beta_i}^{\beta_{i+1}} p(x)dx = \frac{1}{r} \tag{9.11}$$
$$\beta_0 = -\infty, \qquad \beta_r = \infty$$

holds for any $i \in \{0, 1, \ldots, r-1\}$, where r is the number of symbols and by p(x) we denote the probability density function of measured values.

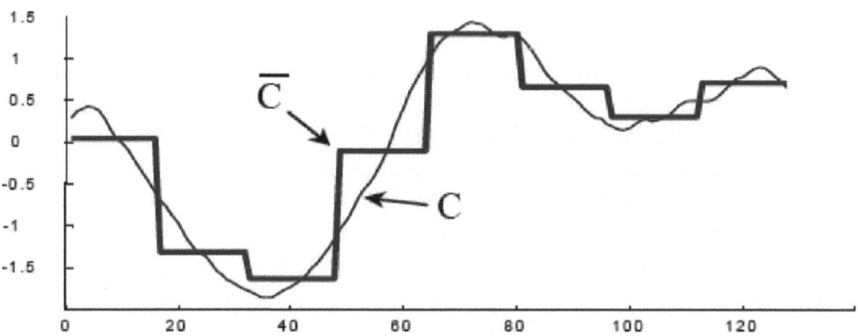

Fig. 9.4. PPA sequence approximating an original signal [43]

Table 9.1. The breakpoints when measurements subject to Gaussian distribution N(0,1)

Symbol number	β_1	β_2	β_3	β_4	β_5	β_6	β_7	β_8	β_9
3	−0.43	0.43							
4	−0.67	0	0.67						
5	−0.84	−0.25	0.25	0.84					
6	−0.97	−0.43	0	0.43	0.97				
7	−1.07	−0.57	−0.18	0.18	0.57	1.07			
8	−1.15	−0.67	−0.32	0	0.32	0.67	1.15		
9	−1.22	−0.76	−0.43	−0.14	0.14	0.43	0.76	1.22	
10	−1.28	−0.84	−0.52	−0.25	0	0.25	0.52	0.84	1.28

Specifically, for sensor measurements subject to the Gaussian distribution N(0,1), the values of these breakpoints are given in Table 9.1 where the number of symbols ranges from 3 to 10.

Once the breakpoints for separation of regions are fixed, the symbolic sequence $\hat{C} = [\hat{c}_1, \hat{c}_2, \ldots, \hat{c}_w]$ can be obtained from the intermediate PAA sequence $\tilde{C} = [\tilde{c}_1, \tilde{c}_2, \ldots, \tilde{c}_w]$ by using the following rule:

$$\hat{c}_i = alpha_j \qquad iff \quad \beta_{j-1} \leq \bar{c}_i \leq \beta_j \tag{9.12}$$

where $alpha_j$ denotes the symbol assigned to the jth region bounded by β_{j-1} and β_j.

9.3.3 Temporal Abstraction

Temporal abstraction is an artificial intelligence technique first proposed by [45] to solve data interpretation tasks. The goal is to derive high level generalization from time-stamped representations of time series to evolve toward interval-based representations. Basically this is achieved by aggregating adjacent events exhibiting a common behavior overtime into a generalized concept. Through temporal abstraction, large amounts of temporal data in primary, longitudinal signals can be compressed into compact, abstract, and more meaningful descriptions in the form of series of symbolic values.

Basic temporal abstraction seems sufficient to derive symbolic time series data in the context of this chapter. The ontology for basic temporal abstraction is depicted in Fig. 9.5 which includes state abstraction and trend abstraction. The former focuses on the measured values themselves to extract intervals associated to qualitative concepts such as low, normal, and high, while the latter considers differences between two neighboring records to detect specific patterns like increase, decrease, and stationarity in the series. If differences between consecutive measurements are treated as data for a new

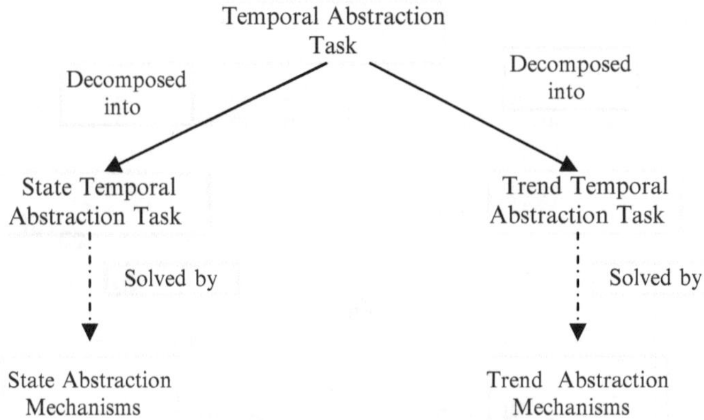

Fig. 9.5. Ontology of basic temporal abstraction

series, trend abstraction is equivalent to applying state abstraction to the secondary series of differences derived from the original series of measured values.

The regions associated with qualitative concepts are to be defined beforehand through discretization of the range of measured values for state abstraction or the range of difference values for trend abstraction. The essential thing to do in abstraction is merging adjacent entities falling in the same region into a cluster and summarizing behaviors in this cluster with a concept (symbol) corresponding to the region. Then, arranging concepts of clusters according to the order of their appearances produces a required symbolic series. This new series is compact, abstract, and contains more meaningful information than the primary one.

Temporal abstraction was applied successfully to intelligent analysis of longitudinal data series gathered from monitoring of chronic patients, as reported in [6]. This work, for instance, analyzed and abstracted body temperature profiles in terms of the concepts of low, normal, high, and very high. At the same time the trend for temperature changes were identified as stationarity, increase, or decrease. A simple example given in [6] to illustrate abstractions of body temperatures is shown in Fig. 9.6.

9.3.4 Phase-Based Pattern Identification

In some cases a longitudinal signal consists of a series of phases and every phase has its physical significance to identify its property (pattern) alone. This motivates us to divide the whole signal profile into pieces of subsignals and each of which corresponds to a phase. Since subsignals are assumed to be relatively short and simple, conventional signal processing methods like Fourier or wavelet transforms can be applied to them for extracting features and classifying their patterns. Further, arranging patterns of subsignals in order of their appearances creates a symbolic series as compact and abstract

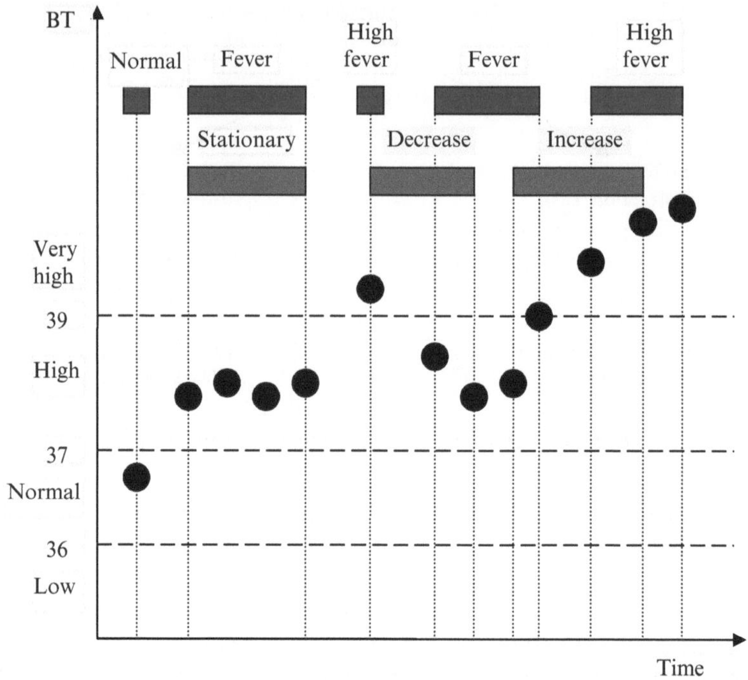

Fig. 9.6. A simple example for temporal abstractions on body temperatures [6]

representation for the overall signal. Each symbol in this series signifies the pattern of a subsignal corresponding to a phase. We refer the method stated above as *phase-based pattern identification.*

This method is used in one of our AI projects related to stress medicine [35], where RSA signals obtained from patients are employed to classify their stress levels. A patient is usually tested through a series of 40–80 breathing cycles (including inhalation and exhalation). Every respiration cycle lasts on average 5–15 s and corresponds to either a normal breathing pattern or one of the dysfunctional patterns. The patterns of breathing (also called RSA patterns) are identified from RSA measurements in the respective respiration periods. Further patterns from consecutive breathing cycles constitute a symbolic time series, which is to be investigated to find information reflecting stress levels of patients.

An overview of the stress medicine project is depicted in Fig. 9.7. First the RSA signal measured during the whole test period is decomposed into a collection of subsignals. By subsignal i in Fig. 9.7 we denote the phase of the signal recorded for the ith breathing cycle. Each subsignal i is delivered to the block "signal classifier," where wavelet analysis and CBR are applied to decide upon its pattern [35]. The identified patterns are then composed into a symbolic series in terms of their appearance order for classifying categories concerning stress levels.

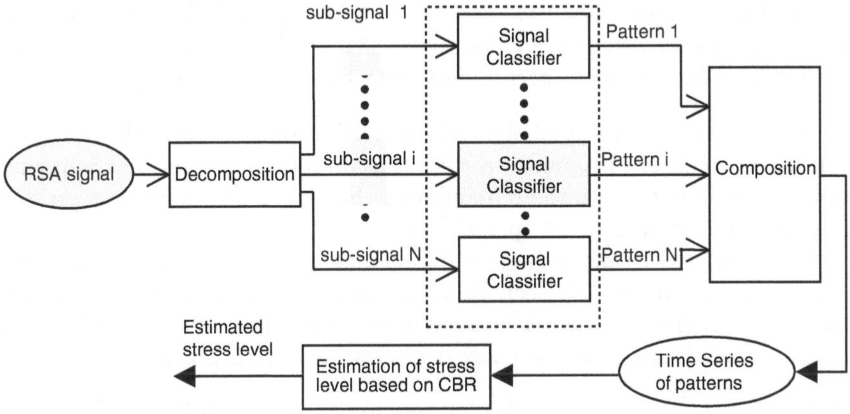

Fig. 9.7. An overview of the stress medicine project

9.4 Knowledge Discovery on Symbolic Time Series Representations

Once a numerical signal has been converted into a symbolic series, it is crucial to get awareness of which parts in the series make sense while others are ignorable. This entails discovering knowledge about key sequences for an underlying domain. Key sequences help better interpret and characterize time series at hand to capture real nature of dynamic systems

Indeed the value of knowledge about key sequences has been made obvious in many application scenarios. For instance, in health monitoring of engineering equipments, original sensor readings can be converted into discrete symbols [41], and some critical changes in series of measurements like swell, sag, impulsive transients, might be signs indicating a present or potential anomaly. In telecommunications, useful information can be obtained from sequences of alarms produced by switches for analysis and prediction of network faults. In defense, sequences of deployments/actions of enemies would possibly betray their tactical intentions. Finally, in a medical scenario, a data sequence of symptoms exhibited on a patient may help to forecast a disease that follows the emerging symptoms.

9.4.1 Problem Statements

To clearly present the proposed knowledge discovery approach, we now give descriptions of the various terms and concepts that are related. We start from formal definitions of symbolic time series, sequences, and time series databases, and then we precisely formulate the problem this section aims to tackle.

Definition 1. *A symbolic time series is a series of symbols signifying events that occurred sequentially overtime,* $X = [x(1), x(2), \ldots, x(i), \ldots, x(w)]$*, where* i *indexes an time segment corresponding to symbolic value* $x(i)$*.*

In a general sense, a symbolic series in Definition 1 can be either a conversion from a numerical signal or a recording of discrete events that happened sequentially in nature. In the remaining of this chapter we refer time series to only symbolic ones given no special notes.

Moreover, every time series has an inherent class. The previous time series data are assumed to have been classified and they are stored in a database together with their associated classes to facilitate data mining. A definition of time series database in the context of classification is given as follows:

Definition 2. *A time series database is a set of pairs* $\{(X_i, Z_i)\}_{i=1}^{K}$*, where* X_i *denotes a time series,* Z_i *the class assigned to* X_i*, and* K *is the number of time series cases in the database.*

With a time series database at hand, the data mining process involves analyzing sequences that are included in the database. A sequence of a time series is formally described in Definition 3.

Definition 3. *A sequence* S *of a time series* $X = [x(1), x(2), \ldots, x(w)]$ *is a list consisting of elements taken from contiguous positions of* X*, i.e.,* $S = [x(k), x(k+1), \ldots, x(k+m-1)]$ *with* $m \leq w$ *and* $1 \leq k \leq w - m + 1$*.*

Usually there is a very large amount of sequences included in the time series database. But only a part of them that carry useful information for estimating classes are in line with our interest. Such sequences are referred to as indicative sequences and defined in the following:

Definition 4. *A sequence is regarded as indicative given a time series database provided that:*

(1) *It appears in a sufficient amount of time series profiles of the database*
(2) *The discriminating power of it, assessed upon the database, is above a specified threshold*

A measure for discriminating power together with the arguments that lie behind this definition will be elaborated in Sect. 9.4.2. The intuitive explanation is that an indicative sequence is such a one that, on one hand, appears frequently in the database, and on the other hand, exhibits high cooccurrence with a certain class.

Obviously, if a sequence is indicative, another sequence that contains it as subsequence may also be indicative for predicting the class. However, if these both are indicative of the same class, the second sequence is considered as redundant with respect to the first one because it conveys no more information. Redundant sequences can be easily recognized by checking possible inclusion between sequences encountered. The goal here is to find sequences that are not only indicative but also nonredundant and independent of each other.

Having given necessary notions and clarifications we can now formally define our problem to be addressed as follows:

Given a time series database consisting of time series profiles and associated classes, find a set of indicative sequences $\{S_1, S_2, \ldots, S_p\}$ *that satisfy the following two criteria:*

(1) *For any* $i, j \in \{1, 2, \ldots, p\}$ *neither* $S_i \subseteq S_j$ *nor* $S_j \subseteq S_i$ *if* S_i *and* S_j *are indicative of a same class*
(2) *For any sequence* S *that is indicative,* $S \in \{S_1, S_2, \ldots, S_p\}$ *if* S *is not redundant with respect to* S_j *for any* $j \in \{1, 2, \ldots, p\}$

The first criterion above requests compactness of the set of sequences $\{S_1, S_2, \ldots, S_p\}$ in the sense that no sequence in it is redundant by having a subsequence indicative of the same class as it. A sequence that is both indicative and nonredundant is called a key sequence. The second criterion further requires that no single key sequence shall be lost, which signifies a demand for completeness of the set of key sequences to be discovered.

9.4.2 Evaluation of Single Sequences

This section aims to evaluate individual sequences to decide whether one sequence can be regarded as indicative. The main thread is to assess the discriminating power of sequences in terms of their cooccurrence relationship with possible time series classes. In addition we also illustrate the importance of sequence appearing frequencies in the case base for ensuring reliable assessments of the discriminating power.

We assume that given a sequence S there are a set of probable consequent classes $\{C_1, C_2, \ldots, C_k\}$. The strength of the cooccurrence between sequence S and class $C_i (i = 1 \ldots k)$ can be measured by the probability, $p(C_i|S)$, of C_i conditioned upon S. Sequence S is considered as discriminative in predicting outcomes as long as it has a strong cooccurrence with either of the possible classes. The discriminating power of S is defined as the maximum of the strengths of its relations with probable consequent classes. Formally this definition of discriminating power PD is expressed as:

$$PD(S) = \max_{i=1\ldots k} P(C_i|S) \tag{9.13}$$

In addition we say that the class yielding the maximum strength of the cooccurrences, i.e., $C = \arg_{i=1\ldots k} \max P(C_i|S)$, is the class that sequence S is indicative of.

The conditional probabilities in (9.13) can be derived according to the Bayesian theorem as:

$$P(C_i|S) = \frac{P(S|C_i)P(C_i)}{P(S)} \tag{9.14}$$

As the probability $P(S)$ is generally obtainable by

$$P(S) = P(S|C_i)P(C_i) + P(S|\overline{C}_i)P(\overline{C}_i) \tag{9.15}$$

(9.14) for conditional probability assessment can be rewritten as

$$P(C_i|S) = \frac{P(S|C_i)P(C_i)}{P(S|C_i)P(C_i) + P(S|\overline{C}_i)P(\overline{C}_i)} \tag{9.16}$$

Our aim here is to yield the conditional probability $P(C_i|S)$ in terms of (9.16). As $P(C_i)$ is a priori probability of occurrence of C_i which can be acquired from domain knowledge or approximated by experiences with randomly selected samples, the only things that remain to be resolved are the probabilities of S in (time series) cases having class C_i and in cases not belonging to class C_i, respectively. Fortunately such probability values can be easily estimated by resorting to the given case base. For instance we use the appearance frequency of sequence S in class C_i cases as an approximation of $P(S|C_i)$, thus we have:

$$P(S|C_i) \approx \tfrac{N(C_i,S)}{N(C_i)} \tag{9.17}$$

where $N(C_i)$ denotes the number of cases having class C_i in the case base and $N(C_i,\ S)$ is the number of cases having both class C_i and sequence S. Likewise the probability $P(S|\overline{C}_i)$ is approximated by

$$P(S|\overline{C}_i) \approx \tfrac{N(\overline{C}_i,S)}{N(\overline{C}_i)} \tag{9.18}$$

with $N(\overline{C}_i)$ denoting the number of cases not having class C_i and $N(\overline{C}_i, S)$ being the number of cases containing sequence S but not belonging to class C_i.

The denominator in (9.16) has to stay enough above zero to enable reliable probability assessment using the estimates in (9.17) and (9.18). Hence it is crucial to acquire an adequate amount of time series cases containing S in the case base. The more such cases available the more reliably the probability assessment could be derived. For this reason we refer the quantity $N(S) = N(C_i, S) + N(\overline{C}_i, S)$ as evaluation base of sequence S in this chapter.

At this point we realize that two requirements have to be satisfied for believing a sequence to be indicative of a certain class. Firstly the sequence has to possess an adequate evaluation base by appearing in a sufficient amount of time series cases. Obviously a sequence that occurred randomly in few occasions is not convincing and can hardly be deemed significant. Secondly, the conditional probability of that class under the sequence must be dominatingly high, signifying a strong discriminating power. These explain why indicative sequence is defined by the demands on its appearance frequency and discriminating power in Definition 4.

In real applications two minimum thresholds need to be specified for the evaluation base and discriminating power, respectively, to judge sequences as indicative or not. The values of these thresholds are domain dependent and are to be decided by human experts in the related area. The threshold for discriminating power may reflect the minimum probability value that suffices to predict a potential outcome in a specific scenario. The threshold for the evaluation base indicates the minimum amount of samples required to fairly approximate the conditional probabilities of interest. This threshold value can be estimated in terms of the distribution of cases in classes in the case library as well as their prior probabilities. It is shown in the following.

Let $\delta > 0$ be the smallest distance for the denominator in (9.16) to remain sufficiently away from zero, we demand

$$\frac{N(C_i,S)}{N(C_i)}P(C_i) + \frac{N(\overline{C}_i,S)}{N(\overline{C}_i)}P(\overline{C}_i) \geq \delta \tag{9.19}$$

Further the above relation has to hold for every class C_i to ensure reliable assessments of conditional probabilities for all the classes given sequence S. Next the lower bound for the left side of inequality (9.19) is yielded by

$$\frac{N(C_i,S)}{N(C_i)}P(C) + \frac{N(\overline{C}_i,S)}{N(\overline{C}_i)}P(\overline{C}_i) \geq \frac{N(C_i,S)P(C_i) + N(\overline{C}_i,S)P(\overline{C}_i)}{\max[N(C_i),N(\overline{C}_i)]}$$

$$\geq \frac{[N(C_i,S) + N(\overline{C}_i,S)] \bullet \min[P(C_i),P(\overline{C}_i)]}{\max[N(C_i),N(\overline{C}_i)]} = \frac{\min[P(C_i),P(\overline{C}_i)]}{\max[N(C_i),N(\overline{C}_i)]}N(S) \tag{9.20}$$

Since this lower bound not being less than δ is a sufficient condition for satisfaction of inequality (9.19), we simply impose constraints on the evaluation base N(S) as given by

$$N(S) \geq \frac{\max[N(C_i),N(\overline{C}_i)]}{\min[P(C_i),P(\overline{C}_i)]} \bullet \delta \qquad \forall i \tag{9.21}$$

Herewith it is clearly seen that the threshold value for the evaluation base can be defined as the minimum number of N(S) that satisfies all the constraints in (9.21) for every class C_i. Finally only those sequences that pass thresholds for both discriminating power and evaluation base are evaluated as indicative ones.

9.4.3 Searching for Key Sequences

With the evaluation of sequences being established, we now turn to exploration of qualified sequences in the problem space. The goal is to locate all key sequences that are nonredundant and indicative. We first detail a sequence

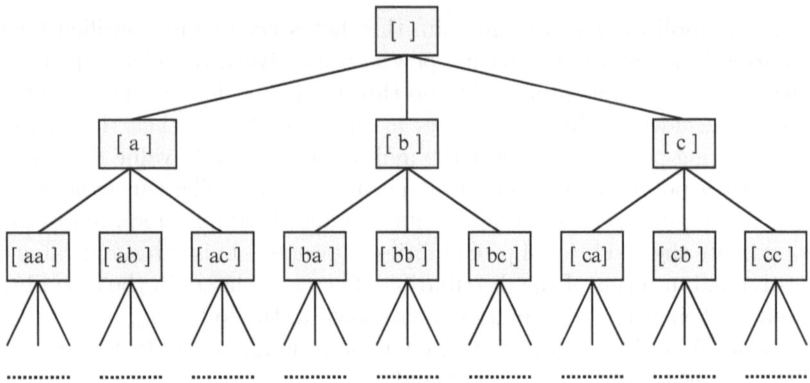

Fig. 9.8. The state space for sequences with three symbols

search algorithm for this purpose in this section and then we demonstrate simulation results on a synthetic case base with the proposed algorithm in Sect. 9.4.4

Discovery of key sequences can be considered as a search problem in a state space in which each state represents a sequence of symbols. Connection between two states signifies an operator between them for transition, i.e., addition or removal of a single symbol in time sequences. The state space for a scenario with three symbols *a, b, c* is illustrated in Fig. 9.8, where an arc connects two states if one can be created by extending the sequence of the other with a following symbol.

A systematic exploration in the state space is entailed for finding a complete set of key sequences. We start from a null sequence and generate new sequences by adding a single symbol to parent nodes for expansion. The child sequences are evaluated according to evaluation bases and discriminating powers. The results of evaluation determine the way to treat each child node in one of the following three situations:

(a) If the evaluation base of the sequence is under a threshold required for conveying reliable probability assessment, terminate expansion at this node. The reason is that the child nodes will have even smaller evaluation bases by appearing in fewer cases than their parent node.
(b) If the evaluation base and discriminating power are both above their respective thresholds, do the redundancy checking for the sequence against the list of key sequences already identified. The sequence is redundant if at least one known key sequence constitutes its subsequence while both remaining indicative of the same consequent. Otherwise the sequence is considered nonredundant and hence is stored into the list of key sequences together with the consequent it indicates. After that this node is further expanded with the hope of finding, among its children, qualified sequences that might be indicative of other consequents.

(c) If the evaluation base is above its threshold whereas the discriminating power still not reaching the threshold, continue to expand this node with the hope of finding qualified sequences among its children.

The expansion of nonterminate nodes is proceeded in a level-by-level fashion. A level in the search space consists of nodes for sequences of the same length and only when all nodes at a current level have been visited does the algorithm move on to the next level of sequences having one more symbol. This order of treating nodes is very beneficial for redundancy checking because a redundant sequence will always be encountered later than its subsequences including the key one(s) during the search procedure.

From a general structure, the proposed sequence search algorithm is a little similar to the traditional breadth-first procedure. However, there are still substantial differences between both. The features distinguishing our search algorithm are (1) it does not attempt to expand every node encountered and criteria are established to decide whether exploration needs to be proceeded at any given state and (2) it presumes multiple goals in the search space and thus the search procedure is not terminated when a single key sequence is found. Instead the search continues on other prospective nodes until none of the nodes in the latest level needs to be expanded. A formal description of the proposed search algorithm is given as follows:

Algorithm for finding a complete set of key sequences

1. Initialize the *Open* list with an empty sequence.
2. Initialize the Key_List to be an empty list.
3. Remove the most left node t from the Open list.
4. Generate all child nodes of t
5. For each child node, $C(t)$, of the parent node t
   a) Evaluate $C(t)$ according to its discriminating power and evaluation base;
   b) If the evaluation base and discriminating power are both above their respective thresholds, do the redundancy checking for $C(t)$ against the sequences in the *Key_list*. Store $C(t)$ into the *Key_list* if it is judged as not redundant. Finally put $C(t)$ on the right of the *Open* list.
   c) If the evaluation base of $C(t)$ is above its threshold but the discriminating power is not satisfying, put $C(t)$ on the right of the *Open* list.
6. If the *Open* list is not empty go to step 3, otherwise return the *Key_list* and terminate the search.

Finally it bears mentioning that finding key sequences in our context differs from those [2, 13, 46] in the literature of sequence mining. Usually the goal in sequence mining is merely to find all legal sequential patterns with their frequencies of appearances above a user-specified threshold. Here we have to consider the cause-outcome effect for classification purpose. Only those nonredundant sequences that are not only frequent but also possess strong discriminating power will be selected as the results of search.

Table 9.2. Sequences discovered on a synthetic case base

Sequence discovered	Discriminating power%	Evaluation base	Dominating consequent
[a d c]	76.70	103	Class 1
[b c a]	78.22	101	Class 1
[d e b]	73.39	124	Class 2
[e a e]	83.18	107	Class 3

9.4.4 Simulation Results

To verify the feasibility of the mechanism addressed above we now present the simulation results on a synthetic case base. A case in this case base is depicted by a time series of 20 symbols and one diagnosis class as the outcome. A symbol in a time series belongs to {a, b, c, d, e} and a diagnosis class is either 1, 2, or 3. The four key sequences assumed are [a d c], [b c a], [d e b], and [e a e]. The first two sequences were supposed to have strong cooccurrences with class 1 and the third and fourth exhibit strong cooccurrences with classes 2 and 3, respectively. Each time series in the case base was created in such a way that both sequences [a d c] and [b c a] had a chance of 80% of being reproduced once in the time series cases of class 1 while sequences [d e b] and [e a e] were added into class 2 and class 3 cases, respectively, with a probability of 90%. After stochastic reproduction of these key sequences, the remaining symbols in the time series of all cases were generated randomly. The whole case base consists of 100 instances for each class. Presuming such time series cases to be randomly selected samples from a certain domain, a priori probability of each class is believed to be one-third.

The sequence search algorithm was applied to this case base to find key sequences and potential cooccurrences hidden in the data. The threshold for the discriminating power was set at 70% to ensure an adequate strength of the relationships discovered. We also specified 50 as the threshold of the evaluation base for reliable assessment of probabilities. The sequences found in our test are shown in Table 9.2 below.

As seen from Table 9.2 we detected all the four key sequences previously assumed. They were recognized to potentially cause the respective consequents with the probabilities ranging from 73.39 to 83.18%. These relationships with a degree of uncertainty are due to the many randomly generated symbols in the case base such that any sequence of symbols is more or less probable to appear in time series of any class. But such nondeterministic property is prevalent in many real-world domains.

9.5 Utility of Key Sequences in Case-Based Classification

The key sequences discovered help us better focus on the most important dynamic patterns while ignoring trivial randomness in examining a time series. They are treated as significant features in capturing dynamic system behaviors. Rather than enumerating what happened in every consecutive time segment, we can now more concisely represent a time series case in terms of occurrences of key sequences in it. Let $\{S_1, S_2, \ldots, S_p\}$ be the set of key sequences. We have to search for every $S_i (i = 1 \ldots P)$ in a time series X to detect all possible appearances. Then case index for X can be established according to the results of key sequence detection. In the following four alternate ways to index X based on key sequences are suggested.

9.5.1 Naive Case Index

A naive means of indexing a time series case X is to depict it by a vector of binary numbers each of which corresponds to a key sequence. A number in the vector is unity if the corresponding sequence is detected in X and zero otherwise. This means that, by the naive method, the index of X is given by

$$Id_1(X \mid S_1, \ldots, S_P) = [b_1, b_2, \ldots, b_P] \tag{9.22}$$

where

$$b_i = \begin{cases} 1 \; if & S_i \; is \; subsequence \quad of \quad X \\ 0 & otherwise \end{cases} \tag{9.23}$$

This index has the merit of imposing low demand in computation. It also enables the similarity between two cases to be calculated as the proportion of the positions where their indexing vectors have identical values. Suppose two time series cases X_1 and X_2 which are indexed by binary vectors $[b_{11}, \ldots, b_{1P}]$ and $[b_{21}, \ldots, b_{2P}]$, respectively, the similarity between them is simply defined as

$$Sim_1(X_1, X_2) = 1 - \frac{1}{P} \sum\nolimits_{j=1}^{P} |b_{ij} - b_{2j}| \tag{9.24}$$

9.5.2 Case Index Using Sequence Appearance Numbers

With a binary structure the case index in Sect. 9.5.1 carries a little limited content and would be usable only in relatively simple circumstances. A main reason is that the index cannot reflect how many times a key sequence has appeared in a series of consideration. To incorporate that information, an alternate way is to directly employ the numbers of appearances of single key sequences in describing time series cases. By doing this we acquire the second method of indexing time series X by an integer vector as

$$Id_2(X \mid S_1, \ldots, S_P) = [f_1, f_2, \ldots, f_P] \tag{9.25}$$

where f_i denotes the number of occurrences of sequence S_i in series X.

Further, considering the case index in (9.25) as a state vector, we use the cosine matching function [44] as the similarity measure between two time series cases X_1 and X_2. Thus we have

$$Sim_2(X_1, X_2) = \frac{\sum_{j=1}^{P} f_{1j} f_{2j}}{\sqrt{\sum_{j=1}^{P} f_{1j}^2}\sqrt{\sum_{j=1}^{P} f_{2j}^2}} \tag{9.26}$$

with f_{1j}, f_{2j} denoting the numbers of occurrences of key sequence S_j in X_1 and X_2, respectively.

9.5.3 Case Index in Terms of Discriminating Power

Although the case index in (9.25) can distinguish two cases having a same key sequence but with different numbers of appearances, it still might not be an optimal representation to capture the exact nature of the problem. Recall that the value of a key sequence is conveying a degree of confidence in the sense of discriminating power for predicting a potential consequent, a time series X would be more precisely characterized by the discriminating powers of the appearances of single key sequences. Intuitively two times of occurrences of a key sequence would give a stronger discriminating power than occurring just once, but not twice in the quantity of the strength. From view of this we suggest indexing X as a vector of real numbers, representing discriminating powers for the appearances of single key sequences, as follows:

$$Id_3(X \mid S_1, \ldots, S_P) = [g_1, g_2, \ldots, g_P] \tag{9.27}$$

with

$$g_i = \begin{cases} DP(f_i * S_i) & if \quad f_i \geq 1 \\ 0 & if \quad f_i = 0 \end{cases} \tag{9.28}$$

By $DP(f_i * S_i)$ we denote the discriminating power by sequence S_i appearing f_i times in X.

Let C be the class that the key sequence S_i is indicative of. We define the discriminating power $DP(f_i * S_i)$ as the probability for class C given f_i appearances of sequence S_i. This probability can be obtained by applying the Bayes theorem in a sequential procedure. Assuming a two class problem without loss of generality, this procedure is depicted here by a series of equations as follows:

$$P(C|S_i) = \frac{P(S_i|C)P(C)}{P(S_i|C)P(C) + P(S_i|\overline{C})P(\overline{C})} \tag{9.29}$$

$$P(C|2 * S_i) = \frac{P(S_i|C)P(C|S_i)}{P(S_i|C)P(C|S_i) + P(S_i|\overline{C})P(\overline{C}|S_i)} \tag{9.30}$$

$$P(C|t * S_i) = \frac{P(S_i|C)P\left(C|(t-1) * S_i\right)}{P(S_i|C)P\left(C|(t-1) * S_i\right) + P(S_i|\overline{C})P\left(\overline{C}|(t-1) * S_i\right)} \tag{9.31}$$

$$DP(f_i * S_i) = P(C|f_i * S_i) = \frac{P(S_i|C)P\left(C|(f_i - 1) * S_i\right)}{P(S_i|C)P\left(C|(f_i - 1) * S_i\right) + P(S_i|\overline{C})P\left(\overline{C}|(f_i - 1) * S_i\right)} \tag{9.32}$$

where the probabilities $P(S_i|C)$ and $P(S_i|\overline{C})$ can be estimated according to (9.17) and (9.18), respectively. The probability updated in (9.29) represents the probability for class C given one appearance of S_i, which is further updated in (9.30) by the second appearance of S_i producing a higher probability considering both occurrences. Generally, the probability $P(C|t * S_i)$ is yielded by updating the prior probability $P\left(C|(t-1) * S_i\right)$ with one more occurrence of S_i in (9.31). Finally we obtain the ultimate probability assessment incorporating all appearances, i.e., the required discriminating power, by (9.32).

We now give a concrete example to illustrate how a case index can be built in terms of occurrences of key sequences. Suppose a two class situation in which three key sequences S_1, S_2, and S_3 are discovered. Sequence S_1 appears twice in time series X and S_2 appears once while S_3 is not detected. S_1 and S_2 are both indicative of a certain class C. The a priori probability for class C is 50% and the probabilities of sequences S_1, S_2 in situations of class C and its complementary are shown below:

$$P(S_1|C) = 0.56 \qquad P(S_1|\overline{C}) = 0.24$$
$$P(S_2|C) = 0.80 \qquad P(S_2|\overline{C}) = 0.40$$

With all the information assumed above, the discriminating powers for the appearances of S_1 and S_2 are calculated in the following:

1. Calculate the probability for C with the first appearance of S_1 by

$$P(C|S_1) = \frac{P(S_1|C)P(C)}{P(S_1|C)P(C) + P(S_1|\overline{C})P(\overline{C})} = \frac{0.56 \cdot 0.5}{0.56 \cdot 0.5 + 0.24 \cdot 0.5} = 0.70$$

2. Refine the probability $P(C|S_1)$ with the second appearance of S_1, producing the discriminating power for the appearances of S_1

$$DP(2 * S_1) = P(C|2 * S_1) = \frac{P(S_1|C)P(C|S_1)}{P(S_1|C)P(C|S_1) + P(S_1|\overline{C})P(\overline{C}|S_1)}$$
$$= \frac{0.56 \cdot 0.70}{0.56 \cdot 0.70 + 0.24 \cdot 0.30} = 0.8448$$

   It is clearly seen here that the power of discrimination is increased from 0.70 to 0.8448 due to the key sequence occurring for the second time.

3. Derive the discriminating power for the occurrence of S_2 by calculating the conditional probability for C upon S_2 as

$$DP(1 * S_2) = P(C|S_2) = \frac{P(S_2|C)P(C)}{P(S_2|C)P(C) + P(S_2|\overline{C})P(\overline{C})}$$
$$= \frac{0.80 \cdot 0.50}{0.80 \cdot 0.50 + 0.40 \cdot 0.50} = 0.6667$$

Moreover, because S_3 is not detected in X, there is no discriminating power for it. Hence we construct the index for this time series case as:

$$Id_3(X\,|S_1\,, S_2, S_3) = [0.8448,\ 0.6667,\ 0]$$

Finally, with this case indexing scheme, we use the cosine function again as the similarity measure for case retrieval. So the similarity between two time series X_1 and X_2 is given by

$$Sim_3(X_1, X_2) = \frac{\sum_{j=1}^{P} g_{1j}g_{2j}}{\sqrt{\sum_{j=1}^{P} g_{1j}^2}\sqrt{\sum_{j=1}^{P} g_{2j}^2}} \tag{9.33}$$

where g_{1j} and g_{2j} denote the jth elements in the case indexes (9.27) for X_1 and X_2, respectively.

9.5.4 Case Indexing with Key Sequence Union

In Sect. 9.5.3 cases are indexed according to the discriminating powers of occurrences of single key sequences. Such work could be extended by regarding the key sequences that are indicative of a common class as a collective union. This view motivates us to group occurrences of key sequences in time series X into a set of clusters. For every class C_i there is a cluster V_i corresponding to it. V_i is a collection of events for occurrences of those key sequences that are indicative of class C_i. The discriminating power of cluster V_i is defined as the probability of class C_i in light of the events included in the cluster. Hence we write

$$DP(V_i) = \begin{cases} P(C_i\,|\{e_j\,|e_j \in V_i\}) & if \quad V_i \neq \emptyset \\ 0 & if \quad V_i = \emptyset \end{cases} \tag{9.34}$$

Further, the discriminating powers of clusters of events representing key sequences occurrences are utilized to index a time series case. Hence the index for time series X is given by

$$Id_4(X \mid S_1, \ldots, S_P) = [DP(V_1), DP(V_2), \ldots, DP(V_K)] \tag{9.35}$$

where K denotes the number of classes of interest.

It is clear that the case index in the form of (9.35) is highly concise. It reduces the length of index vector to the number of classes. This is achieved by calculating the discriminating power for a union of key sequences that are consistent. Consequently every component in the vector of (9.35) contains rich information by fusion of occurrences from multiple key sequences. This proposed case index is valuable for further dimensionality reduction particularly under the circumstances when the number of key sequences discovered is still quite large.

Let $V_i = \{e_1, e_2, \Lambda, e_T\}$ be a cluster of events of key sequences occurrences corresponding to class C_i. We now want to obtain the discriminating power of cluster V_i by calculating the conditional probability $P(C_i|e_1, e_2, \Lambda, e_T)$. This probability is yielded by exploiting the events e_j as evidences for probability updating in separate steps. At every step we use a single event to revise prior probabilities according to the Bayes theorem and these updated probability estimates are then propagated as prior beliefs to the next step. The procedure of probability updating using events in cluster V_i is depicted by a series of equations as follows:

$$P(C_i|e_1) = \frac{P(e_1|C_i)P(C_i)}{P(e_1|C_i)P(C_i) + P(e_1|\overline{C}_i)P(\overline{C}_i)} \tag{9.36}$$

$$P(C_i|e_1, e_2) = \frac{P(e_2|C_i)P(C_i\,|e_1\,)}{P(e_2|C_i)P(C_i\,|e_1\,) + P(e_2|\overline{C}_i)P(\overline{C}_i\,|e_1\,)} \tag{9.37}$$

$$P(C_i|e_1, \ldots, e_i) = \frac{P(e_i|C_i)P(C_i\,|e_1\,, \ldots, e_{i-1})}{P(e_i|C_i)P(C_i\,|e_1, \ldots, e_{i-1}) + P(e_i|\overline{C}_i)P(\overline{C}_i\,|e_1, \ldots e_i\,)} \tag{9.38}$$

$$p(C_i|e_1, \ldots, e_T) = \frac{P(e_T|C_i)P(C_i\,|e_1\,, \ldots, e_{T-1})}{P(e_T|C_i)P(C_i\,|e_1, \ldots, e_{T-1}) + P(e_T|\overline{C}_i)P(\overline{C}_i\,|e_1, \ldots\; e_{T-1}\,)} \tag{9.39}$$

where the probabilities $P(e_i|C_i)$ and $P(e_i|\overline{C}_i)$ for $i \in \{1, \ldots, T\}$ can be estimated according to (9.17) and (9.18), respectively, as e_i is considered as the occurrence of a sequence. The probability updated in (9.36) represents the probability for class C_i given event e_1, which is further updated in (9.37) by event e_2 producing a more refined belief considering both e_1 and e_2. Generally the probability $P(C|e_1, \ldots, e_i)$ is yielded by updating the prior probability $P(C|e_1, \ldots, e_{i-1})$ with a new event e_i in (9.38). Finally we obtain the ultimate probability assessment incorporating all available events in (9.39).

At this stage one may question the order in which single events from a cluster are used to refine probability assessments. This seems a fundamental issue and involves allocation of events to different steps of a sequential procedure. Fortunately our study has clarified that the order of events used in probability updating is completely indifferent. The final probability value remains constant as long as each piece of event is assigned to a distinct step. The claims as such are formally based on the following theorems.

Lemma 1. *Let $\{e_1, \ldots, e_T\}$ be a cluster of events representing appearances of certain key sequences in a time series X. The probability for class C given the cluster is not affected if two adjacent events exchange their positions in the order of events used for probability refinements. This means that the relation $P(C|e_1, \ldots e_i, e_{i+1}, \ldots, e_T\} = P(C|e_1, \ldots e_{i+1}, e_i, \ldots, e_T\}$ holds for $i \in \{1, \ldots T-1\}$.*

Proof. For proof of the lemma with the statement that $P(C|e_1,\ldots,e_{i-1},e_i, e_{i+1},\ldots,e_T\} = P(C|e_1,\ldots,e_{i-1},e_{i+1},e_i,\ldots,e_T\}$, we only need to establish the relation for $P(C|e_1,\ldots,e_{i-1},e_i,e_{i+1}\} = P(C|e_1,\ldots,e_{i-1},e_{i+1},e_i)$, which is equivalent to the lemma.

We start to consider the probability $P(C|e_1,\ldots e_i,e_{i+1}\}$ which is acquired by updating the prior belief $P(C|e_1,\ldots,e_i)$ with a new evidence e_{i+1}, hence it can be written as

$$P(C|e_1,\ldots,e_i,e_{i+1}) = \frac{P(e_{i+1}|C)P(C|e_1,\ldots,e_i)}{P(e_{i+1}|C)P(C|e_1,\ldots,e_i) + P(e_{i+1}|\overline{C})P(\overline{C}|e_1,\ldots,e_i)} \tag{9.40}$$

Further the probability $P(C|e_1,\ldots,e_i)$ is formulated by taking $P(C|e_1,\ldots,e_{i-1})$ as its prior estimate such that

$$P(C|e_1,\ldots,e_i) = \frac{P(e_i|C)P(C|e_1,\ldots,e_{i-1})}{P(e_i|e_1,\ldots,e_{i-1})} \tag{9.41}$$

Likewise we obtain

$$P(\overline{C}|e_1,\ldots,e_i) = \frac{P(e_i|\overline{C})P(\overline{C}|e_1,\ldots,e_{i-1})}{P(e_i|e_1,\ldots,e_{i-1})} \tag{9.42}$$

Combining (9.41) and (9.42) into (9.40) gives rise to a transformed formulation as

$$P(C|e_1,\ldots,e_i,e_{i+1}) = \frac{P(e_{i+1}|C)P(e_i|C)P(C|e_1,\ldots,e_{i-1})}{P(e_{i+1}|C)P(e_i|C)P(C|e_1,\ldots,e_{i-1}) + P(e_{i+1}|\overline{C})P(e_i|\overline{C})P(\overline{C}|e_1,\ldots,e_{i-1})} \tag{9.43}$$

Next we express the conditional probabilities $P(e_{i+1}|C)$, $P(e_{i+1}|\overline{C})$, $P(e_i|C)$, $P(e_i|\overline{C})$ with their Bayes forms by

$$P(e_{i+1}|C) = \frac{P(C|e_{i+1})P(e_{i+1})}{P(C)} \tag{9.44}$$

$$P(e_{i+1}|\overline{C}) = \frac{P(\overline{C}|e_{i+1})P(e_{i+1})}{P(\overline{C})} \tag{9.45}$$

$$P(e_i|C) = \frac{P(C|e_i)P(e_i)}{P(C)} \tag{9.46}$$

$$P(e_i|\overline{C}) = \frac{P(\overline{C}|e_i)P(e_i)}{P(\overline{C})} \tag{9.47}$$

where $P(C)$ and $P(\overline{C})$ denote the initial probability estimates for class C and its complementary without any events about key sequences appearances. Using the Bayes forms from (9.44) to (9.47), (9.43) is finally rewritten as

$$P(C|e_1,\ldots,e_i,e_{i+1})$$
$$= \frac{P^2(\overline{C})P(C|e_{i+1})|P(C|e_i)P(C|e_1,\ldots,e_{i-1})}{P^2(\overline{C})P(C|e_{i+1})|P(C|e_i)P(C|e_1,\ldots,e_{i-1}) + P^2(C)P(\overline{C}|e_{i+1})|P(\overline{C}|e_i)P(\overline{C}|e_1,\ldots,e_{i-1})} \tag{9.48}$$

Clearly we see from (9.48) that the order between e_i and e_{i+1} has no effect at all on the probability $P(C|e_1,\ldots,e_i,e_{i+1})$ assessed. It follows that

$$P(C|e_1,\ldots,e_{i-1},e_i,e_{i+1}) = P(C|e_1,\ldots,e_{i-1},e_{i+1},e_i) \tag{9.49}$$

and here from the lemma is proved.

With the lemma justified by the proof above, we further contemplate the implication of it. This leads to a corollary presented below. □

Corollary 1. *Let* $\{e_1,\ldots,e_T\}$ *be a cluster of events representing appearances of certain key sequences in a time series* X*. The probability for* X *in class* C *given the cluster is independent of the order according to which single events* e_1, $e_{2,\ldots}$, e_T, *are used in probability refinements.*

The proof of Corollary 1 *is obvious. According to the lemma, an element in a given order of events can be moved to an arbitrary position by repeatedly exchanging its position with an adjacent one while not affecting the final probability assessments. As this can be done to every piece of event, we enable transitions to any orders of events without altering the estimated value of the probability.*

This corollary is important in providing theoretic arguments allowing for an arbitrary order of sequences to be used in probability fusion based on the Bayes theorem. The connotation is that when a key sequence occurred in the time series does not matter for the case index. Instead only the numbers of appearances of key sequences affect the likelihoods of classes given respective occurrence clusters, which are included as components in the case index vector.

Now let us study an illustrative example to better understand how the above sequential procedure works in derivation of required probabilities using clusters of events as evidences. Consider a time series X *with two probable classes. Suppose that four key sequences* S_1, S_2, S_3, *and* S_4 *are detected in* X, *and* S_1, S_2 *are indicative of class* C *while* S_3 *and* S_4 *are indicative of the complementary of C. The a priori probability of class* C *is 50%, and the probabilities of sequences* S_1, S_2, S_3, *and* S_4 *in situations of class* C *and its complementary are shown below:*

$$\begin{array}{ll} P(S_1|C) = 0.56 & P(S_1|\overline{C}) = 0.24 \\ P(S_2|C) = 0.80 & P(S_2|\overline{C}) = 0.40 \\ P(S_3|C) = 0.35 & P(S_3|\tilde{C}) = 0.62 \\ P(S_4|C) = 0.18 & P(S_4|\tilde{C}) = 0.30 \end{array}$$

Further we assume that sequence S_1 *appears twice in* X *and* S_2, S_3, S_4 *appear once, hence the clusters of key sequence occurrences for* X *are notated as*

$V_1(X) = \{S_1, S_1, S_2\}$ and $V_2(X) = \{S_3, S_4\}$. With these three occurrences detected, the probability of class C is yielded in the following three steps:

Step A1: Update the a priori probability $P(C)$ with the first appearance of S_1 by

$$P(C|S_1) = \frac{P(S_1|C)P(C)}{P(S_1|C)P(C) + P(S_1|\overline{C})P(\overline{C})} = \frac{0.56 \cdot 0.5}{0.56 \cdot 0.5 + 0.24 \cdot 0.5} = 0.70$$

Step A2: Refine the probability updated in step A1 with the second appearance of S_1, thus we have

$$\begin{aligned} P(C|S_1, S_1) &= \frac{P(S_1|C)P(C|S_1)}{P(S_1|C)P(C|S_1) + P(S_1|\overline{C})P(\overline{C}|S_1)} \\ &= \frac{0.56 \cdot 0.70}{0.56 \cdot 0.70 + 0.24 \cdot 0.30} = 0.8448 \end{aligned}$$

Step A3: Refine the probability updated in step A2 with the occurrence of S_2, and we acquire the final probability assessment taking into account all events by

$$\begin{aligned} P(C|S_1, S_1, S_2) &= \frac{P(S_2|C)P(C|S_1, S_1)}{P(S_2|C)P(C|S_1, S_1) + P(S_2|\overline{C})P(\overline{C}|S_1, S_1)} \\ &= \frac{0.80 \cdot 0.8448}{0.80 \cdot 0.8448 + 0.40 \cdot 0.1552} = 0.9159 \end{aligned}$$

Likewise we calculate the probability $P(\overline{C}|S_3, S_4)$ with two steps as follows:

Step B1: Update the prior probability $P(\overline{C})$ with occurrence of S_3

$$P(\overline{C}|S_3) = \frac{P(S_3|\overline{C})P(\overline{C})}{P(S_3|C)P(C) + P(S_3|\overline{C})P(\overline{C})} = \frac{0.62 \cdot 0.5}{0.35 \cdot 0.5 + 0.62 \cdot 0.5} = 0.6392$$

Step B2: Refine the probability updated in step B1 with appearance of S_4

$$\begin{aligned} P(\overline{C}|S_3, S_4) &= \frac{P(S_4|\overline{C})P(\overline{C}|S_3)}{P(S_4|C)P(C|S_3) + P(S_4|\overline{C})P(\overline{C}|S_3))} \\ &= \frac{0.30 \cdot 0.6392}{0.18 \cdot 0.3608 + 0.30 \cdot 0.6392} = 0.7470 \end{aligned}$$

Finally, with the required probabilities at hand, we can establish the case index for the time series X as follows

$$\begin{aligned} & Id_4(X|S_1, S_2, S_3, S_4) = [DP(V_1), DP(V_2)] \\ & = \left[P(C|S_1, S_1, S_2), P(\overline{C}|S_3, S_4)\right] = [0.9159, 0.7470] \end{aligned}$$

For similarity assessment, we first calculate the dissimilarity between two time series X_1 and X_2 as the average of the differences in discriminating powers over all key sequences clusters

$$Dis_4(X_1, X_2) = \frac{1}{K}\sum\nolimits_{j=1}^{K} |DP(V_{1j}) - DP(V_{2j})| \tag{9.50}$$

where V_{1j} *and* V_{2j} *denote the* j*th clusters of key sequences corresponding to class* C_j*, for* X_1 *and* X_2*, respectively. Since the concept of dissimilarity is opposite to that of similarity, the degree of similarity between* X_1 *and* X_2 *is simply defined as unity subtracted by the dissimilarity value*

$$Sim_4(X_1, X_2) = 1 - Dis_4(X_1, X_2) \tag{9.51}$$

9.6 Relation to Relevant Works

Representation and retrieval of sequential sensor measurements as time series cases have received increasing research efforts during the recent years. The primary idea is to convert time-varying profiles into somehow simplified and shorter vectors that still preserve distances between original signals. Fourier transform and wavelet transform are two commonly used methods for such a conversion, and their usages for retrieving similar cases to support clinical decisions and industrial diagnoses have been shown in [33,35,36], respectively.

A more general framework for tackling cases in time dependent domain was proposed by [34], in which temporal knowledge embedded in cases are represented at two levels: case level and history level. The case level is tasked to depict single cases with features varying within case durations, while consecution of cases occurrences have to be captured in the history level to reflect the evolution of the system as a whole. It was also recommended by the authors that, at both of the two levels, the methodology of temporal abstraction [6,45] could be exploited to derive series of qualitative states or behaviors, which facilitate easy interpretation as well as pattern matching for case retrieval.

This chapter would be a valuable supplementary to the framework by Montani and Portinale in the sense that our key sequence discovery approach can be beneficially applied to the series of symbols abstracted from original numerical time series. The point of departure is that, in many practical circumstances, significant transitional patterns in history are more worthy of attentions than the states or behaviors themselves associated with single episodes. It follows that the key sequences discovered will offer us useful knowledge to focus on what are really important in case characterization. Moreover, as the number of key sequences is usually is much smaller than the number of elements in the series, indexing cases in terms of key sequences exhibits a further dimensionality reduction from series obtained via temporal abstraction.

It is worthy noting that the knowledge discovery treated here distinguishes itself from traditional learning included in a CBR cycle. The retain step in CBR typically stores a new case in the library or modifies some existing cases and may contain a number of substeps [1]. Learning therein is therefore case specific with knowledge stemming directly from newly solved cases. Contrarily, in our approach, learning is treated as a background task separated from the retain step and the whole case library is the input to the knowledge discovery process. Some relevant works combining knowledge discovery and CBR systems include: genetic-based knowledge acquisition for case indexing and matching [23], incremental learning to organize a case base [38], exploitation of background knowledge in text classification [53], and analysis of pros and cons for explanations in CBR systems [32].

Finding sequential patterns was widely addressed in the literature of sequence mining [2,13,46], where the goal was merely to find all legal sequential patterns with adequate frequencies of appearances. Identifying key sequences in our context differs from those in sequence mining in that we have to consider the cause-outcome effect for classification purpose. Only those nonredundant sequences that are not only frequent but also indicative in predicting outcomes will be selected.

Finally, time series data mining have gained increasing attention recently. Three embedding methods were proposed by [16] to transform time series data into a vector space for classification purpose. Keogh and his colleagues addressed the issue of dimensionality reduction for indexing large time series databases [25] and also for fast search in these databases [26]. In [52] a family of three unsupervised methods was suggested to identify optimal and valid features given multivariate time series data. Similarity mining in time series was tackled by [21] and various methods for efficient retrieval of similar time sequences were discussed in [9, 17, 37, 51]. Algorithms for mining association rules were handled in [28,40,49] to model and predict time series behaviors in dynamic systems, and the application of association mining to disclose stock prices relations in time series was presented in [20].

9.7 Conclusion

This chapter suggests a novel hybrid methodology combining data symbolization and knowledge discovery for analysis and interpretation of complex, longitudinal signals prevalent in medical and industrial domains. Data symbolization is tasked to transform primary numerical (usually real valued) series of measurements into shorter, more abstract series of symbolic data. The process of knowledge discovery is then applied to the case base of converted symbolic series to find key sequences, which would in turn help better characterizing and indexing primary numerical sensor signals into a concise case structure.

The knowledge discovery approach proposed utilizes the whole case library as available resources and is able to find from the problem space all qualified sequences that are nonredundant and indicative. An indicative sequence exhibits a high cooccurrence with a certain class and is hence valuable in offering discriminative strength for prediction and classification. A sequence that is both indicative and nonredundant is termed as a key sequence.

It is shown that the key sequences discovered are highly usable to characterize time series cases in case-based reasoning. The idea is to transform an original (lengthy) time series into a more concise case structure in terms of the occurrences of key sequences detected. Four alternate ways to develop case indexes based on knowledge about key sequences are suggested. The performance and applicability of these four case indexing methods are being tested in practical case studies related to our medical and industrial projects.

References

1. A. Aamodt and E. Plaza. Case-based reasoning: foundational issues, methodological variations and systems approaches. AI Communications, 7:39–59, 1994.
2. Agrawal, R., Srikant, R.: Mining sequential patterns. In: Proceedings of the 11th International Conference on Data Engineering. (1995) 3–14
3. A. Apostolico, M.E. Bock, and S. Lonardi. Monotony of surprise and large-scale quest for unusual words. In: Proceedings of the 6th International conference on Research in Computational Molecular Biology, Washington, DC, pp. 22–31, 2002.
4. Bar-Shalom, Y. and X. Li, Estimation and Tracking: Principles, Techniques, and Software, Artech House, Boston, 1993.
5. E. Beckenstein, G. Bachman and L. Narici. Fourier and Wavelet Analysis, Springer, 2000.
6. R. Bellazzi, C. Larizza, ans A. Riva. Temporal abstractions for interpreting diabetic patients monitoring data. Intelligent Data Analysis, 2: 97–122, 1998.
7. I. Bichindaritz and E. Conlon. Temporal knowledge representation and organization for case-based reasoning. In Proc. TIME-96, IEEE Computer Society Press, Washington, DC, 1996, pp. 152–159.
8. H.A. Braun et al. Low-Dimensional Dynamics in Sensory Biology 2: Facial Cold Receptors of the Rat. J. of Comp. Neuroscience 7(1), pp. 17–32, 1999.
9. Chan, K.P., Fu, A.W.: Efficient time series matching by wavelets. In: Proceedings of the International Conference on Data Engineering. (1999) 126–133
10. J.P. Crutchfield and K. Young. Inferring statistical complexity. Phys. Rev. Lett., Vol. 63, No. 2, pp. 105–108, 1989.
11. Daw, C.S., Finney, C.E.A.: A review of symbolic analysis of experimental data. Review of Scientific Instruments, 74(2): 915–930, 2003.
12. Fayyad, U., Piatetsky-Shapiro, G., Smyth, P.: From data mining to knowledge discovery. In: Advances in Knowledge Discovery and Data Mining. MIT Press (1996) 1–36
13. Garofalakis, M.N., Rajeev, R., Shim, K.: SPIRIT: Sequential sequential pattern mining with regular expression constraints. In: Proceedings of the 25th International Conference on Very Large Databases. (1999) 223–234
14. J. Godelle and C. Letellier. Symbolic sequence statistical analysis for free liquid jets. Phys. Rev. E 62, Issue 6, pp. 7973–7981, 2000.
15. Gordon, N., A. Marrs and D. Salmond, Sequential Analysis of Nonlinear Dynamic Systems Using Particles and Mixtures, in: Nonlinear and Nonstationary Signal Processing, W. Fitzgerald, R. Smith, A. Walden, and P. Young, ed., Chapter 2, Cambridge University Press, Cambridge, 2001.

16. Hayashi, A., Mizuhara, Y., Suematsu, N.: Embedding time series data for classification. In: Perner, P., Imiya, A. (eds.): Proceedings of the IAPR International Conference on Machine Learning and Data Mining in Pattern Recognition. Leipzig (2005) 356–365
17. Hetland, M.L.: A survey of recent methods for efficient retrieval of similar time sequences. In: Last, M., Kandel, A., Bunke, H. (eds.): Data Mining in Time Series Databases. World Scientific (2004)
18. L.M. Hively, V.A. Protopopescu, and P.C. Gailey. Timely detection of dynamical change in scalp EEG signals. Chaos, Vol. 10, Issue 4, pp. 864–875, 2000.
19. M. Holschneider. Wavelet – An Analysis Tool. Oxford Science publications, 1995.
20. Huang, C.F., Chen, Y.C., Chen, A.P.: An association mining method for time series and its application in the stock prices of TFT-LCD industry. In: Perner, P. (ed.): Proceedings of the 4th Industrial Conference on Data Mining. Leipzig (2004)
21. Huhtala, Y., Kärkkäinen, J., Toivonen, H.: Mining for similarities in aligned time series using wavelets. In: Data Mining and Knowledge Discovery: Theory, Tools, and Technology. SPIE Proceedings Series, Vol. 3695. Orlando, FL (1999) 150–160
22. M.D. Jaere, A. Aamodt, and P. Skalle. Representing temporal knowledge for case-based prediction. In S. Craw and A. Preece, editors, Proceeding of the European Conference on Case-Based Reasoning, 2002, pp. 174–188.
23. J. Jarmulak, S. Craw, and R. Rowe. Genetic algorithms to optimize CBR retrieval. In E. Blanzieri and L. Portinale, editors, Proceedings of the European Conference on Case-Based Reasoning, pages 136–147. Springer, 2000.
24. S. Kadar, J. Wang, and K. Showalter. Noise-supported travelling waves in subexcitable media. Nature 391, pp. 770–772, 1998.
25. Keogh, E., Chakrabarti, K., Pazzani, M., Mehrotra, S.: Locally adaptive dimensionality reduction for indexing large time series databases. In: Proceedings of ACM SIGMOD Conference on Management of Data. Santa Barbara, CA (2001) 151–162
26. Keogh, E., Chakrabarti, K., Pazzani, M., Mehrotra, S.: Dimensionality reduction for fast similarity search in large time series databases. Journal of Knowledge and Information Systems (2001)
27. J.-S. Kim et al. Decreased entropy of symbolic heart rate dynamics during daily activity as a predictor of positive head-up tilt test in patients with alleged neurocardiogenic syncope. Phys. Med. Biol. 45, pp. 3403–3412, 2000.
28. Last, M., Klein, Y., Kandel, A.: Knowledge discovery in time series databases. IEEE Trans. Systems, Man, and Cybernetics — Part B: Cybernetics 31 (2001) 160–169
29. Lee, S.K., White P.R.: The Enhancement of Impulse Noise And Vibration Signals For Fault Detection in Rotating and Reciprocating Machinery, Journal of Sound and Vibration 217 (1998), 485–505.
30. Lin, J.: Feature Extraction of Machine Sound Using Wavelet and Its Application in Fault Diagnosis, NDT&E International 34 (2001), 25–30.
31. S. Lonardi. Global detectors of unusual words: Design, implementation, and applications to pattern discovery in biosequences. Ph.D thesis, Department of Computer Sciences, Purdue University, 2001.
32. D. McSherry. Explaining the Pros andv Cons of conclusions in CBR. In P. Funk and P.A.G. Calero, editors, Proceedings of the European Conference on Case-Based Reasoning, pages 317–330. Springer, 2004.

33. S. Montani, et al. Case-based retrieval to support the treatment of end stage renal failure patients, Artificial Intelligence in Press.
34. S. Montani and L. Portinale. Case based representation and retrieval with time dependent features. In Proceedings of the International Conference on Case-Based Reasoning, pages 353–367, Springer, 2005.
35. Nilsson, M., Funk, P.: A Case-Based Classification of Respiratory Sinus Arrhythmia. In P. Funk and P.A.G. Calero, editors, Proceedings of the European Conference on Case-Based Reasoning, pages 673–685. Springer, 2004.
36. E. Olsson, P. Funk, and N. Xiong. Fault diagnosis in industry using sensor readings and case-based reasoning. Journal of Intelligent & Fuzzy Systems, 15:41–46, 2004.
37. Park, S., Chu, W.W., Yoon, J., Hsu, C.: Efficient search for similar subsequences of different lengths in sequence databases. In: Proceedings of the International Conference on Data Engineering. (2000) 23–32
38. P. Perner. Incremental learning of retrieval knowledge in a case-based reasoning system. In K. D. Ashley and D. G. Bridge, editors, Proceedings of the International Conference on Case-Based Reasoning, pages 422–436. Springer, 2003.
39. Pous, C., Colomer, J., and Melendez, J.: Extending a fault dictionary towards a case based reasoning system for linear electronic analog circuits diagnosis. In: Proceedings of the 7th European Conference on Case-Based Reasoning, Madrid, 2004, pp 748–762.
40. Pray, K.A., Ruiz, C.: Mining expressive temporal associations from complex data. In: Perner, P., Imiya, A. (eds.): Proceedings of the IAPR International Conference on Machine Learning and Data Mining in Pattern Recognition. Leipzig (2005) 384–394
41. Ray, A.: Symbolic dynamic analysis of complex systems for anomaly detection. Signal Processing 84 (2004) 1115–1130
42. R. Schmidt, B. Heindl, B. Pollwein, and L. Gierl. Abstraction of data and time for multiparametric time course prognoses. In: Advances of Case-Based Reasoning, LNAI 1168, Springer-Verlag, Berlin, 1996, pp. 377–391.
43. J. Lin, E. Keogh, S. Lonardi et al. A symbolic representation of time series, with implications for streaming algorithms. In: Proceedings of the 8th ACM SIGMOD Workshop on Research Issues in Data Mining and Knowledge Discovery, pages 2–11, 2003
44. G. Salton. Automatic information organization and retrieval. New York: McGraw-Hill, 1968.
45. Y. Shahar. A framework for knowledge-based temporal abstractions. Artificial Intelligence, 90:79–133, 1997.
46. Srikant, R., Agrawal, R.: Mining sequential patterns: Generalizations and performance improvements. In: Proceedings of the 5th International Conference on Extending Database Technology. (1996) 3–17
47. S.D. Stearns. Digital Signal Processing with Examples in Matlab. CRC Press, Florida, 2003
48. X.Z. Tang, E.R. Tracy, and R. Brown. Symbol statistics and spatio-temporal systems. Physica D, Vol. 102, Issue 3–4, pp. 253–261, 1997.
49. Tung, A.K.H., Lu, H., Han, J., Feng, L.: Breaking the barrier of transactions: Mining inter-transaction association rules. In: Proceedings of ACM Conference on Knowledge Discovery and Data Mining. (1999) 297–301

50. G. Tzanetakis, G. Essl, and P. Cook. Audio Analysis using the Discrete Wavelet Transform. In Proceedings of the WSES International Conference on Acoustics and Music: Theory and Applications (AMTA 2001) Skiathos, Greece, 2001.
51. Wu, Y., Agrawal, D., Abbadi, A. EI: A comparison of DFT and DWT based similarity search in time series databases. In: Proceedings of the 9th ACM CIKM Conference on Information and Knowledge Management. McLean, VA (2000) 488–495
52. Yoon, H., Yang, K., Shahabi, C.: Feature subset selection and feature ranking for multivariate time series. IEEE Trans. Knowledge and Data Engineering 17 (2005) 1186–1198
53. S. Zelikovitz and H. Hirsh. Integrating background knowledge into nearest-neighbor text classification. In S. Craw and A. Preece, editors, Proceedings of the European Conference on Case-Based Reasoning, pages 1–5. Springer, 2002.

10

Prototypes and Case-Based Reasoning for Medical Applications

R. Schmidt, T. Waligora, and O. Vorobieva

Institut für Medizinische Informatik und Biometrie, Universität Rostock

Summary. Already in the early stages of case-based reasoning (CBR) prototypes were considered as an interesting technique to structure the case base and to fill the knowledge gap between single cases and general knowledge. Unfortunately, later on prototypes never became a hot topic within the CBR community. However, for medical applications they have been used rather regularly, because they correspond to the reasoning of doctors in a natural way. In this chapter, we illustrate the role of prototypes by application programs, which cover all typical medical tasks: diagnosis, therapy, and course analysis.

10.1 Introduction

Cases are the most specialised form of knowledge representation. The knowledge of physicians consists of general knowledge they have read in medical books and of their experiences in form of cases they have treated themselves or colleagues have told them about. Not all cases are of the same importance. Some are typical while others are rather exceptional, e.g. a paediatrician does not remember all his patients with measles, but maybe those with serious complications or those where his measles diagnosis was surprisingly wrong. Doctors consider differences between their current patient and typical or known exceptional cases.

We believe that medical case-based reasoning (CBR) systems should take the reasoning of doctors into account [1]. Such systems should not only consist of general medical domain knowledge plus a flat case base, but the case base should be structured by typical case generalisations called prototypes [2].

Though the use of prototypes had been early introduced in the CBR community [3,4], their use is still rather seldom. Later on it fell into oblivion and was brought up again by Bergman in form of generalised cases [5], which are similar but not identical with the idea of prototypes. While generalised cases are general or abstract in contrast to concrete cases, prototypes contain the typical features of a set of cases.

R. Schmidt et al.: *Prototypes and Case-Based Reasoning for Medical Applications*, Studies in Computational Intelligence (SCI) **73**, 285–317 (2008)
www.springerlink.com

However, since doctors reason with typical cases anyway, in medical CBR systems prototypes are a rather common knowledge form, they are used in a variety of applications, e.g. for diabetes [6], for eating disorders [7], and for pulmonology [8]. Prototypical images that can be transformed after certain image processing steps in prototypes are used for the diagnosis of medical images [9].

A prototype is generalised from a set of single cases. The cases in this set are very similar to each other or they belong in some other specific way together and form a sort of class. For example, in a diagnostic system all patients that are diagnosed as measles patient might be grouped together. Usually, prototypes have the same structure as cases but have less and more general features, namely just the typical ones. Sometimes prototypes are defined by medical experts, sometimes they can be found in literature (e.g. the typical symptoms for measles), and sometimes they are computed.

The use of case-oriented generalised knowledge presents the opportunity to structure case bases. Cases can be clustered into groups, prototypical diseases, or schema. Clancey [10] distinguishes between prototypes that represent specific expressions of diseases or therapies and schema that contain essential features of diseases or therapies. As Selz [11] characterises a schema as a description of an entity where at least one part remains vague, the distinction between prototypes and schema seems to be fluid. We only use the term prototype and refer to a hierarchy of prototypes where the most general prototypes that contain the most common features are situated on top and the most specific ones are placed at the bottom.

This notion of prototypes differs from the usual notion of classes and clusters [12] in many ways. Prototypes are not the result of a classification process. Whether a case belongs to a prototype is determined by its features or defined by an expert. There may be a hierarchy of prototypes but there are not relations (similarity, is–a and so on), and the set of cases belonging to a prototype is not represented by its most representative case but by the prototype.

The main purpose of such generalised knowledge is to guide the retrieval and sometimes to decrease the amount of storage by erasing redundant cases. In domains with rather weak domain theories another advantage of case-oriented techniques is their ability to learn from cases. Only gathering new cases may improve the systems ability to find suitable similar cases for current problems, but it does not elicit the intrinsic knowledge of the stored cases. To learn the knowledge contained in cases a generalisation process is necessary. Generally speaking, prototypes fill the knowledge gap between the specificity of single cases and abstract knowledge usually expressed as rules.

In this chapter we present systems we developed during the last 10 years and focus on the role of prototypes within them. We start with a prototype-based system for diagnosis of dysmorphic syndromes. Subsequently we present a system for course analysis and prognosis of the kidney function and finally we present two therapeutic systems, namely one for antibiotic therapy advice and ISOR, a system that deals with therapeutic problems in the endocrine domain.

10.2 Prototype-Based Diagnosis of Dysmorphic Syndromes

In this application, retrieval does not search for former single cases but only for prototypes. Each prototype represents and characterises one specific diagnosis. We assume that this idea is rather typical for diagnostic tasks, because it seems to be reasonable to search for a general description of a disease instead of searching for single patients.

When a child is born with dysmorphic features or with multiple congenital malformations or if mental retardation is observed at a later stage, finding the correct diagnosis is extremely important. Knowledge of the nature and the etiology of the disease enables the pediatrician to predict the patient's future course. So, an initial goal for medical specialists is to diagnose a patient to a recognised syndrome. Genetic counselling and a course of treatments may then be established.

A dysmorphic syndrome describes a morphological disorder and it is characterised by a combination of various symptoms, which form a pattern of morphologic defects. An example is Down syndrome which can be described in terms of characteristic clinical and radiographic manifestations such as mental retardation, sloping forehead, a flat nose, short broad hands, and generally dwarfed physique [13]. The main problems of diagnosing dysmorphic syndromes are as follows [14]:

- More than 200 syndromes are known
- Many cases remain undiagnosed with respect to known syndromes
- Usually many symptoms are used to describe a case (between 40 and 130)
- Every dysmorphic syndrome is characterised by nearly as many symptoms

Furthermore, knowledge about dysmorphic disorders is continuously modified, new cases are observed that cannot be diagnosed (it exists even a journal that only publishes reports of newly observed interesting cases [15]), and sometimes even new syndromes are discovered. Usually, even experts of paediatric genetics only see a small count of dysmorphic syndromes during their lifetime.

So, we have developed a diagnostic system that uses a large case base. Starting point to build the case base was a large case collection of the paediatric genetics of the University of Munich, which consists of nearly 2,000 cases and 229 prototypes. A prototype (prototypical case) represents a dysmorphic syndrome by its typical symptoms. Most of the dysmorphic syndromes are already known and have been defined in literature. And nearly one-third of the prototypes were determined by semiautomatic knowledge acquisition, where an expert selected cases that should belong to same syndrome and subsequently a prototype, characterised by the most frequent symptoms of his cases, was generated. To this database we have added rare dysmorphic syndromes, namely from "clinical dysmorphology" [15] and from the London dysmorphic database [16].

10.2.1 Diagnostic Systems for Dysmorphic Syndromes

Systems to support diagnosis of dysmorphic syndromes have already been developed in the early 1980s. The simple ones perform just information retrieval for rare syndromes, namely the London dysmorphic database [16], where syndromes are described by symptoms, and the Australian POSSUM, where syndromes are visualised [17]. Diagnosis by classification is done in a system developed by Wiener and Anneren [18]. They use more than 200 syndromes as database and apply Bayesian probability to determine the most probable syndromes. Another diagnostic system, which uses data from the London dysmorphic database was developed by Evans [19]. Though he claims to apply CBR, in fact it is again just a classification, this time performed by Tversky's measure of dissimilarity [20]. The most interesting aspect of his approach is the use of weights for the symptoms. That means the symptoms are categorised in three groups – independent of the specific syndromes, instead only according to their intensity of expressing retardation or malformation. However, Evans admits that even features, that are usually unimportant or occur in very many syndromes sometimes play a vital role for discrimination between specific syndromes.

10.2.2 Prototypicality Measures

In CBR usually cases are represented as attribute-value pairs. In medicine, especially in diagnostic applications, this is not always the case, instead often a list of symptoms describes a patient's disease. Sometimes these lists can be very long, and often their lengths are not fixed but vary with the patient. For dysmorphic syndromes usually between 40 and 130 symptoms are used to characterise a patient.

Furthermore, for dysmorphic syndromes it is unreasonable to search for single similar patients (and of course none of the systems mentioned above does so) but for more general prototypes that contain the typical features of a syndrome. To determine the most similar prototype for a given query patient instead of a similarity measure a prototypicality measure is required. One speciality is that for prototypes the list of symptoms is usually much shorter than for single cases.

The result should not be just the one and only most similar prototype, but a list of them – sorted according to their similarity. So, the usual CBR retrieval methods like indexing or nearest neighbour search are inappropriate. Instead, rather old measures for dissimilarities between concepts [20, 21] are applied.

Since our system is still in the evaluation phase, the user has three choices for a prototypicality measure. As humans look upon cases as more typical for a query case as more features they have in common [21], distances between prototypes and cases usually mainly consider the shared features. The first, rather simple measure (10.1) just counts the number of matching symptoms of the query patient (X) and a prototype (Y) and normalises the result by dividing it by the number of symptoms characterising the syndrome. This normalisation is done, because the lengths of the lists of symptoms of the various prototypes vary very much. It is performed by the two other measures too.

The following equations are general (as they were originally proposed) at the point that a general function "f" is used, which usually means a sum that can be weighted. In general these functions "f" can be weighted differently. However, since we do not use any weights at all, in our application "f" means simply a sum.

$$\mathbf{D}(\mathbf{X},\mathbf{Y}) = \frac{\mathbf{f}(\mathbf{X}+\mathbf{Y})}{\mathbf{f}(\mathbf{Y})} \tag{10.1}$$

The second measure (10.2) was developed by Tversky [20]. It is a measure of dissimilarity for concepts. In contrast to the first measure, additionally two numbers are subtracted from the number of matching symptoms. Firstly, the number of symptoms that are observed for the patient but are not used to characterise the prototype (X–Y), and secondly the number of symptoms used for the prototype but are not observed for the patient (X – Y) is subtracted.

$$\mathbf{D}(\mathbf{X},\mathbf{Y}) = \frac{\mathbf{f}(\mathbf{X}+\mathbf{Y}) - \mathbf{f}(\mathbf{X}-\mathbf{Y}) - \mathbf{f}(\mathbf{Y}-\mathbf{X})}{\mathbf{f}(\mathbf{Y})} \tag{10.2}$$

The third prototypicality measure (10.3) was proposed by Rosch and Mervis [21]. It differs from Tversky's measure only in one point: the factor X–Y is not considered:

$$\mathbf{D}(\mathbf{X},\mathbf{Y}) = \frac{\mathbf{f}(\mathbf{X}+\mathbf{Y}) - \mathbf{f}(\mathbf{Y}-\mathbf{X})}{\mathbf{f}(\mathbf{Y})} \tag{10.3}$$

10.2.3 Our System

Our system consists of four steps (Fig. 10.1). At first the user has to select the symptoms that characterise a new patient. This selection is a long and very time consuming process, because we consider more than 800 symptoms. However, diagnosis of dysmorphic syndromes is not a task where the result is very urgent, but it usually requires thorough reasoning and subsequently a long-term therapy has to be started. Secondly, the user can select one of the prototypicality measures explained in Sect. 10.6. In routine use, this step shall be dropped and the measure with best evaluation results shall be used automatically.

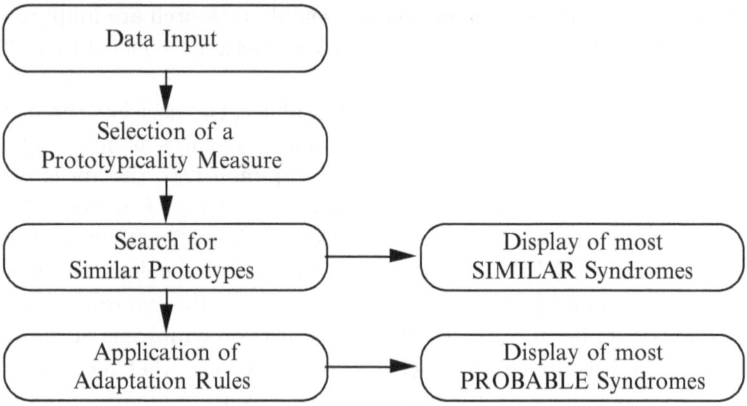

Fig. 10.1. Steps to diagnose dysmorphic syndromes

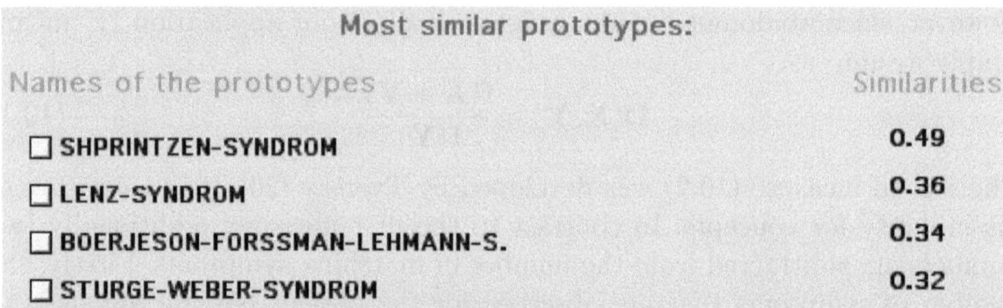

Fig. 10.2. Top part of the listed prototypes after applying a prototypicality measure

In the third step to diagnose dysmorphoic syndromes, the chosen measure is sequentially applied on all prototypes (syndromes). Since the syndrome with maximal similarity is not always the right diagnosis, the 20 syndromes with best similarities are listed in a menu (Fig. 10.2).

In the fourth and final step, the user can optionally choose to apply adaptation rules on the syndromes. These rules state that specific combinations of symptoms favour or disfavour specific dysmorphic syndromes. Unfortunately, the acquisition of these adaptation rules is very difficult, because they cannot be found in textbooks but have to be defined by experts of paediatric genetics. So far, we have got only 18 of them and so far, it is not possible that a syndrome can be favoured by one adaptation rule and disfavoured by another one at the same time. When we, hopefully, acquire more rules such a situation should in principle be possible but would indicate some sort of inconsistency of the rule set.

How shall the adaptation rules alter the results? Our first idea was that the adaptation rules should increase or decrease the similarity scores for favoured and disfavoured syndromes. But the question is how. Of course no medical expert can determine values to manipulate the similarities by adaptation rules and any general value for favoured or disfavoured syndromes would be

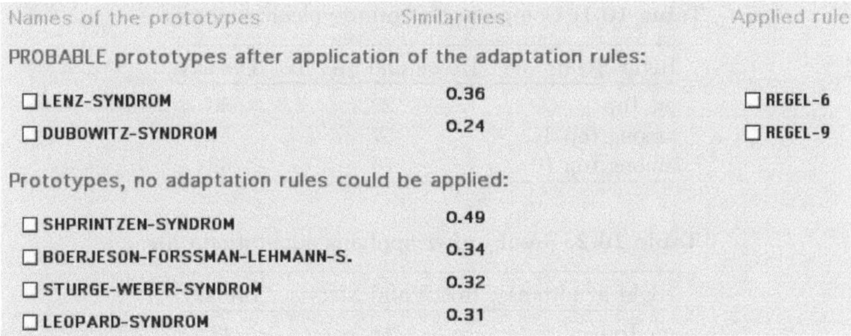

Fig. 10.3. Top part of the listed prototypes after additionally applying adaptation rules

arbitrary. So, instead the result after applying adaptation rules is a menu that contains up to three lists (Fig. 10.3).

On top the favoured syndromes are depicted, then those neither favoured nor disfavoured, and at the bottom the disfavoured ones. Additionally, the user can get information about the specific rules that have been applied on a particular syndrome.

In the example presented in Figs. 10.2 and 10.3, the correct diagnosis is Lenz syndrome. The computation of the prototypicality measure of Rosch and Mervis provided Lenz syndrome as the most similar but one syndrome (here Tversky's measure provides a similar result, only the differences between the similarities are smaller). After application of adaptation rules, the ranking is not obvious. Two syndromes have been favoured, the more similar one is the right one. However, Dubowitz syndrome is favoured too (by a completely different rule), because a specific combination of symptoms makes it probable, while other observed symptoms indicate a rather low similarity.

10.2.4 Results

Cases are difficult to diagnose when patients suffer from a very rare dymorphic syndrome for which neither detailed information can be found in literature nor many cases are stored in our case base. This makes evaluation difficult. If test cases are randomly chosen, frequently observed cases resp. syndromes are frequently selected and the results will probably be fine, because these syndromes are well known. However, the main idea of the system is to support diagnosis of rare syndromes. So, we have chosen our test cases randomly but under the condition that every syndrome can be chosen only once. For 100 cases we have compared the results obtained by both prototypicality measures (Table 10.1).

The results may seem to be rather poor. However, diagnosis of dysmorphic syndromes is very difficult and usually needs further investigation, because often a couple of syndromes are very similar. The first step is to provide the

Table 10.1. Comparison of prototypicality measures

Right Syndrome	Rosch and Mervis	Tversky
on Top	29	40
among top 3	57	57
among top 10	76	69

Table 10.2. Results after applying adaptation rules

Right syndrome	Rosch and Mervis	Tversky
on Top	32	42
among top 3	59	59
among top 10	77	71

Table 10.3. Results after applying some more adaptation rules

Right Syndrome	Rosch and Mervis	Tversky
on Top	36	44
among top 3	65	64
among top 10	77	73

doctor with information about probable syndromes, so that he gets an idea which further investigations are appropriate. That means, the right diagnose among the three most probable syndromes is already a good result.

Obviously, the measure of Tversky provides better results, especially when the right syndrome should be on top of the list of probable syndromes. When it should be only among the first three of this list, both measures provide equal results.

Adaptation Rules

Since the acquisition of adaptation rules is a very difficult and time consuming process, the number of acquired rules is rather limited, namely at first just ten rules. Furthermore, again holds: The better a syndrome is known, the easier adaptation rules can be generated. So, the improvement mainly depends on the question how many syndromes involved by adaptation rules are among the test set. In our experiment this was the case only for five syndromes. Since some had been already diagnosed correctly without adaptation, there was just a small improvement (Table 10.2).

Some More Adaptation Rules

Later on we acquired eight further adaptation rules and repeated the tests with the same test cases. The new adaptation rules again improved the results (Table 10.3). It is obvious that with the number of acquired adaptation rules the quality of the program increases too. Unfortunately, the acquisition of these rules is very difficult and especially for very rare syndromes probably nearly impossible.

10.3 Time Course Prognosis

In this section we present a method for prognosis of temporal courses based on multiparametric numeric values for organ functions.

Since traditional time series techniques [22] work well with known periodicity, but do not fit in domains characterised by possibilities of abrupt changes, much research has been performed in the field of medical temporal course analysis. However, the methods developed so far either require a complete domain theory or well-known standards (e.g. course pattern or periodicity).

An ability of RÉSUMÉ [23] is the abstraction of many parameters into one single parameter and to analyse courses of this abstracted parameter. However, interpretation of these courses requires complete domain knowledge. Haimowitz and Kohane [24] compare many parameters of current courses with well-known standards (trend templates). In VIE-VENT [25] both ideas are combined: Courses of single quantitative measured parameters are abstracted into qualitative course descriptions that are matched with well-known standards.

When we started building a system for course analysis and prediction of the kidney function, we were confronted with a domain where the domain theory is extremely incomplete and no standards were known. So we had to design our own method. For temporal courses, our general idea is to search with CBR retrieval methods [8, 9] for former patients with similar courses and to consider their course continuations as possible prognosis for a query patient.

To make CBR applicable an appropriate case representation has to be found. Usually, a list of attribute-value pairs that contains all case attributes is sufficient. However, for multiparametric time courses the choice of suitable attributes is not obvious. Firstly, not complete courses (they may differ in their length, they may go much further back than it is relevant for the current situation), but only patients' current developments of a certain length should be compared with parts of former patients' courses, which should have about the same length. Secondly, each course consists of a sequence of measured or calculated parameter sets. It cannot be assumed that all parameters are of the same importance, especially the more recent parameter sets are usually more important than older ones.

And even the importance of parameters within the same set may extremely differ. One idea is to look for appropriate weightings for the parameters. However, hundreds of parameters might be involved, much domain knowledge may be required, and weights can be very subjective. Furthermore, it seems to be impossible to visualise a sequence of parameter sets in such a way that a user can rapidly discern the important characteristics.

In the kidney function domain, we chose a different alternative. With the help of medical experts we defined kidney function states based on the most important parameters and subsequently we abstracted each daily parameter set into such a function state. So, courses are represented as a sequence of function states.

10.3.1 Prognostic Model

We propose a prognostic model for multiparametric time courses that combines two abstraction steps with CBR (Fig. 10.4).

The first step is a state abstraction from a set of parameter values to a single function state. Therefore few requirements have to be met. Meaningful states to describe the parameter sets and a hierarchy of these states must exist. Furthermore, knowledge to define the states must be available. These definitions may consist of obligatory or optional conditions on the parameter values. Of course, all obligatory conditions should be met, while for the optional ones some alternatives exist how to determine the appropriate state. One simple idea is to count the met conditions. Additionally, the quality of meeting graduated conditions may be considered (e.g. fuzzy methods may be applied).

The second abstraction means to describe a course of states. An often-realised idea is to use different trend descriptions for different periods of time, e.g. short-term or long-term trend descriptions etc. (e.g. [25]).

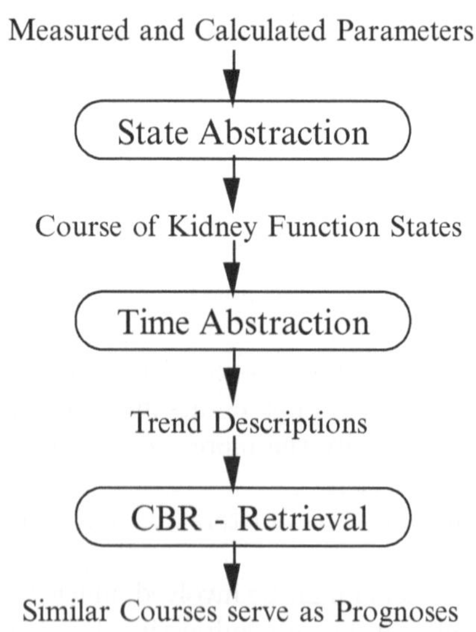

Fig. 10.4. Prognostic model for time course

The lengths of each trend description can be fixed or they may depend on concrete values (e.g. successive equivalent states may be concatenated).

However the trend descriptions may be defined, they can be expressed by four parameters: the length, the first and last state, and an assessment. The lengths and the assessments of the descriptions can vary with domain-dependent demands, while the state definitions and their hierarchy are domain dependent anyway.

The third step means CBR retrieval. Since especially for large case bases a sequential search for similar cases is too time consuming, a few nonsequential retrieval algorithms have been developed in the CBR community. Most of the retrieval algorithms can handle various sorts of attributes, but usually they only work well for those sorts of attributes or problems they have been developed for. So, the choice of the retrieval algorithm should mainly depend on the sort of values of case attributes and sometimes additionally on application characteristics.

The question arises: Of which sort are the four parameters that describe a trend? The states are obviously nominal valued ordered according to their hierarchy. The assessments should have ordered nominal values too, e.g. steady, decreasing etc. Only the lengths should have numeric values. If the time points of the parameter measurements are few integers, they can be treated as ordered nominal values. The proposed retrieval algorithms for ordered nominal valued attributes are CBR-Retrieval-Nets [26], which are based on Spreading Activation [28]. So, if all four parameters have ordered nominal values, the choice of the retrieval algorithm should obviously be CBR-Retrieval-Nets.

However, we made some assumptions that may not necessarily be met in every domain. For example, the lengths may not be transformable into nominal values, the trend assessments may not be just simple nominal values, but more sophisticated descriptions, and there are of course alternatives to describe trends, e.g. even a computed real value might somehow express a trend.

10.3.2 Kidney Function Courses

Up to 60% of the body mass of an adult person consists of water. The electrolytes dissolved in body water are of great importance for an adequate cell function. The human body tends to balance the fluid and electrolyte situation. But intensive care patients are often no longer able to maintain adequate fluid and electrolyte balances themselves due to impaired organ functions, e.g. renal failure, or medical treatment, e.g. parenteral nutrition of mechanically ventilated patients. Therefore physicians need objective criteria for the monitoring of fluid and electrolyte balances and for choosing therapeutic interventions as necessary.

At our ICU, physicians daily get a printed renal report from the monitoring system NIMON [29] which consists of 13 measured and 33 calculated parameters of those patients where renal function monitoring is applied. For example, the urine osmolality and the plasma osmolality are measured parameters that are used to calculate the osmolar clearance and the osmolar excretion. The interpretation of all reported parameters is quite complex and needs special knowledge of the renal physiology.

The aim of our knowledge-based system ICONS is to give an automatic interpretation of the renal state to elicit impairments of the kidney function on time and to give early warnings against forthcoming kidney failures. That means, we need a time course analysis of many parameters without any well-defined standards.

However, in the domain of fluid and electrolyte balance, neither a prototypical approach in ICU settings is known nor exists complete knowledge about the kidney function. Especially, knowledge about the behaviour of the various parameters on time is yet incomplete. So, we combined the idea of RÉSUMÉ [23] to abstract many parameters into one single parameter with the idea of Haimowitz and Kohane [24] to compare many parameters of current courses with well-known standards. Since well-known standards were not available, we used former similar cases instead.

The method in ICONS corresponds to the general method proposed above and shown in Fig. 10.4.

State Abstraction

For the data abstraction we use states of the renal function, which determine states of increasing severity beginning with a normal renal function and ending with a renal failure. Based on the kidney function states (e.g. in Fig. 10.5 a reduced kidney function), characterised by obligatory and optional conditions for selected renal parameters, we first check the obligatory conditions. For each state that satisfies the obligatory conditions we calculate a similarity

Reduced Kidney Function

Obligatory Condtion:	c_kreat40 - 80	
Optional Conditions:		
Retention Rates:	p_kreat_se	<2
	p_urea_se	< 150
Tubular Function:	u_osmol320 – 600	
	u_p_osmol	1.1–1.8
	u_kreat 10 – 40	
	u_p_kreat	20–50
Urine Volume:	urine volume	0.7–3.0
	osmol_ex	800–3000

Fig. 10.5. Definition of the reduced kidney function state. Abbreviations: c, clearance; p, plasma; u, urine; kreat, kreatinin; osmol, osmolality; se, serum; ex, excretion

value concerning the optional conditions. We use a variation of Tversky's [20] measure of dissimilarity between concepts. Only if two or more states are under consideration, ICONS presents them to the user sorted according to their similarity values together with information about the satisfied and not satisfied optional conditions.

The user can accept or reject a presented state. When a suggested state has been rejected, ICONS selects another one. Finally, we determine the central state of occasionally more than one states the user has accepted. This central state is the closest one towards a kidney failure. Our intention is to find the state indicating the most profound impairment of the kidney function.

Temporal Abstraction

First, we have fixed five assessment definitions for the transition of the kidney function state of one day to the state of the, respectively, next day. These assessment definitions are related to the grade of renal impairment:

steady. both states have the same severity value.
increasing. exactly one severity step in the direction towards a normal function.
sharply increasing. at least two severity steps in the direction towards a normal function.
decreasing. exactly one severity step in the direction towards a kidney failure.
sharply decreasing. at least two severity steps in the direction towards a kidney failure.

These assessment definitions are used to determine the state transitions from one qualitative value to another. Based on these state transitions, we generate three trend descriptions. Two trend descriptions especially consider the current state transitions.

short-term trend:=	current state transition; Abbreviation: T1
medium-term trend:=	looks recursively back from the current state transition to the one before and unites them if they are both of the same direction or one of them has a "steady" assessment; Abbreviation: T2
long-term trend:=	characterises the considered course of at most seven days; Abbreviation: T3

For the long-term trend description we additionally introduced four new assessment definitions. If none of the five former assessments fits the complete considered course, we attempt to fit one of these four definitions in the following order:

alternating. at least two up and two down transitions and all local minima are equal.
oscillating. at least two up and two down transitions.

fluctuating. the distance of the highest to the lowest severity state value is greater than 1.

nearly steady. the distance of the highest to the lowest severity state value equals one.

Only if there are several courses with the same trend descriptions, we use a minor fourth trend description T4 to find the most similar among them. We assess the considered course by adding up the state transition values inversely weighted by the distances to the current day. Together with the current kidney function state these four trend descriptions form a course depiction, that abstracts the sequence of the kidney function states.

Looking back from a time point t, these four trend descriptions form a pattern of the immediate course history of the kidney function considering qualitative and quantitative assessments.

Why These Four Trend Descriptions?

There are domain specific reasons for defining the short-, medium-, and long-term trend descriptions T1, T2, and T3. If physicians evaluate courses of the kidney function, they consider at most one week prior to the current date. Earlier renal function states are irrelevant for the current situation of a patient. Most relevant information is derived from the current function state, the current development and sometimes a current development within a slightly longer time period. That means, very long-term trends are of no interest in this domain. In fact, very often only the current state transition or short continuous developments are crucial.

The short-term trend description T1 expresses the current development. For longer time periods, we have defined the medium- and long-term trend descriptions T2 and T3, because there are two different phenomena to discover and for each, a special technique is needed. T2 can be used for detecting a continuous trend independent of its length, because equal or steady state transitions are united recursively beginning with the current one. As the long-term trend description T3 describes a well-defined time period, it is especially useful for detecting fluctuating trends.

Since every abstraction loses some specific information, information about the daily kidney function states is lost in the second abstraction step. The course description contains only information about the current and the start states of the three trend descriptions. The intermediate states are abstracted into trend description assessments.

Example. The following kidney function states may be observed in this temporal sequence (Fig. 10.6):

selective tubular damage, reduced kidney function, reduced kidney function, selective tubular damage, reduced kidney function, reduced kidney function, sharply reduced kidney function

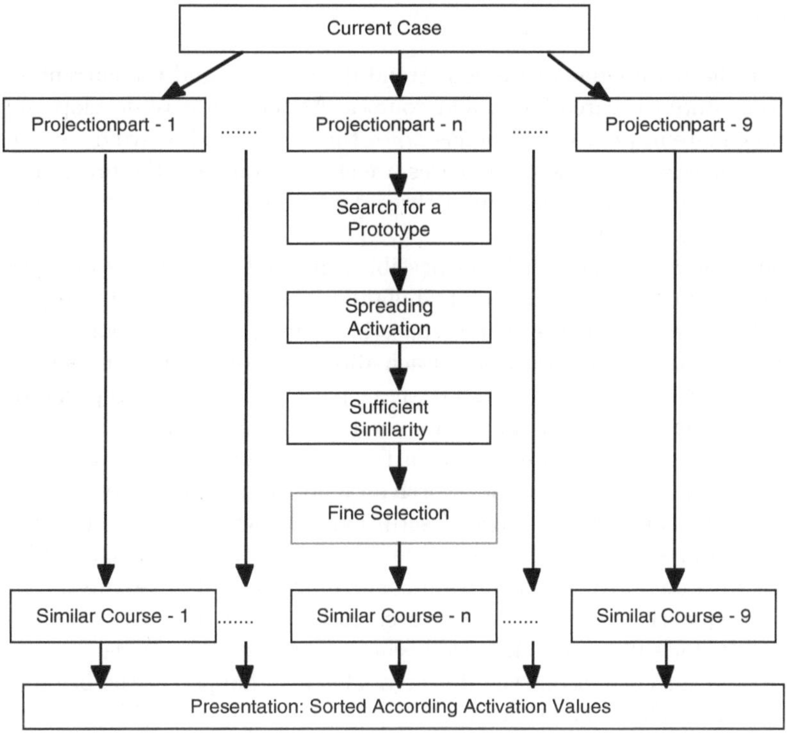

Fig. 10.6. Comparative presentation of a current and a similar course

So we get these six state transitions:
decreasing, steady, increasing, decreasing, steady, decreasing
with these trend descriptions:

current state: sharply reduced kidney function
T1: decreasing, reduced kidney function, one transition
T2: decreasing, selective tubular damage, three transitions
T3: fluctuating, selective tubular damage, six transitions
T4: 1.23

In this example, the short-term trend description T1 assesses the current state transition as "decreasing" from a "reduced kidney function" to a "sharply reduced kidney function." Since the medium-term trend description T2 accumulates steady state transitions, T2 determines a "decrease" in the last four days from a "selective tubular damage" to a "sharply reduced kidney function." The long-term trend description T3 assesses the entire course of seven days as "fluctuating," because there is only one increasing state transition and the difference between the severity values of a "selective tubular damage" and a "sharply reduced kidney function" equals two.

Retrieval

We use the parameters of the four trend descriptions and the current kidney function state to search for similar courses. As the aim is to develop an early warning system, we need a prognosis. For this reason and to avoid a sequential runtime search along the entire cases, we store a course of the previous seven days and a maximal projection of three days for each day a patient spent on the intensive care unit.

Since there are many different possible continuations for the same previous course, it is necessary to search for similar courses and for different projections. Therefore, we divided the search space into nine parts corresponding to the possible continuation directions. Each direction forms an own part of the search space. During the retrieval these parts are searched separately and each part may provide at most one similar case. The similar cases of these parts together are presented in the order of their computed similarity values.

Before the main retrieval, we search for a prototypical case (see Sect. 10.3.3) that matches most of the trend descriptions. Below this prototype the main retrieval starts (Fig. 10.7). It consists of two steps for each part. First we search with an activation algorithm concerning qualitative features.

Subsequently, we check the retrieved cases with a similarity criterion [27] that looks for sufficient similarity, because even the most similar course may differ from the current one significantly. This may happen at the beginning of the use of ICONS, when there are only a few cases known to ICONS, or when the query course is rather exceptional.

If two or more courses are selected in the same projection part, we use the sequential similarity measure of TSCALE [30], which goes back to Tversky [20], concerning the quantitative features in a second step.

Continuation of the example. For the example above, the following similar course (Fig. 10.6) with these transitions is retrieved:

decreasing, increasing, decreasing, steady, steady, decreasing
with these trend descriptions:

current state: sharply reduced kidney function
T1: decreasing, reduced kidney function, one transition
T2: decreasing, selective tubular damage, four transitions
T3: fluctuating, selective tubular damage, six transitions
T4: 1.17

In the lower part of each course the (abbreviated) kidney function states are depicted. The upper part of each course shows the deduced trend descriptions.

T1 describes a "decrease" from a "reduced kidney function" and T2 describes a "decrease" from a "selective tubular damage" to a "sharply reduced kidney function" in the last five days. T3 assesses the considered course as "fluctuating." For T4, a slightly lower value in comparison to the current

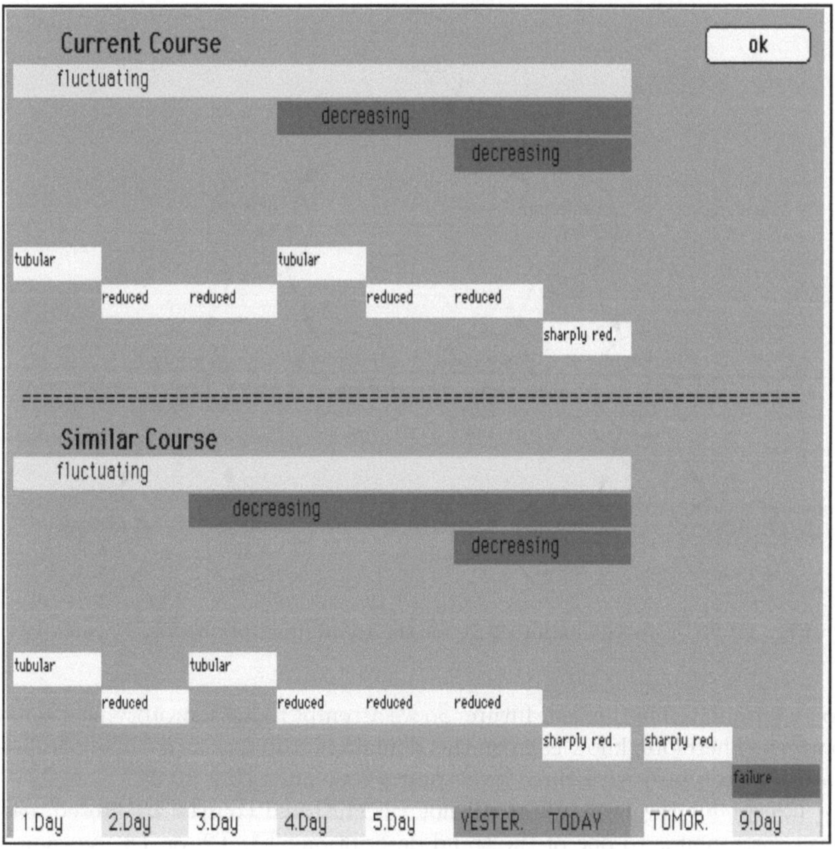

Fig. 10.7. The retrieval procedure

course has been calculated, because the change from a "selective tubular damage" to a "reduced kidney function" state occurs earlier.

After another day with a "sharply reduced kidney function" the patient belonging to the similar course had a kidney failure. The physician may notice this as a warning and it is up to him to interpret it.

This former course was retrieved, because especially the features with the highest weights (the current state and all assessments) equal the features of the query course. As there is no significant difference between both courses, there is no reason for the sufficient similarity criterion to reject this similar course.

10.3.3 Learning a Tree of Prototypes

Prognosis of multiparametric courses of the kidney function for ICU patients is a domain without a medical theory. Moreover, we cannot expect such a theory

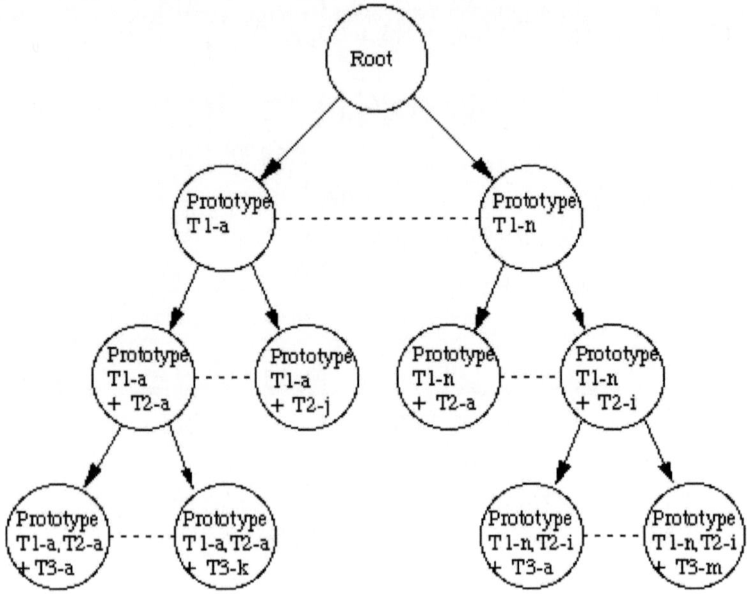

Fig. 10.8. Prototype architecture for the trend descriptions T1, T2, and T3

to be formulated in the near future. So we attempt to learn prototypical course pattern. Therefore, knowledge on this domain is stored as a tree of generalised cases (prototypes) with three levels and a root node (Fig. 10.8).

Except for the root, where all not yet clustered courses are stored, each level corresponds to one of the trend descriptions T1, T2, or T3. As soon as enough courses that share another trend description are stored at a prototype, a new prototype with this trend is created. At a prototype at level 1, we cluster courses that share T1, at level 2, courses that share T1 and T2 and at level 3, courses that share all three trend descriptions T3.

We can do this, because regarding their importance, the short-, medium-, and long-term trend descriptions T1, T2, and T3 refer to hierarchically related time periods. T1 is more important than T2 and T3, and so forth.

We start the retrieval with a search for a prototype that has most of the trend descriptions in common with the query course. The search begins at the root node with a check for a prototype with the same short-term trend description T1. If such a prototype can be found, the search goes on below this prototype for a prototype that has the same trend descriptions T1 and T2, and so forth. If no prototype with a further trend in common can be found, we search for a course at the last accepted prototype.

If no prototype exists that has the same T1 as the query course, we search at the root node, where links to all courses in the case base exist.

Continuation of the example. In the example above, we can create just one prototype at level 1, because at the second level the query course and the similar one, called "similar-1" differ in their length. Although the long-term trend description T3 is equal for both courses, we cannot create a prototype at level 3 because of the strictly hierarchical organisation of the prototype tree. However, learning a prototypical description "fluctuating in seven days from a selective tubular damage to sharply reduced kidney function" which does not consider any more similarities or deviations within this time period would be too general and therefore too impracticable.

Assuming we find another similar course, called "similar-2", for the current case of the example above with the following kidney function states:

reduced kidney function, reduced kidney function, selective tubular damage, selective tubular damage, reduced kidney function, reduced kidney function, sharply reduced kidney function with these trend descriptions:

current state: sharply reduced kidney function
T1: decreasing, reduced kidney function, one transition
T2: decreasing, selective tubular damage, four transitions
T3: oscillating, reduced kidney function, six transitions
T4: 1.33

The current query course, "similar-1", and "similar-2" will be clustered at level 1 to prototype T1-a, defined by T1 as "decreasing, reduced kidney function, one transition". Afterwards at level 2 the current course and "similar-2" will be clustered to a prototype T1-a + T2-a, defined by T1 as "decreasing, reduced kidney function, one transition" plus by T2 as "decreasing, selective tubular damage, four transitions." The attempt to create another prototype at level 3 fails, because the trend descriptions T3 have different assessments and different start states. The result, a tree of prototypes learned from the three courses is shown in Fig. 10.9.

10.3.4 Retrieval Experiments

Since we wished to be convinced that CBR-Retrieval-Nets really are the appropriate retrieval algorithm for our prognostic model, we compared them with an indexing algorithm, which had been developed for nonordered nominal values [31]. The results of this comparison are as follows: The indexing algorithm works faster (Table 10.4), but provides worse results, because stored cases get only activation values for attribute values that exactly match the query case values. The CBR-Retrieval-Nets additionally send smaller activation values to cases with attribute values similar to query case values. Hence, courses can be determined to be similar which have attribute values that slightly deviate from the query case values.

Since one idea of using prototypes is to speed up the retrieval by structuring the case base, we additionally compared both algorithms with and without using prototypes. To decide when a prototype should be generated, a threshold parameter is required. We set this parameter to the value of 2, which

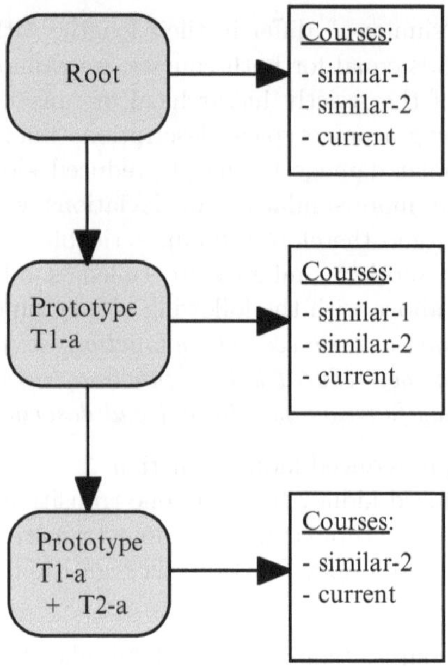

Fig. 10.9. Generated prototype tree from three example courses

Table 10.4. Retrieval times (in seconds) for the retrieval algorithms "CBR-Retrieval-Nets" and "indexing" with and without using prototypes

Courses	Retrieval nets	Retrieval nets, Use of prototypes	Indexing	Indexing, use of prototypes
No. 1	0.163	0.163	0.155	0.155
No. 2	0.284	0.281	0.214	0.218
No. 3	0.316	0.366	0.165	0.213
No. 4	0.455	0.513	0.404	0.452
No. 5	0.514	0.544	0.428	0.506
No. 6	1.328	0.759	0.600	0.717
No. 7	0.649	0.401	0.246	0.347
No. 8	0.685	0.642	0.376	0.469
No. 9	0.550	0.617	0.444	0.551
No. 10	0.386	0.476	0.257	0.394
No. 11	0.537	0.553	0.234	0.367
No. 12	1.396	0.870	0.743	0.890
No. 13	0.577	0.607	0.244	0.332
No. 14	0.518	0.425	0.340	0.494

means, that already two cases with the same trend description are sufficient to generate a prototype. Hence many prototypes were generated.

At first glance the results (Table 10.4) are not very encouraging for using prototypes. However, for the CBR-Retrieval-Nets the time differences between with and without prototypes are very small except for those two courses where the retrieval worked noticeably slower (No.6 and No.12): Here, using prototypes reduces the retrieval by at least a third.

However, so far the determination of the appropriate prototype occurred by sequentially matching the trend description parameters. So, most of the time gained by reducing the number of cases worth to consider is used up to determine the appropriate prototype. This indicates that not only the retrieval algorithm for cases, but also the determination of appropriate prototypes should be organised in a nonsequential way.

10.4 Antibiotics Therapy Advice

We developed an antibiotics therapy advice system called ICONS for patients in an intensive care unit who have caught an infection as additional complication. To speed up the process of finding suitable therapy recommendations, we applied CBR techniques. As information about antibiotics therapy changes in time, new cases are incrementally incorporated into the case base and outdated ones are updated or erased.

10.4.1 Antibiotics Therapy

Severe bacterial infections are still a life-threatening complication in intensive care medicine, correlating with a high mortality [32]. Identification of bacterial pathogens is often difficult. It usually requires at least 24 hours to identify the pathogen that is responsible for an infection and at least another 24 hours to find out which antibiotics have therapeutic effects against the identified pathogen. In order not to endanger the patient, physicians sometimes have to start an antimicrobial therapy before the responsible pathogen and its sensitivities are determined. This sort of antibiotic therapy is called "calculated," in contrast to a "selective" therapy, which is used when microbiological results are already available. For an adequate calculated antibiotic therapy, it is essential to access information about the expected pathogen spectrum and its expected susceptibility, existing contraindications, and the side effects of antibiotics.

The main task of our adviser is to present suitable calculated antibiotics therapy advice for intensive care patients who have caught a bacterial infection as an additional complication. Since, for such critical patients, physicians cannot wait for the laboratory results, we use an expected pathogen spectrum based on medical background knowledge. Each recommended antibiotics therapy should completely cover this spectrum. Furthermore, as advice is needed very quickly we speed up the process of computing recommended antibiotic therapies by using CBR methods (the right path in Fig. 10.10). This means that we search for a previous similar patient and transfer the therapies

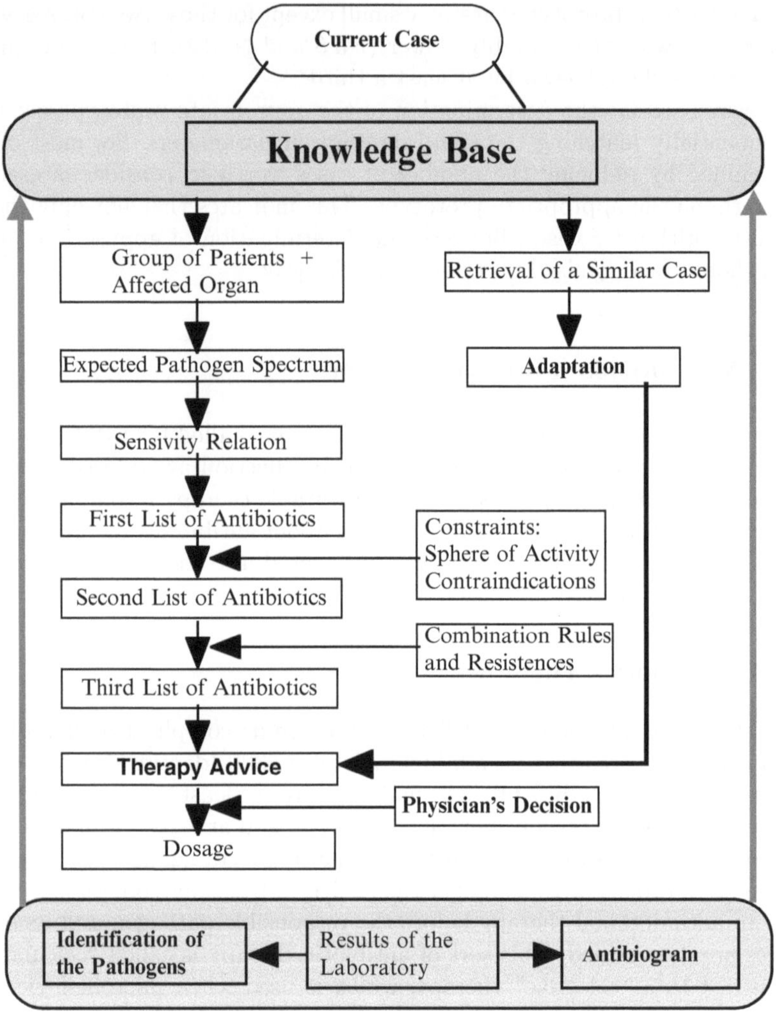

Fig. 10.10. Program flow in ICONS

suggested for his situation to the current patient. These previous therapies are then adapted to take account of any differences between the situations of the previous and current patients.

10.4.2 Strategy for Selecting Recommended Antibiotic Therapies

As ICONS is not a diagnostic system, we do not attempt to deduce evidence for diagnoses based on symptoms, frequencies, or probabilities, but instead pursue a strategy that can be characterised as follows: find all possible solutions, and subsequently reduce them using the patient's contraindications and the requirement to completely cover the calculated pathogen spectrum (establish-refine strategy).

Firstly, we distinguish between different groups of patients (infection acquired in or outside the ward, respectively, the hospital; immunocompromised patients). An initial list of antibiotics is generated by a susceptibility relation, which for each group of pathogens provides all antibiotics that usually have therapeutic effects. This list contains all antibiotics that cover at least a part of the potential pathogen spectrum. We obtain a second list of antibiotics by reducing the first one through applying two constraints: the patient's contraindications and the desired sphere of activity. Using the antibiotics on this second list, we try to find antibiotics that cover the whole pathogen spectrum individually.

Except for some community-acquired infections, monotherapies have to be combined with antibiotics that have synergistic or additive effects. If no adequate single therapy can be found, we use combination rules to generate combinations of antibiotics. Each possible combination must be tested for the ability to cover the expected spectrum completely.

10.4.3 Case-Based Reasoning

In this application, the main argument for using CBR methods is to speed up the process of finding adequate therapies. We shorten the strategy described above for selecting recommended antibiotic therapies by searching for a similar case, retrieving its suggested therapies, and by adapting them according to the contraindications of the current patient.

The retrieval consists of three steps. Firstly we select the part of the case base in which all cases share the following two attributes with the current patient: the group of patients, and the infected organ system. This means a selection of the appropriate prototype tree. Subsequently, we apply the tree-hash retrieval algorithm of Stottler, Henke, and King [33] for nominal valued contraindications and the similarity measure of Tversky [20] for the few integer valued contraindications. Furthermore, we use an adaptability criterion, because not every case is adaptable [34]. The attributes used for the retrieval are the contraindications, which work as constraints on the set of possible antibiotics. It is therefore obvious that we should use only former cases whose contraindications are shared by the current patient. To guarantee this condition the adaptability criterion has to be checked during retrieval.

In ICONS three different sorts of adaptations occur: A CBR adaptation to obtain sets of calculated advisable therapies for current patients, an adaptation of chosen therapies according to laboratory findings, and a periodical update of laboratory information (resistance situation, frequently observed pathogens).

Each contraindication restricts the set of advisable therapies. During the retrieval we require that the retrieved case does not have any additional contraindications besides those of the current case. Otherwise the solution set for the current case would be inadmissibly reduced by the additional contraindications of a previous case.

Because of this criterion, the adaptation of a previous similar case is rather simple. It is simply a matter of transferring the set of advisable therapies and if necessary of reducing this set according to the additional contraindications of the current case.

10.4.4 Prototypes

Since in an incrementally working system the number of cases increases continuously, storing each case would slow down the retrieval time and exceed any space limitations. We therefore decided to structure the case base by prototypes and to store only those cases that differ significantly from their prototype. Like for diagnosis (see Sect. 10.2), we create prototypes that include the most frequent features of the corresponding cases. In diagnostic applications, prototypes correspond to typical diseases or diagnoses. So, for antibiotic therapies, prototypes are expected to correspond to typical antibiotic treatments associated with typical clinical features of patients. However, as the attributes are contraindications that are responsible not for the generation, but for the restriction of the solution set, this is only partly true.

We investigated the growth of a hierarchical prototype structure built up from a randomly ordered stream of cases. The results are presented and discussed in Sect. 10.4.5.

Selection of a Prototype Tree

In ICONS there is not just one prototype tree, but a forest of trees, which are all independent from each other. A specific tree can be generated for each affected organ system combined with each group of patients. So, for nearly 20 organ systems and five patient groups there are nearly 100 possible prototype trees. We generate them dynamically only when required. For example a tree for "community-acquired kidney infections" will be generated as soon as the first data input occurs from a patient who has a kidney infection which he has acquired outside the hospital.

Since all cases within the same prototype tree belong to the same group of patients, and the same organ system is affected, it follows that the same expected pathogen spectrum deduced from background knowledge has to be covered. Cases within the same prototype tree are only discriminated from each other by their contraindications. These are allergies against specific antibiotics, reduced organ functions (kidney and liver), specific diagnoses (e.g. CNS disease), special blood diseases, pregnancy, and the patient's age.

Generating Prototypes

The aim of our concept of prototypical cases is to structure the case base, to keep the prototypes always up to date, and to erase redundant cases. As the prototypes are generated incrementally and as they should always contain the typical features of their cases, we use two threshold parameters:

(1) The parameter "minimum frequency" determines how (relatively) often a contraindication has to occur in the set of cases to be incorporated into the prototype.
(2) The parameter "number of cases" determines the required number of cases that are necessary to fill a prototype or to create an alternative prototype. The lower this threshold the more prototypes are created and the fewer cases are stored.

First, all cases are stored below the prototype they belong to. If the threshold "number of cases" is reached after storing a new case below a prototype, the prototype will be "filled". At this point, every contraindication that occurs in the prototype's cases at least as often as the "minimum frequency" threshold will be included into the prototype. Subsequently, the "filled" prototype can be treated like a case. The same holds for prototypes as for cases: Each contraindication restricts the set of advisable therapies. The contraindications of a prototype are those that occur most often within its cases. So from the viewpoint of frequency they are the typical ones. Those cases that have no additional contraindications in comparison with their prototypes are erased.

When new cases are added later on to an already filled prototype, the observed frequencies may change and consequently the contraindications of the prototype may have to be recomputed. If the contraindications of a prototype change, the suggested antibiotic therapies have to be recomputed, too. In addition, all cases must be inspected again to determine whether they need to be stored.

We create an "alternative" prototype below an already existing prototype if for the latter enough cases exist (which means the threshold "number of cases" is reached) that have at least one contraindication in common, which the already existing prototype does not include. We generate the alternative prototype using those cases that share at least one contraindication not included in the existing prototype. We place this new prototype in the hierarchy directly below the already existing prototype. Alternative prototypes differ from their superior prototypes by their contraindications and therefore also by their sets of advisable antibiotic therapies.

10.4.5 Experimental Results with Prototype Generation Strategies

The general idea of our concept is to keep the prototypes always up to date. They should contain the typical features of their cases. We have tested two contrasting policies for deleting redundant cases and a strategy of keeping all cases. Our evaluation had two aims. First, we wished to find a strategy that best fits the two contrasting aims of finding many adaptable cases or prototypes and requiring little memory. Secondly, we wished to find good settings for the threshold parameters.

Normally, cases without additional features in comparison to their prototype are redundant, because they do not contain any additional information [31]. However, in our application the attributes are contraindications, which are not used to generate a solution, but to restrict a solution set. This means they are applied as constraints. A case with fewer contraindications than its prototype has a greater chance of being adaptable to a query case, because only a case without additional contraindications in comparison to the query case is adaptable.

We have therefore tested two opposing strategies: firstly, deleting cases without additional attributes, and secondly, deleting only cases with additional attributes. Additionally, we have tested a strategy without deleting any cases at all.

The memory size without any stored cases is about 2.248 MB for all three strategies. The argument about the memory might seem to be unreasonable, because the differences between the strategies are only about 40 KB for 75 test cases. However, we performed our tests in just one of about 100 possible parallel sets. 75 cases in each set might lead to differences of up to 4 MB. This leads to the question of whether our system should require about 12 or about 16 MB memory. Certainly, problems should not occur until the number of cases per set exceeds 75.

Without generating any prototypes at all, for 51 of the 75 test cases a similar adaptable case can be found. As prototypes are treated like cases, this number can be exceeded.

Strategy A: Deleting Cases Without Additional Attributes

We have tested the strategies with 75 cases, which were incrementally incorporated into the system. For strategy A, we varied the threshold parameter "number of cases," which indicates how many cases are necessary to generate a prototype. The second threshold parameter "relative frequency" was set to 33%, which means that a contraindication is incorporated into a prototype if at least a third of its cases have this contraindication.

The results (Table 10.5) can be summarised as follows: The more cases necessary to generate a prototype (this is achieved by increasing the value "number of cases") the higher the number of stored cases and the higher the number of retrieved adaptable cases. After a while there is only little to be gained by increasing this threshold parameter any further (fourth setting). A surprise is the big increase in the number of retrieved adaptable cases in the second setting compared with the first one. This cannot be simply explained by the four additionally stored cases, but by the following two phenomena. Firstly, those cases that have no additional information (contraindications) in comparison to their prototype are deleted. This means that the deleted cases would be more likely to be adaptable to future queries.

Secondly, under the second setting the prototypes are generated later and consequently cases are deleted later as well.

Table 10.5. Test results for strategy A

	1. Setting number of cases = 2	2. Setting number of cases = 3	3. Setting number of cases = 4	4. Setting number of cases = 5
memory size	in Mbytes	in Mbytes	in Mbytes	in Mbytes
after 75 cases	2.392	2.390	2.401	2.402
number of prototypes	9	7	8	8
number of stored cases	53	57	62	63
number of deleted cases	22	18	13	12
number of adaptations	12	26	31	31

Table 10.6. Test results for strategy B

	1. Parameter setting: relative frequency = 33 %	2. Parameter setting relative frequency = 25 %
memory size after 75 cases	2.373 MB	2.368 MB
number of generated prototypes	9	6
number of stored cases	20	25
number of deleted cases	55	50
number of adaptations	14	14

One aim of using prototypes is the hope of reducing the memory size. For strategy A this benefit does not occur, because the storage requirement for prototypes is bigger than for cases. This is because prototypes contain some additional information: The intersection of advisable therapies for their cases (cases only contain additional specific therapy suggestions), observed frequencies of contraindications of their cases, etc.

Strategy B: Deleting Cases with Additional Attributes

Our aim with strategy B was to keep in the case base those cases that have a higher chance to be adaptable. These are cases with few contraindications. We therefore adapted a strategy opposite to strategy A, namely deleting cases with additional information (contraindications) to their prototype. As many cases are deleted, we set the threshold parameter "number of cases" to the value two. Here, we varied the second parameter "relative frequency," which determines the frequency with which contraindications have to be observed among the cases to be incorporated into a prototype.

The difference between the results for the two settings for strategy B is rather small (Table 10.6). With a smaller relative frequency (second setting) more contraindications are incorporated into the prototypes. So, fewer stored cases have additional contraindications in comparison to their prototypes and consequently fewer cases are deleted. Furthermore, fewer prototypes are generated, because the prototypes cover more cases. The memory size is nearly the same and the number of retrieved adaptable cases is exactly the same for both settings.

In comparison to strategy A, it is noticeable that about the same number of prototypes have been generated, but much more cases have been deleted. Though those cases which have a bigger chance to be adaptable remain in the case base, the number of retrieved adaptable cases slightly increases in comparison to the first setting of strategy A, but the number is not as high as in the other settings of strategy A.

So, the strategy of keeping those cases that are easily adaptable results in such a small case base that only few adaptable cases can be retrieved.

Strategy C: All Cases Remain in the Case Base

For strategy C no cases are deleted at all. We have evaluated the same threshold parameter settings as for strategy A. It can be seen that many more adaptable cases can be retrieved in comparison to the corresponding settings of strategy C, while the memory requirement increases only slightly (Table 10.7).

Since two cases are sufficient to generate a prototype in the first setting, many prototypes are created and the memory requirement increases correspondingly. It is a little surprising that fewer adaptable cases are retrieved, but this is because a hierarchy with three levels of prototypes has been generated, and since the prototypes are treated as cases, the right prototype on each level has to be determined to be the most similar case.

Really surprising is the big increase of retrieved adaptable cases in the third setting. There are two possible explanations. Firstly, as the number of generated prototypes decreases, the prototype hierarchy is simpler and it is easier to find the appropriate case. Secondly, and probably the main reason, the number of cases which are necessary to generate a prototype is higher (= 4), so that more cases are considered when a generated prototype is filled, and consequently fewer contraindications are incorporated into the prototype. This means the prototypes themselves become more adaptable.

Table 10.7. Test results for strategy C

	1. Setting number of cases = 2	2. Setting number of cases = 3	3. Setting number of cases = 4	4. Setting number of cases = 4
memory size	in Mbytes	in Mbytes	in Mbytes	in Mbytes
after 75 cases	2.439	2.426	2.421	2.419
number of prototypes	19	10	8	7
number of stored cases	75	75	75	75
number of deleted cases	0	0	0	0
number of adaptations	29	32	52	51

However, when the number of generated prototypes decreases, there are fewer cases available to be used for adaptation (fourth setting).

Summary of the Evaluation Results for the Prototype Strategies

Keeping all cases in the case base increases the memory requirement, but increases the number of retrieved adaptable cases dramatically. Considering the number of retrieved adaptable cases, strategy A provides results that are nearly as good as for strategy C, but the achieved reduction is rather small. Keeping more adaptable cases (strategy B) results in a small case base, but only few adaptable cases can be found.

Too many prototypes should be avoided, because a complex hierarchy results in difficulties in finding the desired case. This means the threshold parameter "number of cases" should not be set too low.

The most preferable setting is the third one of strategy C. If the memory limitations become a real problem, strategies that delete redundant cases should be considered. Every stored case increases the memory requirement of our system by approximately 1.7 KB. This might lead to performance problems for much bigger case bases, keeping in mind that our test set of 75 cases covers just one out of a set of more than 80 medical areas.

The best settings, whether all cases are retained (strategy C) or cases without additional information (strategy A) are deleted, are those where the threshold parameter "number of cases" is set sufficiently high. It leads to more retrieved adaptable cases.

10.5 ISOR

ISOR is a CBR system for long-term therapy support in the endocrine domain [35]. It performs typical therapeutic tasks, such as computing initial therapies, initial dose recommendations, and dose updates. Apart from these tasks ISOR deals especially with situations where therapies become ineffective. Causes for inefficacy have to be found and better therapy recommendations should be computed. In addition to the typical CBR knowledge, namely former already solved cases, ISOR uses further knowledge forms, especially medical histories of query patients themselves and prototypical cases (prototypes).

ISOR uses prototypes in two ways, namely in form of guidelines for dose calculations and as generalised solutions for therapy inefficacy.

10.5.1 Computing Initial Doses: Guidelines as Prototypes

For hypothyroidism only one drug exists, namely Levothyroxine. The problem is to calculate effective initial doses (Fig. 10.11). Firstly, a couple of prototypes exist. These are recommendations that have been defined by expert commissions [36]. Though we are not sure whether they are officially accepted, we call

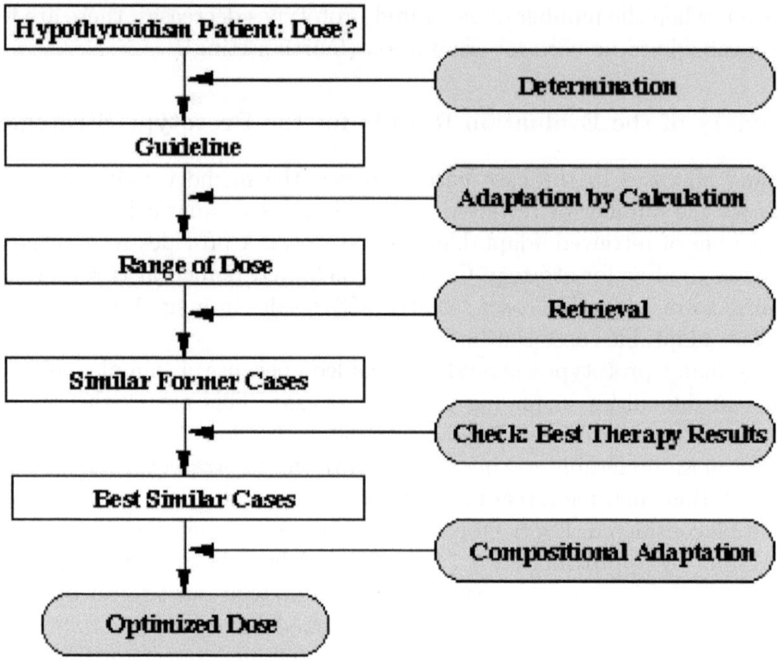

Fig. 10.11. Determination of an initial Levothyroxine dose

them guidelines. The assignment of a patient to a fitting guideline is obvious because of the way the guidelines have been defined. With the help of these guidelines a range for good doses can be calculated.

To compute a dose with best expected impact, we retrieve similar cases whose initial doses are within the calculated ranges. Since cases are described by few attributes and since our case base is rather small, we use Tversky's sequential measure of dissimilarity [20]. On the basis of those retrieved cases that had best therapy results an average initial therapy is calculated. Best therapy results can be determined by values of a blood test after two weeks of treatment with the initial dose. The opposite idea to consider cases with bad therapy results does not work here, because bad results can also be caused by various other reasons.

10.5.2 Generalised Solutions for Therapy Inefficacy

When long-term therapies become inefficient, ISOR searches for reasons and attempts to find better therapies. Solutions are reasons for inefficacy. General solutions like "irregular intake" or "changes of hormonal situation" are used as prototypes on a first level. On a second level these prototypes are more specific. All prototypes are described by three main attributes (problem code, diagnosis, and therapy) and some additional attributes like age, sex, prescribed drug etc. All prototypes have been defined by medical experts.

At first the retrieval searches among the prototypes on the top level and checks which solution might be probable for the query patient. Subsequently prototypes on lower level are considered and finally the single cases, which belong to the retrieved prototype.

10.6 Summary: The Role of Prototypes

The presented systems have one thing in common that distinguishes them from most CBR systems: They use prototypes as a form of knowledge representation that fills the gap between specific cases and general rules. The main purpose of such generalised knowledge is to structure the case base, to guide the retrieval process, and sometimes to decrease the amount of storage by erasing redundant cases.

In domains with rather weak domain theories another advantage of case-oriented techniques is their ability to learn from cases. Only gathering new cases may improve the system's ability to find suitable similar cases for current problems, but it does not elicit the intrinsic knowledge of the stored cases. To learn the knowledge contained in cases a generalisation process is necessary. In our early warning system concerning the kidney function, apart from guiding the retrieval and structuring the case base prototypes mainly serve to learn typical course pattern, because just the relevant kidney parameters are known but no knowledge about their temporal course behaviour exists.

For diagnosis of dysmophic syndromes prototypes correspond directly to the physician's sense of prototypes. As comparisons with single cases are unable to identify typical features, in this application the use of prototypes is not only sensible, but even necessary.

In ISOR, the prototypes for dose calculation are guidelines and the prototypes for therapy inefficacy are similar to those for diagnosis of dysmorphic syndromes. The main difference is that in ISOR all prototypes are defined by medical experts.

Summarising our experiences we would like to make quite clear that the role of prototypes depends on the application and the task. For medical diagnoses they even seem to be necessary because of their correspondence to medical prototypes which guide the physicians diagnoses. In domains with very poor domain theories they may help to learn general knowledge.

References

1. Strube G, Janetzko D (1990) Episodisches Wissen und fallbasiertes Schließen: Aufgaben für die Wissensdiagnostik und die Wissenspsychologie. Schweizerische Zeitschrift für Psychologie 49 (4): 211–221

2. Swanson DB, Feltovich PJ, Johnson PE (1977) Psychological Analysis of Physician Expertise: Implications for Design of Decision Support Systems. In: Shires DB, Wolf H (eds.): Proceedings of MEDINFO 77, North-Holland, Amsterdam, pp 161–164
3. Schank RC (1982) Dynamic Memory: a theory of learning in computer and people. Cambridge University Press, New York
4. Bareiss R (1989) Exemplar-based knowledge acquisition. Academic Press, San Diego
5. Maximini K., Maximini R., Bergmann R.(2003) An Investigation of Generalized Cases. In: Asley, K.D., Bridge, D.G. (eds.): Proc ICCBR 2003, Springer, Berlin Heidelberg New York, pp 261–275
6. Bellazzi R, Montani S, Portinale l (1998) Retrieval in a prototype-based case library: a case study in diabetes therapy revision. In: Smyth B, Cunningham P (eds) Proc European Workshop on Case-Based Reasoning. Springer-Verlag, Berlin Heidelberg New York, pp 64–75
7. Bichindaritz I (1995) Case-based reasoning adaptive to several cognitive tasks. In: Aamodt A, Veloso M (eds) Case-Based Reasoning Research and Development, Proceedings International Conference on Case-Based Reasoning, ICCBR-95, Springer-Verlag, Berlin Heidelberg New York, pp 391–400
8. Turner R (1988) Organizing and Using Schematic Knowledge for Medical Diagnosis. In: Kolodner J (ed) Proceedings of a Workshop on Case-Based Reasoning, Florida, pp 435–446
9. Perner P (2006): A Comparative Study of Catalogue-Based Classification. In: Roth-Berghofer TR et al (eds.): Proc ECCBR, Springer Berlin 301–308
10. Clancey WJ (1985) Heuristic Classification. Artificial Intelligence 27: 289–350
11. Selz O (1913) Über die Gesetze des geordneten Denkverlaufs. Stuttgart
12. Perner P (2004) Are case-based reasoning and dissimilarity-based classification two sides of the same coin?, Journal Engineering Applications of Artificial Intelligence, 15/2: 205–216
13. Taybi H, Lachman RS (1990) Radiology of Syndromes, Metabolic Disorders, and Skeletal Dysplasia. Year Book Medical Publishers, Chicago
14. Gierl L, Stengel-Rutkowski S (1994) Integrating Consultation and Semiautomatic Knowledge Acquisition in a Prototype-based Architecture: Experiences with Dysmorphic Syndromes. Artificial Intelligence in Medicine 6: 29–49
15. Clinical Dysmorphology. www.clindysmorphol.com
16. Winter RM, Baraitser M, Douglas JM (1984) A computerised data base for the diagnosis of rare dysmorphic syndromes. Journal of medical genetics 21 (2): 121–123
17. Stromme P (1991) The diagnosis of syndromes by use of a dysmorphology database. Acta Paeditr Scand 80 (1): 106–109
18. Weiner F, Anneren G (1989) PC-based system for classifying dysmorphic syndromes in children. Computer Methods and Programs in Biomedicine 28 111–117
19. Evans CD (1995) A case-based assistant for diagnosis and analysis of dysmorphic syndromes. International Journal of Medical Informatics 20: 121–131
20. Tversky A (1977) Features of Similarity. Psychological Review 84 (4): 327–352
21. Rosch E, Mervis CB (1975) Family Resemblance: Studies in the Internal Structures of Categories. Cognitive Psychology 7: 573–605

22. Robeson SM, Steyn, DG (1990) Evaluation and comparison of statistical forecast models for daily maximum ozone concentrations. Atmospheric Environment 24 B (2): 303–312
23. Shahar Y (1999) Timing is Everything: Temporal Reasoning and Temporal Data Maintenance in Medicine. In: Horn W, Shahar Y, Lindberg G, Andreassen S, Wyatt J (eds) Proceedings of AIMDM'99, Springer-Verlag, Berlin Heidelberg New York, 30–46
24. Haimowitz IJ, Kohane IS (1993) Automated trend detection with alternate temporal hypotheses. In: Proceedings of the 13th International Joint Conference on Artificial Intelligence, Morgan Kaufmann Publishers, San Mateo, pp 146–151
25. Miksch S, Horn W, Popow C, Paky F (1995) Therapy planning using qualitative trend descriptions. In: Barahona P, Stefanelli M, Wyatt J (eds) Proc AIME'95, Springer-Verlag, Berlin Heidelberg New York, pp 197–208
26. Lenz M, Auriol E, Manago M (1998) Diagnosis and decision support. In: Lenz M, Bartsch-Spörl B, Burkhard H-D, Wess S (eds) Case-Based Reasoning Technology, From Foundations to Applications., Springer-Verlag, Berlin Heidelberg New York, pp 51–90
27. Smyth B, Keane MT (1998) Adaptation-Guided Retrieval: Questioning the Similarity Assumption in Reasoning. Artificial Intelligence 102: 249–293
28. Anderson JR (1989) A theory of the origins of human knowledge. Artificial Intelligence 40, Special Volume on Machine Learning, pp 313–351
29. Wenkebach U, Pollwein B, Finsterer U (1992) Visualization of large datasets in intensive care. Proc Annu Symp Comput Appl Med Care, pp 18–22
30. DeSarbo WS et al. (1992) TSCALE: A new multidemensional scaling procedure based on Tversky's contrast model. Psychometrika 57: 43–69
31. Kolodner J (1993) Case-based reasoning. Morgan Kaufmann Publishers, San Mateo
32. Bueno-Cavanillas A et al. (1994) Influence of nosocomial infection on mortality rate in an intensive care unit. Crit Care Med 22: 55–60
33. Stottler RH, Henke AL, King JA (1989) Rapid retrieval algorithms for Case-Based Reasoning. In: Proceedings of International Joint Conference on Artificial Intelligence, 233–237
34. Smyth B, Keane MT (1993) Retrieving adaptable cases: the role of adaptation knowledge in case retrieval. In: Richter MM et al. (eds) Proceedings of 1st European Workshop on Case-Based Reasoning, pp 76–81
35. Schmidt R, Vorobieva O (2006) Case-Based Reasoning Investigation of Therapy Inefficacy. Knowledge-Based Systems 19 (5): 333–340
36. Working group for paediatric endocrinology of the German society for endocrinology and of the German society for children and youth medicine (1998) 1–15

11

Case-Based Reasoning for Image Segmentation by Watershed Transformation

M. Frucci,[1] P. Perner,[2] and G. Sanniti di Baja[1]

[1]Institute of Cybernetics "E. Caianiello", CNR, Pozzuoli (Naples) Italy
[2]Institute of Computer Vision and Applied Computer Science, Leipzig, Germany

Summary. This chapter introduces a novel image-segmentation scheme based on case-based reasoning. Image segmentation is aimed at dividing an image into a number of different regions in such a way that each region is homogeneous with respect to a given property, but the union of any two adjacent regions is not. To reach this goal, a number of different approaches have been suggested in the literature, among which we consider here watershed-based segmentation. The basic idea of this segmentation scheme is to identify in the gray-level image a suitable set of seeds from which to perform a growing process. The growing process groups to each seed all pixels that are closer to that seed more than to any other seed, provided that a certain homogeneity condition is satisfied. Since any segmentation method includes some parameters, whose values depend on the image characteristics, CBR can be profitably used to improve the performance of the adopted segmentation method and to ensure that good segmentation results are obtained even if the segmentation method is applied to images with different characteristics. In practice, CBR will select from a case-base the cases having image characteristics similar to those of the current input image, and will apply to the current image the segmentation parameters associated to the most similar case. Image characteristics will be computed in terms of mean features on the whole image, and a proper similarity measure will be used to select in the case-base the most similar case.

11.1 Introduction

Image segmentation is a process of dividing an image into a number of different regions such that each region is homogeneous with respect to a given property, but the union of any two adjacent regions is not. This process has received much attention in the literature, being a necessary preliminary step for any image analysis task, and a number of surveys of different approaches have been published (see, e.g., [1–7]).

Image thresholding is a well-known technique for image segmentation. Because of its wide applicability to many areas of digital image processing, a large number of thresholding methods have been proposed over the

M. Frucci et al.: *Case-Based Reasoning for Image Segmentation by Watershed Transformation*, Studies in Computational Intelligence (SCI) **73**, 319–353 (2008)
www.springerlink.com

years (see, e.g., [8–17]). Image thresholding has low-computational complexity, which makes it an attractive method, but does not take into account spatial information and is mostly suitable for images where the gray-levels constitute well-defined peaks, separated by not too broad and flat valleys. Actually, some of the limitations of this approach can be overcome by combining rule-based methods with learning methods, such as case-based reasoning (CBR) (see e.g., [18]). Based on a rule set, the histogram of the gray-levels is properly smoothed and the right number of peaks can be selected more easily and in a reliable manner. CBR ensures the incremental learning of the rule set with the proper parameters.

Another common approach to image segmentation is based on feature space clustering, which has sometimes been regarded as the multidimensional extension of the concept of thresholding. Clustering schemes using different kinds of features (multispectral information, mean/variation of gray-level, texture, color) have been suggested (see, e.g., [19–29]). This approach can be successfully used if each perceived region of the image constitutes an individual cluster in the feature space. This requires a careful selection of the proper features, which depends on image domain.

Segmentation can also be accomplished by using region-based methods, or edge-detection-based methods, or methods based on a combination of those two approaches (see, e.g., [30–43]). Region-based methods imply the selection of suitable seeds from which to perform a growing process. In general, region merging and region splitting are accomplished to obtain a meaningful number of homogeneous regions. Seed selection and homogeneity criterion play a critical role for the quality of the obtained results. Edge-detection-based methods follow the way in which human observers perceive objects, as they take into account the difference in contrast between adjacent regions. Edge detection does not work well if the image is not well contrasted, or in the presence of ill-defined or too many edges.

Watershed-based segmentation (see, e.g., [44]) exploits both region-based and edge-detection-based methods. The basic idea of watershed-based segmentation is to identify in the gray-level image a suitable set of *seeds* from which to perform a growing process. If the main feature taken into account is gray-level distribution, the seeds are mostly detected as the sets of pixels with locally minimal gray-level (called *regional minima*). The growing process groups to each seed all pixels that are closer to that seed more than to any other seed, provided that a certain homogeneity in gray-level is satisfied. Thus, watershed-based segmentation limits the drawbacks of region-based and edge-detection-based methods. In fact, the seeds from which to perform region growing are generally detected in the gradient image of the input gray-level image, where the edges are enhanced. In turn, the problem generally affecting segmentation by edge detection, i.e., that the edges seldom constitute closed curves surrounding the regions of interest, is overcome since the regions are determined by the growing process. Since in this chapter we are dealing with watershed-based segmentation, we postpone a more detailed description of it to Sect. 11.4.

Finally, we point out that the literature on segmentation is quite large and includes other approaches besides the ones that we have briefly discussed above. For example, almost all the above approaches can benefit if some fuzziness is taken into account. When this is done, segmentation is referred to as fuzzy segmentation (see, e.g., [45–49]).

To overcome the drawbacks of the algorithms mentioned above, learning methods are applied to image segmentation. These learning methods are applied to learn the mapping between image features and semantically meaningful parts, to learn the parameters of the segmentation algorithm or to learn the mapping between rank performance of the segmentation algorithm and the image features.

There are statistical learning methods, machine learning methods, neural-net-based learning methods, and learning methods using a combination of different techniques. The main drawbacks of these methods are (1) the need of a sufficiently large training set and (2) the need of training again the whole model, when new data come in. Therefore, it seems to be useful to use CBR for a flexible image segmentation system, since CBR can be used as a reasoning approach as well as an incremental knowledge-acquisition approach. A CBR framework has, indeed, been successfully used for the high-level unit of an image interpretation system [50–52] and has shown to outperform with respect to other approaches.

This chapter proposes a novel image-segmentation scheme based on CBR. The watershed transformation is used for image segmentation and CBR is used to select the segmentation parameters according to the characteristics of the current image. CBR should ensure that we obtain good segmentation results, independently of the images to which we apply to our segmentation algorithm. We assume that images having similar image characteristics will show similarly good segmentation results when the same segmentation parameters are used. Therefore, CBR will select cases having similar image characteristics from a case base and apply to the current case the segmentation parameters associated to the most similar case. In Sect. 11.2 we describe related work on CBR image processing and interpretation. Then, we describe in Sect. 11.3 the general framework for case-based image segmentation. Section 11.4 describes our approach for CBR image segmentation. It explains in Sect. 11.4.1, the watershed algorithm and in Sect. 11.4.2 the merging strategy for reducing the oversegmentation that occurs when applying the watershed transformation. Section 11.4.3 describes the introduction of a similarity-based control scheme for oversegmentation reduction that makes the control process more flexible. The case description and its evaluation are given in Sect. 11.4.4. The similarity determination between the current case and the cases in the case base is described in Sect. 11.4.5. Sections 11.4.6 and 11.4.7 give some starting ideas for the automatic evaluation of the segmentation results, and for case generalization and similarity learning, respectively. Final remarks are given in Sect. 11.5. Section 11.6 concludes the work.

11.2 Related Work

Several systems have been developed that apply CBR for image retrieval and interpretation at the symbolic level. Usually, the images are not processed and the symbolic terms are user-specified. Early examples are [53] for retrieval of radiological images, [54] for the detection of coronary heart disease and [55] for the diagnosis of breast cancer in histopathology.

In [50], Grimnes and Aamodt presented a system that integrates CBR into a task-oriented model-based system for interpreting abdominal CT images. A CBR unit is used, where each case is an individual image segment. These cases are associated with labels that, in turn, are used as indices for a case base consisting of organs. The case base of organs is used by another CBR unit for organ interpretation. The system is based on a propose-critique-modify learning cycle.

A completely different application of CBR to image processing was described in [56] by Ficet-Cauchard et al. CBR was applied for the development of the image processing steps of formerly not solved image processing problems, by using experiences and plan adaptation. In [51], Jarmulak presented a system employing a tree-based retrieval strategy for ultrasonic B-scans that are one-dimensional signals. In [57], Micarelli et al. applied CBR to scene recognition.

In [58], Perner used CBR for segmentation of brain CT images, involving more complex case representations, reasoning and learning strategies, and data mining techniques for pattern recognition. In [59], Perner proposed a system that uses CBR at three different levels. At the low-level stage, image segmentation was optimized by taking into account different image-acquisition conditions and image quality. At the intermediate-level stage, the case representation to be used by the high-level unit was extracted. At the high-level stage, image interpretation was dynamically adapted. The system worked on different case representations, such as graph representation for high-level image description, and raw image matrix for low-level image representation. Therefore, the system used two different CBR strategies for reasoning and learning: reasoning used structural similarity and learning used digital image distance. Different learning strategies in a hierarchy of structural cases were presented by Perner [52, 60]. In [61], Perner compares CBR and dissimilarity classification methods, which has become important in pattern recognition. The application of case-based image interpretation to health monitoring and biotechnology is described in [62]. Learning case representations and improving system performance by controlling the similarity measure is described in [63]. Recent research has focused on mining raw information into more general cases, [64], and making object recognition more robust against model variation [65].

Learning in pattern recognition for image segmentation can be distinguished into three different tasks:

1. Learning the mapping function from the image features to the labels for the regions in the image [19–29]
2. Learning the mapping function from the image quality to the rank performance of an algorithm [66]
3. Learning the mapping function from image features to segmentation parameters [58]

The above learning tasks are classification tasks having different aims. In the first task, the image is partitioned into small regions that are described by image features. Generally, these regions are labeled by the operator, either manually or interactively, based on an image analysis algorithm. The task is to learn a classifier that can map the image features of the regions to the labels.

Tasks 2 and 3 are control tasks solved by classification. The basic idea is that there is a strong correlation between image characteristics, such as contrast, noise and illumination, and the performance of the algorithms that are applied to the image. Applying the same segmentation algorithm to images sharing the same characteristics is expected to produce results with the same performance.

11.3 The General Framework of a Case-Based Image Segmentation Approach

The segmentation problem can be seen as a classification problem for which we have to learn the best classifier. Depending on the segmentation task, the output of the classifier can be the labels for the image regions, the segmentation algorithm selected as the most adequate, or the parameters for the selected segmentation algorithm. In any case, the final result is a segmented image. The learning of the classifier should be done on a sufficiently large test data set, which should represent the entire domain well enough in order to be able to build up a general model for the segmentation problem. However, often it is not possible to obtain a sufficiently large data set and, therefore, the segmentation model does not fit the entire data set and needs to be adjusted to process new data. We note that a general model does not guarantee the best segmentation for each image; rather, it guarantees an average best fit over the entire set of images.

Another aspect of the problem is related to the changes in image quality caused by variations in environmental conditions, image devices, etc. Thus, the segmentation performance needs to be adapted to changes in image quality. All this suggests to use CBR as a basic methodology for image segmentation, since CBR can be seen as a method for problem solving as well as a method to capture new experiences. It can be seen as a learning and knowledge discovery

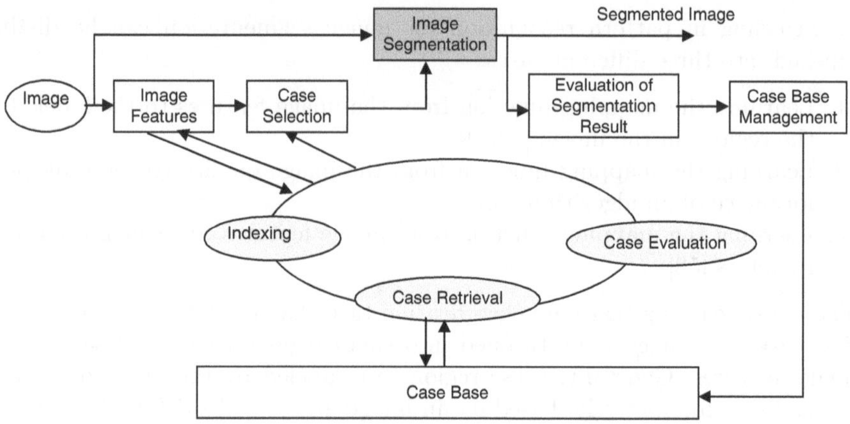

Fig. 11.1. Scheme of case-based image segmentation

approach. The CBR process consists of six phases: extracting the case description, indexing, retrieval, learning, adaptation, and application of the solution.

The CBR process for image segmentation is shown in Fig. 11.1. The actual image characteristics are described by mean features computed on the whole image. These features are used for indexing the case base and for retrieval of a set of cases close to the current problem, based on a proper similarity measure.

To adjust the segmentation parameters, indexing should find out images sharing the same segmentation parameters. Among the cases close to the current problem, the closest one is selected and its associated solution is given as control input to the image segmentation unit. The image segmentation unit takes the current image and processes it according to the current control state. Finally, the output is the segmented image.

A case consists of a description of the image characteristics and the solution. The description of the image characteristics can take into account nonimage and image information. Based on image description, we can reduce our complex solution space to a subspace of relevant cases, where variation in image quality among the cases is limited.

The solution of a case can be one of the outputs of the classifiers described above. Suppose that our aim is to control the parameters of the segmentation unit, then the solution is the set of parameters applicable for the segmentation of the current image. The solution is given as input to the segmentation unit and the current image is processed by the segmentation unit, based on the selected parameters.

If we want to control the selection of the best algorithm in a set of possible algorithms, then the solution given to the image segmentation unit would be the selected algorithm. If we want to label the regions, then the output would be the labeled regions.

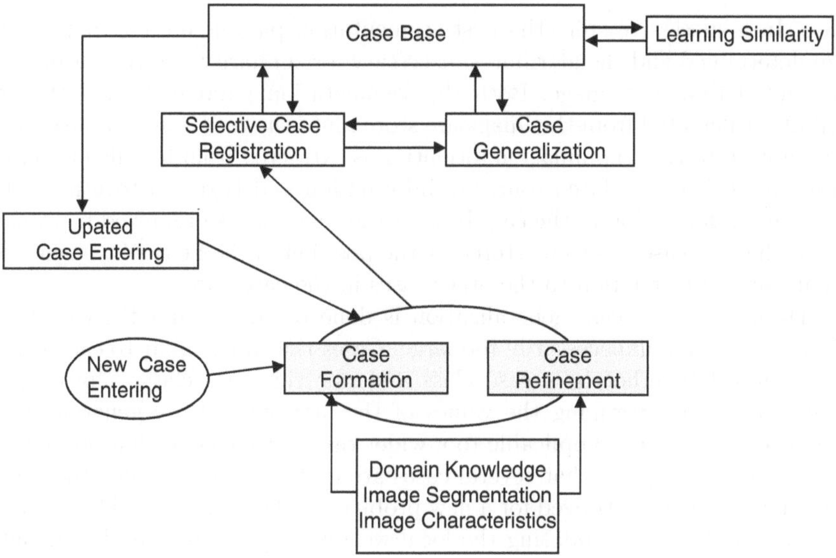

Fig. 11.2. Case base maintenance

After the image has been processed, the segmentation result is evaluated. This means that the segmentation quality is judged either by an expert or automatically. Depending on the obtained evaluation, the knowledge containers (case description, similarity solution) are modified to ensure a better segmentation result by processing again the same image. This task is done by the case base maintenance unit.

The case base maintenance unit is shown in Fig. 11.2. Differently from a conventional segmentation process, CBR also includes the evaluation of the segmentation result and takes it as a feedback to improve the system performance semi- or automatically. Usually this is an open problem in many segmentation applications.

When the evaluation of the segmentation result is done manually, the expert compares the original image with the labeled image on display. If the expert detects significant differences in the two images, the result is tagged as incorrect and the case base management will start. The proposed method is close to the critique-modify framework described in [50].

The evaluation procedure can also be done automatically. However, there is no general procedure available and evaluation can be done automatically only in a domain-dependent fashion.

Case-based maintenance is done for several purposes (1) to enter a new case, when no similar cases are available in the case base, (2) to update an existing case by case refinement, and (3) to obtain case generalization.

Once an incorrect result is observed for an input image either by the user or by the automatic evaluation procedure, the case is tagged as a bad case. In a successive step, the best segmentation parameters for that image are determined and the attributes, necessary for similarity determination, are computed from the image. Both the segmentation parameters and the attributes calculated from the image are stored into the case base as a new case. In addition to that, nonimage information is extracted from the file header or from any other associated source of information and is stored together with the other information in the case base. If the case base is organized hierarchically, the new case has to be stored at the position in the hierarchy suggested by its similarity-relation to the other cases in the case base.

During storage, case generalization is done to ensure that the case base does not become unnecessarily too large. Cases that are similar to each other are grouped together in a case class and, for this case class, a prototype is computed by averaging the values of the attributes. Case generalization ensures that a case is applicable to a wider range of segmentation problems.

It can also happen that several cases are quite similar to each other and, therefore, they get retrieved for a new problem at the same time. Then, learning the similarity by updating the local weights associated to each attribute should be done.

11.4 Our Approach for Case-Based Image Segmentation

We take a case-based approach for image segmentation that controls the parameters of a given segmentation algorithm. Such an approach has been proposed for the first time by Perner [58] with reference to on an histogram-based segmentation algorithm. Here, we consider a watershed-based image segmentation algorithm (WTS) and control by CBR the merging process after the watershed transformation has been applied to the image. This is a crucial step when using WTS.

To our knowledge, watershed-based segmentation was introduced in [67]. Since then, a number of papers have been published dealing with the use of watershed transformation for different applications (see, e.g., [68–81]), or improving the basic algorithm and suggesting solutions to the main problems affecting the watershed partition (see, e.g., [82–110]). In fact, the watershed partition may result to be characterized by a number of regions which is either too large (oversegmentation) or too small (undersegmentation) with respect to the expected result. Oversegmentation mainly occurs because even objects that, in a continuous image, a human observer would classify as homogeneous (e.g., with respect to texture or gray-level distribution), in the digital image consist of a large number of homogeneous parts, and each of these parts constitutes a region of the partition. Undersegmentation is not equally frequent. It mostly occurs when the input gray-level image has low contrast. In the following, we will focus on oversegmentation reduction.

In the next sections we will describe the watershed transformation and our method to reduce oversegmentation in the watershed partitioned image.

11.4.1 Watershed Transformation

The *landscape* paradigm can be conveniently used to give an easily understandable explanation on how the watershed transformation works. A gray-level bidimensional image I can be interpreted as the top surface of a three-dimensional binary landscape, which can be obtained by considering, for each pixel of I with planar coordinates (x,y), the relative gray-value as representing the third coordinate z, that is the height in the landscape in position (x,y). In this landscape, the bottom of each valley (called *pit*) corresponds to a connected set of pixels of I characterized by locally minimal gray-level, while the top of each hill (called *peak*) corresponds to a connected set of pixels of I characterized by locally maximal gray-level.

If the pits of the valleys are pierced and the landscape is slowly immersed into water, the landscape will be flooded and its valleys will be transformed into lakes. Of course, the first valleys that will be transformed into lakes are those with the lowest pits, since these are reached first by the increasing level of water. When the level of water of a lake reaches the edge separating the catchment basin of that lake from an adjacent catchment basin, the water of that lake would overflow into the adjacent basin. To prevent this, a dam is built wherever water of certain lakes could overflow into adjacent catchment basins. This is done until the level of water reaches the highest peak(s) of the landscape. At this stage of the process, the top surface of the flooded landscape coincides with the desired watershed partition. The top lines of the built dams constitute the closed watershed lines, each of which surrounding a catchment basin. In this way, a tessellation of the input image into regions is achieved, where each region corresponds to a catchment basin.

To implement the watershed transformation, the sets of pixels characterized by locally minimal value (*seeds*) should be identified in the gray-level image. Actually, the seeds are identified in the gradient image of the gray-level image, which is computed, for example, by means of the Sobel operator. In this way the seeds are identified in an image that effectively accounts for the homogeneity criterion, since it enhances the edges in the zones with higher variations of gray-value [82]. From the seeds, an iterated growing process is started, which can be accomplished in two different ways, known as watershed by topographical distances and watershed by immersion. Both strategies are deeply discussed in [98].

From an operative point of view, we point out that since the catchment basins must be separated from each other by a leak-proof set of watershed lines, the catchment basins must be 4-connected (8-connected) if watershed lines are defined as sets of 8-connected (4-connected) pixels. In fact, if the same connectedness type is used for both the catchment basins and the watershed lines, topological paradoxa would be originated. Namely, a closed watershed line would not separate the surrounded catchment basin from the adjacent ones. The regions of the obtained partition can be individually identified by assigning to each catchment basin a distinct label. A unique value is in turn assigned to all the watershed lines.

Fig. 11.3. A gray-level image, *left*, the gradient image, *middle*, and the watershed partition, *right*

In Fig. 11.3, a gray-level image, the corresponding gradient image, and the watershed partition into 3,237 regions, obtained by using all seeds detected in the gradient image, are shown from left to right. The watershed lines (in black) are superimposed onto a lighter version of the input image only for better visualization of the lines. This input image is used as a running example in the following. It can be observed that the image is noticeably oversegmented. Oversegmentation is due to the large number of detected seeds, which exceeds the number of perceived regions.

To reduce oversegmentation, a careful selection of the seeds to be used for region growing is necessary: Only seeds corresponding to significant regions should be used. In principle, seed selection can be achieved by using a filter to remove irrelevant minima. However, a priori knowledge on the class of images is necessary to design the proper filter.

A more general method, introduced in [110], is based on the iterated computation of the watershed transform coupled with the use of two techniques, called *flooding* and *digging*, which can be employed to cause disappearance in the gradient image of those seeds that are recognized as corresponding to nonsignificant regions. Only significant basins should be preserved in the final watershed partition (i.e., their seeds should be regarded as relevant) and nonsignificant basins should be removed by aggregating them to significant ones (i.e., their seeds should be regarded as irrelevant). The whole process (i.e., flooding, digging, and watershed transformation) is iterated until all basins result to be significant. Of course, the definition of significant region is crucial to obtain a meaningful partition.

11.4.2 Oversegmentation Reduction Based on Region Significance (Control Parameters)

In [110], the significance of a catchment basin was defined by taking into account the portion of the landscape where the basin is placed, i.e., it was evaluated with respect to the adjacent basins. Let us consider the basin X and let Y be one of the basins adjacent to X. The pixel p at the minimal height along the ridge separating X from Y is called the *relative local overflow* of X

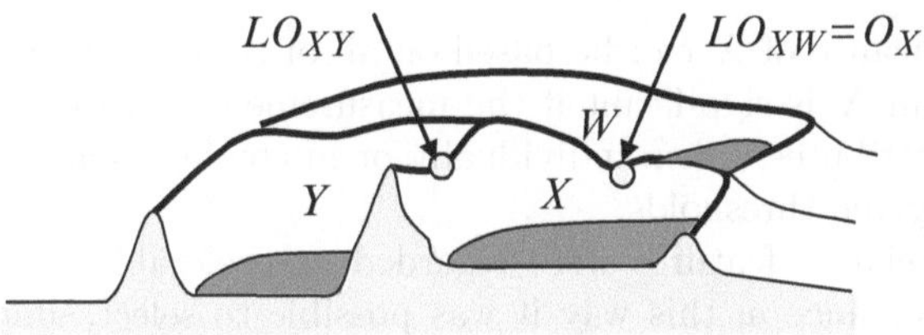

Fig. 11.4. Local overflows for the basin X

Fig. 11.5. Depth, volume, and surface of the lake created in the catchment basin X

with respect to Y and is denoted by LO_{XY}. The local overflow pixel is the one where the dam separating X from Y should start to be built, to prevent overflow from X to Y. Once the local overflow pixels have been identified in correspondence with all the basins adjacent to X, the pixel among them with the smallest value is called the *overflow* of X and is denoted by O_X, see Fig. 11.4.

For a basin X we can consider both absolute and relative features. Relative features take into account X and any of its adjacent basins. Absolute features take into account X and all the adjacent basins. The absolute (relative) features that can be used to characterize X can be obtained in terms of measurements performed on the set of pixels of X having gray-values less than the (relative local) overflow. This set of pixels is the lake formed when the water reaches the (relative local) overflow pixel and is denoted by (L_{XY}) L_X. In the following, we use Z to denote X or XY, depending on whether we are interested in relative or absolute attributes of the lake created in X when the water reaches the relative local overflow or the overflow. Moreover, let us denote by R_X the gray-level of the pit of the basin X. With reference to Fig. 11.5 we can define for X:

- The *depth* of L_Z

$$\text{Depth } [X,O_Z] = \max_{p \in L_Z} \{O_Z - p\} = O_Z - R_X \tag{11.1}$$

- The *volume* of L_Z

$$\text{Volume}\,[X,O_Z] = \sum_{p \in L_Z} (O_Z - p) \tag{11.2}$$

- The *surface* of L_Z, as the number of pixels in the lake L_Z.

The significance of X can be based on all or some of the above measurements. A basin X is significant if the measurements regarding its (absolute or relative) attributes, taken individually or in combination, have values not smaller than given thresholds.

In [110], relative features were regarded as preferable to define the significance of X, since in this way it was possible to select, among all basins adjacent to X, the ones more adequate to absorb X, if X was not significant. Moreover, a relative *similarity parameter* SA_{XY} was also introduced, as the absolute value of the difference in altitude between the pits of X and the adjacent basin Y:

$$SA_{XY} = |R_X - R_Y| \tag{11.3}$$

The similarity between X and Y increases when SA_{XY} decreases.

The relative depth D_{XY} and the similarity parameter SA_{XY} were, then, used to evaluate the relative significance of X with respect to Y. Precisely, a basin X was termed significant with respect to Y if the following holds.

$$SA_{XY} > At \text{ OR } D_{XY} > Dt \tag{11.4}$$

where At and Dt are threshold values, computed automatically by using statistics on the initial watershed partition of the gray-level image.

By taking into account that X can be adjacent to more than one region, three cases are possible for the classification of X:

- X is significant with respect to each adjacent region Y. Then, X is termed strongly significant (and the corresponding seed is relevant).
- X is not significant with respect to every adjacent region Y. Then, X is termed nonsignificant. (The corresponding seed is irrelevant and X has to be absorbed by the adjacent regions.)
- X is significant in correspondence of some adjacent regions only. Then, X is termed partially significant. (The seed is irrelevant and X has to be merged, but only with proper regions, selected among those with respect to which X is nonsignificant.)

Two different techniques were adopted to remove the irrelevant seeds, depending on whether X is nonsignificant or is partially significant. When X was nonsignificant, flooding was accomplished by setting all pixels of X with gray-level lower than the value q of the overflow pixel, to value q. In this way, when watershed transformation was newly applied, X resulted as merged to the adjacent regions with relative local overflow equal to the overflow q. In turn, when X was partially significant, digging was performed to open a canal connecting the pit of X with the pit of each basin Y with respect to which X was not significant. The canal between X and any such a basin Y was identified as the minimal length path linking the pits of X and Y, and passing through the local overflow pixel common to X and Y. The gray-level of all the pixels in the path was set to the lower value between those of the pits of X and Y. When the watershed transformation was newly applied, only the pit of the basins Y were detected as seeds and the desired merging was obtained. The watershed lines of X, which were already detected as separating X from regions W with respect to which X was significant, were not altered.

When flooding and digging were performed by using the thresholds At and Dt, the gradient image resulted to be modified with respect to the initial gradient image. In fact, a certain number of pixels had their gray-values increased by flooding, while other pixels had their gray-values decreased by digging. This modification of the gradient image caused a smaller number of seeds to be detected and, accordingly, a smaller number of regions to be identified by the growing process. Thus, a new watershed partition, less fragmented than the original one, was achieved. In general, flooding, digging, and watershed transformation had to be applied for a number of times. In fact, the regions of the new watershed partition were not necessarily all significant with respect to the threshold At and Dt. When, after k applications of flooding, digging, and watershed transformation, all the obtained regions were strongly significant with respect to the threshold At and Dt, the obtained kth watershed partition was possibly still oversegmented. Then, new values for the two thresholds At and Dt were computed on the kth watershed partition. Of course, if the newly computed threshold values were smaller than or equal to the previous ones, no merging would have been possible and the kth watershed partition would have been the final one. Otherwise, if at least one of the new two thresholds was larger than the old ones, the whole process was repeated. To guarantee that the process terminates, the maximum number of repetitions of the whole process was fixed to five. In general, the process terminated after at most three repetitions of the process.

The values At and Dt of the two thresholds involved in the significance criterion were computed in [110] by taking into account the initial watershed transform of the gradient image. Let M be the number of sets each of which including all basins characterized by pits having the same height h, with h $= 1, 2, \ldots M$, in the initial watershed partition. For the ith set, the maximal values of the depth and the similarity parameter ($max\text{-}depth_i$ and $max\text{-}sim_i$) were computed. Then, the two thresholds At and Dt were assigned, respectively, the minimum of $max\text{-}depth_i$ and the minimum of $max\text{-}sim_i$, computed for i $= 1, 2, \ldots, M$. The same criterion was used also to compute the new values of the thresholds, by using the watershed partition resulting when region merging was no longer possible with the initial thresholds.

The complete scheme of the segmentation algorithm [110] is given in Fig. 11.6. More details can be found in [110].

In Fig. 11.7, the watershed partition of the running example is given to show the performance of flooding and digging to reduce over-segmentation With respect to the initially detected 3,237 basins, only 82 basins are found in the final image. As in Fig. 11.3, the watershed lines (in black) are superimposed onto a lighter version of the input image only for better visualization of the lines.

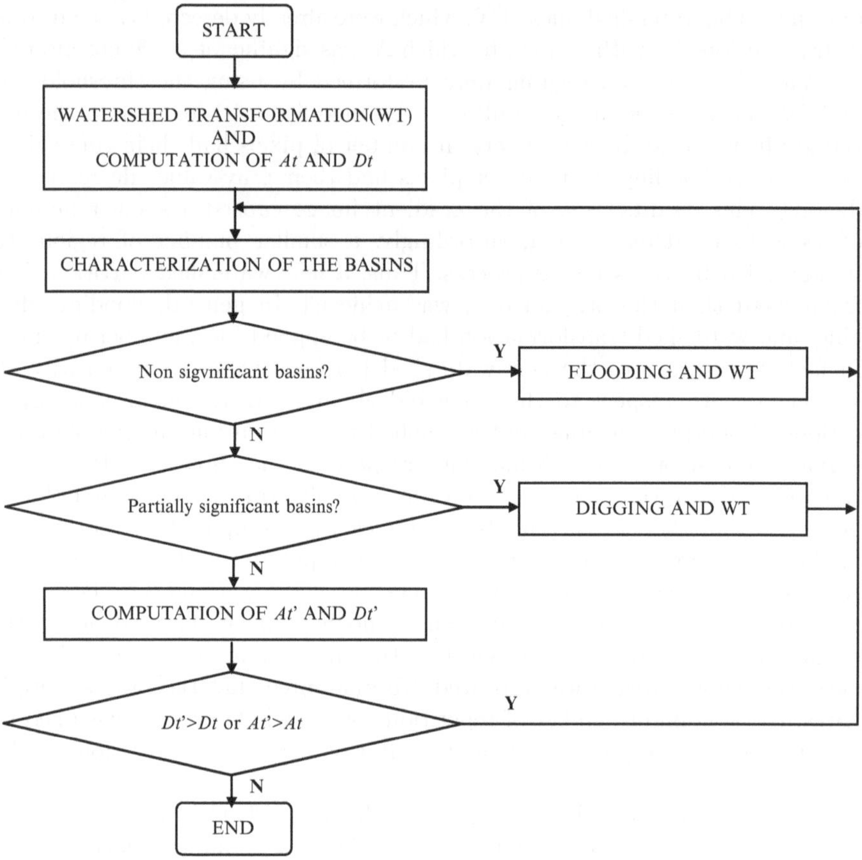

Fig. 11.6. The segmentation algorithm introduced in [110]

Fig. 11.7. The original image, watershed partition, *left*, over-segmentation reduction by using algorithm [110], *right*

11.4.3 From Crisp Rules to Similarity

The algorithm in [110] is based on a crisp rule to decide when to do the merging process:

$$\text{if } SA_{XY} > At \text{ OR } D_{XY} > Dt, \text{ then } X \text{ is significant} \qquad (11.5)$$

This crisp rule based on an OR condition has been applied to all images regardless of the actual image characteristics.

To improve the performance of the segmentation algorithm [110], we should not use a crisp test to decide about merging. In fact, according to rule (11.5) it is enough that one of the two measures overcomes the relative threshold, in order a region be classified as significant with respect to an adjacent region. We think that better results could be achieved if we require that both measures SA_{XY} and D_{XY} are taken into account, possibly giving different weights to their contributions. We also think that the weights should be determined by analyzing the image characteristics.

We introduce a similarity-based control scheme that gives us the freedom to weight the influence of the two different parameters SA_{XY} and D_{XY}.

The suggested scheme takes into account the relative measure of the actual value of SA_{XY} with respect to the threshold At and weights this value by a factor a. The same is done for D_{XY} and Dt, whose ratio is weighted by a factor b. Depending on image characteristics, we weight the influence of region similarity and of depth by means of the two weights a and b, and introduce a threshold T as in the following rule:

$$\text{if } \frac{1}{2}(a \cdot \frac{SA_{XY}}{At} + b \cdot \frac{D_{XY}}{Dt}) \geq T, \quad \text{then } X \text{ is significant} \qquad (11.6)$$

If at least one of the values SA_{XY}/At and D_{XY}/Dt is larger than 1, then rule (11.5) would classify the region X as significant with respect to the adjacent region Y. If $a = b = 1$ and the threshold T is set to 0.5, rule (11.6) would also classify X as significant with respect to Y. If both SA_{XY}/At and D_{XY}/Dt have value larger than 1, then the threshold T in rule (11.6) can be set to 1 to classify X as by rule (11.5).

Table 11.1 shows some combinations of values for a, b, and T and the relative interpretations.

Different combinations of a, b, and T were used to segment the running example. The obtained segmentation results were judged, based on a comparison between the relative numbers and positions of the obtained regions and the number and positions of the regions in a manually segmented image.

Table 11.2 shows some results obtained by using the new similarity-based algorithm on the running example. The best segmentation result obtained by the new algorithm is shown in Fig. 11.8 right, where the watershed lines (in black) are superimposed onto a lighter version of the input image only for better visualization of the lines. This result was obtained by selecting

Table 11.1. Some combinations of a, b, and T, and relative interpretation

a	b	T	Interpretation
1.5	0.5	1	Region similarity is weighted more than depth
1	1	0.5	Region similarity and depth are equally weighted
0.75	1.25	0.7	Depth is weighted more than region similarity
0.75	1.25	1.35	Region similarity is weighted less than depth, and SA_{XY} and D_{XY} are quite larger than the relative thresholds At and Dt
1	1	0.95	Region similarity and depth are equally weighted, and SA_{XY} and D_{XY} can be smaller than the relative thresholds At and Dt

Table 11.2. Selections of a, b, and T for the running example

a	b	T	# regions	Evaluation
1	1	0.9	45	Undersegmented: the dome results to be merged with the sky
1	1	0.8	56	Undersegmented: the dome results to be merged with the sky
1	1	0.7	62	Undersegmented: part of the church to the left is merged to the sky
1	1	0.6	76	A good result with small oversegmentation
1	1	0.5	89	Quite good, even if includes a few small nonmeaningful regions
0.75	1.25	1.3	32	Extremely undersegmented
0.75	1.25	0.9	47	Undersegmented: the dome results to be merged with the sky
0.75	1.25	0.8	56	A very good result. The best one
0.75	1.25	0.7	64	A very good result with small oversegmentation
0.75	1.25	0.6	78	A very good result with small oversegmentation
1.5	0.5	1.	29	Extremely undersegmented
1.5	0.5	0.9	30	Extremely undersegmented
1.5	0.5	0.5	75	Over and undersegmented: the dome is merged to the sky and nonmeaningful regions are detected
1.5	0.5	0.4	80	Over and undersegmented: the dome is merged to the sky and nonmeaningful regions are detected
1.5	0.5	0.3	83	A very good result with small oversegmentation
1.5	0.5	0.2	104	Oversegmented

Fig. 11.8. Watershed partition, *left*, oversegmentation reduction by the algorithm [110], *middle*, and segmentation by the new algorithm, *right*

$a = 0.75$, $b = 1.25$, and $T = 0.8$, i.e., by weighting more D_{XY} than SA_{XY}. Only 56 regions are detected and oversegmentation is really very limited. We note that this result is significantly better than the segmentation obtained by the algorithm [110] resulting in 82 regions.

Rather good results are also obtained by the new algorithm when still selecting $a = 0.75$ and $b = 1.25$, but by using different values for T. As expected, the number of regions decreases when the value of T increases. Undersegmented results are obtained for T ranging from 1.3 to 0.9, and slightly oversegmented results are obtained for T smaller than 0.8.

Reasonably good results are also obtained with different selections of the weights. For example, for $a = 1$, $b = 1$, the value $T = 0.6$ produces a result that is only slightly oversegmented. For $a = 1.5$, $b = 0.5$, the value $T = 0.3$ also produces a slightly oversegmented result.

We tested the new algorithm on different images and noted that there are cases in which both the algorithm in [110] and the new algorithm show a similar behavior. This happens when the values of the parameters are $a = 1$, $b = 1$, and $T = 0.5$, i.e., when similarity and depth have the same influence and, due to the value chosen for the threshold, it is enough that at least one of the two measures overcomes the threshold to classify a region as significant. In all the other cases, the similarity-based algorithm behaves better, or even produces for the first time a reasonable result. For illustrative purpose, some examples are shown in the Appendix. Original images taken from our case base, the watershed partitions, the segmented images by using the algorithm [110] and the segmentation results obtained by using the new algorithm are given. The numbers of regions corresponding to the segmented images and, for the new algorithm, the values of a, b, and T selected for each image, are also indicated to help the reader to appreciate the differences among the results of the three algorithms, since it may result difficult to clearly distinguish the watershed lines superimposed on the input images.

Of course, an objective and automatic evaluation procedure to compare the achieved results with those expected by the user is desirable.

11.4.4 Case Description

A case generally consists of nonimage information, the parameters describing the image characteristics, and the solution (i.e., the values of the segmentation parameters). The segmentation parameters are, in our case, the weights a and b and the threshold T.

Nonimage Information

Nonimage information includes different issues, depending on the application. For example, in case of motion analysis [111], nonimage information includes the camera position, the relative movement of the camera, and the object category. For brain/liquor determination in CT-images [58], nonimage information includes patient-specific parameters (like age and sex), slice thickness, and the number of slices. Nonimage information is recorded in the header of the CT image file, so that it can be automatically accessed. Young patients have smaller liquor areas than elderly patients, and the CT images accordingly show different image characteristics. The anatomical structures (and hence again the image characteristics) also differ between women and men. The number of slices may vary from patient to patient because of this biological diversity, and so may the starting position of the slices. Therefore, the numerical values are mapped onto three intervals: bottom, middle, and top slices. These intervals correspond to the segments of the head with different characteristics. The intervals can easily be computed by dividing the number of slices by three. The remaining uncertainty in position can be ignored.

In this chapter we tried to solve an open problem since we are not dealing with images from one specific domain, as it was, for example, in the CT image analysis [58]. We are actually dealing with different kinds of images such as those coming from biological applications, landscape images, and face images. A straightforward approach to separate our case base might be by the type of images. However, this information is not generally annotated to the images and, hence, has to be given manually. We point out that in case of image databases having annotated text information [112] associated to the images, we would be able to access this information directly from the database.

In the next sections we study how far we would come to determine image similarity, only based on low-level image features.

Image Information

Image information should describe the characteristics of an image in terms of contrast, noise, and illumination. It should reflect the idea that images, classified as similar by the above features, should be segmented equally well by applying the segmentation algorithm, which produces the best segmentation for any of them. In other words, it should reflect the link between the behavior of the segmentation algorithm and the image characteristics. Therefore the question is: What is the right description of the image characteristics?

A straightforward description of an image is in terms of statistical features. The selected statistical features should be computed and used to cluster various images into groups of similar images. For each group, the relative features, stored as image description into the case base, could then be used to check image similarity, when an input image is presented to the system.

Another approach to determine image similarity could be based on a direct comparison of the images, see e.g., [113, 114]. Finally, an image may also be characterized by its texture, so that texture features might be useful to provide a description and for checking similarity among images.

An Example of Case Description

In this work, we consider statistical features, texture features, and a combination of both. This choice is motivated by the fact that the gradient images that are the input to the watershed-based segmentation algorithm differ in contrast, steepness of the edges, and number of edges per area.

The statistical measures are mean, variance, skewness, kurtosis, variation coefficient, energy, entropy, and centroid [115] (see Table 11.3).

The texture features are energy, correlation, homogeneity, contrast, entropy, computed from the cooccurrence matrix [116] (see Table 11.4).

We give three case descriptions by using (1) statistical gray-level features, (2) texture features, and (3) a combination of statistical and texture features.

Clustering based on the normalized city-block metric (see Sect. 11.4.5) and the average linkage method [117] were applied to a small data set, including images from different domains, to see how well the three case descriptions separate different cases and form groups of similar cases. The results are show in

Table 11.3. Statistical gray-level features

Feature name	Calculation	Feature name	Calculation
Mean	$\bar{g} = \sum_g g \cdot H(g)$	Variance	$\delta_g^2 = \sum_g (g-\bar{g})^2 H(g)$
Skewness	$g_s = \frac{1}{\delta_g^3} \sum_g (g-\bar{g})^3 H(g)$	Kurtosis	$g_k = \frac{1}{\delta_g^4} \sum_g (g-\bar{g})^4 H(g) - 3$
Variation Coefficient	$v = \frac{\delta}{\bar{g}}$	Entropy	$g_E = -\sum_g H(g) \log_2 H(g)$
Centroid_x	$\bar{x} = \frac{\sum_x \sum_y x f(x,y)}{\sum_x \sum_y f(x,y)} = \frac{\sum_x \sum_y x f(x,y)}{\bar{g} S}$	Centroid_y	$\bar{y} = \frac{\sum_x \sum_y y f(x,y)}{\sum_x \sum_y f(x,y)} = \frac{\sum_x \sum_y y f(x,y)}{\bar{g} S}$
First-order histogram	$H(g) = \frac{N(g)}{S}$ g is the intensity value N(g) is the number of pixels of intensity value g in the image S is the overall number of pixels		

Table 11.4. Texture features

Feature name	Formula
Energy	$E = \sum_{i=0}^{n} \sum_{j=0}^{n} c_{ij} \cdot c_{ij}$
Correlation	$C = \frac{\sum_{i=0}^{n} \sum_{j=0}^{n} (i-u_x) \cdot (j-u_y) \cdot c_{ij}}{s_x \cdot s_x}$
Local homogeneity	$H = \sum_{i=0}^{n} \sum_{j=0}^{n} \frac{1}{1+(i-j) \cdot (i-j)} \cdot c_{ij}$
Contrast	$Con = \sum_{i=0}^{n} \sum_{j=0}^{n} (i-j) \cdot (i-j) \cdot c_{ij}$

with $n = 2 \cdot ldGray - 1$, c_{ij}-Entry of Cooccurrence matrix

$u_x = \sum_{i=0}^{n} \sum_{j=0}^{n} i \cdot c_{ij}$, $u_y = \sum_{i=0}^{n} \sum_{j=0}^{n} j \cdot c_{ij}$

$s_x^2 = \sum_{i=0}^{n} \sum_{j=0}^{n} (i - u_x)^2 \cdot c_{ij}$, $s_y^2 = \sum_{i=0}^{n} \sum_{j=0}^{n} (i - u_y)^2 \cdot c_{ij}$

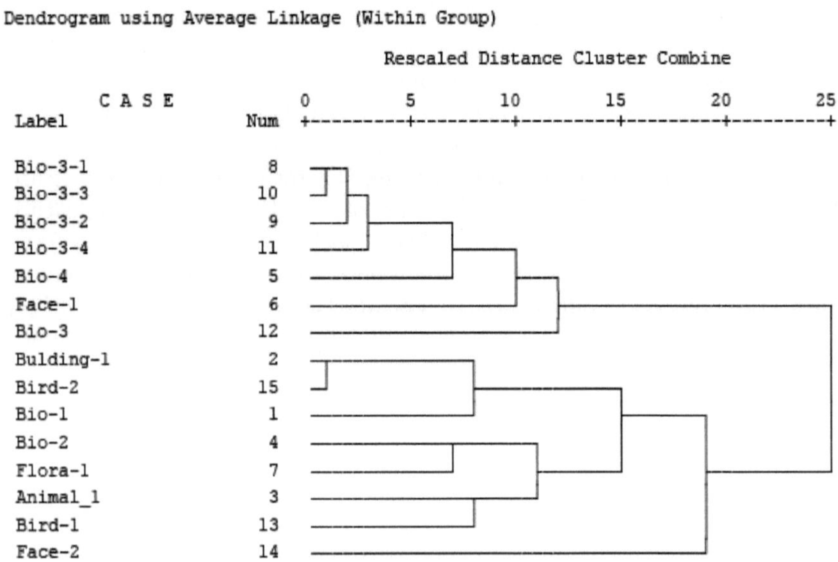

Fig. 11.9. Dendrogram for CBR based on statistical features

Fig. 11.9 for the statistical gray-level features, in Fig. 11.10 for the texture features, and in Fig. 11.11 for the combined feature set. Our expectation was that images, for which we got the best segmentation by using the same values of the parameters, would cluster into groups of similar images. By following this idea, we have to cut the dendrogram in Fig. 11.9 by the cophenetic-similarity value equal to seven. In this way, the first aggregation, where images (image *Bio-1* and image *Flora-1*) having different similarity measures meet, does

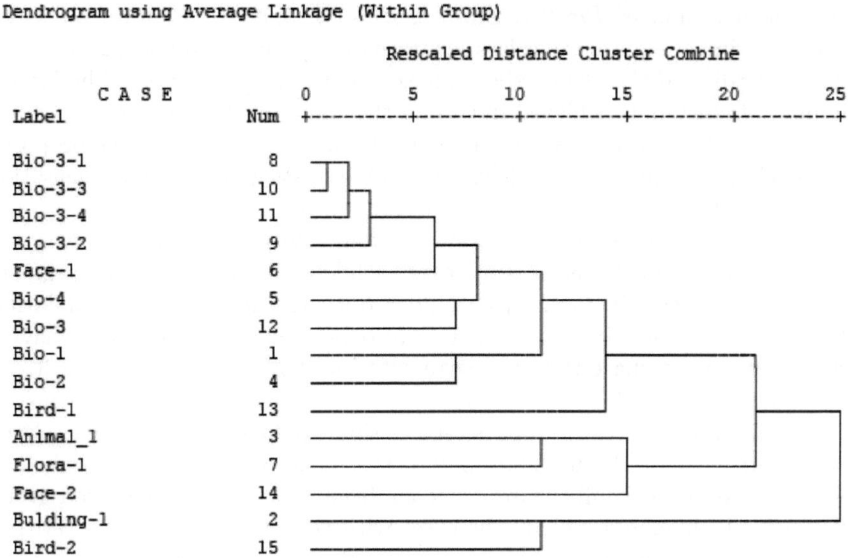

Fig. 11.10. Dendrogram for CBR based on texture features

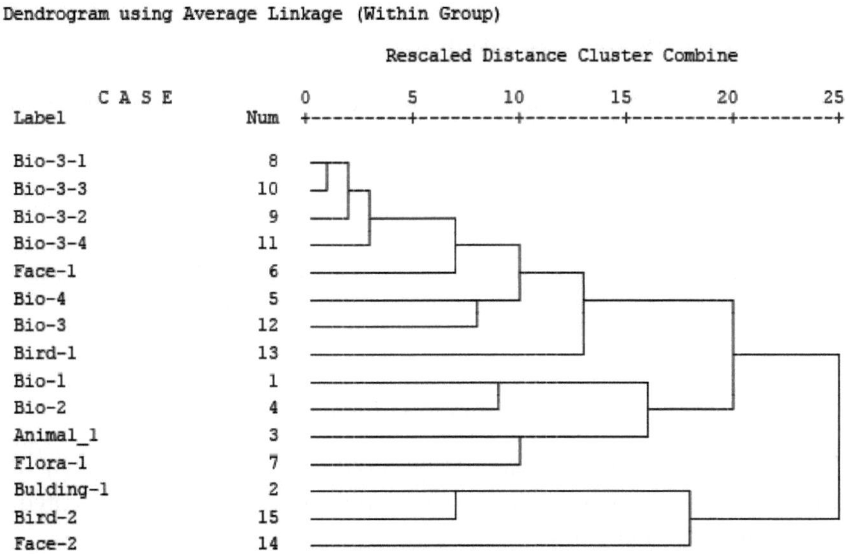

Fig. 11.11. Dendrogram for CBR based on texture and statistical features

not form a group. For the texture features in Fig. 11.10, we have to cut the dendrogram by the cophenetic-similarity value equal to six, to avoid merging of the images *Bio-1* and *Bio-2*. For the combined feature set in Fig. 11.11, the cut-off will be by the cophenetic-similarity value equal to seven, to avoid that images *Building-1* and *Bird-2* form a cluster. In all three approaches, the biological images *Bio-3-1, Bio-3-2, Bio-3-3, and Bio-3-4* form a cluster. The other images get more or less separate into groups with one case member only, regardless if they share the same segmentation parameters. The largest number of clusters is achieved in case of the combined feature set.

The results show that we can distinguish the images based on the proposed low-level image features and, thus, we are able to assign the best segmentation parameters to an input image with specific image characteristics.

From the retrieval point of view, it would be good to have only a few clusters with as many as possible case members sharing the same segmentation parameters. In fact, this would reduce the retrieval and similarity-determination time. In contrast, it is also possible to form groups having cases that do not share the same segmentation parameters, as long as it can support fast retrieval.

The hierarchy of case groups is used to single out cases that are not related to the current case. If the final node of the retrieval hierarchy is used, then searching the most similar case is accomplished within the associated group of cases, which should include the cases with the same image characteristics and the best segmentation parameters. The three dendrograms show that the similarity between all the cases is sensitive enough to achieve this task.

11.4.5 Similarity Determination

Similarity consists of two parts: nonimage similarity, Sim_N, and image similarity, Sim_I. The final similarity is computed by:

$$Sim = \frac{1}{2}(Sim_N + Sim_I) \tag{11.7}$$

In this work, we decided to weigh equally nonimage and image similarity so that they have the same influence on the final similarity. Only when both similarities have high values, the final similarity will be high.

Similarity Measure for Nonimage Information

The Tversky's similarity measure [118] is used for nonimage information. The similarity between a case C_i in the case base and a new case B presented to the system is computed as:

$$SIM_N = S(C_i,B) = \frac{|A_i|}{\alpha\,|A_i| + \beta\,|D_i| + \chi\,|E_i|} \tag{11.8}$$
$$\alpha = 1, \text{ and } \beta = \chi = 0.5$$

where A_i are the features common to both C_i and B, D_i are the features that belong to C_i but not to B, and E_i are the features that belong to B but not to C_i.

Similarity Measure for Image Information

We compute the image dissimilarity between two images A and B in the database of images as the complement to 1 of the distance $dist_{AB}$ between A and B. The distance between A and B is computed as follows:

$$dist_{AB} = \frac{1}{k}\sum_{i=1}^{K} w_i \left| \frac{C_{iA} - C_{imin}}{C_{imax} - C_{imin}} - \frac{C_{iB} - C_{imin}}{C_{imax} - C_{imin}} \right| \qquad (11.9)$$

where C_{iA} and C_{iB} are the values of the ith feature of A and B, respectively, C_{imin} and C_{imax} are the minimum and maximum value, respectively, of the ith feature of all images in the database, and w_i is the weight for the ith feature with $w_1+w_2+\ldots+w_i+\ldots+w_k = 1$. In our case, we assign the same value to all weights.

11.4.6 Automatic Evaluation of the Segmentation Results

The similarity measure developed in [113] can be used for two purposes (1) for image retrieval, based on the image matrix and (2) for the evaluation of the segmentation results.

In this study, we use this measure for the evaluation of the segmentation results. In fact, we are interested in investigating if, by using this measure, we can achieve an objective criterion to compare not only qualitatively different segmentation outcomes. Thus, we use this similarity measure to compare the quality of the obtained segmentation result to the expected result (e.g., the segmentation manually drawn by an expert). We call this image the gold standard.

The algorithm computes the similarity between two image matrixes (see Fig. 11.12). According to the specified distance function, the proximity matrix is calculated, for one pixel at position r,s in image A, to the pixel at the same position in image B and to the surrounding pixels within a predefined window. Then, the minimum distance between the compared pixels is computed. The same process is done for the pixel at position r,s in image B. Afterward, the average of the two minimal values is calculated. This process is repeated until all the pixels of both images have been processed. The final dissimilarity for the whole image is calculated from the average minimal pixel distance. The use of an appropriate window size should make this measure invariant to scaling, rotation, and translation.

We used the above similarity measure to evaluate our segmentation result for the running example. The gold standard was in case A the binarized gradient image of the original image and in case B the manually labeled image. Table 11.5 shows the similarity-values obtained when comparing the original image to the standard watershed-based image segmentation and to the outcome of the algorithm in [110]. The highest dissimilarity was found for the pair-wise computation original-to-standard watershed.

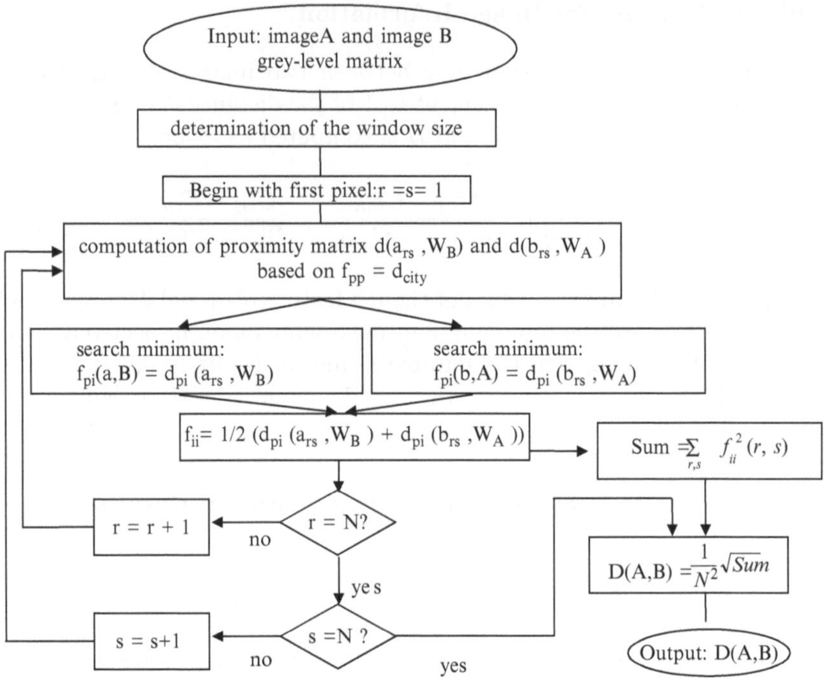

Fig. 11.12. Flowchart of the algorithm [113] for computing the similarity measure

Table 11.5. Evaluation of the segmentation results for the running example, based on the similarity measure in [113], for the partitions obtained by standard watershed and by the algorithm in [110]

	Original-to-original	Original-to-watershed segmentation	Original-to-segmentation by algorithm [110]
Case A	0	0.04656367	0.00808133
Case B	0	0.045201939	0.005551775

Table 11.6 shows the similarity-values obtained when comparing the output of the new algorithm to the original image, in correspondence with different values for the parameters a, b, and T. Based on the similarity-value we should be able to select the best segmentation parameters for the current input image. If this will work, it would also allow us to adjust the segmentation parameters by an automatic optimization procedure where the optimization function is the similarity-value.

The results in Table 11.6 show the best similarity value for the parameter combination $a = 0.75$, $b = 1.25$, and $T = 0.8$. This result confirms our evaluation of the performance, done by visual observation of the results.

Table 11.6. Evaluation of the different segmentation results for the running example, based on the similarity measure in [113], for the new algorithm

Segmentation Results	Parameters $a = 1$, $b = 1$, $T = 0.9$	Parameters $a = 1$, $b = 1$, $T = 0.7$	Parameters $a = 1$, $b = 1$, $T = 0.65$
Case A	0.009195196	0.008050811	0.00805929
Case B	0.006305389	0.005838306	0.00555007
Segmentation Results	Parameters $a = 0.75, b = 1.25, T = 0.8$		Parameters $a = 1.5, b = 0.5$, $T = 0.2$
Case A	0.008032165		0.008313606
Case B	0.005482262		0.006072278

Furthermore the similarity values in Table 11.6 show that values converge to a local minimum when the best possible segmentation is achieved for the chosen parameter combination. It should be possible to guide a search strategy for the automatic selection of the best segmentation-parameter combination.

It is interesting to note that we obtain the same results for the manually labeled image and the binarized gradient image.

11.4.7 Case Generalization and Similarity Learning

The aim of case generalization is to form groups of cases that can be represented by a group representative. This representative should be applicable to a wider range of images for segmentation. Then the cases belonging to the same group have to share the same solution.

Another aim of case generalization is to group cases in a way that it is efficient for retrieval. In this case, it is not necessary that all case members share the same solution.

A possible way to achieve case generalization is by case clustering. In Sect. 11.4.4, we used a conventional hierarchical clustering program, which did not produce many case generalizations on the current set of cases. Only the images *Bio-3-1, Bio-3-2, Bio-3-3*, and *Bio-3-4* can actually be grouped together in a single case class. For this case class, a more general case can be calculated, which would be the mean over all the cases in the group. The other cases remain as individual cases in the case base. Wider generalization will be possible only after more cases will have been inserted into the case base.

In conventional hierarchical clustering, the computed similarity-values are the cophenetic similarities, whose meaning is limited compared to that of absolute similarities. The dendrograms that we have shown are implicit representations of the hierarchy and are computed over the whole case base. In an offline phase, the explicit representation of the hierarchy has to be computed.

If a new case member is inserted in the case base, then clustering has to be done again on the whole case base.

A better way to deal with case generalization and build a higher-order construct (the relation among the clusters) is to do conceptual clustering [119]. During conceptual clustering, the concept hierarchy is built incrementally and, at the same time, the concept description is calculated and explicitly represented in the node of the hierarchy. The similarities are absolute similarities and other measures that describe the concept can be calculated as well.

The image-based similarity measure was described in Sect. 11.4.5. The measure is done by using the city-block metric with equal weights for each attribute. Learning the similarity can be done by learning the weights of the attributes [120]. The weights of each feature w_i are increased by a constant value δ, so that $w_i = w_i \pm \delta$. If the new weights produce an improvement of the retrieval accuracy, then the weights will be updated accordingly; otherwise, the weights will remain unchanged. After all weights have been tested, the constant δ will be divided by 2 and the weight updating procedure is repeated. The process terminates if the difference between the retrieval accuracy of two interactions is less than a predefined threshold.

11.5 Final Remarks

We have shown our approach for watershed-segmentation based on CBR. It is well known that the watershed transformation produces oversegmented images. Oversegmentation can be reduced by applying a merging process, which implies the use of several control criteria, based on characteristics extracted from the initial watershed-segmented image. The control scheme is usually expressed by rules, in the absence of a better theoretical understanding of the control mechanism. The similarity-based control scheme gives a more flexible way to handle the control criteria, depending on the image characteristics and allows understanding the behavioral mechanism by learning the parameters of the controller. Currently, the control scheme of our method is global. The same control scheme is applied to the entire image. We think that a local control scheme could be developed that uses image characteristics of local areas of the image, to control merging in those areas.

The best segmentation-parameter combination should automatically be determined based on a suitable parameter optimization methods [121]. We could show that the similarity measure presented in Sect. 11.4.6 is not only able to detect the best possible segmentation results but can also guide the iterative parameter tuning procedure [122] for the selection of the best parameter combination.

Learning the similarity between the cases, i.e., the weights in formula (11.9) in our case, seems to be a necessary task to face, in order to achieve the aim that images sharing the same image characteristics should have the same image segmentation parameters. These research activities are left for further work.

11.6 Conclusions

The CBR process can be applied to solve all aspects of images segmentation, from choosing the appropriate image segmentation method/parameters for the actual image up to the evaluation of the results, and to provide feedback to the system for performance improvement. Therefore, the model construction aspect for image segmentation can be handled very efficiently based on CBR. CBR is an incremental knowledge-acquisition method as well as a reasoning method. New situations can be captured in an efficient way and the behavior of the segmentation algorithm can be efficiently studied. New situations can be made available for reasoning as soon as they have been captured by the system. This allows the construction of a model for image segmentation that is applicable to wide range of images. We have described how CBR can be applied to watershed-based image segmentation by controlling the merging process. The case image description used for indexing and the similarity measure have been described. The results show that we are able to achieve better results for some groups of images.

More research has to be done on the definition of the proper image description. Image description based on statistical features might properly cover the information about image quality, but this is possibly not enough for watershed-based image segmentation. The introduction of texture features is a promising step in this direction and needs to be further investigated. Moreover, there might be other features, besides statistical and texture features, that could result as more appropriate.

In the future, we plan to investigate more extensively the automatic evaluation of the segmentation results, so as to better judge the quality of the results.

Acknowledgment

This work has been partially supported by the Italian National Research Council, CNR, in the framework of the Short Term Mobility Program 2006.

References

1. K.S. Fu, J.K. Mui, A survey on image segmentation, *Pattern Recognition*, 13, 1, 3–16, 1981.
2. R.M. Haralick, L.G. Shapiro, Image segmentation techniques, *Computer Vision, Graphics, and Image Processing*, 29, 1, 100–132, 1985.
3. N.R. Pal, S.K. Pal, A review on image segmentation techniques, *Pattern Recognition*, 26, 9, 1277–1294, 1993.
4. D.L. Pham, C. Xu, J.L. Prince, Current methods in medical image segmentation, *Annual Review of Biomedical Engineering*, 2, 315–337, 2000.

5. L. Lucchese, S.K. Mitra, Color Image Segmentation: A State-of-the-Art Survey, *"Image Processing, Vision, and Pattern Recognition," Proc. of the Indian National Science Academy (INSA-A)*, New Delhi, India, Vol. 67 A, No. 2, 207–221, 2001.
6. H.D. Cheng, X.H. Jiang, Y. Sun, J. Wang, Color image segmentation: advances and prospects, *Pattern Recognition*, 34, 2259–2281, 2001.
7. J. Freixenet, X. Muñoz, D. Raba, J. Martí, X. Cufí, Yet Another Survey on Image Segmentation: Region and Boundary Information Integration, Proc. 7th ECCV, LNCS 2352, Springer, 408–422, 2002.
8. P.K. Sahoo, S. Soltani, A.K.C. Wong, Y.C. Chen, A survey of thresholding techniques, *Comput. Vis. Graph. Im. Proc.*, 41, 233–260, 1988.
9. A.D. Brink, Grey-level thresholding of images using a correlation criterion, *Pattern Recognition Letters*, 9, 5, 335–341, 1989.
10. H. Luijendijk, Automatic threshold selection using histograms based on the count of 4-connected regions, *Pattern Recognition Letters*, 12, 4, 219–228, 1991.
11. R.J. Whatmough, Automatic threshold selection from a histogram using the "exponential hull", *CVGIP: Graphical Models and Image Processing*, 53, 6, 592–600, 1991.
12. W.-N. Lie, An efficient threshold-evaluation algorithm for image segmentation based on spatial graylevel co-occurrences, *Signal Processing*, 33, 1, 121–126, 1993.
13. L. Wang, J. Bai, Threshold selection by clustering gray levels of boundary *Pattern Recognition Letters*, 24, 12, 1983–1999, 2003.
14. J. Sauvola, M. Pietikäinen, Adaptive document image binarization, *Pattern Recognition*, 33, 2, 225–236, 2000.
15. O. Demirkaya, M.H. Asyali, Determination of image bimodality thresholds for different intensity distributions, *Signal Processing: Image Communication*, 19, 6, 507–516, 2004.
16. M.A. Patricio and D. Maravall, A novel generalization of the gray-scale histogram and its application to the automated visual measurement and inspection of wooden Pallets, *Image and Vision Computing*, 2006 (in press).
17. S. Chen, D. Li, Image binarization focusing on objects, *Neurocomputing*, 69, 16–18, 2411–2415, 2006.
18. P. Perner, An architecture for a CBR image segmentation system, *Journal of Engineering Application in Artificial Intelligence, Engineering Applications of Artificial Intelligence*,12-6, 749–759, 1999.
19. B.J. Schachter, L.S. Davis, A. Rosenfeld, Some experiments in image segmentation by clustering of local feature values, *Pattern Recognition*, 11, 1, 19–28, 1979.
20. M. Celenk, A color clustering technique for image segmentation, *Computer Vision, Graphics, and Image Processing*, 52, 2, 145–170, 1990.
21. J.C. Bezdek, L.A. Hall, L.P. Clarke, Review of MR image segmentation techniques using pattern recognition", *Med. Phys.*, 20, 1033–1048, 1993.
22. P. Schroeter, J. Bigün, Hierarchical image segmentation by multi-dimensional clustering and orientation-adaptive boundary refinement, *Pattern Recognition*, 28, 5, 695–709, 1995.
23. D. Comaniciu, P. Meer, Robust analysis of feature spaces: color image segmentation, *Proc. Society Conference on Computer Vision and Pattern Recognition* (CVPR'97) 750, 1997, 1997.

24. R.P. Velthuizen, L.O. Hall, L.P. Clarke, M.L. Silbiger, An investigation of mountain method clustering for large data sets, *Pattern Recognition*, 30, 7, 1121–1135, 1997.
25. E.J. Pauwels, G. Frederix, Finding Salient Regions in Images: Nonparametric Clustering for Image Segmentation and Grouping, *Computer Vision and Image Understanding*, 75, 1–2, 73–85, 1999.
26. K.B. Eom, Unsupervised segmentation of spaceborne passive radar images, *Pattern Recognition Letters*, 20, 5, 485–494, 1999.
27. J. Cutrona, N. Bonnet, M. Herbin, F. Hofer, Advances in the segmentation of multi-component microanalytical images, *Ultramicroscopy*, 103, 2, 141–152, 2005.
28. K. Hammouche, M. Diaf, J.-G. Postaire, A clustering method based on multidimensional texture analysis, *Pattern Recognition*, 39, 7, 1265–1277, 2006.
29. S. Filin, N. Pfeifer, Segmentation of airborne laser scanning data using a slope adaptive neighborhood, *ISPRS Journal of Photogrammetry and Remote Sensing*, 60, 2, 71–80, 2006.
30. J.-P. Gambotto, A new approach to combining region growing and edge detection, *Pattern Recognition Letters*, 14, 11, 869–875, 1993.
31. Il Y. Kim, H. S. Yang, A systematic way for region-based image segmentation based on Markov Random Field model, *Pattern Recognition Letters*, 15, 10, 969–976, 1994.
32. M.A. Wani, B.G. Batchelor, Edge-Region-Based Segmentation of Range Images, *IEEE Trans on PAMI*, 16, 3, 314–319, 1994.
33. N. Ito, R. Kamekura, Y. Shimazu, T. Yokoyama, Y. Matsushita, The combination of edge detection and region extraction in nonparametric color image segmentation, *Information Sciences*, 92, 1–4, 277–294, 1996.
34. X.M. Pardo, D. Cabello, Biomedical active segmentation guided by edge saliency, *Pattern Recognition Letters, 21, s 6–7, 559–572, 2000.*
35. X. M. Pardo, M.J. Carreira, A. Mosquera, D. Cabello, A snake for CT image segmentation integrating region and edge information, *Image and Vision Computing*, 19, 7, 461–475, 2001.
36. C.D. Kermad, K. Chehdi, Automatic image segmentation system through iterative edge–region co-operation, *Image and Vision Computing*, 20, 8, 541–555, 2002.
37. X. Muñoz, J. Freixenet, X. Cufí, J. Martí, Strategies for image segmentation combining region and boundary information, *Pattern Recognition Letters*, 24, 1–3, 375–392, 2003.
38. M.I. Rajab, M.S. Woolfson, S.P. Morgan, Application of region-based segmentation and neural network edge detection to skin lesions, *Computerized Medical Imaging and Graphics*, 28, 1–2, 61–68, 2004.
39. M. Mueller, K. Segl, H. Kaufmann, Edge- and region-based segmentation technique for the extraction of large, man-made objects in high-resolution satellite imagery, *Pattern Recognition*, 37, 8, 1619–1628, 2004.
40. Y. Zhou, J. Starkey, L. Mansinha, Segmentation of petrographic images by integrating edge detection and region growing, *Computers & Geosciences*, 30, 8, 817–831, 2004.
41. M.I. Rajab, M.S. Woolfson, S.P. Morgan, Application of region-based segmentation and neural network edge detection to skin lesions, *Computerized Medical Imaging and Graphics*, 28, 1–2, 61–68, 2004.

42. T. Chen, D. Metaxas, A hybrid framework for 3D medical image segmentation, *Medical Image Analysis*, 9, 6, 547–565, 2005.
43. I. Dydenko, F. Jamal, O. Bernard, J. D'hooge, I.E. Magnin, D. Friboulet, A level set framework with a shape and motion prior for segmentation and region tracking in echocardiography, Medical Image Analysis, 10, 2, 162–177, 2006.
44. S. Beucher, F. Meyer, 'The morphological approach of segmentation: the watershed transformation', in Dougherty E. (Ed.) *Mathematical Morphology in Image Processing*, Marcel Dekker, New York, 433–481, 1993.
45. P.K. Saha, J.K. Udupa, D. Odhner, Scale-Based Fuzzy Connected Image Segmentation: Theory, Algorithms, and Validation, *Computer Vision and Image Understanding*, 77, 2, 145–174, 2000.
46. Q. Wang, Z. Chi, R. Zhao, Image Thresholding by Maximizing the Index of Nonfuzziness of the 2-D Grayscale Histogram, *Computer Vision and Image Understanding*, 85, 2, 100–116, 2002.
47. G.C. Karmakar, L.S. Dooley, A generic fuzzy rule based image segmentation algorithm, *Pattern Recognition Letters*, 23, 10, 1215–1227, 2002.
48. L. Patino, Fuzzy relations applied to minimize over segmentation in watershed algorithms, *Pattern Recognition Letters*, 26, 6, 819–828, 2005.
49. W. Cai, S. Chen, D. Zhang, Fast and robust fuzzy *c*-means clustering algorithms incorporating local information for image segmentation, *Pattern Recognition*, 2006 (in press).
50. M. Grimnes, A. Aamodt, A two layer case-based reasoning architecture for medical image understanding, in I. Smith & B. Faltings (Eds.) *Advances in Case-Based Reasoning*, Springer Verlag, Berlin, 164–178 1996.
51. J. Jarmulak, Case-based classification of ultrasonic B-Scans: Case-base organisation and case retrieval, in B. Smyth and P. Cunningham (Eds.) *Advances in Case-Based Reasoning*, LNAI 1488, Springer Verlag. Berlin, 100–111, 1998.
52. P. Perner, Different Learning Strategies in a Case-Based Reasoning System for Image Interpretation, in B. Smyth and P. Cunningham (Eds.), *Advances in Case-Based Reasoning*, LNAI 1488, Springer Verlag, Berlin, 251–261, 1998.
53. R. Macura, K. Macura, MacRad: Radiology Image Resource with a Case-Based Retrieval System, in: M. Veloso and A. Aamodt (eds.), *Case-Based Reasoning: Research and Development*, Springer, Berlin, 43–45, 1995.
54. M. Haddad, K-P. Adlassnig, G. Porenta, Feasibility analysis of a case-based reasoning system for automated detection of coronary heart disease from myocardial scintigrams, *Artificial Intelligence in Medicine*, 9, 61–78, 1997.
55. M.C. Jaulent, C. Le Bozec, E. Zapletal, P. Degoulet, Case based diagnosis in histopathology of breast tumours. Medinfo. 9 Pt 1:544–8, 1998.
56. V. Ficet-Cauchard, C. Porquet, M. Revenu, CBR for the reuse of image processing knowledge: A recursive retrieval/adaption strategy, in K.-D. Althoff, R. Bergmann, L.K. Branting (Eds.) *Case-Based Reasoning Research and Development*, Springer, Berlin, 438–453, 1999.
57. A. Micarelli, A. Neri, G. Sansonetti, A case-based approach to image recognition, in E. Blanzieri and L. Portinale (Eds.) *Advances in Case-Based Reasoning*, Springer Verlag, Berlin, 443–454, 2000.
58. P. Perner, An Architecture for a CBR Image Segmentation System, *Journal on Engineering Application in Artificial Intelligence, Engineering Applications of Artificial Intelligence*, 12 (6), 749–759, 1999.

59. P. Perner, CBR Ultra Sonic Image Interpretation. in: E. Blanzieri and L. Portinale (Eds.), *Advances in Case-based Reasoning*, LNAI 1898, Springer Verlag, Berlin, 479–481, 2000.
60. P. Perner, Incremental Learning of Retrieval Knowledge in a Case-Based Reasoning System, in K.D. Ashley and D.G. Bridge (Eds.), *Case-Based Reasoning – Research and Development*, LNAI 2689, Springer Verlag, Berlin, 422–436, 2003.
61. P. Perner, Are case-based reasoning and dissimilarity-based classification two sides of the same coin? *Journal Engineering Applications of Artificial Intelligence*, 5/3, 205–216, 2002.
62. P. Perner, TH. Günther, H. Perner, G. Fiss, R. Ernst, Health Monitoring by an Image Interpretation System - A System for Airborne Fungi Identification, in P. Perner, R. Brause, H-G. Holzhütter (Eds.), *Medical Data Analysis*, LNCS 2868, Springer Verlag, Berlin, 64–77, 2003.
63. P. Perner, H. Perner, B. Müller, Similarity Guided Learning of the Case Description and Improvement of the System Performance in an Image Classification System, in S. Craw and A. Preece (Eds.), *Advances in Case-Based Reasoning*, LNAI 2416, Springer Verlag, Berlin, 604–612, 2002.
64. P. Perner, S. Jähnichen, Case Acquisition and Case Mining for Case-Based Object Recognition, in P. Funk and P.A. González Calero (eds.), *Advances in Case-Based Reasoning*, LNAI 3155, Springer Verlag, Berlin, 616–629, 2004.
65. P. Perner, A. Bühring, Case-Based Object Recognition, in P. Funk and P.A. González Calero (Eds.), *Advances in Case-Based Reasoning*, LNAI 3155, Springer Verlag, Berlin, 375–388, 2004.
66. X. Yong, D. Feng, Z. Rongchun, M. Petrou, Learning-based algorithm selection for image segmentation, *Pattern Recognition Letters*, 26 (8), 1059–1068.
67. S. Beucher, C. Lantuéjoul, Use of watersheds in contour detection, *Proc. Int. Workshop on Image Processing, Real-time Edge and Motion Detection/estimation*, Rennes, France, 12–21, 1979.
68. W.E. Higgins, E.J. Ojard, Interactive morphological watershed analysis for 3D medical images, *Computerized Medical Imaging and Graphics*, 17, 4–5, 387–395, 1993.
69. M. Baccar, L.A. Gee, R.C. Gonzalez, M.A. Abidi, Segmentation of range images via data fusion and morphological watersheds, *Pattern Recognition*, 29, 10, 1673–1687, 1996.
70. J. Sijbers, P. Scheunders, M. Verhoye, A. Van der Linden, D. van Dyck, E. raman, Watershed-based segmentation of 3D MR data for volume quantization, *Magnetic Resonance Imaging*, 15, 6, 679–688, 1997.
71. P.S. Umesh Adiga, B.B. Chaudhuri, An efficient method based on watershed and rule-based merging for segmentation of 3-D histo-pathological images, *Pattern Recognition*, 34, 7,1449–1458, 2001.
72. M.E. Rettmann, X. Han, C. Xu, J.L. Prince, Automated Sulcal Segmentation Using Watersheds on the Cortical Surface, *NeuroImage*, 15, 2, 329–344, 2002.
73. M.M.J. Letteboer, O.F. Olsen, E.B. Dam, P.W.A. Willems, M.A. Viergever, W.J. Niessen, Segmentation of tumors in magnetic resonance brain images using an interactive multiscale watershed algorithm1, *Academic Radiology*, 11, 10, 1125–1138, 2004.
74. Y.-L. Huang, D.-R. Chen, Watershed segmentation for breast tumor in 2-D sonography, *Ultrasound in Medicine & Biology*, 30, 5, 625–632, 2004.

75. J.E. Cates, R.T. Whitaker, G.M. Jones, Case study: an evaluation of user-assisted hierarchical watershed segmentation, *Medical Image Analysis*, 9, 6, 566–578, 2005.
76. R. Rodríguez, T.E. Alarcón, O. Pacheco, A new strategy to obtain robust markers for blood vessels segmentation by using the watersheds method, *Computers in Biology and Medicine*, 35, 8, 665–686, 2005.
77. Z. Wang, C. Song, Z. Wu, X. Chen, Improved watershed segmentation algorithm for high resolution remote sensing images using texture, Proc. IEEE Int Conf. IGARSS '05, 5, 3721–3723, 2005.
78. Y.-M. Li, X.-P. Zeng, A new strategy for urinary sediment segmentation based on wavelet, morphology and combination method, *Computer Methods and Programs in Biomedicine*, 2006 (in press).
79. N. Passat, C. Ronse, J. Baruthio, J.-P. Armspach, J. Foucher, Watershed and multimodal data for brain vessel segmentation: Application to the superior sagittal sinus, *Image and Vision Computing*, 2006 (in press).
80. J. Barraud, The use of watershed segmentation and GIS software for textural analysis of thin sections, *Journal of Volcanology and Geothermal Research*, 154, 1–2, 17–33, 2006.
81. J. Yan, B. Zhao, L. Wang, A. Zelenetz, L. H. Schwartz, Marker-controlled watershed for lymphoma segmentation in sequential CT images, *Medical Physics*, 33, 7, 2452–2460, 2006.
82. F. Meyer, S. Beucher, Morphological segmentation, *Journal of Visual Communication and Image Representation*, 1, 1, 21–46, 1990.
83. P.J. Soille, M.M. Ansoult, Automated basin delineation from digital elevation models using mathematical morphology, *Signal Processing*, 20, 2, Pages 171–182, 1990.
84. Ph. Salembier, Morphological multiscale segmentation for image coding, *Signal Processing*, 38, 3, 359–386, 1994.
85. L. Najman, M. Schmitt, Watershed of a continuous function, *Signal Processing*, 38, 1, 99–112, 1994.
86. F. Meyer, Topographic distance and watershed lines, *Signal Processing*, 38, 1, 113–125, 1994.
87. R. Adams L. Bischof, Seeded Region Growing, IEEE Trans. on PAMI, 16, 6, 641–647, 1994.
88. T. Viero, D. Jeulin, Morphological Extraction of Line Networks from Noisy Low-Contrast Images, *Journal of Visual Communication and Image Representation*, 6, 4, 335–347, 1995.
89. D. Wang, A multiscale gradient algorithm for image segmentation using watersheds, *Pattern Recognition*, 30, 12, 2043–2052, 1997.
90. A. Mehnert, P. Jackway, An improved seeded region growing algorithm, *Pattern Recognition Letters*, 18, 10, 1065–1071, 1997.
91. L. Shafarenko, M. Petrou, J. Kittler, Automatic watershed segmentation of randomly textured color images, IEEE Transactions on Image Processing, 6, 11, 1530–1544, 1997
92. J. Crespo, R.W. Schafer, J. Serra, C. Gratin, F. Meyer, The flat zone approach: A general low-level region merging segmentation method, *Signal Processing*, 62, 1, 37–60, 1997.
93. A.N. Moga, B. Cramariuc, M. Gabbouj, Parallel watershed transformation algorithms for image segmentation, *Parallel Computing*, 24, 14, 1981–2001, 1998.

94. E. Pratikakis, H. Sahli, J. Cornelis, Low level image partitioning guided by the gradient watershed hierarchy, *Signal Processing*, 75, 2, 173–195, 1999.
95. A. Bieniek, A. Moga, An efficient watershed algorithm based on connected components, *Pattern Recognition*, 33, 6, 907–916, 2000.
96. A. Bleau, L.J. Leon, Watershed-Based Segmentation and Region Merging, *Computer Vision and Image Understanding*, 77, 3, 317–370, 2000.
97. J. Weickert, Efficient image segmentation using partial differential equations and morphology, *Pattern Recognition*, 34, 9, 1813–1824, 2001.
98. J.B.T.M. Roerdink, A. Meijster, The watershed transform: definitions, algorithms and parallelization strategies, *Fundamenta Informaticae*, 41, 187–228, 2001.
99. N. Malpica, J.E. Ortuño, A. Santos, A multichannel watershed-based algorithm for supervised texture segmentation, *Pattern Recognition Letters*, 24, 9–10, 1545–1554, 2003.
100. J.-B. Kim, H.-J. Kim, Multiresolution-based watersheds for efficient image segmentation, *Pattern Recognition Letters*, 24, 1–3, pp 473–488, 2003.
101. H.T. Nguyen, M. Worring, R. van den Boomgaard, Watersnakes: Energy-Driven Watershed Segmentation, IEEE Trans. on PAMI, 25, 3, 330–342, 2003.
102. C. Rosito Jung, J. Scharcanski, Robust watershed segmentation using wavelets, *Image and Vision Computing*, 23, 7, 661–669, 2005.
103. C.G. Zhao, T.G. Zhuang, A hybrid boundary detection algorithm based on watershed and snake, *Pattern Recognition Letters*, 26, 9, 1256–1265, 2005.
104. S.E. Hernandez, K.E. Barner, Y. Yuan, Region merging using homogeneity and edge integrity for watershed-based image segmentation, Optical Engineering, 44, 1, 2005.
105. H. Sun, J. Yang, M. Ren, A fast watershed algorithm based on chain code and its application in image segmentation, *Pattern Recognition Letters*, 26, 9, 1266–1274, 2005.
106. M. Frucci, G. Ramella, G. Sanniti di Baja, Using resolution pyramids for watershed image segmentation, *Image and Vision Computing*, 2006 (in press).
107. V. Osma-Ruiz, J.I. Godino-Llorente, N. Sáenz-Lechón, P. Gómez-Vilda, An improved watershed algorithm based on efficient computation of shortest paths, *Pattern Recognition*, 2006 (in press).
108. A. Duarte, Á. Sánchez, F. Fernández, A.S. Montemayor, Improving image segmentation quality through effective region merging using a hierarchical social metaheuristic, *Pattern Recognition Letters, 27, 11 1239–1251, 2006.*
109. C. Rosito Jung, Combining wavelets and watersheds for robust multiscale image segmentation, *Image and Vision Computing*, 2006 (in press).
110. M. Frucci, Over segmentation Reduction by Flooding Regions and Digging Watershed Lines, *International Journal of Pattern Recognition and Artificial Intelligence*, World Scientific, Singapore, 20, 1, 15–38, 2006.
111. F. Kummert, H. Niemann, R. Prechtel, G. Sagerer, Control and explanation in signal understanding environment, *Signal Processing*, 32, 111–145, 1993.
112. J. Hunter, S. Little, A framework to enable the semantic inferencing and querying of multimedia content, *International Journal of Web Engineering and Technology*, 2-2/, 264–286, 2005
113. P. Zamperoni, V. Starovotov, How dissimilar are two gray-scale images, *Proc. 17th DAGM Symposium*, Springer, Berlin, 448–445, 1995.
114. D.L. Wilson, A.J. Baddeley, R.A. Owens, A new metric for grey-scale image comparision, *Interna. Journal of Computer Vision*, 24(1), 1–29, 1997.

115. H. Dreyer, W. Sauer, Prozessanalyse, Berlin, Verlag Technik, 1982.
116. R.M. Haralick, I. Dinstein, K. Shanmugam, Textural features for image classification, *IEEE Transactions on Systems, Man, and Cybernetics*, 3(11), 610–630, 1973.
117. A.K. Jain, R.C. Dubes, Algorithms for clustering data, Prentice-Hall, Inc. Upper Saddle River, NJ, USA, 1988.
118. A. Tversky, Feature similarity, *Psychological Review* 84(4), 327–350, 1977.
119. S. Jänichen, P. Perner, Conceptual clustering and case generalization of 2-dimensional forms, *Computational Intelligence*, 22 (3/4), 177–193, 2006.
120. D. Wettschereck, D.W. Aha, Weighting Features, in M.M. Veloso and A. Aamodt (Eds.), *Case-Based Reasoning Research and Development*, Springer-Verlag, 347–358, 1995.
121. A. Karimi, L. Miskovic, D. Bonvin, Iterative correlation-based controller tuning, *International Journal of Adaptive Control and Signal Processing*, 18 (8), 645–664, 2004.
122. P.E. Gill, W. Murray, M.H. Wright, Practical Optimization, Academic Press, San Diego, USA, 1981.

11.7 Appendix

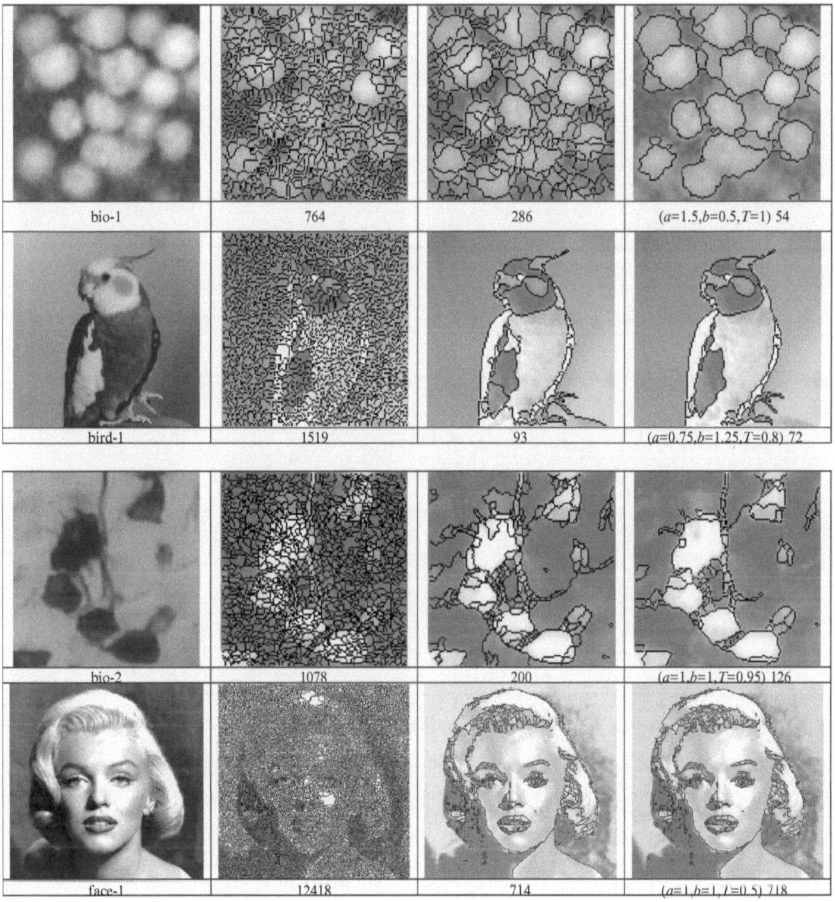

Fig. 11.13. Images bio-1, bird-1, bio-2 and face-1. For each test image, the original image, its watershed partition, the segmentation by using algorithm [110], and the segmentation obtained with the new algorithm are shown from left to right. For visualization purposes, the watershed lines are superimposed on images where the gray-levels have been complemented to 255.

11.7 Appendix

12

Similarity-Based Retrieval for Biomedical Applications

L.G. Shapiro[1], I. Atmosukarto[1], H. Cho[2], H.J. Lin[1], S. Ruiz-Correa[3], and J. Yuen[1]

[1] University of Washington, Seattle, WA 98195, USA
(shapiro@cs,indria@cs,hjlin@u,jenny@cs).washington.edu
[2] Svision, Inc., Bellevue, WA 98006, USA hansang@svisionllc.com
[3] Children's National Medical Center, Washington DC 20010, USA
srcorrea@cnmc.org

Summary. Similarity-based image retrieval, which has become an important area of computer vision, is a part of the case-based reasoning scenario. In similarity-based retrieval, a query image is provided and similar images from a database are retrieved, usually in order of similarity. In this chapter, we discuss the use of similarity-based retrieval for biomedical data. In particular, we describe three different applications that retrieve various types of image and signal data using similarity functions, including brain data (fMRI images and single-unit recording signals), mouse eye data (slit lens images), and skull data (CT scans). We define the similarity measures used in these applications and then discuss a unified query framework for multimedia data in general.

12.1 Introduction

Similarity-based image retrieval is part of the case-based reasoning scenario. It allows for the retrieval of images from a database that are similar in some way to a given query image. It has been used in case-based reasoning systems for both image segmentation and image interpretation. Whereas case-based reasoning focuses on incremental learning of prototypical case classes and index structures and on case mining, computer vision has been studying similarity measures, image indexing, and image features.

Similarity-based retrieval has become an important area of computer vision research. It has reached a state of maturity in which it is starting to be used in real commercial and medical applications. In the commercial domain, similarity measures are being used to organize both personal and institutional databases and to retrieve visually similar images for advertising and marketing. In the medical domain, similarity-based retrieval is used by physicians who want to compare imaging studies from a new patient to those from a database of prior patients to help them determine the diagnosis and potential

L.G. Shapiro et al.: *Similarity-Based Retrieval for Biomedical Applications*, Studies in Computational Intelligence (SCI) **73**, 355–387 (2008)
www.springerlink.com

treatment options. Furthermore, research scientists in the biomedical domain are starting to use similarity measures to organize and retrieve data from large-scale experiments involving multiple types of image and signal data. We have collaborated with three different biomedical research groups who needed similarity-based classification and retrieval systems to aid in their scientific research. In this chapter, we discuss each of these applications, describe the retrieval systems we have developed for them, and suggest the need for a unified query formulation for a general multimedia information retrieval system for biomedical applications.

Section 12.2 provides an account of previous work on content-based retrieval systems. Section 12.3 describes a brain data retrieval system that retrieves neuronal signals and fMRI data from a database of patients who have had surgery for epileptic tumor resection. Section 12.4 describes a skull retrieval system that retrieves skull images from a database of CT images. Section 12.5 describes a mouse eye retrieval system that retrieves images in order to study the cataract development in the eyes. Section 12.6 describes our current work in developing a unified query framework for multimedia biomedical data retrieval. Finally, Sect. 12.7 provides a summary of this chapter.

12.2 Related Work

Content-based image retrieval (CBIR) has been a heavily studied area of computer vision for the past 10 years including both general [48, 49] and medical [31] retrieval systems. The majority of these systems retrieve according to the "query by example" paradigm; that is, the user provides or selects a *query image* and chooses a distance measure (or combination of such measures) that will be used to compare the query image to the images stored in the database. The system retrieves those database images that are judged "similar to" the query image by the distance measure. Usually the retrieved images are returned in order of similarity to the query, and the user has the capability to browse through them in this order. Much of the research work in this area has related to the design of distance measures for accurate retrieval in various application domains. A smaller amount of work has considered the efficiency of retrieval systems, and some indexing methodologies have been developed [3, 6, 8, 9, 46, 47] as well as incremental learning methods for the index structure and the prototypical representation of case classes [17, 36].

In recent years image databases have gone from theory to reality. There are commercial systems, such as IBM's QBIC [14] and Virage's image search engine [4] and research systems including PhotoBook [34], Chabot [32], ImageRover [44], VisualSEEk [50], WebSEEk [51], MARS [28], and our own FIDS [6]. The most common features used for retrieval in these systems are color histograms, texture measures, and simple shape measures. They do not recognize specific objects and are not intended for biological applications.

Region-based systems such as Blobword [7], NETRA [24], and the composite region template (CRT) system [52] perform segmentation based on color-texture properties to identify regions of interest, which can be used in spatial-relationship queries. Some systems employ relevance feedback where the system refines its concept of the user's query according to the user's feedback [29,33,40]. Both regions of interest and relevance feedback are useful in a biological multimedia database system.

There have been a number of systems developed for medical image retrieval [1,2,13,20,30,31,37,46,56,57] as well as for case-based reasoning [15,18,25,35]. Most of these systems were developed for a particular type of retrieval and did not attempt to work with multiple data types. Kelly and Cannon [19] used a global texture signature to characterize images of diseased lungs and a signature distance function to compare them. Chu and Cardenas [12] developed the KMeD (knowledge-based multimedia medical distributed database) system. This large and ambitious project allows querying of medical multimedia data by both image and alphanumeric content and allows for the modeling of both spatial and temporal content of medical objects. Liu et al. [23] have worked on retrieval of brain images with abnormalities, while Perner [35] worked on retrieval of CT brain images based on the image characteristics as well as on patient data for the automatic determination of degenerative brain diseases. Tagare et al. [38,54,55] have studied content-based medical image retrieval for a number of years. Their current research is on high-dimensional indexing in medical image databases applied to cervical and lumbar spine X-ray images, where retrievals are based on shape features. Tang et al. developed a system for histological image retrieval [56], while Zheng et al. designed a system for pathology image retrieval [57]. Shyu et al. [46] developed a structure called the statistical k-d tree to allow for rapid searches in content-based retrieval. More recently they have developed a real time protein structure retrieval system with a multidimensional index [9,47].

12.3 Brain Data Retrieval for the Study of Language Sites in the Brain

During surgery for epileptic tumor resection, a technique called cortical stimulation mapping (CSM) is used to avoid areas on the cortical surface that are essential for language. The technique involves bringing the patient to an awakened state during surgery and presenting text, pictorial, or audio cues of a familiar object for the patient to name. During this process an electrical current is applied to selected *cortical surface sites*, marked by placing small numbered tags on the surface. The electrical current results in a short-term localized disruption of neural function. If this stimulation results in a naming error, then the cortical surface site at which it occurs is marked as essential for language function and is avoided during surgery.

In addition to their clinical use, the CSM language sites are a valuable source of data for understanding language organization in the brain, since their location is highly variable from one patient to the next. Studies have shown that the distribution of sites is correlated with such factors as sex and verbal IQ. The hope is that further insight into language organization can be obtained by correlating these sites, not only with additional demographic data, but also with anatomical features and other measures of language such as functional magnetic resonance imaging (fMRI) and single unit recording (SUR) of individual neurons.

12.3.1 Experimental Procedure

Prior to surgery, each patient undergoes an fMRI scan while being shown the same stimuli as in the later surgical CSM studies (text, audio, pictures). Following each stimulus the scanner generates a time series of three-dimensional volumetric data, where each voxel represents the time course of cerebral blood flow at that voxel. Subsequent statistical processing generates additional three-dimensional volumes, in which each voxel represents the probability that a statistically significant change in blood flow occurred between a stimulus and corresponding control. SUR studies are performed during surgery. As in the CSM studies the locations of the electrodes are marked by numbered tags associated with cortical surface sites. The same stimuli are presented as for fMRI and CSM, but in this case the electrodes record the firing patterns of individual neurons in response to the stimuli. Later processing looks for significant changes in firing rates between stimuli and controls, as well as changes in the firing patterns.

In order to correlate these diverse data sources for a patient, they are registered to a three-dimensional model of the patient's brain, using a structural MRI image volume acquired at the same time as the fMRI volumes. Population data are correlated by warping individual patient brain anatomy to a common "canonical" brain, in the process carrying along the registered functional data. All of the data are stored in several databases that are beginning to be used to help understand the relationships among various factors related to language.

12.3.2 Preprocessing and Feature Extraction

Both the signal data and fMRI data are preprocessed before they can be used in our retrieval system. The raw neuron spike data often contains spikes from two or more separate neurons. This signal data goes through a process called *spike sorting* in which the multiple neuron signals are separated. We have developed a template-extracting method to discriminate the neuron spikes using a rotated principal component analysis algorithm [10]. The raw fMRI data is also noisy. We have developed a method for dynamically detecting the

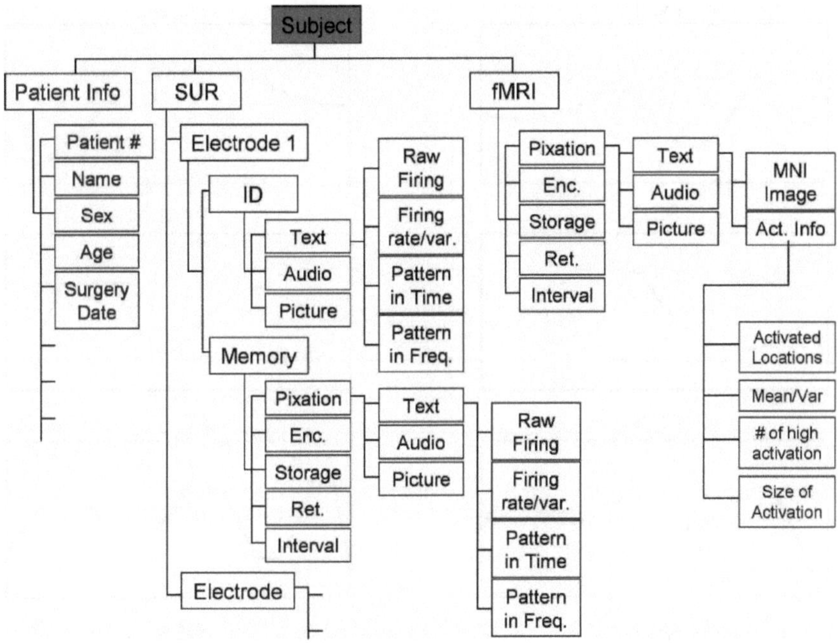

Fig. 12.1. Data structure

activations (areas of high-blood flow) using a maximal overlap wavelet transform to extract hemodynamic responses with minimum shape distortion and a dynamic time-warping algorithm to classify the different types of dynamic waveforms [11].

Figure 12.1 shows the multimedia data that must be stored and retrieved for this application. In addition to standard textual patient information there is SUR data and fMRI data for each patient. The SUR data includes, for each stimulated electron, its ID, the stimuli shown to the patient, the firing pattern in the time and frequency domains, and several numeric statistics that we compute. The fMRI data also has stimuli and statistics and includes the four-dimensional activation data (three-dimensional image over time). The activation level at each voxel over time is the important characteristic of this data.

12.3.3 Similarity-Based Retrieval

Retrieval of signals is based on prior matching of the raw signal data to eight signal templates. The templates come from application of a rotated principal component analysis method [10]. They are shown in Fig. 12.2 for the time domain only. A query signal is decomposed and the two best-matching templates are returned. From these, all signals in the database that match these two templates can be examined.

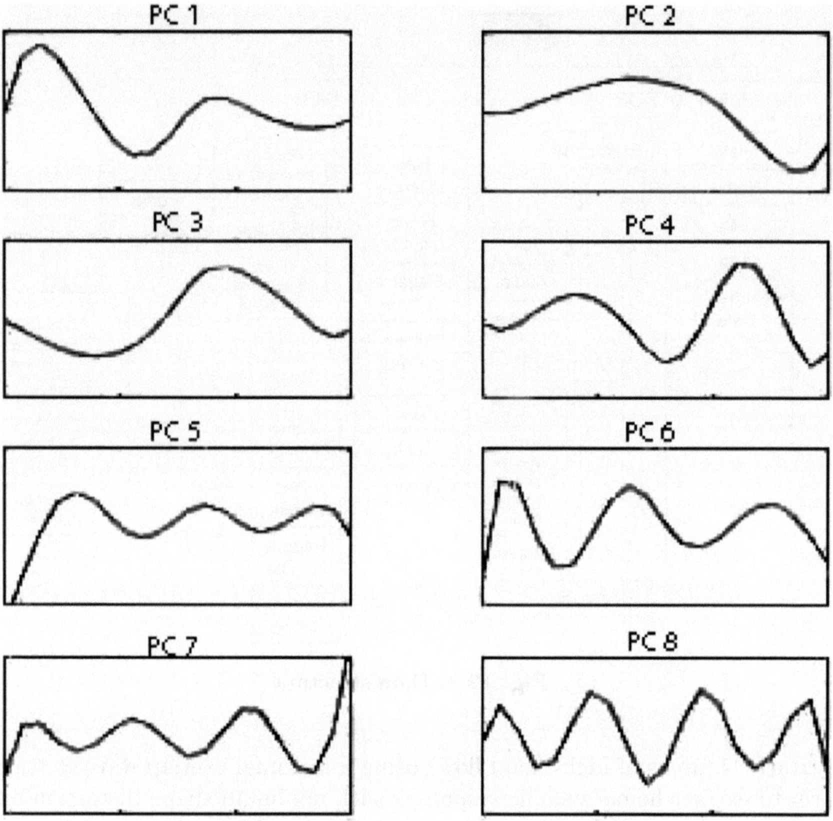

Fig. 12.2. Eight templates in the temporal domain

Figure 12.3 illustrates the results of a query for retrieving firing patterns. The top box shows the patient, microelectrode, and trial protocol information, and the second box gives the firing rate. The third box shows the best and second best template matches to the temporal firing pattern, and the fourth box shows the best and second best template matches to the frequency firing pattern. Users can extend the query by asking for similar results to those retrieved by selecting one of the two options at the bottom of the screen (a) similar firing patterns from a SUR or (b) related fMRI results if they exist.

Figure 12.4 shows the results of a query to display both firing patterns from neurons and the related hemodynamic response from the fMRI image. The top box gives patient information, the closest two activated brain areas in the UW parcellation scheme, and the Brodman area in which the results lie, and the second box gives the average firing rate. The third box shows best template matches to the temporal and frequency firing patterns. The bottom box shows the hemodynamic response that occurred in the most highly activated voxel on the brain surface.

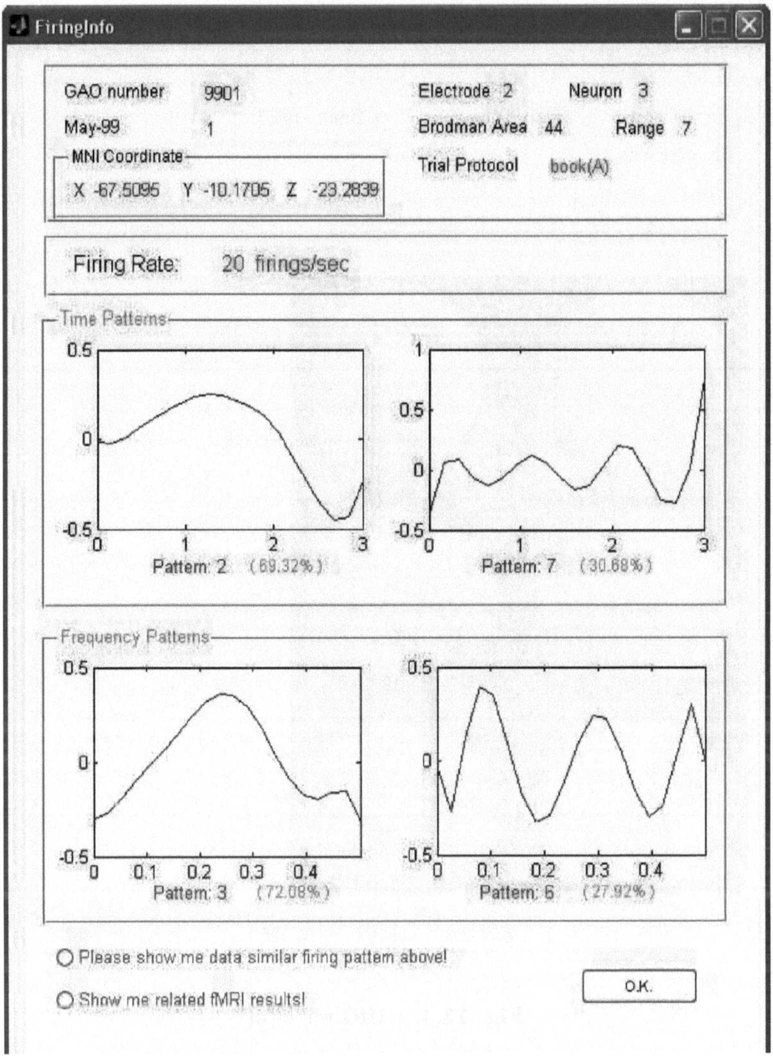

Fig. 12.3. Firing pattern retrieval

Figure 12.5 shows the results of an fMRI activation query. The top box shows the fMRI number, the MNI coordinates of the point with highest activation, and the closest two activated brain areas in the UW parcellaton scheme. The images show the highly activated areas in red.

The query system was custom built for this application. There are three main routes that a query can take (1) query by patient information, (2) query by trial protocols, and (3) query by firing patterns and/or fMRI activations. The third route is content based, while the other two are text based. The

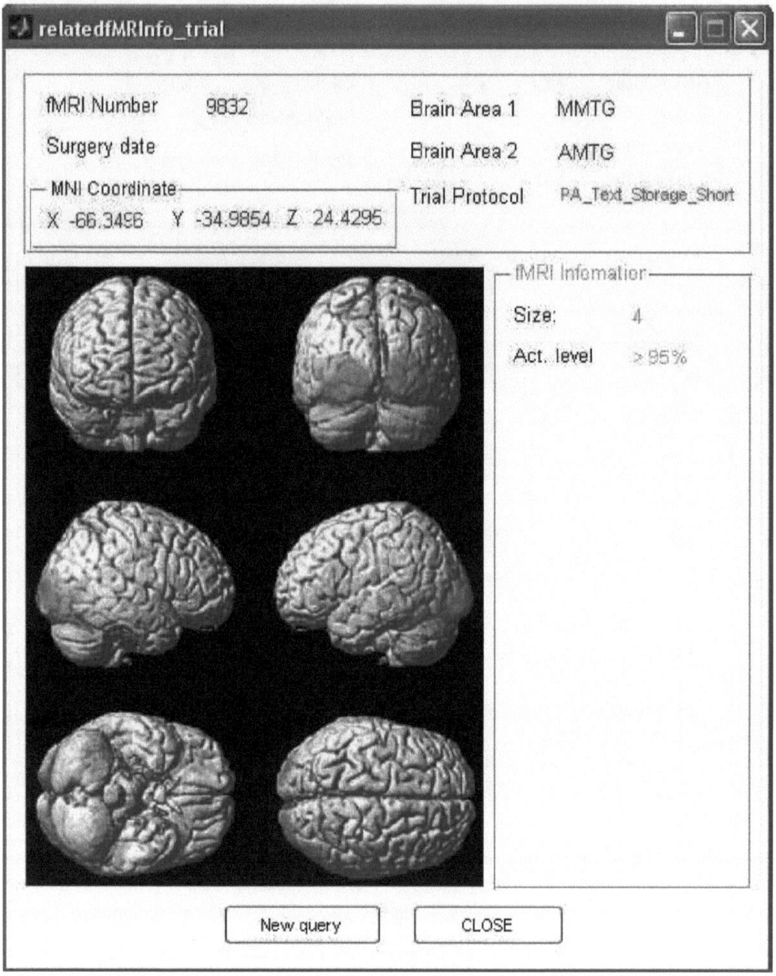

Fig. 12.4. FMRI retrieval

system allows the user to extend the basic query and move between the different query types as needed. A number of different queries have been used to illustrate the system, as summarized below:

1. Query by patient information with SUR characteristics
2. Query by patient information with fMRI characteristics
3. Query by trial protocols of SUR data
4. Query by trial protocols of fMRI data
5. Query by common trial protocols of SUR and fMRI data
6. Query by firing patterns of SUR data

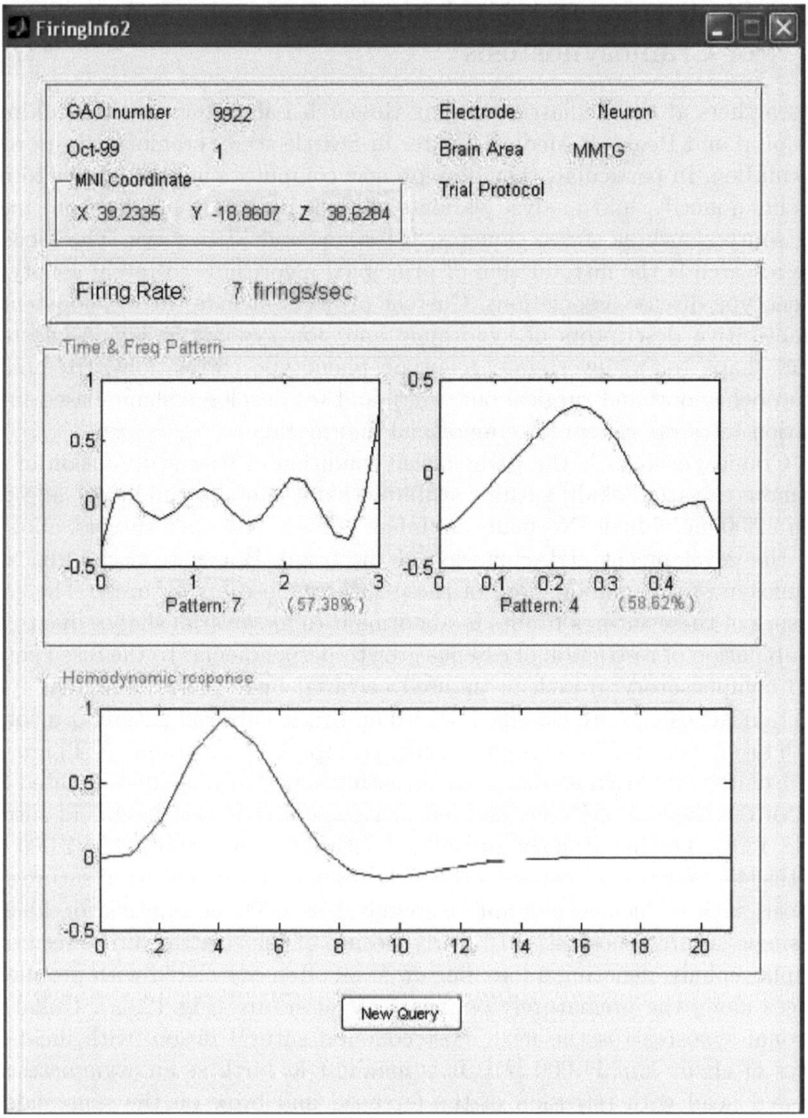

Fig. 12.5. Firing pattern and hemodynamic responses retrieval

7. Query by fMRI activations
8. Query by common features in both SUR and fMRI data

The prototype query system was developed for the scientists studying this particular application. It will become part of a more general system we are developing under National Science Foundation support.

12.4 Skull Data Retrieval for Studying the Effects of Craniosynostosis

Researchers at the Pediatric Imaging Research Laboratory of the Children's Hospital and Regional Medical Center in Seattle study craniofacial disorders in children. In particular, they develop new computational techniques to represent, quantify, and analyze variants of biological morphology from imaging sources such as stereo cameras, CT scans and MRI scans. The focus of the research is the introduction of principled algorithms to reveal genotype–phenotype disease associations. Current projects include the development of quantitative descriptors of syndromic and non syndromic craniosynostotic head shape to (a) investigate genotype–phenotype correlations, (b) predict neurobehavioral and surgical outcomes, and (c) develop a shape-based information retrieval system for craniofacial information.

Craniosynostosis is the pathological condition of premature fusion of one or more calvarial (skull) sutures or fibrous skull joints in childhood, affecting 1 in 2,500 individuals. Normally, an infant is born with open sutures, allowing for the development and expansion of the brain. However, in children with craniosynostosis, one or more of these sutures close prematurely. The early closure of these sutures results in abnormalities in calvarial shapes due to the combination of restriction of osseous growth perpendicular to the fused suture and compensatory growth in unfused calvarial bone plates. Different types of craniosynostosis are classified according to the calvarial suture(s) involved with fusion detected in three-dimensional computed tomography (CT) images.

Isolated craniosynostosis, or single-suture synostosis, includes isolated fusion of the sagittal, metopic, and left or right unicoronal or lambdoid sutures (Fig. 12.6). The incidence of an isolated suture fusion is about 1 in 2,000 live births [45]. Sagittal synostosis is the most common form of isolated suture synostosis with an incidence of approximately 1 in 5,000, accounting for 40–60% of single-suture synostosis [21]. Early closure of the sagittal suture results in scaphocephaly, denoting a long narrow skull often associated with prominent ridges along the prematurely ossified sagittal suture (Fig. 12.6a). Unilateral coronal synostosis is the next most common sutural fusion, with incidence rates of about 1 in 11,000 [21]. It is manifest at birth as an asymmetrically skewed head with retrusion of the forehead and brow on the same side as the fused suture and with compensatory bulging of the forehead on the side opposite the fused suture (Fig. 12.6b). Metopic synostosis is less common and affects 1 in 15,000 individuals [21]. The premature fusion of the metopic suture produces trigonocephaly, denoting a triangular shaped head with prominent frontal crest (Fig. 12.6c). The degree of skull shape deformity and changes in frontal and occipital bulging and biparietal narrowing can vary significantly between individuals even in the same disease class. In most cases of single-suture synostoses, a single surgery, i.e., cranioplasty, is required to release the fused suture and reshape the deformed calvaria. This surgery is preferentially performed within the first year to capitalize on the malleability of the infant's skull and to minimize secondary facial deformation [26, 27].

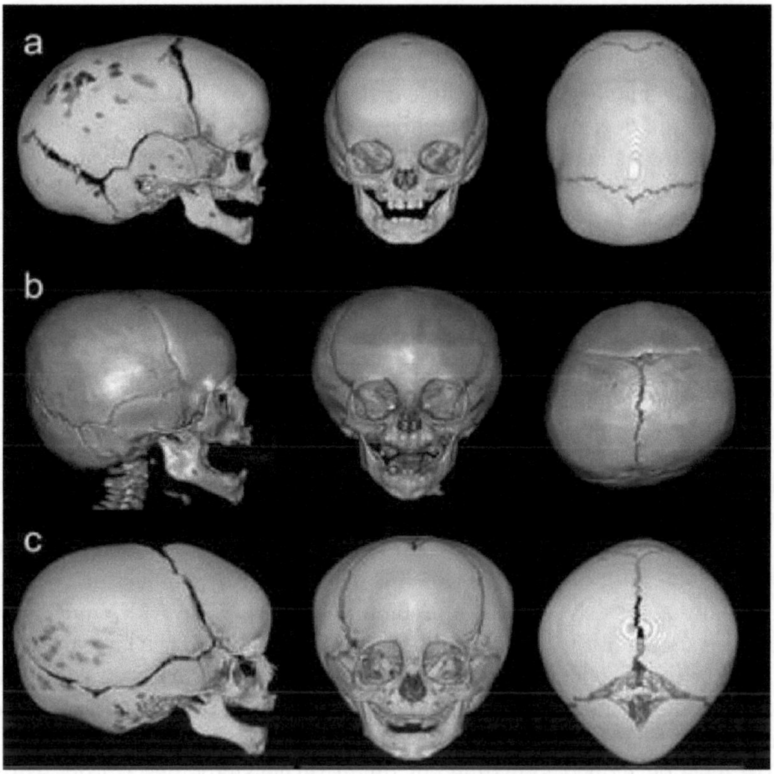

Fig. 12.6. Lateral, frontal, and top views of (**a**) a patient with sagittal synostosis, (**b**) a patient affected with unicoronal synostosis, and (**c**) a patient affected with metopic synostosis

Whether or not different calvarial shapes directly affect the severity of the cases, neuropsychological development, complications during cranioplasty, and postsurgical long-term outcomes for cases of single-suture synostoses remain largely unknown. In addition, the assessment of surgical "success" has been subjective, and no standardized method has been established for the post-surgery evaluation. However, surgeons and craniofacial experts often use cases of similar skull shapes from their past experience as guidelines in the preparation and evaluation of the reconstruction of the skull. This "case-based" study of reasoning makes it possible to reuse prototype images and diagnoses of previously resolved cases to assess the possible surgical complications and the postsurgery outcomes of the new cases. Therefore, the case-based clinical decision support technique produces a need to retrieve similar images of shapes in patients with single-suture synostosis objectively and reproducibly. In addition, medical images, especially the diagnostic images, are produced in ever-increasing quantities. Searching through a large collection of digital images by keyword indexing, or simply by browsing, may be time-consuming and may not satisfy physicians' needs for retrieving cases of similar shapes. Therefore, it is critical to design a shape-based image retrieval system to not only ease the management of clinical data but to aid the radiologists and surgeons in the decision making analysis of the reconstruction of the skull.

In order to design an image-based clinical decision support system for surgeons and craniofacial experts, it is critical to first develop shape descriptors that will enable objective and reproducible detection and quantification of similarities and differences in the three-dimensional skull shapes. We have developed several different quantitative shape descriptors for our retrieval system.

12.4.1 Shape Descriptors

Our shape descriptors were computed from CT image slices obtained with skull imaging. To standardize our computations, we used a calibrated lateral view of a three-dimensional reconstruction of the skull to select three CT planar slices defined by internal brain landmarks (Fig. 12.7) These planes are parallel to the skull base plane, which is determined by connecting the frontal nasal suture anteriorly and the opisthion posteriorly. The A, F, and M planes we use are shown in Fig. 12.7. The A-plane is at the top of the lateral ventricle, the F-plane is at the Foramina of Muntro, and the M-plane is at the level of the maximal dimension of the fourth ventrical. Oriented outlines were extracted from the CT image at each of these planes.

We have developed four different quantitative shape descriptors, three of which are numeric [41–43] and one of which is symbolic [22]. The numeric shape descriptors range from a single number per planar slice to a large matrix of numbers. The symbolic shape descriptor (SSD) is a vector of probabilities obtained from a bag-of-words approach to shape description. The shape descriptors we have developed to characterize skull morphology in craniosynostosis are summarized below.

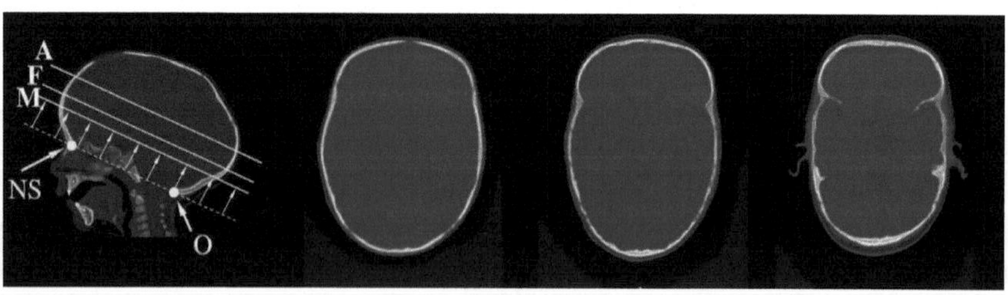

Fig. 12.7. All shape descriptors are constructed using the CT image slices extracted from skull imaging. These bone slides are defined by selecting the base plane on the three-dimensional image and finding internal anatomical landmarks on cerebral ventricles

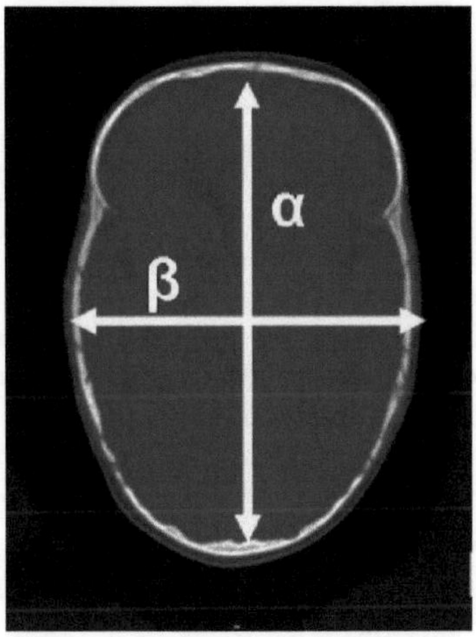

Fig. 12.8. The scaphocephaly severity index is computed as the head width to length ratio β/α as measured on a CT plane extracted from skull shape imaging

1. *Scaphocephaly severity index* (*SSI*). SSI is similar to the cephalic index used as the current clinical standard to measure the variation in head shapes in patients with isolated craniosynostosis. However, instead of measuring the ratio of the head width to the length by the measurer's subjective judgment, SSI takes the ratio, β/α, computed at an outline extracted from CT image slices from skull imaging (Fig. 12.8) [43]. The SSI assumes that a skull shape can be approximated as an ellipse that has eccentricity

$$e = \sqrt{1 - \frac{\beta^2}{\alpha^2}} \tag{12.1}$$

Note that if $e = 0$, $\beta/\alpha = 1$ and the outline shape would be a perfect circle. An eccentricity value close to 1 would suggest a very narrow outline shape. To increase the sensitivity of the shape description using SSI, users can combine the SSI measurements from multiple outlines extracted from skull images to represent the three-dimensional data. Depending on the performance evaluation metric, either individual or combined SSI values can be used as shape descriptors. This measurement is very simple and efficient to compute, but it disregards a broad range of shape variations that may be of importance in capturing skull morphology.
2. *Cranial spectrum* (*CS*). CS is a Fourier-based shape representation that describes a skull shape with the magnitude of the Fourier series coefficients of a periodic function [42]. The planes extracted from CT images are oriented outlines that have directions defined by their corresponding tangent (T) vectors (Fig. 12.9b). An oriented outline can be represented

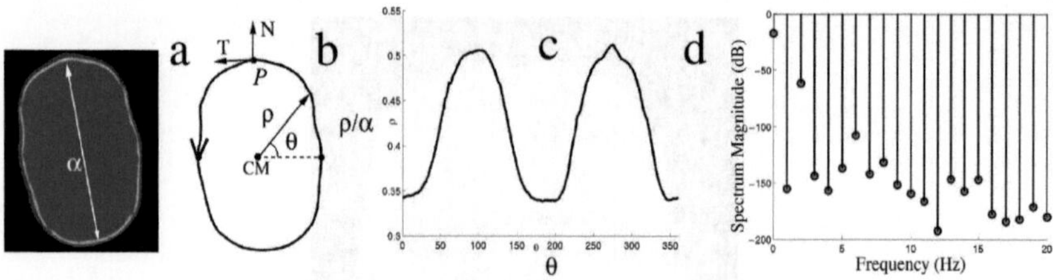

Fig. 12.9. (**a**) Bone CT slice at the level of the A-plane; (**b**) oriented outline counter clockwise direction; (**c**) same outline represented in polar coordinates (ρ, θ); and (**d**) 21 components of the corresponding cranial spectrum. Key: α (maximum outline length), T (tangent vector), N (normal vector), and (CM) center of mass

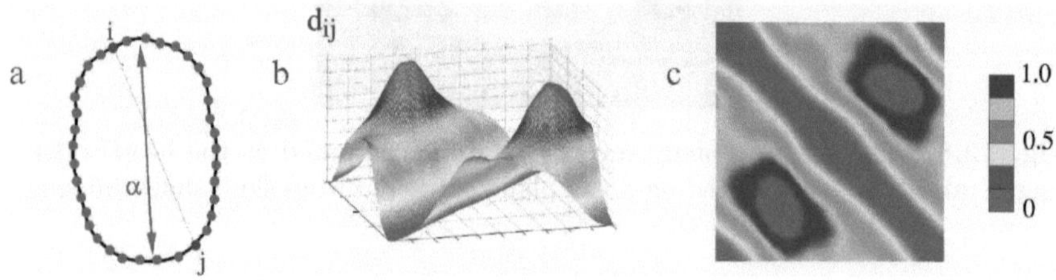

Fig. 12.10. (**a**) Oriented contour represented as a sequence of N evenly spaced points. (**b**) Cranial image. (**c**) Top view and normalized distance scale

as a periodic function by using polar coordinates with the center of mass as the origin of the coordinate system (Fig. 12.9c). This periodic function is then decomposed into a weighted sum of basis outlines by Fourier series. The coefficients from this decomposition then constitute the resulting shape descriptor. This representation encompasses shape information that cannot be captured by the SSI ratios. It is also closely related to traditional DFT-based descriptors [39]. The aim of Fourier analysis as applied here is to decompose an outline shape into a weighted sum of basis outlines, where each basis stratifies particular geometric features of a shape.

3. *Cranial image (CI).* The CI descriptor is a matrix representation of pair-wise normalized square distances computed for all the vertices of an oriented outline that has been discretized into N evenly spaced vertices [41]. Let D be a symmetric matrix with elements $D_{ij} = d_{ij}/\alpha$, for $i, j = 1 \cdots, N$, where d_{ij} is the Euclidean distance between vertices i and j, α is the maximum length of the contour (Fig. 12.10), and N is an arbitrary number defined by the user. Since the outline is oriented, the vertices can be sequentially ordered up to the selection of the first vertex. As a consequence, the matrix D is defined up to a periodic shift along the main diagonal. The definition of CI can be extended to incorporate an arbitrary number of oriented outlines by computing inter and intra-oriented outline distances for each of the vertices of all of the outlines representing a skull.

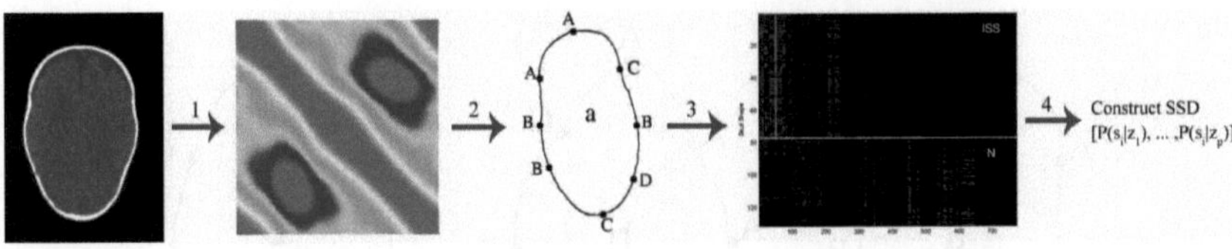

Fig. 12.11. Construction of the SSD representation of an extracted skull outline. 1 = cranial image representation, 2 = k-means clustering, 3 = cooccurrence matrix construction, 4 = PLSA analysis

4. *Symbolic shape descriptor* (*SSD*). The SSD algorithm was developed to model morphological variations that characterize different synostotic skull shapes [22]. It uses a *bag of words* (BOW) approach to capture the local shape variations. Even though this BOW approach has the advantage of producing a simple data representation, it creates a high-dimensional feature space and subsequently hinders efficient statistical analysis and image retrieval. In order to overcome these challenges, we use *probabilistic latent semantic analysis* (PLSA) to capture the cooccurrence information and relationship between elements in the BOW in order to reduce the high-computational complexity from the BOW representation.
   The construction of the SSDs from skull imaging involves several steps, as illustrated in Fig. 12.11. In brief, the cranial image distance matrices are computed for all skull outlines in the training set. K-means clustering of the rows of the distance matrices are applied and yields a set of clusters of the outline points. A symbolic cluster label is assigned to each outline vertex. A cooccurrence matrix is then computed for all training data using the strings of symbols constructed from the cluster labels of the vertices. Finally, PLSA is employed to reduce the dimension of the cooccurrence matrix and to construct the SSD.
   The input of the SSD algorithm is a set of skull shapes $\mathcal{S} = \{S_1, \cdots, S_M\}$. Each skull shape is represented by L oriented outlines, and each outline is discretized into N evenly spaced vertices. For the sake of simplicity and without loss of generality, we assume that $L = 1$. The feature generation algorithm is as follows:
   (a) For each shape S_j in $\mathcal{S}$ and each vertex v_i of S_j, compute the vector of distances from all other vertices of S_j to v_i. This vector is the same as the ith row of the cranial image matrix descriptor (Fig. 12.10).
   (b) Cluster all vectors in S by the k-means clustering algorithm with user-selected k and assign each cluster a symbolic label. Each vertex receives the label of its cluster.
   (c) Compute a *BOW* representation of the skull outlines in $\mathcal{S}$. More specifically, the symbols associated with the vertices of an oriented outline are used to construct strings of symbols or *words*. The string size is fixed at some integer $1 \leq W \leq N$ and is specified by the user. For instance, when $W = 3$, each word contains a string of three symbols.

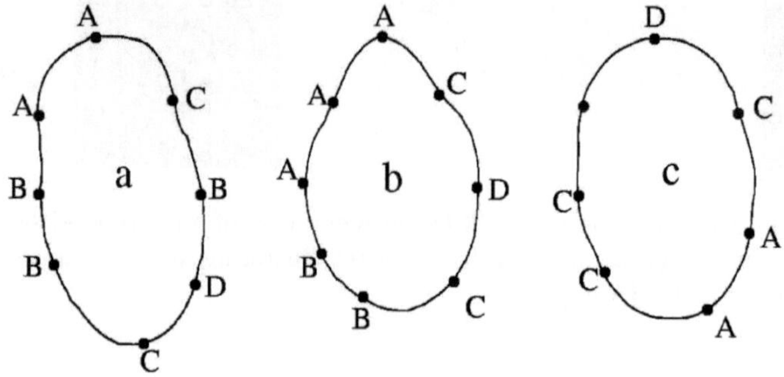

Fig. 12.12. Symbolic labels are assigned to the vertices of the oriented outlines after applying k-means clustering to their numeric attributes. Local geometric aspects are represented by forming strings of these symbols. Oriented outlines of (**a**) sagittal, (**b**) metopic, and (**c**) normal head shapes, computed at the level of the A-plane

A *BOW* representation for the outline in Fig. 12.12a is the unordered set s=$\{'CAA','AAB','ABB','BBC','BCD', 'CDB','DBC','BCA'\}$. These strings represent the *local* geometric properties in the skull shapes.

(d) Compute a $M \times V$ cooccurrence matrix of counts $n(s_i, w_j)$, denoting the number of times the word w_j occurs in the *BOW* s_i associated with the skull outline S_i in the training sets.
(e) Apply PLSA to the cooccurrence matrix $\mathcal{S}$ [16]. PLSA is a latent variable model which associates an unobserved class variable $z_k \in z_1, \ldots, z_P$ with each observation, an observation being the occurrence of a string of symbols from a particular *BOW* s_j. PLSA was originally applied to document retrieval for which a particular object is a document and the string of symbols is a word in that document.
(f) Use the class-conditional probabilities $P(s|z)$ estimated in the previous step to construct the SSDs for the outlines in $\mathcal{S}$. More specifically, for each outline S_i in $\mathcal{S}$, form its corresponding *SSD* as the P-dimensional vector $[P(s_i|z_1), \cdots, P(s_i|z_P)]$.

12.4.2 Shape-Based Retrieval

We have implemented a prototype system for shape-based image retrieval of skull CT imaging for craniofacial medicine. The system of database images supports retrieval based on their shape similarity to a query image. The shapes are represented by either numeric or symbolic features extracted using the four shape descriptors described above. A cosine similarity measure is incorporated for the retrieval of similar cases. The proposed system can help physicians and craniofacial experts with such tasks as diagnosis, intervention, and neurodevelopmental prediction.

The user is given a graphical user interface to the retrieval system. All the feature extraction and computation, feature classification, and feature vector

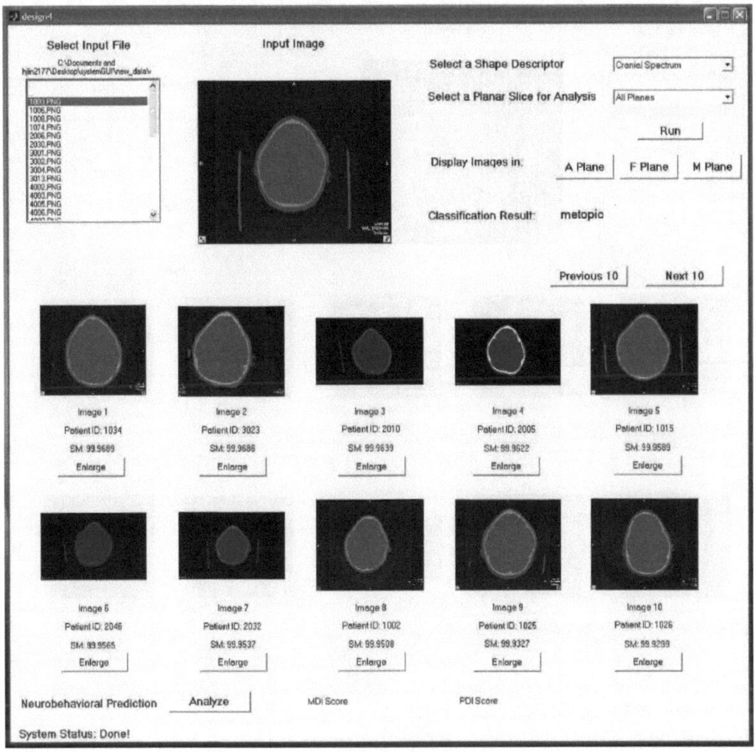

Fig. 12.13. An example of the system retrieving database images using the cranial spectrum representation on all three planar slices

comparison functions are done in the background. The queries to the image data can be specified using a query image to retrieve images that share similar features in the system. Figure 12.13 shows the system retrieving database images that are most similar to a query image using the cranial spectrum representation on all three planar slices. Figure 12.14 shows an example of the system retrieving database images that are most similar to a query image using the SSD on all three planar slices. The system is currently undergoing a rigorous evaluation that compares the results it retrieves to those selected by a craniofacial expert.

12.5 Mouse Eye Image Retrieval for the Study of Cataract Development

Researchers in the Eye Lab at the University of Washington are studying cataract formation in the eye, using mice as the subjects of their experiments. The mice are of several different strains with different genetic factors including a control group of normal mice. The eyes of the mice are photographed at

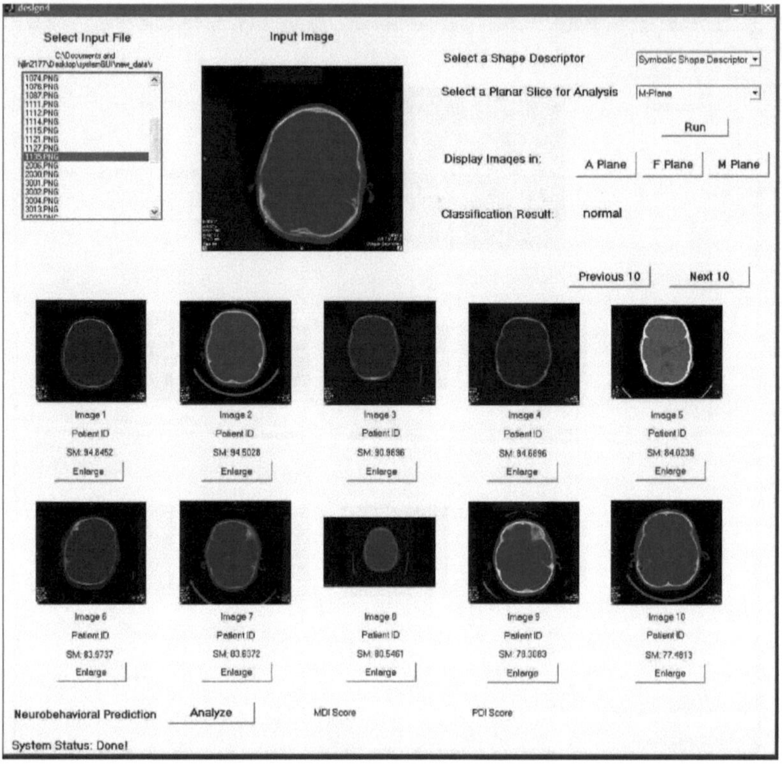

Fig. 12.14. An example of the system retrieving database images using the SSD representation on the M-plane

regular intervals using a slit-lens technique. Thus each mouse has associated subject data and genotype that need to be correlated with the progression of opacification observed in a sequence of slit lamp images. A current study will relate genotype (the mutation in a particular strain of mouse) to phenotype (the presentation of the opacity with respect to pattern and intensity of light scattering) using these images. The experiment involves eight different mutations in two different strains that are imaged approximately once per week. The number of images is enormous. Organizing the images through classification and clustering procedures as well as through the particular mice they came from is essential for the success of the project.

The set of cataract classes used in our work contains three classes (1) WT, which stands for wild type and has no laboratory-induced cataract, (2) secreted protein acidic and rich in cysteine (SPARC) knockout (a matricellular protein), and (3) the synaptic cleft (SC1) knockout. A knockout means the gene coding for a specific protein has been truncated or replaced so that the functional protein is no longer expressed. Figure 12.15 shows typical images for each class.

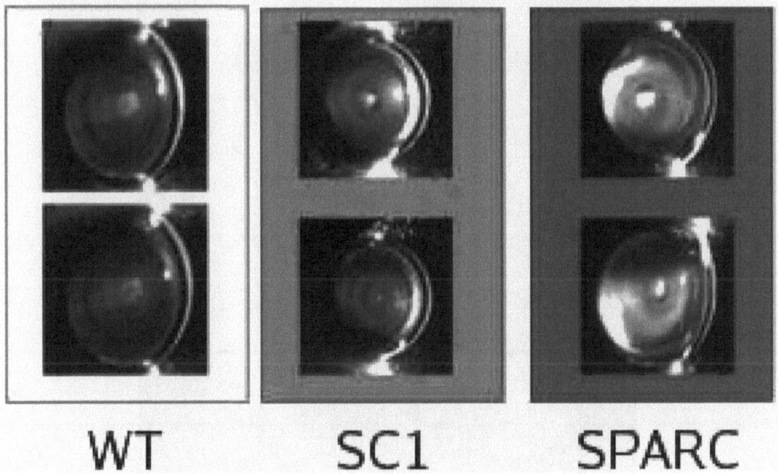

Fig. 12.15. Classes of our study: wild type (WT), synaptic cleft 1 protein knockout (SC1) [53], and secreted acydic rich cysteine (SPARC) knockout [5]

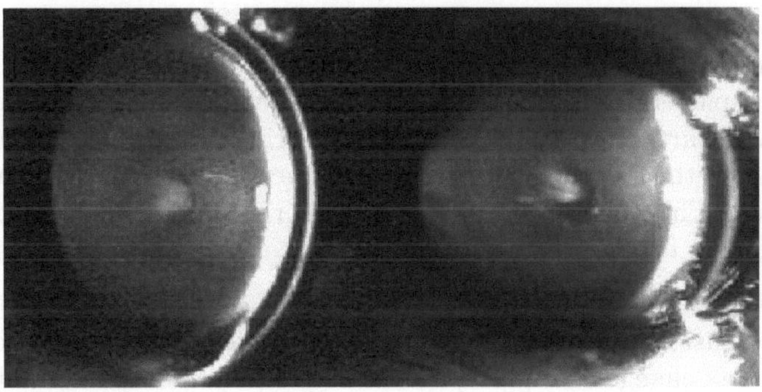

Fig. 12.16. The shape of the eye in an image varies depending on both the mouse and the angle of incidence of the slit-lamp ray with the surface of the eye. In addition to the shape, the mean intensity of the image varies due to the illumination variations not associated with the type of cataract

The original high-resolution images are resized to 300 by 300 pixel versions. Some of the properties of the images in the datasets are:

- The eyes in the images are approximately the same size and are approximately centered at the same location in each image.
- The illumination during the image capture process varies among the different experiments (Fig. 12.16).
- There are artifacts caused by the illumination that are independent from the cataracts (Fig. 12.17).

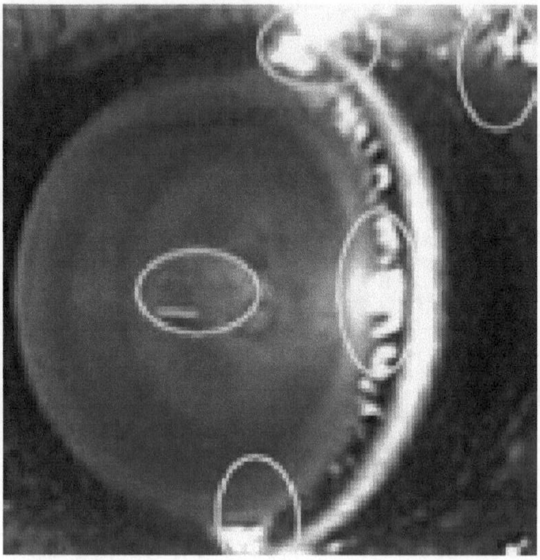

Fig. 12.17. Some artifacts caused by the camera are marked by yellow ellipses

- The pattern in the center of the eye is directly related to the cataract in the lens.
- A ring pattern can be observed in the WT class. Depending on the cataract, partial or total occlusion of this ring pattern is observed.
- Because of the way the images are taken, the ring pattern is not circular but elliptical. This makes the task of detecting rings more difficult (Fig. 12.16).
- The images contain, besides the eye, adjacent parts of the mouse such as eyelashes, which are not of interest.

The occlusion or cloudiness of the lens caused by the presence of certain cataracts changes the perception of the cell layers in the lens. Therefore, the pattern of the lens rings such as the relative colors between two rings will change, making this an important feature in the characterization of the cataract class.

The angle of incidence of the slit-lamp light on the cornea is a reason for the elliptical shape of the ring pattern shown in the medical image. In order to encapsulate the largest amount of information, images with ring patterns as close to a circle as possible are preferred. However, not all the given images have this property, since they are selected manually and the image of the mouse lens may only approximate a sphere. Fitting circular arcs to the rings in the images will not necessarily give a good approximation.

The numerous noise factors in the image make this problem nontrivial. Because the conditions in which the image is taken are not completely controlled, we cannot assume constant illumination or even constant shape (Fig. 12.17).

12.5.1 Feature Extraction

We have developed three different features that can discriminates between the different known classes. The features are (1) ring pattern, (2) intensity profile, and (3) histogram features. The features we have developed are summarized below.

Ring Pattern

Rings, or elliptical layers of cells, are present in both the normal and cataractous lens, unless the lens cells are disrupted. Visualization of the layers of lens cells depends on the magnification and contrast. In the normal lens, the contrast between adjacent cells or layers of cells is small (observation of the layers of cells in a normal lens requires high magnification in a microscope). The contrast increases when an increase in opacity occurs in some layers and not in adjacent layers. The increased contrast allows the rings to be observed at the magnification of the slit lamp. The pattern of rings can be correlated with the formation of cataracts. The proposed approach in identifying and quantifying these characteristics of the rings consists of a five-step process:

1. Ring enhancement.
2. Isolation of an elliptical sector of the lens.
3. Transformation of the elliptical sector into a rectangular image containing only pixels corresponding to the lens.
4. Compression of the rectangular image representation into a one-dimensional array of mean intensities.
5. Extraction of feature vector values from the one-dimensional array produced in the previous step.

First the ring pattern is enhanced by a local histogram equalization transform as shown in Fig. 12.18. Once the rings have been enhanced, the parameters of the ring pattern have to be extracted. Let the coordinates of the center of the eye be (c_x, c_y). We will consider the region that corresponds to

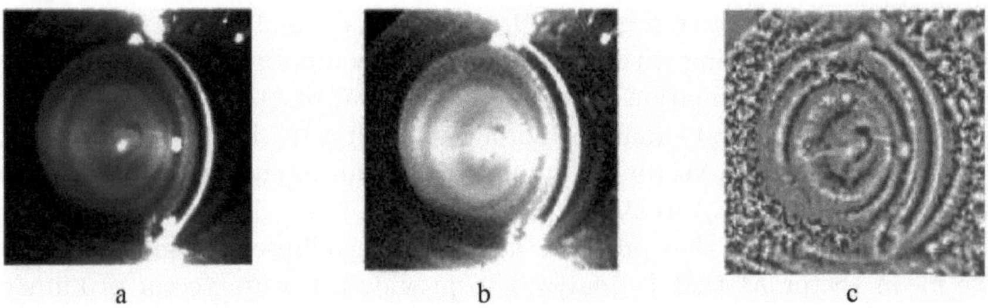

Fig. 12.18. The original image (**a**), the original image after a histogram equalization (**b**), and the original image after a local equalization (**c**)

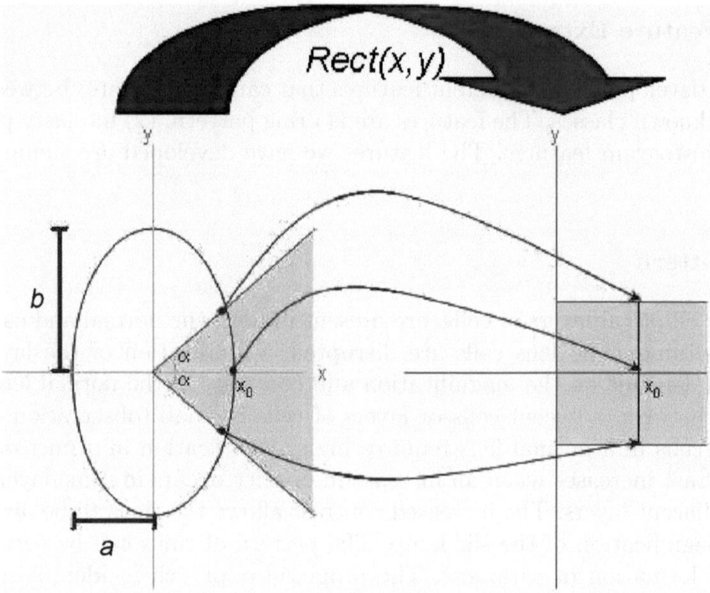

Fig. 12.19. The region inside an elliptical sector is mapped to a rectangular sector using the $Rect(x, y) = (c_x + a, y)$ transformation

the elliptical sector centered at (c_x, c_y) and between the angles α and $-\alpha$, where $0 < \alpha < \frac{\pi}{2}$.

A ring is modeled as an elliptical arc with axes of length a and b parallel to the x and y axes, respectively, and centered at (c_x, c_y). Every point (x, y) that lies in this elliptical arc is mapped to a point on a vertical line by the transformation $Rect(x, y) = (c_x + a, y)$ as shown in Fig. 12.19.

The *Rect* transformation is applied to every ellipse centered at (c_x, c_y) with a fixed $c = \frac{a}{b}$ and $0 < a < w - c_x$, where w and h are the width and height of the image, respectively, to form a rectangular version of the elliptical-form eye. The resulting image of the *Rect* mapping is cropped to remove the cornea and the area outside the lens. The elliptical-to-rectangular transformation has three degrees of freedom: c_x, c_y, and c. The value of c_y is restricted to $\frac{h}{2}$, working with two degrees of freedom for each image.

The *Rect* transformation converts the elliptical rings into vertical lines for easier and more accurate analysis. For a given center (c_x, c_y) and length–width ratio c, it produces a vector $M(c_x, c)$ of the mean intensities for each of the columns in $Rect(c_x, c_y)$ as shown in Fig. 12.20.

The (c_x, c_y) and c that produce the best-fitting ellipse are used to generate the mean vector M that is analyzed to provide a feature vector of numeric attributes that can be used for pattern recognition. The vector we use contains the following ring attributes:

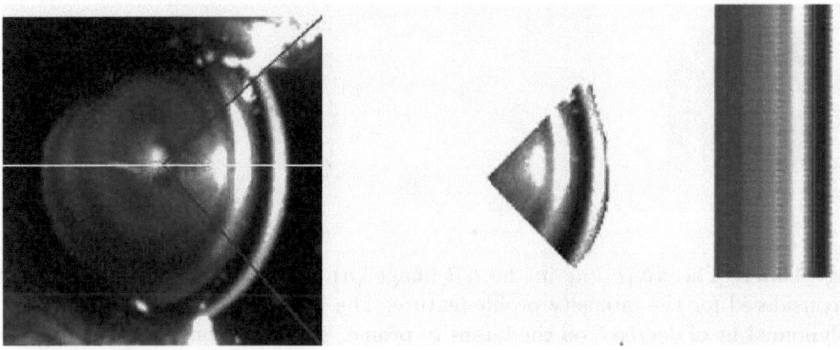

Fig. 12.20. An iterative process selects a center point, isolates an elliptical sector of the lens and maps it to a rectangular version. The variance at each column of the rectangular mapping is computed, and the case where the sum of the column variances is the lowest is selected as the best ellipse sector isolation; its center corresponds to the best center of the lens

- Number, maximum, minimum, mean, and variance of maxima of the function
- Number, maximum, minimum, mean, and variance of minima of the function
- Maximum, minimum, mean, and variance of the distance between pairs of consecutive maxima
- Maximum, minimum, mean, and variance of the distance between pairs of consecutive minima
- Maximum, minimum, mean, and variance of the difference between each pair of consecutive maxima and a minima
- Number of consecutive pairs of maxima and minima (regions formed by peaks and valleys)

These attributes capture several characteristics of the rings: Globally, the attributes include the number of rings and the distribution of their properties such as mean intensity, width, and others; locally, the attributes describe the width and opacity of each ring and provide a comparison of these characteristics with the characteristics of other rings in the same image.

Intensity Profile

In order to capture the level of cloudiness of the lens, the intensity profile of the row of pixels in the middle of the image is also considered. Due to the different opacities in each layer of cells of the lens, a characteristic distribution of intensities exists for each class. With this information, a plot of intensity vs column (with the row being constant) is created. Call this function $I(x)$, where the domain is the columns in the image, and the range is $\{0, 255\}$

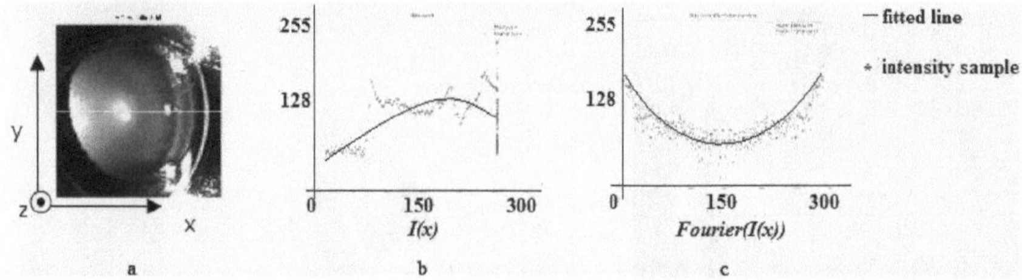

Fig. 12.21. The *white line* in the *left* image (**a**) indicates the row of pixels that is considered for the intensity profile feature. The image in the *middle* (**b**) shows a polynomial fit of degree 5 on the intensity profile. The image on the *right* (**c**) shows a polynomial fit on the output of the Fast Fourier Transform of the intensity profile

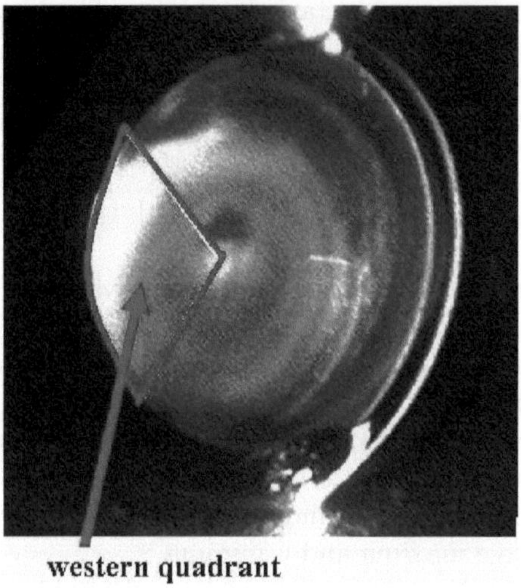

Fig. 12.22. Western quadrant of the lens

representing the possible values for intensity. One way of quantifying $I(x)$ is to fit a polynomial to it and use the values of its coefficients as features.

However, the large number of peaks and valleys creates too much noise to produce an accurate fit as shown in Fig. 12.21b. Instead, a Fast Fourier Transform $Fourier_I(x)$ as observed in Fig. 12.21c is applied. A polynomial of degree 5 is fit to this function, resulting in a complex function. The 12 coefficients of this polynomial (six real and six imaginary) are concatenated to form a feature vector.

Histogram of Western Quadrant of the Lens

The difference in opacity, particularly in the western quadrant (as shown in Fig. 12.22) of the lens layers is characteristic for each cataract class. A histogram of the western quadrant of the lens is created and is fitted to a one-dimensional Gaussian using Maximum Likelihood. The variance of the Gaussian is included in the feature vector. Because of the variations in illumination caused by external sources in the image, only the variance, which is related to the distribution of intensities, and not the mean, which is directly associated with the change in external illuminations, is considered. Figure 12.23 shows the histogram of the images corresponding to each of the three classes.

Segmented Least Squares Fitting on the Intensity Profile

While the intensity profile feature can characterize some classes, the function is assumed to be continuous. Therefore, noncontinuous fluctuations are approximated as smooth curves. Some classes such as SPARC and WT have similar polynomial fits but differ in the number of noncontinuous changes (Fig. 12.24). To encapsulate the amount of "continuity," we fit a piecewise linear function on $I(x)$. The standard least squares linear fitting algorithm fits a line to a set of n two-dimensional points $P = \{(x_1, y1), (x_2, y_2), ..., (x_n, y_n)\}$. The line with the minimum error is $y = ax + b$, where

$$a = \frac{n \sum_i x_i y_i - (\sum_i x_i)(\sum_i y_i)}{n \sum_i x_i^2 - (\sum_i x_i)^2}$$

$$b = \frac{\sum_i y_i - a \sum_i x_i}{n}$$

The error of the fit is:

$$Error(L, P) = \sum_{i=1}^{n} (y - i - ax_i - b)^2$$

The objective of the segmented least squares approximation is to find the partition of consecutive points that minimizes the error of the fit, which corresponds to the sum of the errors for each linear segment used in the fit with an additional cost C for each segment used.

$$E_{fit} = \sum_{p_i \in P} Error(l_i, p_i) + C$$

where p_i is a set of consecutive points in $I(x)$, l_i is the least squares linear fit on this set, and C is the cost for each extra segment. We used values of C on the order of 500.

After fitting this piecewise linear function to $I(x)$, we only consider the left side of the eye for this feature extraction. This is done by considering the segments up to the middle of the image and discarding the first segments that have a very small value for their slope and intersection at the origin. These segments correspond to the dark section of the image, not a part of the mouse lens. The features extracted are:

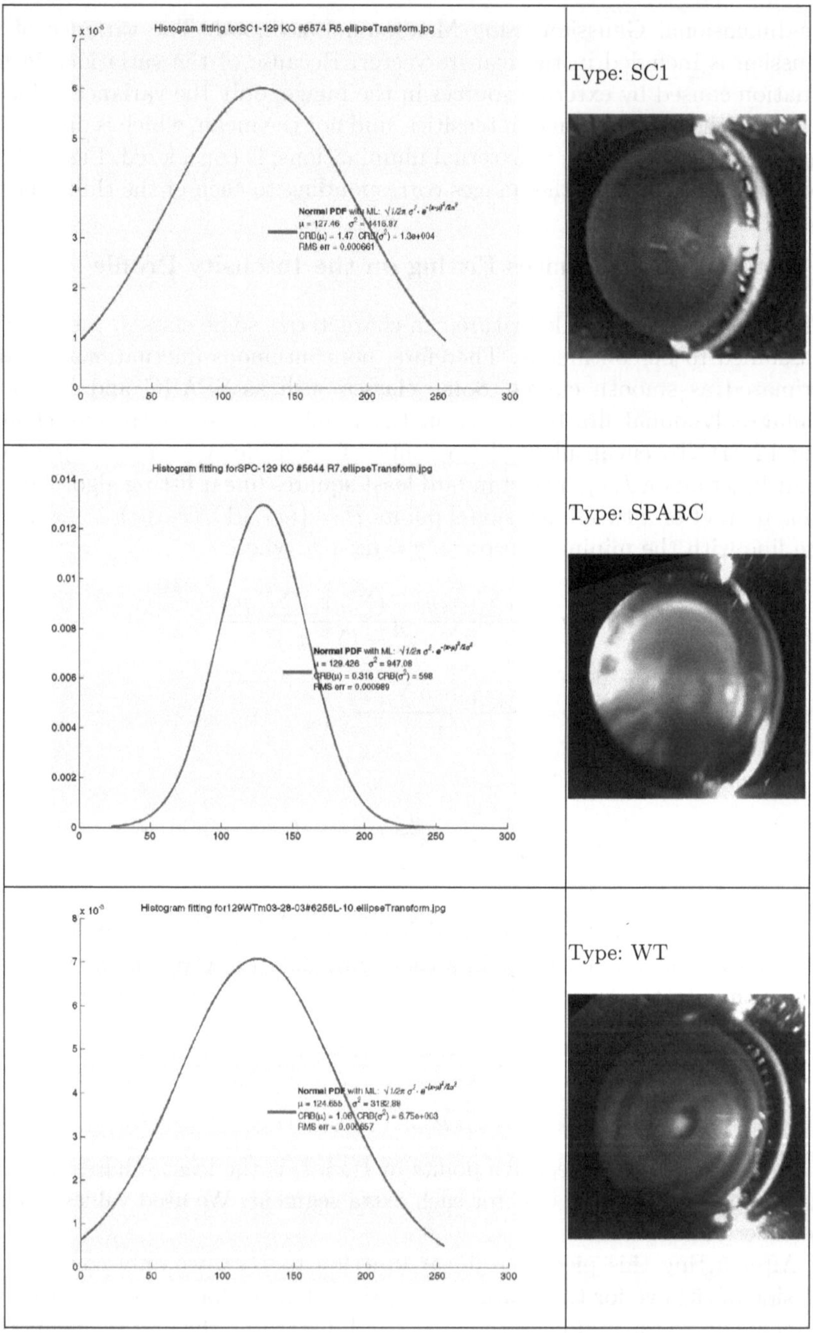

Fig. 12.23. The histogram of the western quadrant of the eye in three different classes. The variance of pixel intensities is smaller in SPARC type images and larger in SC1 type images, while the WT type images tend to have a variance value in between

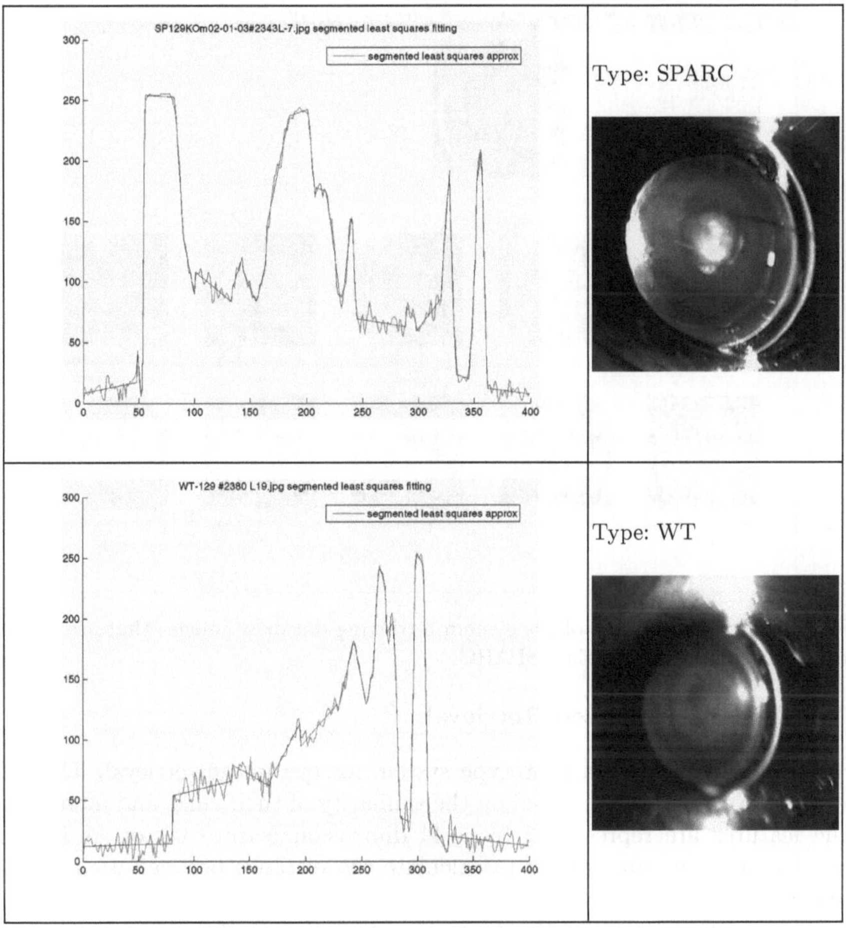

Fig. 12.24. The noncontinuous changes in the intensity profile are not well encapsulated when fitting the intensity profile to a continuous function. The graph on the *top* is the intensity profile for the SPARC type lens on the *top right*; the *bottom* graph corresponds to the intensity profile for the WT type, shown on the *bottom right*. The noncontinuous fluctuations of intensities in the image are observed as sharp edges in the intensity profile that are characteristic in SPARC images, while their presence is not as prominent in the WT class

- Number of segments
- Means of the slope and the intersection at the origin (a, b) of each of the segments
- Variances of the slope and the intersection at the origin (a, b) of each of the segments

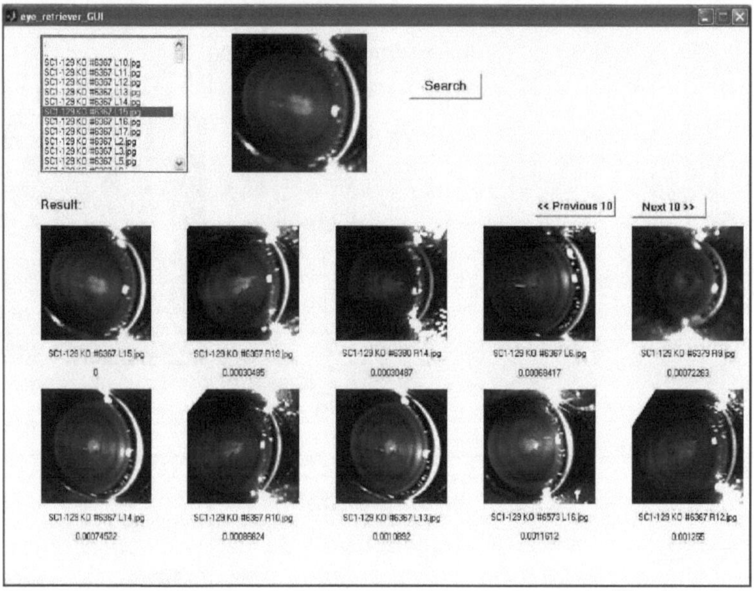

Fig. 12.25. An example of the system retrieving database images that are similar to the query image from class SPARC

12.5.2 Similarity-Based Retrieval

We have implemented a prototype system for eye image retrieval. The system retrieves eye images based on the similarity of their rings and intensities. The features are represented by a 44 dimension feature vector. A Euclidean distance measure is used to calculate the distance between each pair of images.

The user is presented with a graphical user interface of the retrieval system. Feature extraction and classification are done in the background. Figure 12.25 shows the system retrieving database images that are most similar to the query image. This system will allow the scientists doing the basic research to test their theories of genotype–phenotype relationships.

12.6 A Unified Query Framework for Multimedia Data

We wish to develop a unified query framework for multimedia biomedical data. The unified query framework must allow the specification of queries on alphanumeric data, text data, and multiple kinds of image and signal data. Our query framework will be an extended SQL that can handle probabilistic queries through *similarity measures*, which compare a query object to a database object of the same or comparable type and return a value between 0 (no match) and 1 (perfect match), which can be loosely interpreted as a probability.

A query will have three integral parts, which are common to all database systems and can be expressed in SQL or other query languages (1) the query specifications, (2) the result specifications, and (3) the matching requirements. The query specifications will include the objects that are provided to the query. These can include many different data types, such as text, numeric, date, matrix, table, signal, two-dimensional image, three-dimensional image, three-dimensional mesh, and three-dimensional image over time (video). The result specifications will include the objects that are to be returned by the query, which can include the same variety of data types. The matching requirements will specify constraints (predicates) on the returned objects, including both hard constraints that must be satisfied and soft constraints whose satisfaction is probabilistic in nature and involves the execution of similarity measures. For soft constraints, the data will be retrieved in ranked order, according to the probability of a match. The main focus of our work is on these similarity measures and probabilistic retrievals.

12.7 Summary

We have described three separate content-based retrieval systems for biomedical applications. The first retrieves neuronal signals and fMRI data from a database of patients who have undergone epileptic tumor resection. The second retrieves skull images from a database of CT images from patients with both normal and abnormal craniofacial structure. The third retrieves slit-lens mouse eye images from a database of images from normal and DNA-modified mice. All three use very specialized image and signal features in their similarity measures. Our goal in our future work is to develop a set of feature extractors that can be used to construct useful similarity measures as part of a unified system for multimedia biomedical data retrieval.

12.8 Acknowledgments

This research was supported by the National Science Foundation under grant DBI-0543631, NIHI NIDCR grant R01 DE 013813-05, NIH grant DC02310, and NEI grant EY04542.

References

1. A.M. Aisen, L.S. Broderick, H. Winer-Muram, C.E. Brodley, A.C. Kak, C. Pavlopoulou, J. Dy, C.R. Shyu, and A. Marchiori. Automated storage and retrieval of thin section CT images to assist diagnosis: System description and preliminary assessment. *Radiology*, 228:265–270, 2003.
2. S. Atnafu, R. Chbeir, and L. Brunie. Content-based and metadata retrieval in medical image database. In *Proceedings of the IEEE Symposium on Computer-Based Medical Systems*, pages 327–332, 2002.

3. Z. Aung and K.-L. Tan. Rapid 3d protein structure database searching using information retrieval techniques. *Bioinformatics*, 20(7):1045–1052, 2004.
4. J.R. Bach, C. Fuller, A. Gupta, A. Hampapur, B. Horowitz, R. Humphrey, R. Jain, and C.F. Shu. The virage search engine: An open framework for image management. In *Proceedings of SPIE Storage and Retrieval of Image and Video Databases*, 1996.
5. J.A. Bassuk, T. Birkebak, J.D. Rothmier, J.M. Clark, A. Bradshaw, P.J. Muchowski, C.C. Howe, J.I. Clark, and E.H. Sage. Disruption of the sparc locus in mice alters the differentiation of lenticular epithelial cells and leads to cataract formation. *Experimental Eye Research*, 68(3):321–331, 1999.
6. A. Berman and L.G. Shapiro. A flexible image database system for content-based retrieval. *Computer Vision and Image Understanding*, 75:175–195, 1999.
7. C. Carson, S. Belongie, H. Greenspan, and J. Malik. Region-based image querying. In *Proceedings of IEEE Workshop on Content-Based Access of Image and Video Libraries*, pages 42–49, June 1997.
8. K. Chakrabarti and S. Mehrotra. The hybrid tree: An index structure for high dimensional feature space. In *International Conference on Data Engineering*, pages 440–447, 1999.
9. P.-H. Chi, G. Scott, and C.-R. Shyu. A fast protein structure retrieval system using image-based distance matrices and multidimensional index. *International Journal of Software Engineering and Knowledge Engineering*, 15(4), 2005.
10. H. Cho, D. Corina, G.A. Ojemann, J. Schoenfield, L. Zamora, and L. Shapiro. A new neuron spike sorting method using maximal overlap discrete wavelet transform and rotated principal component analysis. In *Proc. IEEE Engineering in Medicine and Biology Society Annual International Conference*, volume 3, pages 2921–2924, 2003.
11. H. Cho, G.A. Ojemann, D. Corina, and L.G. Shapiro. Detection of neural activity in event-related fmri using wavelet and dynamic time warping. In *Proc. IEEE Applications of Digital Image Processing XXVI*, volume 5203, pages 638–647, 2003.
12. W.W. Chu, C.C. Hsu, A.F. Cardenas, and R.K. Taira. Knowledge-based image retrieval with spatial and temporal constructs. *IEEE Transactions on Knowledge and Data Engineering*, 10(6):872–888, 1998.
13. T. Deselaers, D. Keysers, and H. Ney. Fire - flexible image retrieval engine: Image clef 2004 evaluation. In *Proceedings of the CLEF 2004 Workshop*, pages 535–544, 2004.
14. M. Flickner, H. Sawhney, W. Niblack, J. Ashely, Q. Hyang, B. Dom, M. Gorkani, J. Hafner, D. Lee, D. Petkovic, D. Steele, and P. Yanker. The QBIC project: Querying images by content using color, texture, and shape. In *Proceedings of SPIE Storage and Retrieval of Image and Video Databases*, pages 173–181, 1993.
15. M. Haddad, K-P. Adlassnig, and G. Porenta. Feasibility analysis of a case-based reasoning system for automated detection of coronary heart disease from myocardial scintigrams. *Artifical Intelligence in Medicine*, 9:61–78, 1997.
16. T. Hofmann. Unsupervised learning by probabilistic latent semantic analysis. *Machine Learning*, 42:177–196, 2001.
17. S. Janichen and P. Perner. Conceptual clustering and case generalization of two-dimensional forms. *Computational Intelligence*, 22(3/4), 2006.
18. MC Jaulent, C. Le Bozec, E. Zapletal, and P. Degoulet. Case-based diagnosis in histopathology of breast tumors. *Medinfo*, 9(1), 1998.

19. P. Kelly, M. Cannon, and D. Bush. Query by image example: the CANDID approach. In *Proceedings of the SPIE Conference on Storage and Retrieval for Image and Video Databases III*, pages 238–248, 1995.
20. P. Korn, N. Sidiropoulos, C. Faloutsos, E. Siegal, and Z. Protopapas. Fast and effective retrieval of medical tumor shapes. *IEEE Transactions on Knowledge Data Engineering*, 10(6):889–904, 1998.
21. E. Lajeunie, M. Le Merrer, C. Marchac, and D. Renier. Genetic study of scaphocephaly. *American Journal of Medical Genetics*, 62(3):282–285, 1996.
22. H.J. Lin, S. Ruiz-Correa, L.G. Shapiro, A.V. Hing, M.L. Cunningham, M.L. Speltz, and R.W. Sze. Symbolic shape descriptors for classifying craniosynostosis deformations from skull imaging. In *Engineering in Medicine and Biology Society 2005. IEEE EMBS 2005. 27th Annual International Conference of the*, pages 6325–6331, 2005.
23. Y. Liu, N. Lazar, and W. Rothfus. Semantic-based biomedical image indexing and retrieval. In *International Conference on Diagnostic Imaging and Analysis*, 2002.
24. W.Y. Ma and B.S. Manjunath. NETRA: A toolbox for navagating large image databases. In *Proceedings of the IEEE International Conference on Image Processing*, pages 568–571, 1997.
25. R. Macura and K. Macura. Macrad: Radiology image resource with a case-based retrieval system. In M. Veloso and A. Aamodt, editors, *Case-Based Reasoning: Research and Development, Springer*, pages 43–54, 1995.
26. J.L. Marsh, A. Jenny, Galic M, Picker S, and M.W. Vannier. Surgical management of sagittal synostosis. a quantitative evaluation of two techniques. *Neurosurgery Clinics of North America*, 2(3):629–640, 1991.
27. J.L. Marsh and M.W. Vannier. Cranial base changes following surgical treatment of craniosynostosis. *The Cleft Palate Journal*, 23:9–19, 1986.
28. S. Mehrotra, Y. Rui, M. Ortega, and T.S. Huang. Supporting content-based queries over images in MARS. In *Proceedings of the IEEE International Conference on Multimedia Computing and Systems*, 1997.
29. T.P. Minka and R.W. Picard. Interactive learning with a society of models. In *Proceedings of CVPR-96*, pages 447–452, 1996.
30. A. Mojsilovis and J. Gomes. Semantic based image categorization, browsing and retrieval in medical image databases. In *IEEE. International Conference on Image Processing*, 2000.
31. H. Muller, N. Michous, D. Bandon, and A. Geissbuhler. A review of content-based image retrieval systems in medical applications–clinical benefits and future directions. *International Journal of Medical Informatics*, 73:1–23, 2004.
32. V.E. Ogle and M. Stonebraker. Chabot: Retrieval from a relational database of images. *IEEE Computer*, 28:40–48, 1995.
33. J. Peng, B. Bhanu, and S. Qing. Probabilistic feature relevance learning for content-based image retrieval. *Computer Vision and Image Understanding: Special Issue on Content-Based Access of Image and Video Libraries*, 75:150–164, 1999.
34. A. Pentland, R.W. Picard, and S. Sclaroff. Photobook: Content-based manipulation of image databases. In *Proceedings of SPIE Storage and Retrieval of Image and Video Databases II*, pages 34–47, 1994.
35. P. Perner. An architecture for a cbr image segmentation system. *Journal on Engineering Application in Artifical Intelligence*, 12(6):749–759, 1999.

36. P. Perner. Case-base maintenance by conceptual clustering of graphs. *Engineering Applications of Artificial Intelligence*, 19(4):381–295, 2006.
37. E.G.M. Petrakis. Content-based retrieval of medical images. *International Journal of Computing Research*, 11(2):171–182, 2002.
38. X. Qian and H.D. Tagare. Optimally adapted indexing trees for medical image databases. In *IEEE International Symposium on Biomedical Imaging*, pages 321–324, 2002.
39. C. Rao. Geometry of circular vectors and pattern recognition of shape of a boundary. *Proc. Nat. Acad. Sci.*, 95:12783, 2002.
40. Y. Rui, T.S. Huang, M. Ortega, and S. Mehrotra. Relevane feedback: A power tool for interactive content-based image retrieval. *IEEE Transactions on Circuits and Systems for Video Technology: Special Issue on Segmentation, Description, and Retrieval of Video Content*, 8(5):644–655, 1998.
41. S. Ruiz-Correa, R.W. Sze, H.J. Lin, L.G. Shapiro, M.L. Speltz, and M.L. Cunningham. Classifying craniosynostosis deformations by skull shape imaging. In *Proceedings of the 18th IEEE International Symposium on Computer-Based Medical Systems*, pages 335–340, 2005.
42. S. Ruiz-Correa, R.W. Sze, J.R. Starr, A.V. Hing, H.J. Lin, and M.L. Cunningham. A fourier-based approach for quantifying sagittal synostosis head shape. In *American Cleft Palate Craniofacial Association*, 2005.
43. S. Ruiz-Correa, R.W. Sze, J.R. Starr, H.T. Lin, M.L. Speltz, M.L. Cunningham, and A.V. Hing. New scaphocephaly severity indices of sagittal craniosynostosis: A comparative study with cranial index quantifications. *Cleft Palate Craniofacial Journal*, 43(2):211–221, 2006.
44. S. Sclaroff, L. Taycher, and M. L. Cascia. Imagerover: A content-based image browser for the world wide web. In *Proceedings of the IEEE Workshop on Content- Based Access of Image and Video Libraries*, pages 2–9, June 1997.
45. A. Shuper, P. Merlob, M. Grunebaum, and S.H. Reisner. The incidence of isolated craniosynostosis in the newborn infants. *American Journal of Diseases of Children*, 139(1):85–86, 1985.
46. C.-R. Shyu, C.E. Brodley, A.C. Kak, A. Kosaka, A.M. Aisen, and L.S. Broderick. Assert: A physician-in-the-loop content-based image retrieval system for hrct image databases. In *Computer Vision and Image Understanding {Special Issue on Content-Based Retrieval from Image Databases}*, pages 111–131, 1999.
47. C.-R. Shyu, P.-H. Chi, G. Scott, and D. Xu. Proteindbs: A real-time retrieval system for protein structure comparison. *Nucleic Acids Research*, 32:W572–W575, 2004.
48. A.W.M. Smeulders, M. Worring, S. Santini, A. Gupta, and R. Jain. Content-based image retrieval at the end of the early years. *IEEE Transactions on Pattern Analysis and Machine Intelligence*, 22(12):1349–1380, 2000.
49. A.W.M. Smeulders, T.S. Huang, and T. Gevers. Special issue on content-based image retrieval. *International Journal of Computer Vision*, 56:5–6, 2000.
50. J.R. Smith and S.F. Chang. Visually searching the web for content. *IEEE Multimedia Magazine*, 4(3):12–20, 1997.
51. J.R. Smith and S.F. Chang. Visually searching the web for content. *IEEE Multimedia Magazine*, 4(3):12–20, 1997.
52. J.R. Smith and C.S. Li. Image classification and querying using composite region templates. *Computer Vision and Image Understanding: Special Issue on Content-Based Access of Image and Video Libraries*, 75(1–2):165–174, 1999.

53. M.M. Sullivan and E.H. Sage. Hevin/sc1, a matricellular glycoprotein and potential tumor suppressor of the sparc/bm-40/osteonection family. *International Journal of Biochemistry and Cell Biology*, 36(6):482–490, 2004.
54. H.D. Tagare. Increasing retrieval efficiency by index tree adaptation. In *IEEE Workshop on Content-Based Access of Image and Video Libraries*, pages 28–35, 1997.
55. H.D. Tagare, C.C. Jaffe, and J. Duncan. Medical image databases: a content-based retrieval approach. *Journal of the American Medical Informatics Association*, 4(3):184–198, 1997.
56. H. Tang, R. Hanka, and H. Ip. Histological image retrieval based on semantic content. *IEEE Transaction on Information Technology in Biomedicine*, 7(1):26–36, 2003.
57. L. Zheng, A.W. Wetzel, J. Gilbertson, and M. Becich. Design and analysis of a content based pathology image retrieval system. *IEEE Transaction on Information Technology in Biomedicine*, 7(4):249–255, 2003.

13

Medical Imagery in Case-Based Reasoning

D.C. Wilson[1] and D. O'Sullivan[2]

[1] Department of Software and Information Systems
University of North Carolina at Charlotte
9201 University City Boulevard
Charlotte, NC 28223-0001, USA
davils@uncc.edu

[2] Mobile Emergency Triage (MET) Research Group
School of Management
University of Ottawa, ON K1N 6N5, Canada
dympna@management.uottawa.ca

Summary. Image processing is an important focus area within case-based reasoning. CBR systems have been developed both to support image-centric functionality such as segmentation, as well as domain-specific imagery applications. In particular, case-based medical applications employ significant imagery elements to support tasks such as diagnosis and treatment planning. Whereas previous surveys have focused either on imagery or on medicine, this chapter takes a look specifically at their conjunction, providing a novel perspective and overview of the main issues and research work in case-based reasoning involving medical imagery.

13.1 Introduction

Image processing has long been an area of significant interest for case-based reasoning researchers. Research efforts have investigated CBR as applied to diverse imagery tasks including: ultrasonic interpretation [59], submarine classification [4, 19], face recognition [49], architectural support [14], remotely sensed data [71], weather prediction [36], protein crystallization [39], satellite imagery [37], and video retrieval [11]. Further, case-based reasoning has been applied in sketch-based retrieval of architectural data [15, 27], as well as for prediction in spatial GIS applications [33, 34, 40–42]. In addition to domain-specific applications, case-based approaches have also been applied to more general image-centric functionality, such as segmentation [55] and recognition [46].

In particular, though, the medical domain has been strongly coupled with image processing in case-based reasoning, for example in radiology [44] and computer tomography [26, 56]. While the two areas of image processing and medical applications have both been identified as primary topics in CBR research [35, 62], the crossover between the two has not typically been addressed

D.C. Wilson and D. O'Sullivan: *Medical Imagery in Case-Based Reasoning*, Studies in Computational Intelligence (SCI) **73**, 389–418 (2008)
www.springerlink.com

as a whole. This chapter looks specifically at medical imagery applications in case-based reasoning. Several very good surveys have been published, some quite recently, on integrating imagery with case-based reasoning [24,56,61,62], as well as on case-based reasoning in medicine [10, 35, 50, 65]. Our goal is to examine the conjunction of the two areas in light of ongoing developments in medical information systems, in order to provide a novel perspective and overview of the main issues and research work on medical imagery in case-based reasoning. This chapter begins with an overview of image storage and retrieval in general, as well as for medical applications in particular in Sections 13.2 and 13.3. Section 13.4 goes on to introduce the notion of "Media CBR" and related issues for cases that incorporate media elements to varying degrees. Finally, Section 13.5 provides a survey of case-based reasoning work for medical imagery, setting the systems in the context of a comparative framework.

13.2 Image Storage and Retrieval

Image retrieval systems are typically characterized by one of two main approaches, they either support keyword-based annotation and indexing (*query-by-text*) or a content-based approach where the retrieval of images is on the basis of features automatically extracted from the images themselves (*query-by-example*). Each of the two approaches can be practically applied to different image domains, however, it is recognized that neither of these approaches are fully adequate for answering the complete range of user search questions [48]. Both approaches must contend with the so-called *sensory gap* [70], which describes the loss between the actual structure in the world and the representation in a digital image. More precisely, the sensory gap deals with issues of resolution.

The keyword-based approach is a *human-centric* approach that depends on images being accompanied by textual descriptions, which typically incurs a significant knowledge engineering effort. The indexes for such large image collections are time consuming to create and maintain, particularly since entries are not grounded in how the collections are being used (which exacerbates indexing subjectivity). Also, keyword indexing for images only provides hit-or-miss type searching as the range of successful queries is limited to the interpretation of the indexer.

Content-based approaches are *computer-centric* and focus on the automatic extraction of low-level visual features such as colors and shapes from imagery. Substantial research efforts within the computer vision community have been focused on retrieving specific images from a large database by querying the properties of these images [13, 28, 45, 51, 63]. Some notable prototypes for intelligent image retrieval have been developed, including [12,17,18,21,54,66]. Most of these efforts address the problem in the context of general-use applications, where the images stored in the database display substantial differences in their low-level properties, such as: color (histogram matching), texture (image coarseness and contrast matching), and composition (dividing an image into homogeneous color/texture regions and analyzing the relative positions of those regions). Content-based approaches must address the so-called *semantic gap* [70]. General semantic layers are insufficient to model the full complexity of medical knowledge, and the semantic gap refers to transformation from an image to a representation by features. More specifically, the semantic gap deals with the problem of information loss in such transformations.

Addressing the semantic gap is a difficult problem. Baseline approaches may employ combinations of low-level features, such as may be represented in the MPEG-7 standard. In general, however, it requires significant effort in knowledge representation and knowledge acquisition through feature extraction, and classification. Research in semantic-inference [58,60] tries to address this problem by enabling incremental development of fully automatic image retrieval systems, beginning with human-defined and automatically extracted features and progressing to more specific and refined extraction procedures for domain problems.

In our own research [52,53], we identify an additional *task-based* approach to image analysis and retrieval that leverages an associative context, grounded by domain tasks. Images that are grouped for a common purpose may be considered to have a tacit relationship, even though the image contents or annotations themselves would not indicate such.

13.3 Imagery in Medical Application Context

The medical domain, in particular, incorporates a broad range of imagery applications to support laboratory technicians, physicians, or patients in a variety of tasks, such as diagnosis or treatment planning. In order to motivate our survey of medical imagery in CBR, we first take a look at some of the issues in how imagery is applied in the medical application domain.

Medical diagnosis and decision making involves interplay between vast numbers of medical knowledge resources [2,3,69,73]. This can range from implicit knowledge held by healthcare workers to experiential and data-induced knowledge. Systems that can simultaneously access and combine relevant information from these various knowledge resources are crucial to the diagnostic process and subsequently the efficient treatment of patients. From a decision support viewpoint, healthcare workers need complete, contextually relevant information that is consistent with the patient's current medical state and that is appropriately presented at the correct level of abstraction.

For many years, the medical industry has managed imagery in the context of "picture archiving and communication systems" (PACS) [8,47,64,68]. More recently, episodic records of patient interactions are coming to reside in "electronic health records" (EHRs) [16,31,32], which incorporate the traditional

PACS information. The Healthcare Information and Management Systems Society has defined EHR [31]:

> The EHR is a longitudinal electronic record of patient health information generated by one or more encounters in any care delivery setting. Included in this information are patient demographics, progress notes, problems, medications, vital signs, past medical history, immunizations, laboratory data, and radiology reports. The EHR automates and streamlines the clinician's workflow. The EHR has the ability to generate a complete record of a clinical patient encounter, as well as supporting other care-related activities directly or indirectly via interface—including evidence-based decision support, quality management, and outcomes reporting.

The medical community has recognized the need for both context-based and content-based image retrieval, and a good overview from that perspective can be found in [48]. Moreover, the medical community has recognized the value in capturing and linking tacit knowledge to explicit knowledge [6,20,67], supporting the development of task-based approaches.

For integrated medical systems to achieve significant success, it is necessary to provide a level of standardization for image diagnosis and retrieval. There is significant ongoing work in developing standardized vocabularies and medical ontologies for representation. In particular, for medical imagery, it is necessary to have a vocabulary that makes the meaning and the visual appearance of image features clear and universally understood among medical practitioners. The BI-RADS Atlas [1] provides a leading example in standardized classification for radiological image diagnosis in mammographic studies.

13.4 Media CBR and Imagery Cases

The different medical contexts for imagery help to define different roles that imagery can play as part of case structures in a CBR system. The range of image contribution to a case will depend on the domain task, and it can range from being the sole/primary source of case knowledge to being a contributing part of a larger whole. Many CBR medical applications have incorporated nonimage context, such as image metadata or patient characteristics, and nonimage context has been recognized as part of case representation for individual images [61]. The widespread deployment of EHR type systems argues for the incorporation of such imagery as a significant part of overall patient media structures. Here we provide a more general account of how imagery can play a role in case composition, by looking at the role of media in cases. We present a view of media case composition directly applicable to imagery, which extends a characterization originally developed by the authors in the context of textual case-based reasoning [72].

One aspect of CBR imagery research focuses on transforming or augmenting knowledge-poor images [56, 61] in order to generate (hybrid-) structured representations that can be used by more traditional knowledge-based CBR methods. Such support for imagery CBR is especially important when the raw case information is composed entirely of image data. However, there are many situations in which image information plays an important, but ancillary role in case composition and reasoning. For example, a medical diagnosis may be made primarily based on demographic information and nonimage laboratory analysis (e.g., blood testing), but an X-ray may still provide useful support for that diagnosis.

Thus, techniques from content-based image retrieval, which have been used to support image-as-case CBR, can be applied in situations where image information represents only a small, but useful part of the overall reasoning context. This has led us to view media CBR along a continuum from "media-light," where media information offers limited reasoning support but does not require sophisticated processing, to "media-heavy," where media information is the focus of reasoning but requires much more specialized treatment. For media-light contexts, we suggest that relatively simple and general content-based techniques can be used to provide a local measure of similarity which, although relatively weak in itself, is strong enough to enhance reasoning within a larger knowledge-based context. Such measures provide standard similarity metrics, such as nearest neighbor methods, with support for "media features," providing they are used with appropriate contextual support from other case content.

13.4.1 Defining Media CBR

We propose a working definition of media CBR that will allow us to make finer distinctions about media in cases. While the definition itself is very simple, it requires making some subtle distinctions.

First, the definition is made in terms of raw case context, that is, the unmodified media component(s) that will in some (often significantly modified or augmented) form be used for reasoning. These represent the raw experiences that a media-enabled CBR system works with. For example, the definition would be applied to medical imagery prior to going through a specialized feature extraction/indexing process (manual or automatic) to support retrieval. If we did not make this distinction, some media-focused CBR systems might not be considered media systems at all, since their reasoning processes use only the extracted knowledge-structure counterparts as a proxy for the actual media.

Second, we presume that there is a sound reason for having the media as part of the experiential context: Either the media itself is an end (as in retrieving a particular image) or the media represents useful knowledge that, if possible to do so, is either difficult or impractical to fully encode across the semantic gap. Thus we presume that there is some value in the media itself, and that it is not simply a poor representational choice.

With these conditions in mind, we define media CBR:

> *Media CBR is CBR that will make use of media to enable or to enhance its reasoning process.*

13.4.2 Semistructured Cases

In making finer distinctions about media CBR, it is important to understand the roles that media can play in case composition. Some media CBR systems work with cases that correspond in their entirety to a media unit such as an image or audio segment. However, a case may also contain both raw media and knowledge-rich components. For example, in a diagnosis situation, the physician may represent structured demographic information about the patient (e.g., age, gender) while incorporating radiological imagery that illustrates aspects of the medical condition. Here, the media and knowledge-based components together make up the raw case experience. Thus our definition of media CBR does not presume that the media involved in reasoning comprises the whole of a case, in order to allow for such "semistructured" cases.

Semistructured cases are simple or complex cases that are composed in part of well-defined, knowledge-rich (having a rigorous semantic interpretation in the system context) structured features, as well as of ill-defined, knowledge-poor unstructured features such as raw imagery. We view this distinction on the knowledge-level, instead of the representation or implementation levels, so the fact that an image may be represented as color properties for measuring similarity does not change our view of the knowledge it affords as an image. We recognize the semantic gap as part of the similarity measure, but presume that the media itself is available upon retrieval.

A raw media-case, then, can be viewed as a case with a single "media" feature that contains the entire representation of the media element. This enables us to view case composition as a continuum of case types from fully media (single media component) to semistructured (some media and some knowledge-rich components) to fully structured (complete set of knowledge-rich components).

13.4.3 Media-Heavy/Light CBR

Given our working definition of media CBR and a view of semistructured cases, we make the following finer-grained distinctions. In media CBR, when the importance of media (in the raw experiential context) as a basis for reasoning exceeds the importance of the available knowledge-rich reasoning context, it is considered *media-heavy CBR* (MH-CBR). When the importance of media as a basis for reasoning is overshadowed by the available knowledge-rich reasoning context, it is considered *media-light CBR* (ML-CBR). There is a continuum between media-light and media-heavy CBR from the standpoint of an individual media component. When little additional case context is available, it is very important to make distinctions based on the media component in question, which will require greater discrimination power. As additional case context becomes available, it is less important to make fine distinctions based on the particular feature in question.

13.4.4 Media CBR and Semistructured Cases

In order to round out our discussion of media CBR, we consider some of the issues involved in working with semistructured cases.

The degree of media of a CBR system is closely related both to the type of case structuring and to the composition and media processing power of the system's indexing/similarity metric. For purposes of indexing and retrieval, media CBR systems can follow two related routes to varying degrees. First, cases can be transformed from more raw media to more structured representations, enabling the use of traditional knowledge-based methods. Second, similarity metrics can be made more complicated, perhaps involving multiple stages.

There is a complementary relationship in moving between the extremes of case composition. A great deal of media CBR can be viewed as enabling existing knowledge-based CBR methods by transforming full-media, fully unstructured image-cases into more structured, knowledge-rich cases. From our point of view, the complementary goal is also important; that is, enabling the augmentation of fully structured, knowledge-rich cases with fully media, knowledge-poor information by extending the functionality of standard similarity metrics to deal with media components.

The idea of moving from a single-featured media-case to a set of more knowledge-based features is compelling. We can view the process of transforming a media case into a more structured case as happening in two possible ways. First, media can be transformed into or augmented with knowledge-rich features, either manually or automatically. Second, larger media components could be transformed into multiple smaller components, each containing a portion of the original and providing a more narrowly defined contextual interpretation of the media therein. For example, an image may be segmented into regions of interest or a video may be segmented into scenes. Both help to fill-in otherwise empty contextual portions of case knowledge and can strengthen reasoning. The former enables more powerful reasoning, but the latter may require less processing.

This gives rise to two interesting and related issues. The first is whether it would be possible for a set of many contextualized media features to perform as well as some knowledge-rich representations. This question recalls the relative roles of deep and surface features in analogical reminding (e.g., [23, 30]). A related issue is whether it is possible to reduce the amount of effort required to transform a case or to apply a complex similarity measure by judging when

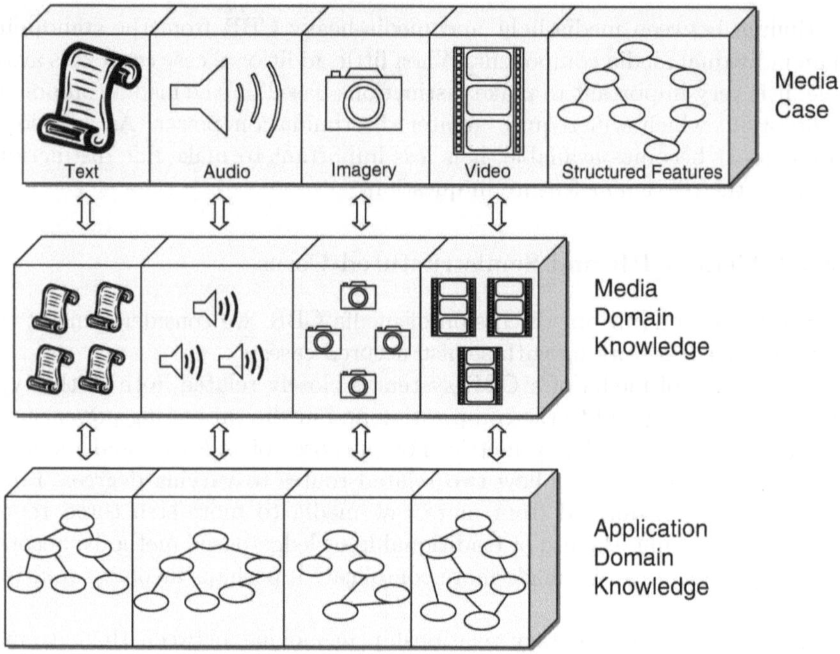

Fig. 13.1. Media case development

enough new contextual knowledge has been gained to make reasoning practicable, and discontinuing the knowledge-building effort at that point.

Figure 13.1 provides a general view of a multiple-media case, incorporating imagery, text (or markup), audio, and video. It illustrates two aspects of media feature processing (1) segmentation of raw media features according to media-specific knowledge and (2) semantic mapping of media features according to application domain knowledge. Given the background context in image processing and medical imagery, as well as our development of media CBR notions, we turn our attention to a survey of medical imagery research in case-based reasoning.

13.5 Case-Based Medical Imagery Applications

We survey eight approaches involving medical imagery in case-based reasoning. Tables 13.1–13.3 provide a framework summary of the systems in terms of primary system and metric characteristics. The systems are categorized according to a number of common image features as well as know techniques for image analysis, manipulation, and retrieval. Table 13.1 outlines each CBR image system we describe in the course of the chapter, the number of cases employed by the specified application (if available) and a brief representation of how each case is represented by the particular system. Table 13.2 outlines

Table 13.1. CBR medical image systems, cases, and case representation

Author and year	Cases	Case representation
Alexandrini et al, 2003	n/a	Case can consist of papers, printed films, clinical patient information such as anamnesis, '1qdiagnosis, prescriptions in written text form, patient data, images, waveforms, and structured reports stored DICOM format, and SNOMED terms
Balaa et al., 2003	130	140 attributes
Berger, 1992	n/a	25 features: polygons which define the geometry of tissue in the cross section for the case
Grimnes and Aamodt, 1996	n/a	Two Layers: 1st layer: case is description of a segment possibly together with tentative pathologic/anatomical hypothesis as well as any previously rejected hypotheses with justifications. Second layer: case is set of segments with diagnostic segment hypotheses together with problem description not pertaining to single segments only, as well as any previously rejected segment hypotheses with justifications
Galushka et al., 2005	n/a	18 RGB and 18 textural features extracted from each region of interest to form cases in two separate case bases
Golobardes et al., 2002	216	18 features representing microclacifications
Haddad et al., 1997	100	Each image consists of six planes; each plane is divided into 12 segments. For each segment, a value of relative thallium activity obtained by polar map analysis
Jurisica and Glasgow, 2003	n/a	11 features
Lobozek et al., 1998	35	A case is described as a collection of macroscopic areas, each of them associated to a collection of histologic areas. Each type of area is defined by a set of about ten-specific features
Perner et al., 1999	n/a	Image and nonimage features
Perner et al., 2003	n/a	Four features: color, shape, contour, object
Wilson et al., 2006	1,000	Clinical data and medical image annotations
Yearwood and Pham, 2000	n/a	A case is made up of relevant clinical data along with 11 attributes for each vertebrate line of alignment. Also six values corresponding to the intervertebral spacings and five curvature values are included

whether the specified application performs case adaption and/or retrieval, the type of media CBR implemented by the system, and the image segmentation and interpretation capabilities of the systems. Table 13.3 illustrates the

Table 13.2. Case adaption, retrieval, media CBR type, image segmentation, and interpretation

Author year	Case adaption	Case retrieval	Media CBR	Image segment	Image interpret
Alexandrini et al., 2003	No	Yes	Light	No	No
Balaa et al., 2003	No	Yes	Light	No	No
Berger, 1992	Yes	Yes	Light	No	Yes
Grimnes and Aamodt, 1996	Yes	Yes	Light	Yes	Yes
Galushka et al., 2005	Yes	Yes	Heavy	Yes	Yes
Golobardes et al., 2002	No	Yes	Heavy	Yes	Yes
Haddad et al., 1997	Yes	Yes	Heavy	Yes	Yes
Jurisica and Glasgow, 2003	No	Yes	Heavy	Yes	Yes
Lobozek et al., 1998	No	Yes	Light	No	No
Perner et al., 1999	No	Yes	Heavy	Yes	Yes
Perner et al., 2003	No	No	Light	No	Yes
Wilson et al., 2006	No	Yes	Light	No	No
Yearwood and Pham, 2000	No	Yes	Light	No	No

Table 13.3. Pattern recognition, content-based features, explicit, and implicit image features

Author year	Pattern recog	Color	Shape	Edge	Texture	Impl feat	Expl feat
Alexandrini et al., 2003	No	No	No	No	No	No	Yes
Balaa et al., 2003	Yes	No	No	No	No	No	Yes
Berger, 1992	No	No	Yes	No	No	Yes	Yes
Grimnes and Aamodt, 1996	No	No	No	No	No	Yes	Yes
Galushka et al., 2005	No	Yes	No	No	Yes	Yes	No
Golobardes et al., 2002	No	No	No	No	No	Yes	No
Haddad et al., 1997	No	No	No	No	No	Yes	No
Jurisica and Glasgow, 2003	No	No	Yes	Yes	No	Yes	Yes
Lobozek et al., 1998	No	No	No	No	No	No	Yes
Perner et al., 1999	No	Yes	No	No	No	Yes	Yes
Perner et al., 2003	No	Yes	Yes	No	No	Yes	No
Wilson et al., 2006	No	No	No	No	No	No	Yes
Yearwood and Pham, 2000	No	No	No	Yes	No	Yes	Yes

pattern recognition and content-based retrieval capabilities of each system in terms of image similarity matching performed using color, shapes, edge detection, and textural analysis. Finally the table outlines whether images are compared and retrieved using implicit and/or explicit features.

13.5.1 FM-Ultranet

FM-Ultranet [7] is used in the domain of ultrasonography and aims to detect malformations and abnormalities of fetus through ultrasonographical examinations. Ultrasonographists decide whether a fetal abnormality is a dangerous malformation and if a pregnancy should be terminated or if the abnormality is just a particularity without importance. To come to this decision, an ultrasonographist may use several different knowledge sources, medical literature, expert professional, and tacit knowledge and/or the advice and second opinion of a colleague expert. FM-Ultranet attempts to automatically relate these knowledge sources using CBR to represent and leverage medical literature and expert tacit knowledge and by providing a mechanism to electronically link and network ultrasonography clinicians.

The case base consists of 130 cases which are arranged in a hierarchical and object oriented structure. Each case is composed of about 140 attributes. These 140 attributes are organized in 39 concepts and subconcepts to represent characteristics in a medical sense. Most of the attributes store knowledge about the anatomical structure of the fetus, like the urinary tract and the morphology of the head, thorax, or rachis. The case representation also includes clinical data of the mother (medical history), ultrasound imaging, bibliographic references, outcome of the pregnancy, the status of the child at birth, the result of the autopsy, and the international classification of diseases code.

According to the authors, ultrasound scans interpretation and diagnosis can largely be improved through comparison of past existing similar cases and with guidance of an expert. Therefore a CBR approach is employed for the task. The FM-Ultranet tool collects the clinical data at the ultrasonographer's workplace. Data collection is done easily from a single page that offers all relevant entries for a sane fetus. In the case of an abnormal value the related detailed descriptions are provided. Instead of creating a report manually, the examiner annotates picture information where necessary to the filled attributes and then generates his case documentation report with a single click.

This tool is built on top of the CBR works engine, which is used to retrieve the most similar past cases from a reference case base as well as from the ultrasonographist's local case base. Case data is utilized to support the decision process, i.e., the ultrasonographist can compare the found cases with his present examination. No adaptation of cases is performed.

The system retrieves similar cases/images by calculating similarity between attributes in the concepts. This is calculated mathematically or compared through a look up table, depending on the attribute type. No content-based image processing is performed by the application. Multimedia support is also provided by making the database available for consultation by other ultrasonographists and offering the possibility to submit cases to experts for advice. Finally a report of system findings is generated when the detection CBR process is completed.

The system was evaluated during a 3 months trial by 11 medical doctors and three experts, in hospitals in Nmes in France and Lige in Belgium. The application aimed to help ultrasonographists to improve their diagnostic skills by providing comparison with existing clinical cases, exploitation of their own past experience, and a means to contact external experts. With a unanimous "yes" answer to the question on usefulness and with 45% of the testing population having acquired new knowledge, the FM-Ultranet system was deemed to have met ultrasonographers needs.

13.5.2 Roentgen

Roentgen [9] is an application of case-based planning in the radiation therapy domain for thorax cancer. In radiation therapy, beams of radiation high-energy photons or particles are directed at cancerous tissue in the patient's body. Since many vital tissues are highly sensitive to radiation extreme care must be used when deciding how to position the radiation beams, what energies to use, and what the beam cross sections should be. Radiation therapy plans are developed by a dosimetrist, an oncology specialist. The dosimetrist uses past planning experience to develop a plan for a patient once the physician has determined what region of the body is to be treated and the amount of dose to be delivered. The physician decides what plan to use based on the work of the dosimetrist.

Roentgen employs an archive of past therapy cases to suggest therapy plans for new patients. It processes the images from the previous cases using content-based image techniques. The image processing is based on implicit geometric image features. Each case/image in the system contains polygons which define the outline of each tissue in the anatomical cross section for the case. For each of the important tissues – target, spinal cord, lungs, body outline – there is a corresponding list of coordinates which defines a polygon representing the tissue. Furthermore, a list of features which capture the essential geometric information which characterizes a patient with respect to the physics of radiation therapy is stored. For thorax cancer patients therefore the application computes the following geometric features for the lungs, spinal cord, and body outline:

Area: The area of the polygon. For the body outline, this is the absolute area and for other tissues it is the area as a proportion of the total body area.
Eccentricity: This feature parallels the subjective perception of how "elongated" the corresponding tissue is.
Orientation: This feature is associated with the subjective perception of the direction in which a tissue points.
Rho: The distance from the target centroid to the centroid of the polygon of the tissue.
Theta: The angle formed by the vector from the target centroid to the tissue centroid with the positive X-axis.

Just the first three parameters are computed for the target. All five parameters are computed for the other tissues. In addition, two other parameters are computed which are concerned with the geometry of the body outline. They are:

Apx: The posterior to anterior extent of the body outline in centimeters.
Rlrx: The relative right to left extent of the body outline as a proportion of the anterior to posterior extent.

Roentgen is integrated into the therapy design task. When designing a new therapy plan the dosimetrist can search using Roentgen for an initial plan suggestion. Each entry for a therapy plan contains information describing the geometry (polygons which define the outline of each tissue in the cross section for the case) and dose requirements of the particular plan. Since the dose distribution produced by a plan is determined by the patient's internal geometry, a good match should resemble the new patient geometrically. In addition, the doses delivered to important tissues in the patient by the retrieved plan should meet the dose constraints for the new patient. Roentgen addresses these two requirements by calculating a matching score which is the product of a geometry subscore and a dose requirement subscore. The retrieval and recommendation of new plans is therefore calculated using both implicit image features (geometry) and explicit patient features (dosage).

Roentgen supports case-based therapy planning in following ways. Firstly it retrieves the case which best matches the geometry and treatment constraints of the new patient from a knowledge base of previous cases. It adapts this plan by tailoring it to the specific details of the current patient. The effect of applying the plan to the current patient is evaluated and if detrimental effects are discovered the plan can be repaired to avoid faults in treatment processes. This final plan is then recommended by the system to the human planner. Roentgen solves problems in a domain dominated by spatial reasoning by applying theories of constraint satisfaction.

13.5.3 Image Creek

This work attempts the task of medical image understanding, in particular computer tomography (CT) imagery. Cases in the ImageCREEK application [26] combine lower level segment identification with a higher level interpretation and understanding of the image as a whole. In order to do this radiologist knowledge in the form of high-level interpretation of image content is captured and combined with low-level structure analysis. Therefore cases consist of both image and nonimage attributes.

The ImageCreek architecture is designed and implemented within the Creek system for knowledge-intensive case-based problem solving and learning. Cases, as well as general domain knowledge and information are captured in the frame-based representation language CreekL, where a knowledge model represented in CreekL is viewed as a dense semantic network. The case-based method of Creek relies heavily on an extensive body of general domain knowledge in its problem understanding, similarity assessment, case adaptation, and learning. The underlying case-based interpreter in Creek contains a three-step process of (1) activating relevant parts of the semantic network, (2) explaining derived consequences and new information, and (3) focusing toward a conclusion that conforms to the task goal. This "activate-explain-focus" cycle, is a general mechanism that has been specialized for each of the four major reasoning tasks of the CBR cycle (retrieve, reuse, revise, retain).

Using the case-based activate-explain-focus cycle a novel architecture for medical image understanding was proposed. The architecture incorporated information about how radiologists diagnose patients as well the mechanics of current image processing algorithms. Therefore a two-layered architecture, corresponding to two case bases storing these two different kinds of experience and supporting two different kinds of solutions was developed. One layer lays on top of the other and both employ a "propose-verify-critique-modify" framework. Specifically the two layers of the application compose the Segment ImageCreek and the Wholistic ImageCreek layers. These two layers operate as follows:

The Segment ImageCreek layer works with image segments (i.e., subsets of the image sharing some similarity) in isolation. In the case base, a case is a description of a segment possibly together with a tentative pathologic/anatomical hypothesis as well as any previously rejected hypotheses with justifications. A segment hypothesis is such a description and some of the hypotheses may be pathological labels; others may be normal anatomical labels and some may neither be pathological nor anatomical due to imaging or therapeutic artifacts. In this layer only one segment description at a time is examined and a segment hypothesis is suggested for each single segment.

The Wholistic ImageCreek layer works with the entire image in question. An overall image interpretation is achieved using different segments with suitable segment hypotheses that fit in with each other and with the general problem description context. In the case base, a case is a set of segments with diagnostic segment hypotheses together with the problem description not pertaining to single segments only, as well as any previously rejected segment hypotheses with justifications. In this layer the idea is to look at the broader aspects and the totality of the all the segment hypotheses in light of the problem description context and the findings not pertaining to a particular segment only.

Image interpretation is performed using both implicit and explicit image features. Implicit features used are image segments which are acquired using the publicly available Khoros image processing environment, for segmenting images from CT scanners. A segment, in this context, refers to an area of the image corresponding to an anatomical object or a significant part of one, for example an area picturing a liver. Explicit features used in the ImageCreek application correspond to the high-level interpretation of image content provided by radiographers.

13.5.4 SCINA

Haddad et al. [29] developed the SCINA application to perform Myocardial perfusion scintigraphy, a noninvasive diagnostic method for the evaluation of patients with suspected or proven coronary artery disease (CAD). The SCINA application interprets medical images using implicit image features extracted by image segmentation.

A case library was compiled of 100 patients who underwent both stress perfusion scintigraphy and coronary angiography to document or exclude the presence of significant CAD. The study group was obtained by a retrospective search in databases of our clinic and included 77 men and 23 women with a mean age of 59. Each patient scintigraphic image consists of six planes; with each plane divided into 12 segments. For each segment, a value of the relative thallium activity obtained by polar map analysis is determined. This processing is performed using OSIRIS software and polar map data were extracted from the six short axis and saved as 2 x 6, 4 x 6, 6 x 6, and 6 x 12 matrices of integer numbers. Therefore, a case in the case library consists of the following features: name, sex, and age of the patient, 12, 24, 36, or 72 integer values obtained by polar map analysis of the scintigraphic image one binary value representing the result of the visual analysis of the coronary angiography (1 or 0 for the presence/absence of CAD).

SCINA uses case-based reasoning for image interpretation by deriving an assessment concerning the presence of CAD from scintigraphic image data. The full case-based cycle of retrieve, reuse, revise, and retain is implemented. The retrieval, which is performed by nearest neighbor match, considers the integer values obtained by polar map analysis. Those values from segments of planes placed in the middle are weighted high, those from the edges are weighted low. In order to define the similarity metric, two observers independently processed scintigraphic images of ten patients using OSIRIS software. Respective polar map data were compared and three different intervals were used as tolerance ranges: + 1 standard deviation (SD), + 2 standard deviations, and + 2.5 standard deviations of the difference in segmental tracer uptake. Numeric values for the difference between the current case and the case in the library that are within the tolerance range were considered a complete or partial match depending on the similarity metric.

Two different adaptation strategies were implemented and tested. The first adaptation strategy was developed based on a mapping between the main coronary vessels and the polar map according to a schematic model of the coronary circulation. If a segmental activity value on the polar map of the target case was significantly lower than the corresponding value of the most similar case, the score of the respective vessel was increased to indicate more severe CAD in this vascular territory and vice versa. Thus, the model of the coronary anatomy together with scintigraphic data from the current case and the case library were used to adapt the output of the CBR application.

For the second adaptation strategy, the binary number reflecting the presence or absence of CAD of the five retrieved cases with the highest similarity scores were used to derive an assessment concerning the presence of CAD in the current patient. This adaptation algorithm uses the differences in the input domain between retrieved cases and the current case to calculate a data point in the output domain that corresponds to the angiographic evidence of CAD.

SCINA is implemented in the CBR programming shell ESTEEM. The final prototype of SCINA reported the sensitivity and specificity for detection of coronary heart disease at 98% and 70%, respectively, suggesting the diagnostic accuracy is feasible for clinical use.

13.5.5 $TA3_{IVF}$

Jurisica and Glasgow [38] describe the application of automated image analysis to evaluate morphology and developmental features of oocytes and embryos in the domain of in vitro fertilization (IVF). An important aspect of this research is in its focus on a domain where processing only symbolic information is not sufficient and where one representation formalism is not adequate to satisfy diverse users and tasks.

In the application a case is composed of 11 features extracted from the medical images. The features are represented by numeric attributes. The aim of the work is to combine image analysis techniques with case-based reasoning to provide an application that can: serve as a feature extraction technique, serve as an indexing approach and serve as an image analysis tool. Case retrieval of IVF images is implemented by the application.

The system uses implicit image features for image processing. This is performed using computer-based morphometry to precisely and objectively identify developmental features of oocytes and embryos. In order to perform morphological analysis of oocyte and embryo images the authors chose to use deformable models, which are model-based techniques for image analysis. They are used in medical image analysis for image segmentation, matching and deriving image object size, shape, and location. In this system the available models are recognized images and the task of deformable models is proper image alignment recognition. For example, the system has a repository of models of embryos with different quality. Then the task is to align the new embryo with existing models.

Deformable models are also used to perform automated image-feature extraction from the images of the embryos. This information can be used to classify the quality of the embryo and analyze its morphology, which in turn can be used in conjunction with other clinical attributes during decision making. Finally deformable models are applied to identifying dynamic properties of images. Namely, sequences (or groups) of models may be compared to another group to identify dynamic or evolving images (e.g., development of an embryo over a period of time).

Image preprocessing isolates the region of interest (ROI) in the image and attempts to standardize images with regards to lighting and size. The image processing system is divided into the analysis and retrieval modules. The analysis module is used to extract features from the raw image to create a case. This is performed by evaluating the morphology and developmental features of oocytes and embryos (including cell number, fragmentation, cellular appearance, zona thickness, etc.). The retrieval module uses variable context retrieval. Each case is assumed to have only one attribute, which is a two-dimensional array of values. Contexts are represented by two arrays indicating a maximum and a minimum. Thus, the domain of a context is a continuous range of values. Post-processing then implements the classification strategy.

13.5.6 BIOGEFA Airborne Fungi

Perner et al. [57] have developed a case-based system for the detection and identification of airborne fungi using an image acquisition and interpretation system. The images used originate from microscope enhanced pictures. Cases in the case base are image descriptions which are automatically extracted from the images themselves. Each case consists of the solution which is the type of fungi spores and the features describing the visual properties of the object. The features are color, shape, special properties inside the objects such as structure inside, size, and appearance of the cell contour. The image processing therefore makes use of implicit image features.

The fungal strains have a high-biological variability, i.e., dissimilarity between the features of individual fungi is quite extensive. A strain cannot be generalized to a few cases because of this variability. Therefore a case-based reasoning approach for the image interpretation rather than a generalized approach is undertaken. Similarity between an actual case and cases in case base is determined using Euclidean distance. The initial case base is a flat case base. An index structure is incrementally learnt as soon as new cases are input into the case base using a decision tree. The reasoning system also allows for the learning of a more compact case description.

13.5.7 Image Segmentation System

Perner [56] proposes a system that uses CBR to optimize image segmentation at the low-level unit according to changing image acquisition conditions and image quality. The system has been used to detect degenerative brain disease in particular Alzheimer disease in CT images of a patient and was implemented at the Radiology Department at the University of Halle.

Cases are comprised of image features as well as nonimage information about the image acquisition and the patient. Image information is described by statistical measures of the gray level like: mean, variance, skewness, kurtosis, variation coefficient, energy, entropy, and centroid. For CT-images, nonimage information consists of patient-specific parameters, slice thickness, and scanning sequence. This information is contained in the header of the CT image file.

The case-based reasoning unit for image segmentation consists of a case base, in which formerly processed cases are stored by their original images, their nonimage and their image segmentation parameters. The task is now to find the best segmentation for the current image by looking up the case base for similar cases, CBR is used to select the segmentation parameter according to the current image characteristics. This is done using both implicit image features (segmentation calculated using gray-level histograms) and explicit nonimage features (patient data).

In an offline phase, the best segmentation parameters for the image are determined and the attributes, which are necessary for similarity determination, are calculated from the image. The similarity measures for the image and nonimage information are combined to calculate an overall similarity measure. Both, the segmentation parameters and the attributes calculated from the image, are stored into the case base as new case. During storage, case generalization will be done to ensure that the case base will not become too large.

The result of the segmentation process is observed by the user. The original image can be compared with the labeled image on display. If the user detects deviations of the marked areas in the segmented image from the objects area in the original image, which should be labeled then the result will be evaluated as a bad result and case base management will start. This will also be done if no similar case is available in the case base. The evaluation procedure can also be done automatically. However, the drawback is that there is no general procedure available, so it must be developed in a domain dependent fashion. Therefore, an automatic evaluation procedure would constrain the usage of the system.

13.5.8 CyclopsDistMedDB

Alexandrini et al. [5] argue that CBR technology has proven its usefulness in supporting reasoning for medical domains, and they advocate a research focus on the integration of such "CBR-modules" into systems supporting medical processes and healthcare organizations' workflow. In this way, CBR-modules can be accessed at the place where diagnosis is performed and in the time when it is performed. In the proposed application, a case can be represented by a number of heterogeneous components including, papers, printed films, clinical patient information such as anamnesis, diagnosis, prescriptions in written text form, patient data, images, waveform. This information for each case is converted to a structured report in a DICOM format and assigned a number of SNOMED terms that accurately describe its content. This application integrates medical image data into a case with other patient information and retrieves image data using only explicit patient information and not by using any image processing techniques.

The CBR component being developed by the authors will be integrated in an existing transparent gateway for distributed healthcare information access to enable the collaboration of several health care organizations. The authors have previously developed a number of computer-based medical systems which are used in hospitals and clinics in Germany and Brazil. Each of these systems are connected using CyclopsDistMedDB, a transparent gateway for distributed data access in DICOM format. CyclopsDistMedDB is Client/Server based and is installed on every computer within the partner hospitals, so during every process step (registration of the patient, anamnesis, diagnosis, etc.) all information stored for a patient can be accessed and specialized tools can be triggered. CyclopsDistMedDB supports collaboration between several healthcare organizations by allowing users to send DICOM images or DICOM structured reports via e-mail. The CBR component described in this article is an attempt to extend these collaboration possibilities by allowing access to shared database from several incorporated healthcare organizations.

The CBR approach used by the authors is used for case storage and retrieval. The authors envisage the retrieval component being used in the following way: A physician wants to find all relevant information for a patient, e.g., he or she uses a DICOM structured report of a current patient's case as query and can retrieve suitable old reports in text form, relevant passages from medical text books, and DICOM structured reports of other similar cases. This helps to shorten examination time, because the physician can access all relevant information using their computer. Furthermore the approach lends itself to the storage and preservation of information, for example information from old medical texts can be retrieved and can be converted to a new pattern like a DICOM structured report. The system also has additional benefits for new or inexperienced health professionals as they can use the database to find similar cases to compare with patients they are currently diagnosing. The application also provides a shared database of knowledge that can be accessed by experts looking for solutions in the case of difficult or unusual cases.

13.5.9 IDEM

The IDEM system [43] is a Web-based application for querying and browsing histopathology images by exploring a collection of illustrated medical cases. In the histopathology domain physicians usually write a description of a case in natural language terms in a standardized report. By analyzing reports, a case representation strategy was developed. A case is described as a collection of macroscopic areas, each of them associated to a collection of histologic areas where each histologic area can contain several histologic areas as well as cytological descriptions. Each type of area is defined by approximately ten features. As well as defining each case in terms of semantic information from the reports, the cases are also defined in terms of a tree structure by translating data from the physician reports. Each node of a tree corresponds to the description of an important macroscopic or histologic area, according to the diagnostic conclusion. Therefore in this research only explicit image features are used to perform similarity matching and no image processing based on implicit features is performed.

A similarity metric was constructed where cases are compared using their composite features both in terms of their semantic and structural characteristics. Semantic characteristics are given by the set of features extracted from physician reports. Structural characteristics correspond to the tree structure of the cases. A global similarity is calculated as the arithmetic mean of both characteristics. CBR retrieval is implemented to be used both in a querying and browsing module, as follows. No case adaptation occurs within the application.

Querying with CBR. Target documents are retrieved using the similarity metric. The query functionality aims at retrieving the images and diagnosis attached to a case of the base that are most similar to a new case. The new case has the same format than the cases in the base. The system output consists of two main parts corresponding to the description of the most relevant case and to the justification of its relevancy

Browsing with CBR. The CBR approach allows an expert proximity between cases to be defined and justified. A browsing path is created in the case base by creating hypertext links between the nearest cases using the outline tree structure. As a result of the tree structure the similarity measure can provide a quantification of the browsing process. The user can always know to what extent the case he/she watches is close to the previous one through the similarity links.

13.5.10 CaB-CS

Golobardes et al. [25] are concerned with computer aided diagnosis of breast cancer using mammographic images. Diagnosis is performed using microcalcifications from the mammographic images. Cases in the application are represented by 18 microclacification features which appear on mammographic images. Some of these features include the area, perimeter, and compactness of the microclacification. The image processing method uses only these implicit features.

The diagnostic procedure can be broken down into four main steps. These are (1) digitizing the mammographic image, (2) processing the image, (3) performing microcalcification identification and feature extraction, and finally (4) using machine learning techniques to diagnose the processed mammogram automatically.

The first step is straightforward and the radiological mammographic image is digitized. In steps two and three the microcalcifications are analyzed and characterized through the extraction of features and visibility descriptors. This is performed by means of different image processing techniques such as gray-level image analysis, signal processing algorithms, or morphological methods. The images are then segmented and the digitization and segmentation processes transform the original gray-level image into a binary image, where the background tissue is removed and clustered microcalcifications appear. Finally, in the fourth step (the focus of the article), machine learning techniques are applied to diagnose the images where images can be diagnosed as malign, benign, or unclassified.

The two machine learning techniques used in the work are case-based reasoning and genetic algorithms. Firstly, both algorithms are compared to results obtained by human experts and statistical models. Secondly the algorithms are compared to six other well-known machine learning techniques.

The case-based classified system CaB-CS is used in this research. In this system the reuse phase of the CBR cycle is simplified as it classifies a new case using the same class of the most similar retrieved one. In CaB-CS, the notion of similarity between two cases is computed using different similarity measures. In the research outlined in this chapter a number of different similarity functions are used for diagnosis. The different similarity functions used can be classified in two groups: functions based on distance metrics and functions based on spheres.

For functions based on distance metrics the authors utilized Minkowski's metric and Clark's distance similarity metrics. For functions based on spheres the authors used the Proximity Sphere and the MinMax Sphere which compute the similarity between two cases using local similarity measures. They also used the Mean Sphere which computes similarity using a global similarity measure.

An evaluation of the application consists of two main phases. Firstly, the case-based reasoning and genetic algorithm approaches are compared with previous results obtained by human experts and statistical models. Secondly the results achieved by the case-based reasoning and genetic algorithms are compared with six classifier schemes provided by different machine learning theories. These classifiers are instance-base learning (IB1 and IBk with k=3), statistical modeling, Naive Bayes, tree induction (C4.5), rule learning (PART), and support vector machines (SVM). The algorithms were compared in terms of classification accuracy, sensitivity, and specificity.

In the first phase of the evaluation, results demonstrated that the accuracy achieved by case-based reasoning and genetic algorithms overcome the accuracy obtained by the human experts and the statistical model. The highest accuracy achieved was 77.14 and was calculated by a case-based classifier using Clark's distance metric. However, human experts and the statistical model showed better sensitivity and specificity than those calculated by the case-based and genetic algorithms. This is as a result of human experts not classifying uncertain examples.

In the second phase of the evaluation the average accuracy rate across the case-based and genetic algorithm approaches as well as the six other machine learning techniques tended to be very similar. This was also the case for sensitivity and specificity scores with all of the algorithms tending to be more specific than sensitive.

13.5.11 Telemedicine for Cooperative Medical Diagnosis

Yearwood and Pham [74] are concerned with the domain of telemedicine (specifically teleradiology) and providing medical services to people living in remote areas with a lack of medical specialists. They identify a number of obstacles to implementing applications in the domain of teleradiology. Firstly expensive equipment is required to collect images digitally rather than on film. Secondly there are problems associated with the storage of large amounts of digital data produced by the different medical imaging techniques. Thirdly preserving the integrity and diagnostic quality of images sent over networks and transmission lines is a difficult task and finally it is also necessary to have appropriate equipment at the other end to present a diagnostic quality display of images.

To solve some of these issues the authors have developed a case-based environment that allows physicians to exchange ideas and to collaborate and cooperate in diagnosis with centrally located medical specialists. Specifically the remote database allows access to medical images, retrieval of relevant medical cases to support diagnosis, and communication among participants through telepointers and image annotation by free-hand drawing. In this article the author describe research performed using images that depict cervical spine injuries.

A case in the case base is represented by relevant patient clinical data along with 11 extracted features. Six distance values corresponding to the intervertebral spacings and five curvature values are also stored.

Feature extraction can be carried out on clients for the purpose of real-time collaboration and discussion. Features are extracted using content-based image techniques including edge detection, corner identification, and contour following. For the particular problem of cervical spine trauma a fixed set of features is extracted and the cases in the case base all contain these features. The features extracted are those identified by radiologists to confirm the normality of the spine and to identify any deviation from normal, including lines of alignment, bony contours, craniocervical junction, disc spaces, facet joint spaces, vertebral spacing, and prevertebral soft tissues. In total a case is made up of relevant clinical data along with 11 extracted features. Six distance values corresponding to the intervertebral spacings and five curvature values are also stored.

The authors presented their case-based component of the application in the design phase. It was envisaged that the already developed feature extraction application would integrate with a hospital PACs system as well as a case-based reasoner. The application would use case query language (CQL) which is the query language of the CBR works shell, to query for a target case. This language would permit querying on individual features or queries based on a combination of features in a similarity measure.

13.5.12 Tissue Classification

Galushka et al. [22] propose a case-based monitoring system for leg ulcers. In the article they concentrate on the first stage of the monitoring process which is that of tissue classification. The wound healing process is assessed by examining the base of wounds, which consists of many different tissue types such as granulation, slough, necrotic, and epithelial tissues. Muscle, tendon, or haematoma may also be present. The color of the wound bed gives an indication of these tissue types and thus the particular phase of healing. In order to automatically monitor the wound healing process, it is important to assess different tissue types continuously, and compare the process on consecutive visits. Therefore the importance of color to tissue classification is obvious but the researchers also felt that texture may also have an important part to play and classification based on this feature was also examined. Cases in the case base were represented in terms of color and textural feature and image analysis was based only on these implicit image features.

The extraction of color features from the imagery was performed using RGB histograms. The extraction consisted of two steps. Firstly in the signal processing step the histogram was smoothed by a low-pass filter to remove high-frequency fluctuations. In the second step the three highest peaks in the histogram are detected which correspond to the highest intensity level elements contained in a ROI in the image.

In extracting textural features three features were used: angular second-moment feature, contrast feature, and correlation feature. The angular second-moment feature is a measure of the homogeneity of an image (or ROI). The contrast is a measure of the contrast or amount of local variation present in an image. The correlation is a measure of gray-tone linear dependencies within the image.

Using a CBR methodology, the classification procedure consists of three steps: feature extraction, retrieval, and adaptation. Each image is split into regions of cells by applying a grid structure during a pre-processing step. Each cell element is 10x10 pixels in size and equates to a ROI. 18 RGB and 18 textural features are extracted from each ROI to form cases in two separate case bases (one based on RGB features, the other on textural features). The ten closest cases, which correspond to the regions of interest with the most similar feature values, are selected from the case base during the retrieval process. Retrieval is carried out by k nearest neighbor algorithm ($k = 10$) and the majority class was used to classify the target ROI.

In an evaluation, classification using RGB features produced both very high-average accuracies, whereas using textural features produced much lower accuracy values. The results indicate that RGB features are better used in multiclass tissue classification, while textural feature are less effective for this type of classification. This is significant as textural features had previously proven useful in wound image segmentation; however it is less accurate in this application due to the small size of the ROI's (10x10 pixels). The results also show that the CBR approach to tissue classification has advantages over other classifiers reported in the literature, such as logistic regression, ANN, and SVM.

13.5.13 MEDIC

O'Sullivan et al [52,53] have developed a clinical decision support system that integrates patient medical imagery with other patient data within an EHR system. Cases in the application are therefore represented by textual patient data such as demographics, clinical information, and laboratory test results as well as any medical imagery and associated medical image annotations. No image processing using implicit features is performed; rather images are retrieved using explicit image data in the form of patient information and image annotations.

The authors state that medical diagnosis and decision making involves interplay between vast numbers of medical knowledge resources. This ranges from explicit patient information to implicit knowledge held by caregivers to experiential and data-induced knowledge. However many decision support systems fail to capture and take advantage of these complementary knowledge sources. The authors state that CBR provides excellent methods and opportunities for combining and representing different types of medical knowledge and patient data in such applications. This is because one of the intuitively attractive features of CBR in medicine is that the concepts of patient and disease lend themselves naturally to a case representation. Also medical practitioners logically approach diagnosis from a case-based standpoint (i.e., previous patient interactions are as strong a factor as individual symptoms in making a diagnosis).

The application allows doctors to wirelessly input, query, and compare electronic patient records including associated medical imagery on any mobile or desktop device. The type of functionality provided includes a user interface that allows caregivers to input patient information in a straightforward manner and dedicated multimedia annotation tools for medical imagery that support communication and collaboration between different caregivers as well as management of medical image resources. All important patient information including medical images and annotations, endoscopies, and physician dictations are integrated into encapsulated cases in a multimedia case base of patient profiles. This knowledge base is used to facilitate decision support for all healthcare personnel as similar patient cases can be quickly and easily retrieved for comparison of patient information. As well as capturing and recording all patient information, the application attempts to leverage insights from a caregiver's decisions and rationale as they diagnose and treat patients. As a caregiver interacts with a patient profile during a diagnosis the system records their actions. This enables the capture of human expertise and proficiency in diagnosing and treating the particular illness which in turn allows us to understand why relevant information was accessed and investigated. Once this expert domain knowledge is captured, it is combined with relevant patient medical data and stored in the case base of encapsulated patient cases. This information can then be used to filter, retrieve, and display the most relevant similar patient case histories from huge medical repositories for comparison of diagnoses and treatment procedures with newly presenting patients.

The application attempts to overcome difficulties associated with the semantic gap in image retrieval by uniting information about underlying visual data with more high-level concepts provided by healthcare professionals as they interact with medical imagery. For example, capturing a measure of human expertise and proficiency involved in making a diagnosis from an X-ray gives an insight into why relevant information was selected (e.g., the highlighting a particular body organ) and also how it was employed in the context of the diagnosis (e.g., inferred from added annotations). This is implemented using a set of image annotation tools. The tools can accept and integrate with a number of common medical image formats (e.g., DICOM, X-ray) and allow caregivers to formulate aspects of their task in diagnosing patients. Annotation information is collected implicitly to shield caregivers from the burden of explicit knowledge engineering. From their perspective, the image interaction tools support them in carrying out their task (e.g., producing a report on the current patient) by making it easier for them to select and highlight relevant features, to store insights and to summarize aspects of their work. However from a system perspective the tools monitor and record the clinician's actions and ultimately capture contextual diagnostic knowledge to improve the ability of the application to recommend other profiles for comparing diagnostic information and treatment procedure.

Case retrieval within the system is taking place in the context of an overall workflow. Some of the most important steps in this workflow (which may or may not be relevant to all patients) are: entering preliminary patient details, recording results of initial examinations, inputting symptoms, uploading and annotating medical imagery, recording diagnoses, and recommending treatments. Each of these steps is stored as a textual feature within a patent case and each case is indexed along each of these constituent features. Using these textual indices textual queries imputed by a caregiver about current patients are matched against previous patient contexts. The retrieval system employs indexes in separate spaces across constituent segments of the patient profile. When a caregiver enters a query, the parameters and any weights specified by caregivers are combined and compared to local features from other previous patient profiles in the case base. A weighted average is used to compute similarity between the current patient and other patients in the central database.

13.6 Conclusion

In case-based reasoning research, image processing and the medical domain are strongly coupled. While the two areas of image processing and medical applications have both been identified as primary topics in CBR research, this crossover has not typically been addressed as a whole. We have examined the conjunction of the two areas in light of ongoing developments in medical information systems, in order to provide a novel perspective and overview of the main issues and research work on medical imagery in case-based reasoning. We expect that this survey will provide case-based reasoning researchers with a good reference point and perspective for new medical imagery developments. Medical informatics in general, and imagery in particular, is an increasingly important application domain, and we look forward to significant advances from the case-based reasoning community.

References

1. *Breast Imaging Reporting and Data System Atlas (BI-RADS Atlas).* The American College of Radiology, 2003.
2. Syed Sibte Raza Abidi. Knowledge management in healthcare: Towards "knowledge-driven" decision-support services. *International Journal of Medical Informatics*, 63:5–18, 2001.
3. Syed Sibte Raza Abidi. Medical knowledge morphing: Towards case-specific integration of heterogeneous medical knowledge resources. In *Proceedings of the Eighteenth IEEE Symposium on Computer-Based Medical Systems*, 2005.
4. David Aha and Patrick Harrison. Case-based sonogram classification. Technical Report Technical Report NRL/FR/5510-94-9707, Also listed as NCARAI TR: AIC-93-041, Naval Research Center, Navy Center for Applied Research in Artificial Intelligence, Washington, DC, 1994.
5. Fábio Alexandrini, Dirk Krechel, Kerstin Maximini, and Aldo von Wangenheim. Integrating cbr into the health care organization. In *Proceedings of the 16th IEEE Symposium on Computer-Based Medical Systems*, pages 130–135. IEEE Computer Society, 2003.
6. Yu-N Cheah and Syed Sibte Raza Abidi. Augmenting knowledge-based medical systems with tacit healthcare expertise: Towards an intelligent tacit knowledge acquisition info-structure. In *Proceedings of the 14th IEEE Symposium on Computer-Based Medical Systems*, pages 264–269, 2001.
7. Z.E. Balaa, A. Strauss, P. Uziel, K. Maximini, and R. Traphöner. FM-Ultranet: a decision support system using case-based reasoning, applied to ultrasonography. In *Proceedings of the ICCBR-03 Workshop on CBR in the Health Sciences*, 2003.
8. E. Bellon, M. Feron, T. Deprez, H. Pauwels, M. Vanautgaerden, A. De Deurwaerder, R. Reynders, W. Reviers, B. Draelants, P. Suetens, G. Marchal, and B. Van Den Bosch. Integrating images into a central medical information system. *Studies in Health Technology and Informatics*, 2002.
9. J. Berger. Roentgen: Case-based reasoning and radiation therapy planning. In *Proceedings of the 16th Annual Symposium on Computer Applications in Medical Care*, 1992.

10. Isabelle Bichindaritz. Case-based reasoning in the health sciences. *Artificial Intelligence in Medicine*, 36(2):121–125, 2006. Guest Editorial.
11. Robin Burke and Alex Kass. Supporting learning through active retrieval of video stories. *Expert Systems with Applications*, 9(3):361–378, 1995.
12. C. Carson, S. Belongie, H. Greenspan, and J. Malik. Region-based image querying. In *Proceedings of the IEEE Workshop on Content-Based Access of Image and Video Libraries*, pages 42–49, San Juan, Puerto Rico, 1997.
13. S.D. Cohen and L.J. Guibas. Shape-based indexing and retrieval; some first steps. In *Proceedings of the 1996 ARPA Image Understanding Workshop*, volume 2, pages 1209–1212, 1996.
14. Carl-Helmut Coulon. Image retrieval without recognition. In *First European Workshop on Case-Based Reasoning (EWCBR'93), Posters and Presentations*, volume 2 of *SEKI-Report 93-12-2*, pages 399–402, 1993.
15. Ellen Yi-Luen Do and Mark D. Gross. Reasoning about cases with diagrams. In *American Society of Civil Engineers (ASCE) 3rd Congress on Computing in Civil Engineering*, pages 314–320, 1996.
16. Electronic health records overview. National Institutes of Health National Center for Research Resources, April 2006.
17. D.A. Forsyth et al. Finding pictures of objects in large collections of images. In *Proceedings of the ECCV 96 Workshop on Object Representation*, 1996.
18. M. Flickner et al. Query by image and video content: The QBIC system. *IEEE Computer*, 28(9):23–32, 1995.
19. G. Fala and J. Walker. Using case-based reasoning to automate acoustic submarine classification: Detailed summary of technical progress. ONR Computer Science Division Program Summary, Fiscal Year 1992, 1993.
20. Gabby Fennessy. Knowledge management in evidence-based healthcare: Issues raised when specialist information services search for the evidence. *Health Informatics Journal (2001)* 7(1):4–7, 2001.
21. C. Frankel, M. Swain, and W. Athitsos. Webseer: An image search engine for the World Wide web. Technical Report TR-96-14, Department of Computer Science, University of Chicago, 1996.
22. Mykola Galushka, H. Zheng, David Patterson, and L. Bradley. Case-based tissue classification for monitoring leg ulcer healing. In *Proceedings of the 18th IEEE Symposium on Computer-Based Medical Systems*, pages 353–358. IEEE Computer Society, 2005.
23. D. Gentner, M. Rattermann, and K. Forbus. The roles of similarity in transfer: Separating retrievability from inferential soundness. *Cognitive Psychology*, 25:524–575, 1993.
24. J. Glasgow and I. Jurisica. Integration of case-based and image-based reasoning. In *Proceedings of the AAAI-98 Workshop on Case-Based Reasoning*, 1998.
25. Elisabet Golobardes, Xavier Llorà, Maria Salamó, and Joan Martí. Computer aided diagnosis with case-based reasoning and genetic algorithms. *Knowledge-Based Systems*, 15(1-2):45–52, 2002.
26. M. Grimnes and A. Aamodt. A two layer case-based reasoning architecture for medical image understanding. In *Proceedings of the Third European Workshop on Case-Based Reasoning*, pages 164–178, 1996.
27. M. Gross, C. Zimring, and E. Do. Using diagrams to access a case library of design. In *Proceedings of the 1994 International Conference on Artificial Intelligence in Design*, pages 129–144. Kluwer, 1994.

28. V.N. Gudivada and V.V. Raghavan. Design and evaluation of algorithms for image retrieval by spatial similarity. *ACM Transactions on Information Systems*, 13(2):115–144, 1995.
29. M. Haddad, K.P. Adlassnig, and G. Porenta. A feasibility analysis of a case-based reasoning system for automated detection of coronary heart disease from myocardial scintigrams. *Artificial Intelligence in Medicine*, 9(1):61–78, 1997.
30. Kristian J. Hammond, Colleen M. Seifert, and Kenneth C. Gray. Functionality in analogical transfer: A hard match is good to find. *Journal of the Learning Sciences*, 1(2):111–152, 1991.
31. Thomas Handler, Rick Holtmeier, Jane Metzger, Marc Overhage, Sheryl Taylor, and Charlene Underwood. HIMSS electronic health record definitional model version 1.0, November 2003.
32. Health informatics — requirements for an electronic health record architecture. Technical Report ISO/TS 18308:2004, International Organization for Standardization, 2004.
33. A. Holt and G.L. Benwell. Case-based reasoning and spatial analysis. *Journal of the Urban and Regional Information Systems Association*, 8(1):27–36, 1996.
34. A. Holt and G.L. Benwell. Applying case-based reasoning techniques in GIS. *The International Journal of Geographical Information Science*, 13(1):9–25, 1999.
35. Alec Holt, Isabelle Bichindaritz, Rainer Schmidt, and Petra Perner. Medical applications in case-based reasoning. *The Knowledge Engineering Review*, 20(3):289–292, 2006.
36. E. Jones and A. Roydhouse. Intelligent retrieval of historical meteorological data. *AI Applications*, 8(3):43–54, 1994.
37. B. Jose, B. P. Singh, S. Venkataraman, and R. Krishnan. Vector based image matching for indexing in case based reasoning systems. In *Proceedings of GWCBR-96*, 1996.
38. I. Jurisica and J. Glasgow. Extending case-based reasoning by discovering and using image features in IVF. In *Proceedings of the ACM Symposium on Applied Computing*, 2003.
39. I. Jurisica, P. Rogers, J. Glasgow, S. Fortier, J. Luft, D. Bianca, and G. DeTitta. Image-feature extraction for protein crystallization: Integrating image analysis and case-based reasoning. In *Proceedings of the Thirteenth Annual Conference on Innovative Applications of Artificial Intelligence (IAAI-2001)*, pages 73–80, 2001.
40. Daniel Kaster, Heloisa Vieira Rocha, and Claudia Bauzer Medeiros. Case-based reasoning applied to environmental modeling with GIS. In *Proceedings of GIScience 2000*, 2000.
41. A. Khattak and H. Renski. PLAN<>HOV: A case-based reasoning planning tool for high-occupancy-vehicle lane analysis in a GIS environment. *Transportation Research Record*, 1682:18–27, 1999.
42. Yu Ting Hung Ko-Wan Tsou, Yao-Lin Chang. A case-based urban planning support system using an integrated combination of geographical information systems and remote sensing. In *Proceedings of the 21st Asian Conference on Remote Sensing*, 2000.
43. Christel LeBozec, Marie-Christine Jaulent, Eric Zapletal, Didier Heudes, and Patrice Degoulet. IDEM: A web application of case-based reasoning in histopathology. *Computers In Biology And Medicine*, 28(5):473–487, 1998.

44. R.T. Macura and K.J. Macura. MacRad: Case-based retrieval system for radiology image resource. In *Proceedings of ICCBR-95*, 1995.
45. R. Mehrotra and J. Gray. Similar-shape retrieval in shape data management. *IEEE Computer*, 28(9):57–62, 1995.
46. Alessandro Micarelli, Alessandro Neri, and Giuseppe Sansonetti. A case-based approach to image recognition. In *Proceedings of EWCBR-2000*, pages 443–454, 2000.
47. Craig A. Morioka, Suzie El-Saden, Gary Duckwiler, Qinghua Zou, Rene Ying, Alex Bui, David Johnson, and Hooshang Kangarloo. Workflow management of HIS/RIS textual documents with PACS image studies for neuroradiology. In *Proceedings of the 2003 American Medical Informatics Association Annual Symposium*, pages 475–479, 2003.
48. Henning Müller, Nicolas Michoux, David Bandon, and Antoine Geissbuhler. A review of content-based image retrieval systems in medicine - clinical benefits and future directions. *International Journal of Medical Informatics*, 73(1):1–23, 2004.
49. A.D. Narasimhalu. CAFIIR: An image based CBR/IR application. In *Proceedings of the 1993 AAAI Spring Symposium on Case-Based Reasoning and Information Retrieval*, pages 70–77, 1993.
50. Markus Nilsson and Mikael Sollenborn. Advancements and trends in medical case-based reasoning: An overview of systems and system development. In *Proceedings of the Seventeenth International FLAIRS Conference*, pages 178–183, 2004.
51. V.E. Ogle. Chabot: Retrieval from a relational database of images. *IEEE Computer*, pages 23–32, September 1995.
52. Dympna OSullivan, Michela Bertolotto, David Wilson, and Eoin McLoughlin. Fusing mobile case-based decision support with intelligent patient knowledge management. In *Proceedings of the Eighth European Conference on Case-Based Reasoning Workshop on CBR in the Health Sciences.*, 2006.
53. Dympna OSullivan, Eoin McLoughlin, Michela Bertolotto, and David C. Wilson. Mobile case-based decision support for intelligent patient knowledge management. In *Proceedings of the Eleventh International Symposium on Health Information Management Research*, 2006.
54. A. Pentland, R.W. Picard, and S. Scarloff. Photobook: Content-based manipulation of image databases. *International Journal of Computer Vision*, 18(3):233–254, 1996.
55. P. Perner. An architeture for a cbr image segmentation system. *Engineering Applications of Artificial Intelligence*, 12(6):749–759, 1999.
56. P. Perner. An architeture for a CBR image segmentation system. *Engineering Applications of Artificial Intelligence*, 12(6):749–759, 1999.
57. P. Perner, T. Gunther, and H. Perner. Airborne fungi identification by case-based reasoning. In *Proceedings of the ICCBR-03 Workshop on CBR in the Health Sciences*, 2003.
58. Petra Perner. Image mining for the construction of semantic-inference rules and for the development of automatic image diagnosis systems. In Xingquan Zhu and Ian Davidson, editors, *Knowledge Discovery and Data Mining: Challenges and Realities with Real World Data*. IDEA Group Inc.
59. Petra Perner. Case-based reasoning for image interpretation in non-destructive testing. In *Proceedings of the First European Workshop on Case-Based Reasoning*, pages 403–410, 1993.

60. Petra Perner. Mining knowledge in medical image databases. In *Data Mining and Knowledge Discovery: Theory, Tools, and Technology—Proceedings of SPIE*, volume 4057, 2000.
61. Petra Perner. Why case-based reasoning is attractive for image interpretation. In *Proceedings of the Fourth International Conference on Case-Based Reasoning*, pages 27–44, 2001.
62. Petra Perner, Alec Holt, and Michael Richter. Image processing in case-based reasoning. *The Knowledge Engineering Review*, 20(3):311–314, 2006.
63. R.W. Pickard and T.P. Minka. Vision texture for annotation. *Multimedia Systems*, 3(1):3–14, 1995.
64. Osman Ratib, Michael Swiernikb, and J. Michael McCoyb. From PACS to integrated EMR. *Computerized Medical Imaging and Graphics*, February 2003.
65. Rainer Schmidt, Stefania Montani, Riccardo Bellazzi, Luigi Portinale, and Lothar Gierl. Cased-based reasoning for medical knowledge-based systems. *International Journal of Medical Informatics*, 64(2-3):355–367, 2001.
66. S. Sclaroff, L. Taycher, and M. La Cascia. Imagerover: A content-based image browser for the world wide web. In *Proceedings of the IEEE Workshop on Content-Based Access of Image and Video Libraries*, pages 2–9, 1997.
67. M. Shepherd, S.R.R. Abidi, Q. Gao, Z. Chen, Q. Qi, and G. Allen Finley. Accessing tacit knowledge and linking it to the peer-reviewed literature. In *Communications of the Association for Information Systems*, volume 17, pages 873–889, 2006.
68. E.L Siegel, J.M. Denner, S.M. Pomerantz, B. Reiner, and Z. Protopapas. The utility of a PACS in the medical media department. In *Proceedings of the Fourth International Conference on Image Management and Communications*, pages 130–131, 1995.
69. Ida Sim, Paul Gorman, Robert A. Greenes, R. Brian Haynes, Bonnie Kaplan, Harold Lehmann, and Paul C. Tang. Clinical decision support systems for the practice of evidence-based medicine. *Journal of the American Medical Informatics Assocation*, 2001.
70. A.W.M. Smeulders, M. Worring, S. Santini, A. Gupta, and R. Jain. Content-based image retrieval at the end of the early years. *IEEE Transactions on Pattern Analysis and Machine Intelligence*, 22(12), 2000.
71. S. Venkataraman, R. Krishnan, and K. Rao. A rule-rule-case based system for image analysis. In *First European Workshop on Case-Based Reasoning (EWCBR'93), Posters and Presentations*, volume 2 of *SEKI-Report 93-12-2*, pages 410–414, 1993.
72. David C. Wilson and Shannon Bradshaw. CBR textuality. *Expert Update*, 3(1):28–37, 2000.
73. Jeremy C. Wyatt. Management of explicit and tacit knowledge. *Journal of the Royal Society of Medicine*, 94(1):6–9, 2001.
74. John Yearwood and Binh Pham. Case-based support in a cooperative medical diagnosis environment. *Telemedicine Journal*, 6(2):243–250, 2000.

14

Instance-Based Relevance Feedback in Image Retrieval Using Dissimilarity Spaces

G. Giacinto and F. Roli

Department of Electrical and Electronic Engineering – University of Cagliari, Italy

Summary. High-retrieval precision in content-based image retrieval can be attained by adopting relevance feedback mechanisms. The user marks all the retrieved images as being either relevant or not, then the search engine exploits this relevance feedback to adapt the search to better meet user's needs. The main difficulties in exploiting relevance information are (a) the gap between user perception of similarity and the similarity computed in the feature space used for the representation of image content and (b) the availability of few training data (users typically label a few dozen of images. At present, SVM are extensively used to learn from relevance feedback due to their capability of effectively tackling the above difficulties. As the performances of SVM depend on the tuning of a number of parameters, in this chapter a different approach to relevance feedback is proposed. First, images are represented in the dissimilarity space made up of the dissimilarities from the set of relevant images. Then a relevance score is computed in terms of the distance from the nearest nonrelevant image, and the distance from the nearest relevant one. Images are ranked according to this score and the top k images are displayed. Reported results show that the performances of the proposed approach are comparable to the highest performances that can be attained by SVM by suitably tuning the learning parameters.

14.1 Introduction

Search engines are becoming increasingly popular as far as the amount of information available in digital form makes it nearly impossible to users to browse inside predefined categories, as each category may include thousands of items. On the other hand, search engines employ techniques from the artificial intelligence domain that allows retrieving data not only according to the associated metadata used to describe them, but also in term of their content [16]. To this end, a suitable model of the content is needed. While search engines for textual documents have reached a good level of maturity (e.g., the search engine employed by Google), search engines for images are still far to be accepted by the average user. The main reason is that images

G. Giacinto and F. Roli: *Instance-Based Relevance Feedback in Image Retrieval Using Dissimilarity Spaces*, Studies in Computational Intelligence (SCI) **73**, 419–436 (2008)
www.springerlink.com

convey a vast amount of information (images have been used as the first mean of communication among humans) [33]. Thus every description of an image is inherently subjective and partial, and search for images based on keywords may fits users' needs only partially. This fact explains why researchers are investigating techniques to retrieve images from databases where the search is based on image content. This research topic is usually referred to as content-based image retrieval (CBIR) and it is currently attracting many researchers from different fields [16,33].

As it is very difficult to capture the complex semantics of an image, the vast majority of CBIR techniques relies on the representation of images by low-level features, e.g., color, texture, shape, etc., [8,16,33]. As a consequence, content-based queries are typically expressed by visual examples in order to retrieve from the database all images that are "similar" to the examples. It is easy to see that the effectiveness of CBIR techniques strongly depends on the choice of the set of visual features, and on the choice of the "metric" used to model the user's perception of image similarity. A number of metrics have been proposed in the literature to adequately measure (dis)similarities in a given feature space [16,30,33].

However, no matter how suitable for the task at hand the features and the similarity metric have been designed, the set of retrieved images often fits the user's needs only partly. It is easy to see that different users may categorize images according to different semantic criteria [3,41]. Thus, if we allow different users to mark the images retrieved with a given query as relevant or nonrelevant, different subsets of images will be marked as relevant. Accordingly, the need for mechanisms to adapt the CBIR system response based on some feedback from the user is widely recognized [16,33,41].

This issue has been studied thoroughly in the text retrieval field, where the relevance feedback concept has been introduced [29]. However, as far as effective document models have been devised, relevance feedback became less attractive so that they are rarely used in the text retrieval field. On the other hand, relevance feedback has been recognized to be necessary in CBIR due to the inherent subjectivity in measuring the degree of relevance of an image with respect to a given query.

A number of relevance feedback techniques have been proposed in the literature to date [41]. Early works on relevance feedback have been formulated in terms of the optimization of one or more CBIR components, e.g., the formulation of a new query and/or the modification of the similarity metric to take into account the relevance of each feature to the user query [15,25,28]. Other CBIR systems employ parametric similarity metrics whose parameters are computed from relevance feedback [31,32].

More recently, relevance feedback has been formulated in terms of a classification problem [37–40]. This formulation requires a careful design, as the number of training samples is typically small (the user is asked to mark as being relevant or not a number of images in the order of few dozens), while the number of features used to represent image content can be large. In addition, the problem can be either formulated as a two-class problem (relevant vs. nonrelevant), or as a $(1 + x)$-class problem. This second formulation, a.k.a. "biased learning," takes into account that the total number of classes in the image database in unknown, but the user is only interested in one class [40]. Two learning techniques are widely used in this context: support vector machines (SVM), and discriminant analysis (DA).

SVM are quite popular in the pattern recognition field as they can handle the above two issues (i.e., small training sets and high dimensionality), and the reported results are quite good if compared to other techniques. However, it should be noted that the choice of the most appropriate SVM parameters for the problem at hand is far from being trivial. Different values of the learning parameters as well as different choices of the kernel may lead to different performances. This aspect is almost neglected in research papers, but it needs to be tackled in order to use SVM in an operational environment.

Discriminant analysis on the other hand has been formulated in a "biased" environment in account of the lack of knowledge on the number of classes. Reported results are quite good even if, as for SVM, a number of learning parameters must be set appropriately in order to get valuable results.

In this chapter, we present a relevance feedback mechanism based on the representation of the images in the database in terms of their (dis)similarities from a *representation set*. The use of the dissimilarity representation of objects has been recently studied in the pattern recognition field where it provided alternative solutions to a number of pattern recognition problems [10, 22, 23]. In particular, it is suited in cases when it is difficult to provide a good feature representation of data, while it can be easier to provide a set of dissimilarities from a set of representative objects. It has also been shown that the dissimilarity representation of data can be viewed as an application of case-based reasoning [26].

As the low-level feature representation of images does not always allow for capturing the user concept of similarity, the dissimilarity representation may represent an approach to exploit relevance feedback [5, 12, 20]. It has been shown that the dissimilarity representation of images allows bridging the gap between low-level (feature) representation of images and the user's perception of similarity, as low level features are used an intermediate step between images and relevance feedback computation.

In order to exploit the benefit of the dissimilarity representation of images, we compute a relevance score for each image of the database in terms of the kth distance from the set of relevant images, and the kth distance from the set of nonrelevant images retrieved so far. For this reason we called this approach "instance based" as, for each image of the database, its relevance score depends on the distance from one relevant image, and the distance from one nonrelevant image.

This chapter is organized as follows. In Sect. 14.2, a brief review of the related works on relevance feedback is presented. In Sect. 14.3, the dissimilarity representation of images in the context of CBIR systems is proposed.

The proposed relevance feedback mechanism based on the nearest neighbor paradigm is described in Sect. 14.4. Experimental results on an image data set are reported in Sect. 14.5. Reported results show that the performances of the proposed method can be compared to other relevance feedback mechanisms described in the literature. Conclusions are drawn in Sect. 14.6.

14.2 Relevance Feedback for CBIR

It is well known that information retrieval system performances can be improved by user interaction mechanisms. This issue has been studied thoroughly in the text retrieval field, where the relevance feedback concept has been introduced [29]. Techniques developed for text retrieval should be suitably adapted to CBIR, on account of differences in both feature number and meaning, and in similarity measures [17, 27, 28].

Basically, relevance feedback strategies are motivated by the observation that the user is unaware of the distribution of images in the feature space, nor of the feature space itself, nor of the similarity metric. Therefore, relevance feedback techniques proposed in the literature involve the optimization of one or more CBIR components, e.g., the formulation of a new query and/or the modification of the similarity metric to take into account the relevance of each feature to the user query.

Query reformulation is motivated by the observation that the image used to query the database may be placed in a region of the feature space that is "far" from the one containing images that are relevant to the user. A query shifting technique for CBIR based on the well-known Rocchio formula, developed in the text retrieval field has been proposed in [27].

Relevance feedback is used in many CBIR systems to optimize a parametric similarity metric. A linear combination of different similarity metrics, each suited for a particular feature set, has been proposed in [32]. Relevance feedback information is then used to modify the weights of the combination to reflect different feature relevance. Santini and Jain also proposed a parameterized similarity measure updated according to feedback from the user [31]. Rather than modifying the similarity metric, Frederix et al. proposed a transformation of the feature space by a logistic regression model so that relevant images represented in the new feature space exhibit higher similarity values [11]. A probabilistic feature relevance scheme has been proposed in [25], where a weighted Euclidean distance is used. Theoretical frameworks involving both the computation of a new query and the optimization of the parameters of similarity metric have been also proposed [15, 28].

A different perspective has been followed in [6] where relevance feedback technique based on the Bayesian decision theory was first proposed. The probability of all images in the database of being relevant is estimated, and images are presented to the user according to the estimated probability. Bayesian decision theory also inspired a query shifting approach aimed at computing a new query whose k-nearest neighbors belongs to the *relevant* region of the feature space [13].

More recently, relevance feedback has been formulated in terms of a classification problem [37–40]. The problem has been either formulated as a two-class problem (relevant vs. nonrelevant), or as a $(1+x)$-class problem. This second formulation, a.k.a. "biased learning," takes into account that the total number of classes in the image database in unknown, but the user is only interested in one class [40]. Two learning techniques are widely used in this context: SVM, and discriminant analysis (DA). Reported results showed that these approaches allow for attaining better results than those provided by early relevance feedback approaches. A number of papers proposed different approaches based on the above paradigms. Advanced techniques as well as an extended overview of the two techniques can be found in [35] and [34].

The use of the dissimilarity representation for relevance feedback in CBIR has been first proposed in [12]. Other researchers independently proposed different approaches based on the dissimilarity representation of data where SVM are used as the final classification tool [5, 20]. Reported results showed that this representation of images allows bridging the gap between low-level (feature) representation of images and the user's perception of similarity.

14.3 Dissimilarity Representation of Images

Dissimilarity representation of data has been proposed in the pattern recognition field as an alternative approach in representing data w.r.t. the feature representation of data [23]. Instead of representing patterns in terms of a feature vector, patterns are represented by a vector of (dis)similarities from a set of *representative* data. This approach is well suited for those applications where it is difficult to provide a suitable representation of patterns in terms of a feature vector. On the other hand, a set of dissimilarities may be more easily available as it is argued that the notion of proximity between patterns is more fundamental than that of features [23].

The dissimilarity representation of data is defined w.r.t. a "representation set"

$$R = \{p_1, p_2, \ldots, p_n\}$$

made up of n objects that are used as a reference for all other objects of interest. An object x can be represented in terms of dissimilarities as a vector

$$[d(x, p_1), d(x, p_2), \ldots, d(x, p_n)]$$

where $d(\,,\,)$ is a distance measure between pair of objects. This distance may be computed using some intermediate feature representation of patterns.

14.3.1 Dissimilarity Evaluation and Prototype Selection

The process usually employed by humans in assigning a class label to an object involves the use of some measure of "similarity" between objects. An object is thus assigned a label according to the class label of the most "similar" patterns whose label is known. Such measures of similarity may depend or not from some quantitative measure made on the objects. These measures are usually called "features." It is easy to see that the definition is, suitable features should depend on the notion of similarity between the objects belonging to the domain of interest. However, as in many applications it is hard to easily extract a set of effective features, it is common to extract a large number of candidate features, and then select the more significant subset for the task at hand. For this selection process to be effective the number of labeled prototypes should be large w.r.t. the number of extracted features. Unfortunately in many application domains the use of a large number of features is accompanied by a relatively small number of labeled prototypes. As a consequence, there are a number of difficulties in solving the pattern recognition problem by a statistical formulation. The dissimilarity representation of data aims at coming back to the roots of pattern recognition and machine learning, by emphasizing the role of similarity between patterns w.r.t. the feature representation of patterns [23, 26]. This focus on dissimilarity representation however does not exclude the use of feature spaces where patterns can be represented. On the other hand, it focus on formulating the problem in terms of dissimilarity between patterns, regardless the way such dissimilarities are computed [23]. Thus, dissimilarities between patterns may be computed using some feature spaces, but the problem itself is not formulated in some feature spaces, but in the dissimilarity space.

In the field of image retrieval for large image databases, usually a large number of low-level features are extracted as the semantic content of images typically exhibit a high variability (they are usually referred to "broad domain" database). In addition, as different users typically have different goals, the extraction of suitable features is not an easy task. To this end research in the field of CBIR focused on defining suitable techniques for manipulating the feature space (feature selection, feature weighting, feature space transformations, etc.). On the other hand, the use of a dissimilarity space may represent an alternative solution, as the low-level feature spaces can be used to compute similarities, which are then used to build a new space. Some researchers recently proposed to use a "manipulation space," i.e., a space where the dissimilarity between images are visualized in a 2D space [20]. In this way, the user may provide the feedback to the system by marking those images that are relevant to the query. However, it is worth noting that while in the feature space relevant images may be represented as "distant" points, in the dissimilarity space these points should be represented as"close" to each other. This effect can be explained by observing that two similar images should exhibit similar distances from some of the prototypes of the representative set. Thus the images are close each other in the dissimilarity space.

From the above discussion it is clear that a key role is played by the selection of the prototypes used to build the representative set [20, 23, 24]. It has been shown that in a number of pattern recognition applications the selection of prototypes can be performed randomly. On the other hand, a number of selection techniques can also be used for prototype selection. In the field of image retrieval, prototypes can be selected by browsing the image collection [20]. First images are clustered according to some criteria, and then one representative for each cluster are showed to the user. The user then selects the images that are more relevant to the query, and they are used as prototypes to build the dissimilarity space. Prototypes may also be chosen by employing the well-known editing techniques proposed both to speed-up nearest neighbor and case-based techniques, and to increase their performance by eliminating noisy or redundant cases [7, 18]. However, as the use of nearest neighbor and case-based techniques for relevance feedback in image retrieval is still at an early stage, such techniques are worth to be considered for further improvements.

In conclusion, it can be said that typically dissimilarities are computed in some feature spaces, while prototypes may be chosen in a number of ways. The key concept in the dissimilarity representation is that two similar objects are close to each other in the dissimilarity space as they typically exhibit similar distance from at least some of the prototypes used to build the dissimilarity space.

The use of "manipulation spaces," where objects are displayed in a 2D space, may help in devising user-defined dissimilarities. The user may use this kind of visualization for "moving" relevant images close to each other. Then this visual movement should be transformed in quantitative dissimilarities to be used for further processing. These kinds of tools are currently at an early stage. For example, in [19] one of such tools have been proposed. In this case, the movement of images is used to compute weights for a weighted distance metric in some low-level feature space.

Thus, in the image retrieval field, the use of some low-level feature space is recognized as a useful tool for representing image content. However these features cannot be used directly to measure the similarity between images, but some processing is needed in order to bridge the gap between the low-level representation and the user perception of similarity. The dissimilarity representation is one approach that allows bridging the gap. This representation may also allow for using some user-defined similarity measure that does not depend on low-level feature space. However, the definition of such a measure is not an easy task, and it is argued that such a definition cannot be used as an alternative to low-level image representation, but as an additional measure to better estimate similarity between images.

14.3.2 The Proposed Dissimilarity Representation of Images for Relevance Feedback

In the proposed relevance feedback mechanism for image retrieval, we define the set R as the set made up of all the relevant images that the user has marked during the relevance feedback iterations. Thus the dimensionality of data strictly depends on the number of relevant images retrieved so far. In addition, the image representation depends on the user concept of similarity.

Let us consider an image database whose images I are represented in a d-dimensional low-level feature space, e.g., color, texture, etc. Let us assume that a dissimilarity metric $d(I_j, I_k)$ has been defined in such a feature space. In the following, we will neither make any assumption about the feature space, nor about the similarity metric employed. If R is the set of relevant images the user has marked during the first k relevance feedback iterations, and n is the size of R, the proposed dissimilarity representation I_{diss} of an image I from the database is the following:

$$I_{diss} = [d(I, I_{r,1}), d(I, I_{r,2}), \ldots, d(I, I_{r,n})] \tag{14.1}$$

where I_r represents an image belonging to R. It is worth noting that (14.1) is also used to represent the relevant images $I_1, I_2, \ldots, I_{Nr}$. This representation allows using the Euclidean distance measure to compute the dissimilarity between pairs of images [23].

The proposed dissimilarity representation strictly depends on the set of images that the user has marked to be relevant. Thus it can be argued that using this representation, an image I_{diss} will be as much as relevant as it is *near* to the relevant images and, at the same time, *far* from the nonrelevant ones.

In the following all the proposed formulas refer to the dissimilarity representation of images. Thus, for the sake of clarity, we will omit the subscript "*diss*."

14.4 Instance-Based Relevance Estimation

The proposed mechanism has been inspired by classification techniques based on the "nearest case," which are used in pattern recognition and machine learning for classification and outlier detection [1,2,4,9,36]. The present section illustrates the rationale behind the use of the nearest case paradigm, and provides the details of the techniques that we propose to measure the relevance of images.

Nearest neighbor techniques, as used in statistical pattern recognition, case-based reasoning, or instance-based learning, are effective in all applications where it is difficult to produce a high-level generalization of a "class" of objects. Relevance learning in content base image retrieval may well fit into this definition, as it is difficult to provide a general model that can be adapted to represent different concepts of similarity. In addition, the number of available cases may be too small to estimate the optimal set of parameters for such a general model. On the other hand, it can be more effective to use each "relevant" image as well as each "nonrelevant" image, as "cases" or "instances" against which the images of the database should be compared [14]. Consequently, we assume that an image is as much as relevant as much as its dissimilarity from the nearest relevant image is small. Analogously, an image is as much as nonrelevant as much as its dissimilarity from the nearest nonrelevant image is small. As this assumption may not hold in a feature space representation, images are represented in terms of dissimilarities, as illustrated in Sect.14.3.

14.4.1 Relevance Score Computation

The degree of relevance can be computed as follows. Let us recall that the nearest neighbor (NN) classifier is derived from the local estimation of densities in the neighborhood of the test pattern [9]. Such a local density can be written as [36]

$$p_{NN}(I) = \frac{1/N}{V(\|I - NN(I)\|)} \tag{14.2}$$

where N is the number of training patterns, I the test image, and NN denotes the nearest neighbor of I. Thus we can compute the local density of relevant images in I as

$$p_{NN}^{r}(I) = \frac{1/N}{V(\|I - NN^{r}(I)\|)} \tag{14.3}$$

where NN^r is the nearest relevant image of I. Analogously the local density of nonrelevant images can be computed as

$$p_{NN}^{nr}(I) = \frac{1/N}{V(\|I - NN^{nr}(I)\|)} \tag{14.4}$$

where NN^{nr} is the nearest nonrelevant image of I.

These densities can be used to estimate the degree of relevance of an image as

$$\begin{aligned} relevance(I) = P(relevant|I) &= \frac{p_{NN}^{r}}{p_{NN}^{r} + p_{NN}^{nr}} \\ &= \frac{\|I - NN^{nr}(I)\|}{\|I - NN^{r}(I)\| + \|I - NN^{nr}(I)\|} \end{aligned} \tag{14.5}$$

The relevance score computed according to (14.5) is then used to rank the images and the first k are presented to the user. It is worth noting that this relevance score can be thought of as an estimation of the posterior in I, as it is computed from an estimation of densities. However, as the estimation of densities cannot be deemed reliable as they are based on a very small training set, we will refer to this measure of relevance as a "relevance score."

14.4.2 Stabilization of the Relevance Score

The proposed score suffers from two problems. First of all, let us consider the training set size. Typically the number of images presented to the user by the retrieval system is in the order of few dozens (e.g., 20 images). When the user marks the first set of images retrieved by the system in response to the query, typically very few of them are relevant, the remaining being nonrelevant. It is easy to see that in these cases the density of relevant images computed according to (14.2) is small almost elsewhere, so that the proposed score output large values for those images *far* from the nonrelevant images (i.e., for those images where the density of nonrelevant images is small too).

To solve this problem, we propose to use the distance of I from a modified query vector computed according to [13]. This modified query vector is aimed at moving the search toward regions of the original feature space where it is more likely to find relevant images. We call this new query vector as "Bayesian query shifting" (BQS) as its formulations is derived from the Bayes decision theory:

$$Q_{BQS} = \mathbf{m}_R + \frac{\sigma}{\|\mathbf{m}_R - \mathbf{m}_N\|}\left(1 - \frac{k_R - k_N}{\max(k_R, k_N)}\right)(\mathbf{m}_R - \mathbf{m}_N) \qquad (14.6)$$

where $\mathbf{m}_R$ and $\mathbf{m}_N$ are the mean vectors of relevant and nonrelevant images, respectively, σ is the standard deviation of the images belonging to the neighborhood of the original query, and k_R and k_N are the number of relevant and nonrelevant images, respectively. The new query Q_{BQS} lies on the line connecting the two means, in the $\mathbf{m_R} - \mathbf{m_N}$ direction, the magnitude of the shift depending on the proportion of relevant and nonrelevant images retrieved. It is easy to see that the larger the number of nonrelevant images retrieved, the larger the magnitude of the shift. For more details about this technique, the reader is referred to [13]. It has been shown that this new query vectors allows attaining good performances in terms of retrieval precision, especially when the number of relevant images is small.

Let us denote with d_{BQS} the distance of image I from Q_{BQS}

$$d_{BQS}(I) = \|I - Q_{BQS}\| \qquad (14.7)$$

In order to combine this distance with the relevance score, we need to transform the distance into a score in [0,1]. We used a Gaussian model and denoted the resulting score as *locality(I)*:

$$locality(I) = \frac{1 - e^{1 - d_{BQS}(I) \big/ \max_I d_{BQS}(I)}}{1 - e} \qquad (14.8)$$

In order to compute the stabilized score, let r and n be the number of relevant and nonrelevant images retrieved after the latter iteration, respectively. The stabilized score can be computed as follows:

$$relevance(I)_{stab} = \left(\frac{n/k}{1+n/k}\right) \cdot locality(I) + \left(\frac{1}{n/k+1}\right) \cdot relevance(I) \tag{14.9}$$

The weights of the combination are both equal to 1/2 when no relevant image is retrieved in the latter iteration, while the weights of *locality(I)* decreases as the number of relevant images increases. The weights of *locality(I)* goes to zero when all the retrieved images are relevant.

The second problem in nearest neighbor density estimation is related to the reliability of the estimation with small sample size. It is worth recalling that the previous issue was related to the cases when the number of relevant images is small, while in this paragraph we are addressing the issue of small training set size, i.e., the fact that the number of retrieved images is small. In these cases the use of the first nearest neighbor can hardly be considered a reliable estimation of the local density. In particular this problem is more severe near the boundary between relevant and nonrelevant images. To solve this problem, we propose to use the kth distance instead of the distance from the first nearest neighbor in (14.3)–(14.5) [4, 36].

14.5 Experimental Results

In order to test the proposed method and compare it with other methods described in the literature we used a subset of the Corel data set (Fig. 14.1). This data set is currently used for assessing and comparing relevance feedback techniques.

The data set extracted from the Corel collection is available at the KDD-UCI repository (http://kdd.ics.uci.edu/databases/CorelFeatures/CorelFeatures.data.html). We used a subset made up of 19,511 images, manually

Fig. 14.1. Some examples of the images contained in the Corel data set

subdivided into 42 semantic classes. For each image, the four sets of features available at the Web site have been considered, i.e., Color Histogram (32 features), Color Histogram Layout (32 features), Color Moments (nine features), and Cooccurrence Texture, (16 features). More details on the feature extraction process can be found in [21]. In particular, the similarity between pairs of images is computed according to the Manhattan distance for the first two sets of features, while the Euclidean distance is used for the latter two sets of features. A linear normalization procedure has been performed, so that each feature takes values in the range between 0 and 1.

500 images have been randomly extracted and used as query. The top 20 nearest neighbors of each query are returned. Relevance feedback is performed by marking images belonging to the same class of the query as relevant, and all other images in the top 20 as nonrelevant. This experimental set up affords an objective comparison among different methods and is currently used by many researchers. Performances are computed in terms of the retrieval precision, i.e., the average percentage of relevant images among the top 20 images retrieved by the systems.

The proposed relevance feedback technique has been implemented by considering the second-NN distance. In particular, we report results attained by the *relevance* score computed according to (14.5), and the ones attained by the *stabilized relevance* score computed according to (14.9). Thus it is possible to compare the behavior of the "pure" instance-based technique (14.5) with the instance-based technique combined with the *locality* term (14.9).

For the sake of comparison, retrieval performances obtained with three methods recently described in the literature, are also reported, namely the BQS and SVM. The BQS technique has been illustrated in Sect. 14.4 as it is also used in the present chapter in combination with the instance-based rule. The SVM have been trained using the sets of relevant and nonrelevant images as a training set of a two-class classification problem, and images have been ranked according to the SVM output. As SVM training requires choosing the kernel and the learning parameters, we used two commonly used kernels, namely the linear and the Gaussian kernel. In particular, for the SVM with Gaussian kernel, reported results are the best ones attaining with different values of the learning parameters, while in the case of linear kernel no parameter optimization is required.

Figures 14.2–14.5 show the results attained with the four sets of features of the Corel data set. The retrieval performances attained after the query image is presented to the system are quite low, thus showing that the chosen feature sets are not suited for the task at hand. In addition, different sets of features provided different results, thus confirming that the choice of the set of features is a key aspect of CBIR systems.

It is easy to see that each of the considered relevance feedback techniques allow improving the retrieval precision. The only exception is the linear SVM, which is not suited for the Color Histogram and Color Histogram Layout sets of features. In fact the retrieval performances not only does not improve with

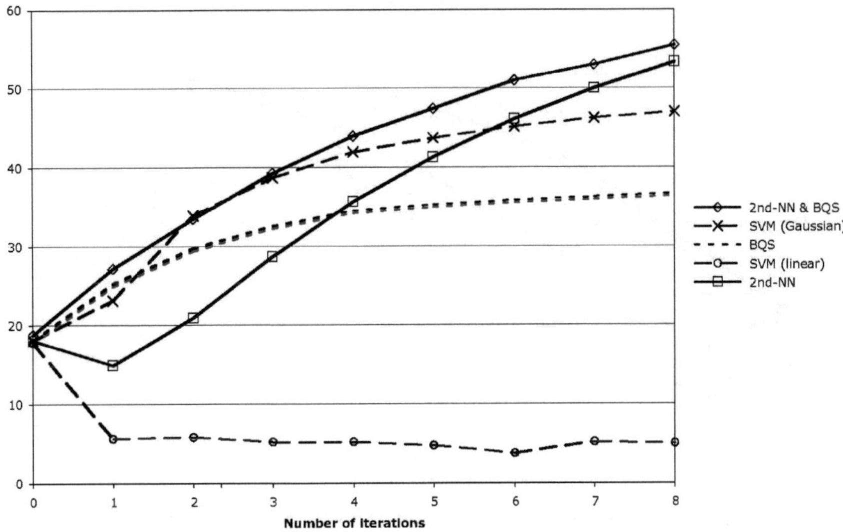

Fig. 14.2. Average percentage precision for the Color Histogram feature set. Eight relevance feedback iterations were performed

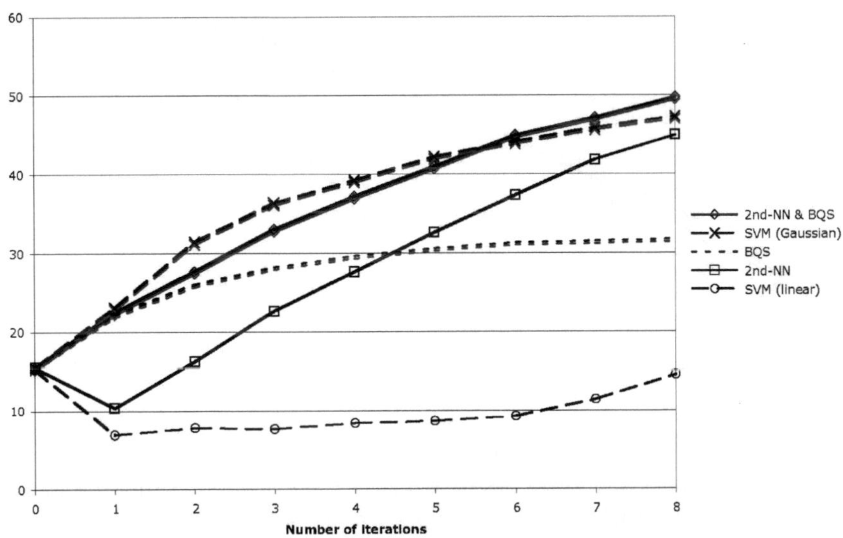

Fig. 14.3. Average percentage precision for the Color Histogram Layout feature set. Eight relevance feedback iterations were performed

respect to the first retrieval, but also they get worse. On the other hand, linear SVM provided performance improvements on the other two sets of features.

The proposed instance-based technique in the dissimilarity space combined with BQS, provided performances higher than those provided by the SVM with Gaussian kernel. This superiority is more evident in Figs. 14.3–14.5. It is

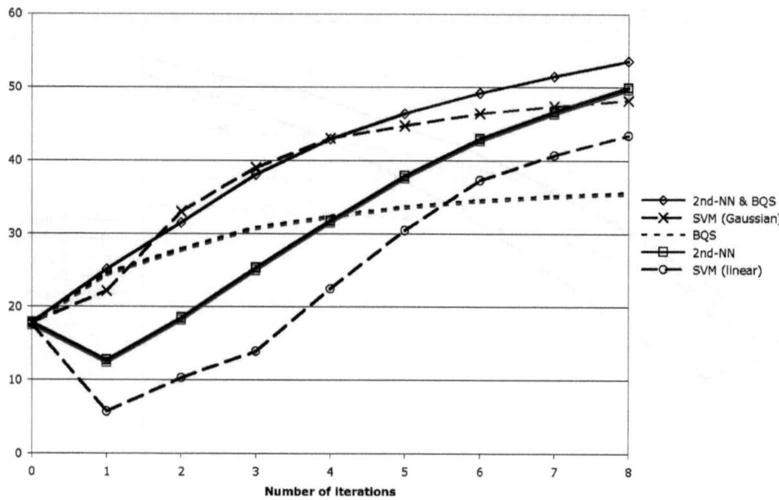

Fig. 14.4. Average percentage precision for the Color Moments feature set. Eight relevance feedback iterations were performed

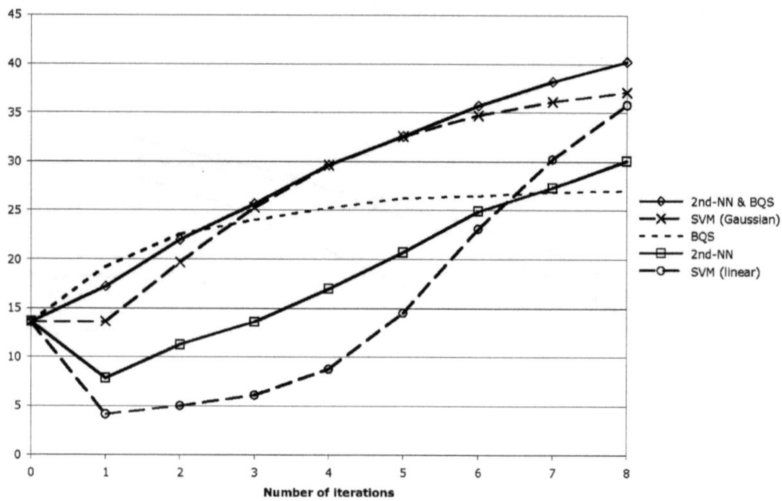

Fig. 14.5. Average percentage precision for the Cooccurrence texture feature set. Eight relevance feedback iterations were performed

worth recalling that the reported performances of the SVM with Gaussian kernel are the best one attained with different parameter values. On the other hand, the proposed mechanism does not require setting any parameter. Thus the proposed technique allows attaining good performances without requiring parameter tunings.

If we analyze the results, we see that the precision decreases in the first iteration when the "pure" second-NN technique, and the linear SVM are used.

This effect is motivated by the fact the in the first iteration very few relevant images are usually retrieved. As a result SVM and second-NN assign high scores to images that are *different* form the nonrelevant images. As the problem is not a two-class problem but a $(1 + x)$ problem, thus it follows that an image that is different from some nonrelevant images is not necessarily a relevant image. If the iterations following the first are considered, the precision attained by the second-NN technique increases in all of the four considered sets of features. In the case of the linear SVM, the precision increases only in two out of the four sets of features, namely Color Moments and Cooccurrence texture.

On the other hand, the performances of BQS allows attaining high performance increases in the first iteration, while the increase is modest in the following iterations. This behavior may be due to the locality of BQS that does not allow exploring new regions of the feature space.

Gaussian SVM and the stabilized second-NN (second-NN & BQS) provide the highest performances. In all the considered feature sets the precision of the stabilized second-NN is always higher than the one provided by the Gaussian SVM. The only exceptions are the iterations from 2 to 5 in the Color Histogram Layout feature set, and the iterations 2 and 3 in the Color Moments feature set where the performances of the Gaussian SVM are slightly higher than those provided by the stabilized second-NN.

However, it is worth recalling that reported results for the Gaussian SVM are related to the best one attained in a number of trials. In a number of experiments with different values of the learning parameters, the performances of the Gaussian SVM after the first relevance feedback iteration were usually smaller than those attained in the first retrieval (i.e., the k nearest neighbors of the query). As the estimation of the optimal or suboptimal parameters for the Gaussian SVM in relevance feedback is beyond the state of the art, it can be concluded that the proposed mechanism is a robust technique in different image representation contexts.

14.6 Conclusions

The dissimilarity representation of data is receiving increasing attention in a number of applications. Recent research in relevance feedback for CBIR proposed some techniques based on this representation of data. This representation allows for avoiding the direct use of low-level feature spaces that may not be suited to the user's needs.

The dissimilarity representation proposed in this chapter use the set of relevant images retrieved so far as the representation set. Then, images are ranked according to a relevance score computed by a combination of two terms, namely a *relevance* term based on the distances from the second relevant neighbor and the second nonrelevant neighbor, and a *locality* term computed in terms of a shifted query in the original feature space. These two terms plays the role of an exploration term and exploitation term, respectively.

Reported results on four feature sets of the Corel image database showed the superiority of the proposed method with respect to state-of-the-art relevance feedback techniques. In particular it has been pointed out that the proposed technique does not require any parameter setting, while other relevance feedback techniques require a long phase for tuning the parameters. Thus it can be concluded that the proposed relevance feedback mechanism is a robust tool that allows attaining good performances in a number of different image representations.

As far as the computational complexity of the proposed technique is concerned, a large number of distances are to be computed. Nevertheless, the response time between two consecutive feedbacks is far below the classic limit of 1.0s for the user's flow of thought to stay uninterrupted. Thus, despite the computational complexity of the algorithm, the response time on a typical PC configuration can be considered acceptable for a large database. However, the response time of the implemented algorithm could be further improved by using some editing techniques for decreasing the number of distances to be evaluated.

Acknowledgments

This work was partially supported by the Italian Ministry of University and Research within the framework of the project "Similarity-based Methods for Computer Vision and pattern recognition: Theory, Algorithms, Applications."

The authors wish to thank Dr. Michael Ortega for providing the images of the Corel Data set, and Dr. Petra Perner for her useful comments and suggestions.

References

1. Aha DW, Kibler D, Albert MK (1991) Instance Based learning Algorithms. Machine Learning 6:37–66
2. Althoff K-D (2001) Case-Based Reasoning. In Chang S.K. (ed.) Handbook on Software Engineering and Knowledge Engineering, World Scientific, 549–588
3. Bhanu B, Dong D (2001) Concepts Learning with Fuzzy Clustering and Relevance Feedback. In: Perner, P. (Ed.): Machine Learning and Data Mining in Pattern Recognition. LNAI 2123, Springer-Verlag, Berlin 102–116
4. Breunig M, Kriegel H-P, Ng R, Sander J. (2000) LOF: indentifying density-based local outliers. In Proc. of the ACM SIGMOD 2000 Int. Conf. on management of data
5. Bruno E, Loccoz N, Maillet S (2005) Learning user queries in multimodal dissimilarity spaces. Proc. of the 3rd Int'l Workshop on Adaptive Multimedia Retrieval

6. Cox I.J, Miller ML, Minka TP, Papathomas TV, Yianilos PN (200) The Bayesian image retrieval system, PicHunter: theory, implementation, and psychophysical experiments. IEEE Trans. on Image Processing 9:20–37
7. Dasarathy DV (Ed.) (1991) Nearest Neighbor Norms: NN Pattern Classification Techniques. IEEE Press
8. Del Bimbo A (1999) Visual Information Retrieval. Morgan Kaufmann Pub. Inc., San Francisco, CA
9. Duda RO, Hart PE, Stork DG (2001) Pattern Classification. John Wiley and Sons, Inc., New York
10. Duin RPW, de Ridder D, Tax DMJ (1997) Experiments with object based discriminant functions: a featureless approach to pattern recognition. Pattern Recognition Letters 18:1159–1166
11. Frederix G, Caenen G, Pauwels EJ (2000) PARISS: Panoramic, Adaptive and Reconfigurable Interface for Similairty Search. Proc. of ICIP 2000 Intern. Conf. on Image Processing. WA 07.04, vol. III, 222–225
12. Giacinto G, Roli F (2003) Dissimilarity Representation of Images for Relevance Feedback in Content-Based Image Retrieval. In: Perner P. (Ed.) Machine Learning and Data Mining in Pattern Recognition. LNAI 2734, Springer-Verlag, Berlin 202–214
13. Giacinto G, Roli F (2004) Bayesian Relevance Feedback for Content-Based Image Retrieval. Pattern Recognition 37:1499–1508
14. Giacinto G, Roli F (2005) Instance-Based Relevance Feedback for Image Retrieval. In Saul L.K., Weiss Y., and Bottou L.: Advances in Neural Information Processing Systems 17, MIT Press 489–496
15. Ishikawa Y, Subramanys R, Faloutsos C (1998) MindReader: Querying databases through multiple examples. In Proceedings. of the 24th VLDB Conference 433–438
16. Lew MS, Sebe N, Djeraba C, Jain R (2006) Content-Based Multimedia Information Retrieval: State of the Art and Challenges. ACM Trans. On Multimedia Computing, Communications and Applications 2:1–19
17. McG Squire D, Müller W, Müller H, Pun T (2000) Content-based query of image databases: inspirations from text retrieval. Pattern Recognition Letters 21:1193–1198
18. McKenna E, Smyth B (2000) Competence-Guided Case-Base Editing Techniques in Blanzeri and Portinale (Eds.) Proceedings of the 5th European Workshop on Advances in Case-Based Reasoning, LNCS 1898, Springer-Verlag: 186–197
19. Moghaddam B, Tian Q, Lesh N, Shen C, Huang TS (2004) Visualization and User-Modeling for Browsing Personal Photo Libraries. International Journal of Computer Vision 56:109–130
20. Nguyen GP, Worring M, Smeulders AWM (2006) Similarity learning via dissimilarity space in CBIR. Proc. of the 8th ACM Int'l workshop on Multimedia Information retrieval 107–116
21. Ortega M, Rui Y, Chakrabarti K, Porkaew K, Mehrotra S, Huang TS (1998) Supporting ranked boolean similarity queries in MARS. IEEE Trans. on KDE 10:905–925
22. Pekalska E, Duin RPW (2002) Dissimilarity representations allow for building good classifiers. Pattern Recognition Letters 23:943–956
23. Pekalska E, Duin RPW (2005) The dissimilarity representation for pattern recognition: foundations and applications. World Scientific Publishing

24. Pekalska E, Duin RPW, Paclick (2006) Prototype selection for dissimilarity-based classifiers. Pattern Recognition 39:189:208
25. Peng J, Bhanu B, Qing S (1999) Probabilistic feature relevance learning for content-based image retrieval. Computer Vision and Image Understanding 75:150–164
26. Perner P (2002) Are case-based reasoning and dissimilarity-based classification two sides of the same coin? Engineering Applications of Artificial Intelligence 15:193–203
27. Rui Y, Huang TS, Mehrotra S (1997) Content-based image retrieval with relevance feedback in MARS. In Proceedings of the IEEE International Conference on Image Processing, IEEE Press 815–818
28. Rui Y, Huang TS (2001) Relevance Feedback Techniques in Image retrieval. In Lew M.S. (ed.): Principles of Visual Information Retrieval. Springer-Verlag, London, 219–258
29. Salton G, McGill MJ (1998) Introduction to modern information retrieval. McGraw-Hill, New York
30. Santini S, Jain R (1999) Similarity Measures. IEEE Trans. on Pattern Analysis and Machine Intelligence 21:871–883
31. Santini S, Jain R (2000) Integrated browsing and querying for image databases. IEEE Multimedia 7:26–39
32. Sclaroff S, La Cascia M, Sethi S, Taycher L (2001) Mix and Match Features in the ImageRover search engine. In Lew M.S. (ed.): Principles of Visual Information Retrieval. Springer-Verlag, London 219–258
33. Smeulders AWM, Worring M, Santini S, Gupta A, Jain R (2000) Content-based image retrieval at the end of the early years. IEEE Trans. on Pattern Analysis and Machine Intelligence 22:1349–1380
34. Tao D, Tang X, Li X, Rui Y (2006) Direct Kernel Biased Discriminant Analysis: A New Content-based Image Retrieval Relevance Feedback Algorithm. IEEE Trans. on Multimedia 8:716–727
35. Tao D, Tang X, Li X, Wu X (2006) Asymmetric Bagging and Random Subspace for Support Vector Machines-Based Relevance Feedback in Image Retrieval. IEEE Trans. on Pattern Analysis and Machine Intelligence 28:1088–1099
36. Tax D (2001) One-class classification. PhD thesis, Delft University of Technology, The Netherlands
37. Tieu K, Viola P (2001) Boosting Image Retrieval. Proc. IEEE Int'l Conf. Computer Vision and Pattern Recognition, vol. 1, 228–235
38. Tong S, Chang E (2001) Support Vector Machine Active Learning for Image Retrieval. Proc. ACM Int'l Conf. Multimedia 107–118
39. Zhang L, Lin F, Zhang B (2001) Support Vector Machine Learning for Image Retrieval. Proc. IEEE Int'l Conf. Image Processing 721–724
40. Zhou X, Huang TS (2001) Small Sample Learning During Multimedia Retrieval Using Biasmap. Proc. IEEE Int'l Conf. Computer Vision and Pattern Recognition, vol. 1, 11–17
41. Zhou X, Huang TS (2003) Relevance Feedback for Image Retrieval: A Comprehensive Review, ACM Multimedia Systems 8:536–544

MIX
Papier aus verantwortungsvollen Quellen
Paper from responsible sources
FSC® C105338

If you have any concerns about our products,
you can contact us on
ProductSafety@springernature.com

In case Publisher is established outside the EU,
the EU authorized representative is:
Springer Nature Customer Service Center GmbH
Europaplatz 3, 69115 Heidelberg, Germany

Printed by Libri Plureos GmbH
in Hamburg, Germany